Animal Contests

Contests are an important aspect of the lives of diverse animals, from sea anemones competing for space on a rocky shore to male ibex contending for access to females. Why do animals fight? What determines when fights stop and which contestant wins? Addressing fundamental questions on contest behaviour, this volume presents theoretical and empirical perspectives across a range of species.

The historical development of contest research, the evolutionary theory of both dyadic and multiparty contests and approaches to experimental design and data analysis are discussed in the first few chapters. These topics are followed by reviews of research in key animal taxa, from the use of aerial displays and assessment rules in butterflies and the developmental biology of weapons in beetles, through to interstate warfare in humans. The final chapter considers future directions and applications of contest research, making this a comprehensive resource for both graduate students and researchers in the field.

Ian C.W. Hardy is an Associate Professor and Reader in the School of Biosciences at the University of Nottingham, UK.

Mark Briffa is an Associate Professor (Reader) in the School of Marine Science and Engineering at Plymouth University, UK.

Animal Contests

Edited by
Ian C.W. Hardy
University of Nottingham, UK
Mark Briffa
Plymouth University, UK

CAMBRIDGE UNIVERSITY PRESS
Cambridge, New York, Melbourne, Madrid, Cape Town,
Singapore, São Paulo, Delhi, Mexico City

Cambridge University Press
The Edinburgh Building, Cambridge CB2 8RU, UK

Published in the United States of America by Cambridge University Press, New York

www.cambridge.org
Information on this title:
www.cambridge.org/9780521887106

First published 2013

Printed and bound in the United Kingdom by the MPG Books Group

A catalogue record for this publication is available from the British Library

Library of Congress Cataloging in Publication data
Animal contests / edited by Ian C.W. Hardy, University of Nottingham, UK, and Mark Briffa, University of Plymouth, UK.
pages cm
Includes bibliographical references and index.
ISBN 978-0-521-88710-6 (hardback)
1. Competition (Biology) I. Hardy, Ian C.W. (Ian Charles Wrighton), 1965– editor of compilation. II. Briffa, Mark., editor of compilation.
QH546.3.A55 2013
577.8′3 – dc23 2013003979

ISBN 978-0-521-88710-6 Hardback

To Lidia, Fenna and Zenta
and to Kate, Ben and Fergus

Contents

Contributors

Troy A. Baird
Department of Biology, University of Central Oklahoma, Edmond, Oklahoma, USA.

Tim P. Batchelor
School of Marine Science and Engineering, Plymouth University, Plymouth, UK.

Mark Briffa
School of Marine Science and Engineering, Plymouth University, Plymouth, UK.

David D. Clarke
School of Psychology, University of Nottingham, University Park, Nottingham, UK.

Miranda L. Dyson
Department of Environment, Earth and Ecosystems, The Open University, Milton Keynes, UK.

Ryan L. Earley
Department of Biological Sciences, University of Alabama, Tuscaloosa, Alabama, USA.

Robert W. Elwood
Queen's University Belfast, School of Biological Sciences, Medical Biology Centre, Belfast, Northern Ireland, UK.

Scott A. Field
Skoll Global Threats Fund, San Francisco, California, USA.

Martin P. Gammell
Department of Life and Physical Sciences, Galway-Mayo Institute of Technology, Galway, Ireland.

Marlène Goubault
Institut de Recherche sur la Biologie de l'Insecte, UMR CNRS 7261, Faculté des Sciences et Techniques, Tours, France.

Tim R. Halliday
21 Farndon Road, Oxford OX2 6RT, UK.

Ian C.W. Hardy
School of Biosciences, University of Nottingham, Sutton Bonington Campus, Loughborough, UK.

Yuying Hsu
Department of Life Science, National Taiwan Normal University, Taipei, Taiwan.

Dómhnall J. Jennings
Centre for Behaviour and Evolution, Institute of Neuroscience, Newcastle University, Newcastle-upon-Tyne, UK.

Darrell J. Kemp
Department of Biological Sciences, Macquarie University, North Ryde, New South Wales, Australia.

Hanna Kokko
Centre of Excellence in Biological Interactions, Division of Ecology, Evolution and Genetics, Research School of Biology, Australian National University, Canberra, Australia.

Mike Mesterton-Gibbons
Department of Mathematics, Florida State University, Tallahassee, Florida, USA.

Armin P. Moczek
Department of Biology, Indiana Molecular Biology Institute, Center for the Integrative Study of Animal Behavior, Indiana University, Bloomington, Indiana, USA.

Sophie L. Mowles
School of Biosciences, University of Nottingham, Sutton Bonington Campus, Loughborough, UK.

Geoff A. Parker
Institute of Integrative Biology, University of Liverpool, Liverpool, UK.

John Prenter
Department of Biological Sciences, Macquarie University, North Ryde, New South Wales, Australia.

Sarah R. Pryke
Division of Evolution, Ecology and Genetics, Research School of Biology, The Australian National University, Canberra, Australia.

Michael S. Reichert
Institut für Biologie, Humboldt University, Berlin, Germany.

Tom N. Sherratt
Department of Biology, Carleton University, Ottawa, Ontario, Canada.

Emilie C. Snell-Rood
Department of Ecology, Evolution and Behaviour, University of Minnesota, St Paul, Minnesota, USA.

Foreword
A personal history of the development of animal contest theory and its role in the 1970s

Geoff A. Parker

This book on *Animal Contests* represents a landmark in evolutionary biology that is greater than its immediate title suggests. The adaptive interpretation of fighting behaviour in animals has been a catalyst in the study of evolutionary adaptation: first, it was influential in changing concepts about the mechanism of selection (from implicit group selection to individual selection), and second, it was the focus for shaping our understanding of frequency-dependent optimisation in biology through the evolutionarily stable strategy (ESS) approach. I regard it a great privilege and honour to have been invited by the Editors to write a Foreword, and I would like to use this opportunity to recount some of the history of these roles of contest theory in evolutionary biology, including my personal recollections of the events during that very exciting decade, the 1970s.

Before 1970

The way we now think of animal fighting behaviour owes most to the development of the first theoretical models of animal contests, developed in the 1970s. Before this time, with a few notable exceptions (e.g. Williams 1966, Lack 1968), most researchers in the disciplines of ethology and ecology routinely (and usually implicitly) applied group or species selection interpretations to what they saw. This ethos did not generally apply to evolutionary biologists or population geneticists, whose analyses were usually founded on principles derived from Darwinian natural selection. Fighting or threat behaviour in animals is common, especially male–male combat (interpreted in the context of sexual selection by Darwin 1871), but also in food-fighting, territoriality, social dominance, and in various other contexts, and sometimes in both sexes. It can generally be related to contest competition for some limited, unsharable, fitness-related resource, although this is not always immediately obvious.

Although male–male combat aspects of sexual selection had many followers after Darwin's (1871) treatise and even up to half a century later (e.g. Richards 1927), the influential reviews of Huxley (1938*a*,*b*) had been sceptical about female choice aspects of sexual selection, and unenthusiastic even for male–male combat over females, favouring indirect interpretations based on natural selection. An era followed in which sexual selection was largely ignored. This coincided with the 'advantage to the species' culture that gripped ethology and ecology until the 1970s, broken only by Bateman's (1948) classic sexual selection paper. The prevailing ethological view of contests related essentially that contests had evolved into ritualised threat displays to prevent serious injury (Chapter 2).

This philosophy pervaded the literature for over three decades. For example, Leuthold (1966), in his field observations of the Uganda kob, *Adenota kob thomasi*, concluded that territorial breeding grounds offer ecological advantages, such as providing a social organisation and a spacing mechanism to the population and ensuring maximum efficiency of reproduction. Similarly, Norman Moore, rightly recognised as a distinguished dragonfly researcher, clearly avoided intra-sexual selection as an interpretation of the territorial fighting of male dragonflies. He suggested instead that territorial behaviour had a selective advantage related to 'dispersal of the population' (Moore 1952), and went on to list six functions of male–male fighting: colonisation of new breeding areas, prevention of interference with sexual behaviour, prevention of interference with oviposition, prevention of food shortage, reduction of losses to predators, and reduction of time spent in aggression (Moore 1957). He later refined this further, concluding that territorial behaviour is of selective advantage in helping to maintain the most viable stock in the most suitable habitats, reducing disturbance of mating and oviposition and unnecessary fighting, and causing dispersal of sexually mature males to new habitats (Moore 1964). Although implicit species-advantage interpretations of male territorial guarding were generally then the rule, there were some very notable exceptions; Merle Jacobs' (1955) beautiful work deserves special credit for throughout interpreting dragonfly fighting, mate-guarding and territoriality in terms of individual benefits through intra-sexual selection. Moore cited Jacobs' study, but avoided any mention of sexual selection. This instance gives some indication of the unpopularity of sexual selection and obvious individual-advantage explanations before the 1970s, even for male–male contests at leks.

In most research communities there was thus a curious lack of communication between ethologists/ecologists and evolutionary biologists/population geneticists, which became evident in the group selection debate, beginning in the 1960s with Wynne-Edward's (1962) treatise and *Nature* paper (1963) promoting group selection, and its rejoinders (Maynard Smith 1964, Perrins 1964, Williams 1966). Wynne-Edwards, a field ecologist, interpreted territorial fighting as an adaptive mechanism preventing over-exploitation of resources by a population but, with clarity of vision rare for mainstream ecologists of that time, he realised that such a function ran counter to Darwinian natural selection and proposed an alternative mechanism ('inter-group selection') to account for its evolution. Controversy over the relative importance of group and individual selection continues to this day.

Numerous studies before the 1970s had described fighting behaviour and noted the prevalence of settlements by 'threat displays' without damaging combat; many researchers, including such notable ethologists as Lorenz and Huxley (Chapter 2), stressed the evolution of 'ritualistic display' as a means of avoiding dangerous aggression leading to severe injury, with implicit species-advantage benefits. However, not all researchers adhered to this orthodoxy, and individual-selection interpretations of combat behaviour existed if sought for. For example, Valerius Geist, a field ethologist with extensive experience of ungulate behaviour, reviewed observations of fighting involving weaponry such as horns and tusks, which suggested that when combatants can injure opponents, they do so (Geist 1966). He perceptively concluded that contests could be seen as the interplay between defensive and offensive behaviour:

> . . . horns evolved to function as weapons inflicting damage; as defense organs shielding their owner; as binding organs allowing opponents a secure lock in battle; as display organs having an a priori intimidating effect on certain conspecifics.

This individual-selection interpretation of an arms race between attack and defence still appears valid.

The behavioural ecology revolution

Although the seeds were sown in the 1960s, the 1970s hailed a remarkable renaissance in the study of behavioural adaptation: the 'behavioural ecology revolution' (Parker 2006). It marked a paradigm shift in the way that most ethologists and ecologists interpreted the adaptive value of behaviour. After the long era of group/species selection, behavioural adaptations began to be interpreted differently, as characterised by three features: (i) replacement of the 'survival value to the species' shorthand with adaptive explanations based on advantage to the individual (Williams 1966), (ii) the understanding that the evolutionary interests of interacting individuals could be in conflict, and (iii) the introduction of optimality approaches, including the evolutionarily stable strategy (ESS) approach (Maynard Smith & Price 1973), to predict expected behavioural strategies.

These changes were to revolutionise our understanding of animal behaviour. In particular, they changed completely how we now think about animal contests. The key early contest theory papers are summarised in Table A, and their numbers in the Table are given in the text below, beginning with Maynard Smith & Price's seminal work [1].

Assessment strategy and RHP

My own interest in animal contests began at the University of Bristol while watching the mating behaviour of dung flies, *Scatophaga* (= *Scathophaga*) *stercoraria* L., beginning as an undergraduate final-year project in spring 1965, and later my PhD project, 1965–1968 (Parker 2001, 2006, 2010). In these ubiquitous flies the sexes meet around fresh cattle droppings, the site of mating and oviposition. Males paired

Table A Key animal contest theory publications of the 1970s.

No.	Publication	Receipt, revision and publication dates (where known)	Summary of key elements	Google Scholar citations (17.05.12)
1	Maynard Smith & Price (*Nature*, 246, 15–18)	*Published:* 2 Nov 1973	Introduction of the ESS concept, symmetric Hawk–Dove (= 'Mouse') games with discrete injury costs, showing that an ESS can involve limited escalation. Presented solution for the symmetric war of attrition.	2917
2	Maynard Smith (*Journal of Theoretical Biology*, 47, 209–221)	*Received:* 10 Jan 1974 *Published:* September 1974	Development of ESS concept and full derivation of symmetric war of attrition. First analysis of a game with an 'arbitrary asymmetry' (i.e. not affecting potential gains or contest costs) between two otherwise identical opponents, showing that this can be used as a 'conventional settlement.	1420
3	Parker (*Journal of Theoretical Biology*, 47, 223–243)	*Received:* 18 March 1974 *Published:* September 1974	First investigation of payoff relevant asymmetries between contestants in resource value and 'resource holding power (or potential)' or RHP. Analysis was not formally ESS, but best strategy depended on opponent's best strategy. The proposed 'assessor rule' (see text) was later shown to be an ESS (Hammerstein & Parker 1982).	1097
4	Maynard Smith & Parker (*Animal Behaviour*, 24, 159–175)	*Received:* 8 May 1975 *Revised:* 26 June 1975 *Published:* February 1976	Introduction of symmetric 'Hawks–Doves' game; first ESS analysis of payoff relevant asymmetries (i.e. resource value and RHP), definitions of 'commonsense' and 'paradoxical' strategies, and simulations of a game in which information about RHP is acquired during the contest.	1126
5	Parker (In: *Sexual Selection and Reproductive Competition in Insects*)	*Received:* early Feb 1977 *Published:* April 1979	First formal analyses of inter-locus sexual conflict, an asymmetric game between males and females, and parameter space over which it can occur in various contexts. Simulations of asymmetric war of attrition and asymmetric opponent independent costs game (an arms race game).	802
6	Maynard Smith (*Evolution and the Theory of Games*)	*Preface date:* Nov 1981 *Published:* 1982	Maynard Smith's monograph on ESS, reviewing contest theory and many other areas to which ESS theory can be applied.	7183

to females are attacked continually by the many single males searching for females on and around droppings; should a male's elaborate defences fail, an intense struggle ensues between the two (or more) males (Parker 1970*a*). If a 'take-over' occurs, the new male mates immediately with the female. The fights are impressive, and appear to be costly, not only to the male combatants but also to the female they are competing for. The biologist and novelist Robin Baker and I were fellow students at Bristol during our undergraduate and postgraduate years. During our final year as undergraduates we often discussed the mechanism of selection, and as a result – I like to think to our credit – we became individual selectionists, despite the prevailing culture in ecology and ethology.

My PhD field study of sexual selection and mating behaviour in dung flies caused me to be fascinated by the problem of the evolution of animal contests. I could see that fights were costly, and I liked the Lorenzian idea that individuals fight to gain some resource that increases fitness. I could also see that from an individual selection perspective, self's best strategy depended on the opponent's strategy and on the differences between self and the opponent, and that these features must shape the behaviour we observe in animal fights. It was something I had often talked to

Robin Baker about at Bristol, but I did not start work on this problem until around 1970, a year or so after beginning a lectureship at the University of Liverpool. By mid 1973 I had done much of the work for a theory paper that was published in 1974 [3]. I suggested that the opponents in a contest should respond to (i) asymmetries in the value of the resource to each opponent (I called these resource values 'payoff asymmetries' or 'fitness budgets for fighting') and to (ii) asymmetries in their 'resource holding power' (RHP, which changed later to 'resource holding potential' [4]), a measure of a given opponent's absolute ability to defend the contested resource. I used the term RHP rather than fighting ability directly, because the ability to hold or gain the contested resource and to avoid or to inflict contest costs depended on a suite of aspects: e.g. condition, size, strength, weaponry, position and experience.

My basic idea was that if opponents can assess these asymmetries perfectly, and costs rise continuously during an escalated fight, the contest should be settled conventionally without costly escalation, using the following 'assessor rule' (abbreviated for clarity):

> ... disputes should be decided by each individual's fitness budget available for expenditure during a fight... and on the rate of expenditure of the fitness budget if escalation occurs (determined by the RHPs of the combatants)... strategies ('assessments') will be determined by... which opponent is likely to expend its fitness budget first, should escalation occur. This 'loser' should retreat (before escalation) and the winner should stay in possession of the resource.

To state this rule formally for constant cost rate expenditure, suppose that opponents A and B have resource values V_A and V_B, respectively (V is a measure of benefits of winning, i.e. how much A or B's fitness increases by winning outright, without costs). Should an escalated contest occur, fitness costs increase at constant rates, c_A to A and c_B to B. The assessor rule is that B should retreat immediately without contesting if it assesses that $V_A/c_A > V_B/c_B$. My intuitive reasoning was that A could still achieve a positive gain from the contest after the time that B would sustain a loss from further fighting even if B were to win. I had certainly not proved that this would be an ESS (Maynard Smith and Price's key paper [1] had not been published at the time I was doing this work; see below), although the assessor rule is indeed an ESS if opponents make rare mistakes in assessment of roles A and B, with roles defined as $V_A/c_A > V_B/c_B$ (Hammerstein & Parker 1982, see below).

If assessment is perfect, why should we ever see escalation? I argued that in escalated combats, opponents might sustain damaging injuries rather than continuous costs. The contest could be seen as a series of bouts in which the initial RHP differences between opponents gave an estimate of the probability that the opponent with lower RHP would sustain the injury (the probability that A wins is $RHP_A/(RHP_A + RHP_B)$). The loser sustains an injury that reduces its RHP, which could then affect its decision to continue the contest. I constructed quite a complex model in which there was a normal distribution of RHP in the population of contestants; the highest RHP individuals held a fixed proportion of the resources and individuals with randomly assigned RHP held the remainder. This enabled calculation of the expected search costs incurred in finding an alternative resource (decreasing as an individual's RHP increases), and hence (given a fixed cost of injury to one opponent) what would be the best strategy for each opponent – to persist (if a positive future expectation) or to retreat (if negative future prospects). By simplifying this to a one-round contest, I concluded that

> ... there should be an escalation range of closely matched combatants and that on either side of the range for a given individual, the higher ranking opponent should usually be prepared to escalate and the lower one to withdraw. Much fighting follows this pattern.

This seemed an intuitively satisfying conclusion, and one that has been supported by more recent game-theory models in which information about RHP is acquired during successive bouts in a contest (e.g. the 'sequential assessment game': Enquist & Leimar 1983).

The birth of the ESS

Little did I realise that at the same time as I was working on my 1974 paper [3] John Maynard Smith and George Price were formulating the seminal concept of ESS, and specifically, had been stimulated to derive ESS logic in order to interpret animal contests [1]. I first heard of this from the population geneticist, Brian Charlesworth, then at Liverpool, when mentioning my work on the evolution of fighting behaviour to him in late 1973. I cannot pretend to have been delighted by this news; my paper was already in early draft stages and I had invested much time on it. Little did any of us realise how much Maynard Smith and Price were to change modelling in evolutionary biology and our own work in the future.

The American George Price was a remarkable, tragic genius whose religious convictions drove him to share his home and possessions with the homeless; a venture that ended in his suicide in 1975, aged 52. His poignant story is the subject of an acclaimed monograph by Harman (2010) and a fascinating essay by Schwartz (2000). John Maynard Smith was also a remarkable person (obituary by Charlesworth & Harvey 2005): a major figure, liked and admired by several generations of scientists, and greatly revered for his fundamental insights, his incisive and clear-thinking approaches, and for his help and friendship to young researchers. Price had written a manuscript on contests for *Nature*, proposing that animals had been selected not to escalate in fights as a result of selection on individuals, to avoid receiving damage through retaliation by their opponent. It was written as a rebuttal of the ethological notion that animals do not escalate dangerously in contests for the good of the species. Maynard Smith refereed Price's paper for *Nature*, and was stimulated to formalise the concept of the ESS, which Price had alluded to verbally. Price's original manuscript remained unpublished, but eventually, their joint *Nature* publication appeared in late 1973 [1]. The delays involved resulted in the *Nature* paper being preceded by a similar publication in a collection of essays by Maynard Smith (1972).

This key paper [1] concerned models of 'symmetric contests' (i.e. between identical opponents), and besides introducing the ESS concept, it produced two key results for animal contests. First, using simulations with five strategies ranging from 'total war' to 'limited war', and assuming that escalated combat results in costly injury to one opponent, which then concedes victory, they generated a payoff matrix and deduced that 'limited war' strategies could be an ESS. This formed the basis of the 'hawk–dove' type model (Price's religious sensitivities resulted in 'dove' being replaced by 'mouse'). Second, in a model (later called the 'war of attrition' [4]) that assumed that contest costs increased continuously at a constant rate with time, and strategies consisted of a choice of how long to persist in the contest, they showed that no pure strategy (here a unique persistence time) could be an ESS. The ESS was instead a mixed strategy in which each opponent 'chooses' a time t randomly from the probability distribution $p(t) = (c/V)e^{-ct/V}$, where c is the rate of expenditure of costs and V the resource value (mathematical details of these early models are reviewed in Chapter 2).

It is interesting that Geist (1974) criticised Maynard Smith and Price [1] on the grounds that

> It perpetuates the old ethological myth that animals fight so as not to injure each other, or refuse to strike 'foul blows' and, presumably, kill each other ... They were not aware of the published field studies primarily of large mammals which have shown not only how dangerous combat is, but, more importantly, have also led to new theories of explaining aggressive behaviour on the basis of individual selection.

The new theories related mainly to Geist's offence and defence ideas (1966, 1971), which are not incompatible with game-theoretic approaches, particularly those involving sequential assessment and information acquired in asymmetric contests.

ESS and asymmetric contests

I was struck by the fact that Maynard Smith and Price [1] had ignored asymmetries between opponents; they had assumed opponents to be exactly identical except for their fighting strategies. Therefore I thought that my RHP assessment manuscript [3] still had some value, and the fact that it was based on individual selection principles, and seemed at least not incompatible with the new ESS concept, was some consolation. Rather uncomfortably, I had to modify my Introduction to cover the 1973 ESS paper [1] before I submitted my manuscript in early 1974 to *Journal of Theoretical Biology*. I did not know then that a new Maynard Smith paper [2] was already in press in the same journal. Among other things, this paper [2] dealt with 'arbitrary' (or 'uncorrelated') asymmetries – in which the opponents were exactly identical except for some arbitrary 'label' uncorrelated with RHP or resource value, known to both opponents. It showed that an ESS could consist of a 'conventional settlement' in which one opponent gives up without escalation in response to the arbitrary asymmetry, which could be 'prior resident' and 'interloper', while the other is prepared to escalate. Like Valerius Geist, my interest in contests had been stimulated by watching animals in nature; the notion of opponents equal in RHP and resource value was hard for me to accept in reality. John Maynard Smith was a reviewer of my paper [3]; his review suggested some corrections and additions, and broke the news of the new paper that he had in press [2], also in *Journal of Theoretical Biology*. He kindly sent me a copy of the manuscript, so I made additions to cover it and his various reviewer's suggestions, and also those of the other reviewer (possibly

Robert Hinde). Both recommended acceptance, and John's paper [2] and mine [3] were published back to back in 1974.

After reviewing my RHP paper, John invited me to Sussex for discussions, including an overnight stay at his home in Lewis with himself and his wife, Sheila. He was exceptional for his generous encouragement of young scientists, and I quickly came to admire and respect him greatly, both as a person and as a scientist and thinker. Indeed, he has probably been the scientist I have most admired over the three decades we knew each other; he was witty, amiable, immensely stimulating and great fun. His clear, incisive vision and friendly, good-humoured debate (at its best while relaxing with a glass of whisky) have been very much missed since his death in 2004 by all who knew him; it has been one of life's greatest privileges to have known and worked with him. I am acutely aware, with sadness, that it is he who should be writing this Foreword, not me.

We discussed whether an asymmetry such as prior resident/interloper could be respected in a 'conventional settlement' for cases where the retreating opponent had higher RHP and/or higher resource value than the winner, something we later called a 'paradoxical ESS', as opposed to a 'commonsense ESS' in which the winner has higher RHP and/or resource value. This stimulated a collaboration [4], mainly by post. I would pore over John's letters and the algebra they contained, all handwritten, making desperate attempts to work through the maths and to contribute ideas. However, this was indeed an asymmetric contest, and despite his amiable and genial nature, I felt awestruck by John's intellect. At Sussex I had discussed the fact that male dung fly numbers at a dropping show an ESS distribution such that all males (whatever the length of time they stay) gain an equal payoff rate (Parker 1970*b*). This was similar to the prediction for the symmetric war of attrition, and we thought that although it was an *n*-player game rather than a dyadic contest, it had similarities and resulted in a mixed ESS (however, the central reason why male dung flies depart at different times in an ESS fashion is that they show 'input matching' of the female arrival rate; Parker and Stuart 1976). For some reason I cannot remember, possibly due to a reviewer's comment, 'power' in RHP became changed to 'potential'. We went on to show that in a population where opponents have equal RHP but different resource values, and which starts by ignoring asymmetries and playing the symmetric war of attrition based on the mean resource value, only the commonsense ESS would evolve. Analysis of a simple hawk–dove type contest in which opponents have equal resource values but different RHP (and hence probability of being injured) showed that both paradoxical and commonsense strategies could be ESSs, but if there was imperfect information, the commonsense ESS was more likely, having a larger 'zone of attraction'. My main input to our paper [4] was really only to propose certain analyses relating to the probability of evolution of commonsense and paradoxical ESSs, and the 'information acquired during a contest' model, which had some similarities with the model in the second half of my RHP paper [3]. John approached this with a simple computer simulation. Our joint paper on asymmetric contests [4] became an ISI citation classic in 1989.

I had been preoccupied in my 1974 paper [4] by the fact that territory owners usually win against intruders; this issue took up the first half of the paper. I attempted to explain this in terms of the fact that the owner either had more to gain by winning (e.g. a fixed investment must be paid before resources can be extracted from territories) or had higher RHP (e.g. owners represent the truncated top RHP fraction of the population); they were thus commonsense solutions. However, as a result of our collaboration [4], paradoxical conventions also appeared to be a formal possibility, as did conventions based on an arbitrary asymmetry, such as an owner/intruder asymmetry uncorrelated with payoffs or RHP.

These early theoretical models [1–4, Table A] quickly spawned empirical studies that were interpreted within the new framework. Nick Davies's (1978) beautiful study of territoriality in male speckled wood butterflies, *Pararge aegeria* (Chapter 7), was the first and most notable example. Males meet females in small sunlit areas in woodland, and quickly settle in vacant 'sunspots'. However, if a sunspot is already occupied, any intruding male is quickly expelled by the owner. In a clever experiment, in which a series of two males were both duped into 'thinking' they were owners, Davies showed that there was a dramatic escalation in the contest duration, a result which fits with the theoretical prediction that the arbitrary asymmetry of prior residence in a sunspot is used to settle contests conventionally because of the risk of dangerous escalation if the convention is broken. Although prior residence in sunspots may not be an entirely uncorrelated asymmetry (Austad *et al.* 1979, Stutt & Willmer 1998, but see Davies 1979), Davies' study nevertheless fits the predictions for conventions in asymmetric contests, and was highly influential in stimulating a combination of empirical and theoretical approaches.

In 1975–1976, I became fascinated by the theoretical problem of sexual conflict, i.e. evolutionary conflict of interest between males and females, which could be manifest as behavioural conflict in a specific form of asymmetric contest. After convincing myself that there could indeed be evolutionary conflict (despite the fact the male and female having progeny in common), I wanted to know how this conflict might be resolved. For cases where males provide no parental investment other than sperm, I argued that there would be many instances in which, when a male meets a female, it would pay the male to mate but not the female. I made a computer simulation of an asymmetric ('sexual') war of attrition using six discrete time strategies, in which the male persisted and the female resisted. The result, regardless of starting frequencies of the strategies, was 'conventional'; one sex would give up with minimal time cost while the other would retain the strategies for much longer contests (which sex gave up quickly depended on the conditions). I then went on to examine an asymmetric arms race (the 'opponent-independent costs game') a variant of the war of attrition in which the opponent with the higher armament wins. In the war of attrition, the costs felt by each opponent are determined by the persistence time of the opponent with the lower time bid, while here, costs are developmental, and are felt independently by each opponent in accordance with their own 'bid'; they represent costs paid for armament rather than time in a contest. I could not get this simulation to generate a stable solution; instead, the strategies appeared to change continuously (what I called an 'unresolvable evolutionary chase'). This work, which had been a couple of years in gestation, was submitted as a book chapter [5] early in 1977, and was eventually published in 1979. The long delay in press caused me much angst (see Parker 2010, p. 445). At one of the first conferences on evolutionary game theory (in Bielefeld in late 1978), the game-theorist Reinhardt Selten (later to become a Nobel laureate) pointed out that this game could give a stable solution if opponents made small random deviations in strategy (armament) around their mean level; this related to his classic paper (Selten 1975) in which opponents make slight mistakes causing them to deviate from rationality (now called the 'trembling hand' principle). It struck me that the obvious reason for such random deviations in arms level would be random environmental variation, something I later investigated generally for asymmetric arms races (Parker 1983). This was to be my last work on dyadic contest theory, although I have subsequently worked on various types of biological scramble (n-player games in which gains from a resource can be shared in proportion to the relative effort of each given competitor, e.g. Parker 2000). But I fear it is true (as the distinguished ecological geneticist, Philip Sheppard, told me early in my academic career at Liverpool University) that one's most important contributions are made by the age of 35 and after then it becomes a matter of refinement. I have included this paper [5] in Table A because it has had more impact in contests in sexual conflict theory (Arnqvist & Rowe 2005) than in animal contests in general.

I have outlined the social history of these early papers of the 1970s at some length, probably partly through sheer nostalgia, although I hope more because of their historic significance: it was a remarkably exciting era (Parker 2006, 2010).

The immediate aftermath

Very soon, a series of theoretical developments followed. For example, the assumption of constant cost expenditure rates in the war of attrition was replaced by non-linear cost functions (Norman *et al.* 1977), followed by much more complex versions allowing general cost and reward functions, and discontinuities and gaps in strategy (Bishop & Cannings 1978, Bishop *et al.* 1978).

The hawk–dove model also received early attention. Treisman (1977) was one of the first to consider fights between relatives (see also Grafen 1979). Hammerstein (1981) examined how perfect information about all asymmetries can permit either payoff-irrelevant asymmetries or payoff-relevant asymmetries to be used conventionally. Under hawk–dove rules, i.e. where the contest ends in a costly injury to one opponent, he clarified that the conventional winning role need not necessarily be the role favoured with respect to payoffs, provided the cost of injury is sufficiently high.

However, a collaboration with Dan Rubenstein in 1978–1979 (during a year I spent at King's College Research Centre, Cambridge) led to the proposal that paradoxical conventions could not hold under war of attrition rules, and that the winning role must always be the one favoured with respect to payoffs (Parker & Rubenstein 1981). In a collaboration with an interesting history (Parker 2010, pp. 450–451), Peter Hammerstein and I confirmed this proposal by showing that in an asymmetric war of attrition, when opponents can make (rare) mistakes about role, only one of two roles can be a winning role, in the sense that the contestant in that role usually gains the resource (Hammerstein & Parker 1982). This winning role A is always a commonsense ESS defined by the payoff-relevant asymmetries. It follows the assessor rule (Parker 1974): the individual in role A is able to persist longer than the opponent before his contest costs exceed the value of the resource, i.e. it is 'retreat immediately if you assess yourself to be in role B, defined as $V_A/c_A > V_B/c_B$'. The closer this inequality, the more likely the opponents are to make mistakes in their assessments of roles A and B, which results in greater escalation. We suggested that this model could be applied to contests between dissimilar opponents such as males and females involved in sexual conflict contests. Note that we did not mean to imply (Chapter 2) that in such cases opponents made mistakes about whether they were male or female; rather the suggestion was that the 'sex label' was just one of many cues by which the male or female opponent could assess whether they occupied role A or B (Clutton-Brock & Parker 1995).

Dan Rubenstein and I had also analysed another version of the 'information acquired in a contest' model (Parker & Rubenstein 1981). Enquist and Leimar (1983) took the idea of information acquired in a contest much further in their 'sequential assessment game'; their formulation has much greater biological reality than earlier models. Like the asymmetric war of attrition, it also predicts greater escalation when opponents are more closely matched.

In the two decades following this early explosion in interest, there have been many further developments in dyadic contest theory (Chapter 2) and also in n-player contests (Chapter 3) and the surge of interest spawned Huntingford and Turner's (1987) seminal monograph on animal conflict. My own interest turned mainly to n-player scrambles, either generally (Parker 2000) or in specific contexts, such as parent–offspring conflict (e.g. Parker & Macnair 1979) and sperm allocation (reviewed in Parker & Pizzari 2010). John Maynard Smith retained his interest in contests throughout the rest of his life, sporadically producing further developments. His *magnum opus* (if one can be selected out of many) was his book (Maynard Smith 1982) reviewing ESS theory and posing many new ideas [6]; it still receives huge numbers of citations and although published in 1982, not in the 1970s, is included in Table A for completeness. He became interested in whether biological signals could be uncostly (the 'Philip Sidney game'; Maynard Smith 1991); and afterwards, in the general biology of signalling (his final book was on biological signals; Maynard Smith & Harper 2003).

And beyond

In a history of behavioural ecology (Parker 2006), I now believe that I was wrong to claim:

> Though contest behaviour still attracts both theoretical and detailed empirical research . . . it is now less popular. Its greatest contribution to behavioural ecology probably relates to its role in the development of ESS.

I suspect that this conclusion was influenced not so much because, since the 1970s and early 1980s, I had moved on to other areas of research (contest behaviour always remained an interest), but rather because its importance in the development of the seminal ESS concept and as a catalyst in the groundswell against implicit group-selection made subsequent developments appear less notable. There have been many important theoretical and empirical advances in the past 30 years. Indeed, this book shows that what has actually happened, as with many of the areas in behavioural ecology, is that the topic has matured into a fully fledged subject area in its own right. Its early roles in the group selection debate and the development of ESS in the optimality approach to adaptation do not detract from this – rather, they enhance its status as a new discipline, as this volume surely testifies.

References

Arnqvist G & Rowe L (2005) *Sexual Conflict*. Princeton: Princeton University Press.

Austad SN, Jones WT & Waser PM (1979) Territorial defense in speckled wood butterflies: Why does the resident always win? *Animal Behaviour*, 27, 960–961.

Bateman AJ (1948) Intrasexual selection in *Drosophila*. *Heredity*, 2, 349–368.

Bishop DT & Cannings C (1978) A generalised war of attrition. *Journal of Theoretical Biology*, 70, 85–124.

Bishop DT, Cannings C & Maynard Smith J (1978) The war of attrition with random rewards. *Journal of Theoretical Biology*, 74, 377–388.

Charlesworth B & Harvey P (2005) John Maynard Smith. *Biographical Memoirs of Fellows of the Royal Society*, 51, 253–265.

Clutton-Brock TH & Parker GA (1995) Sexual coercion in animal societies. *Animal Behaviour*, 49, 1345–1365.

Darwin C (1871) *The Descent of Man and Selection in Relation to Sex*. London: Murray.

Davies NB (1978) Territorial defence in the speckled wood butterfly (*Pararge aegeria*): The resident always wins. *Animal Behaviour*, 26, 138–147.

Davies NB (1979) Game theory and territorial behaviour in speckled wood butterflies. *Animal Behaviour*, 27, 961–962.

Enquist M & Leimar O (1983) Evolution of fighting behaviour: Decision rules and assessment of relative strength. *Journal of Theoretical Biology*, 102, 387–410.

Geist, V (1966) The evolution of horn-like organs. *Behaviour*, 27, 175–213.

Geist V (1971) *Mountain Sheep: A Study in Behavior and Evolution*. Chicago, IL: University of Chicago Press.

Geist V (1974) On fighting strategies in animal combat. *Nature*, 250, 354.

Grafen A (1979) The hawk–dove game played between relatives. *Animal Behaviour*, 27, 905–907.

Hammerstein P (1981) The role of asymmetries in animal contests. *Animal Behaviour*, 29, 193–205.

Hammerstein P & Parker GA (1982). The asymmetric war of attrition. *Journal of Theoretical Biology*, 96, 647–682.

Harman O (2010) *The Price of Altruism: George Price and the Search for the Origins of Kindness*. New York, NY: Bodley Head.

Huntingford FA & Turner A (1987) *Animal Conflict*. London: Chapman & Hall.

Huxley JS (1938*a*) Darwin's theory of sexual selection and the data subsumed by it, in the light of recent research. *American Naturalist*, 72, 416–433.

Huxley JS (1938*b*) The present standing of the theory of sexual selection. In: G de Beer (ed.) *Evolution*, pp. 11–41. Oxford: Oxford University Press.

Jacobs ME (1955) Studies on territorialism and sexual selection in dragonflies. *Ecology*, 36, 566–586.

Lack D (1968) *Ecological Adaptations for Breeding in Birds*. London: Methuen.

Leuthold W (1966) Variations in territorial behavior of Uganda kob *Adenota kob thomasi* (Neumann 1896). *Behaviour*, 27, 215–258.

Maynard Smith J (1964) Group selection and kin selection. *Nature*, 201, 1145–1147.

Maynard Smith J (1972) *On Evolution*. Edinburgh: Edinburgh University Press.

Maynard Smith J (1974) The theory of games and the evolution of animal conflicts. *Journal of Theoretical Biology*, 47, 209–221.

Maynard Smith J (1982) *Evolution and the Theory of Games*. Cambridge: Cambridge University Press.

Maynard Smith J (1991) Honest signalling: The Philip Sidney game. *Animal Behaviour*, 42, 1034–1035.

Maynard Smith J & Harper D (2003) *Animal Signals*. Oxford: Oxford University Press.

Maynard Smith J & Parker GA (1976) The logic of asymmetric contests. *Animal Behaviour*, 24, 159–175.

Maynard Smith J & Price GR (1973) The logic of animal conflicts. *Nature*, 246, 15–18.

Moore NW (1952) On the so-called 'territories' of dragonflies (Odonata-Anisoptera). *Behaviour*, 4, 85–100.

Moore NW (1957) Territory in dragonflies and birds. *Bird Study*, 4, 125–30.

Moore NW (1964) Intra- and interspecific competition among dragonflies (Odonata). *Journal of Animal Ecology*, 33, 49–71.

Norman RF, Taylor PD & Robertson RJ (1977) Stable equilibrium strategies and penalty functions in a game of attrition. *Journal of Theoretical Biology*, 65, 571–578.

Parker GA (1970*a*) The reproductive behaviour and the nature of sexual selection in *Scatophaga stercoraria* L. (Diptera: Scatophagidae). IV. Epigamic recognition and competition between males for the possession of females. *Behaviour*, 37, 113–139.

Parker GA (1970*b*) The reproductive behaviour and the nature of sexual selection in *Scatophaga stercoraria* L. II. The fertilization rate and the spatial and temporal relationships of each sex around the site of mating and oviposition. *Journal of Animal Ecology*, 39, 205–228.

Parker GA (1974) Assessment strategy and the evolution of fighting behaviour. *Journal of Theoretical Biology*, 47, 223–243.

Parker GA (1979) Sexual selection and sexual conflict. In: MS Blum & NA Blum (eds.) *Sexual Selection and Reproductive Competition in Insects*, pp. 123–166. London: Academic Press.

Parker GA (1983) Arms races in evolution: An ESS to the opponent-independent costs game. *Journal of Theoretical Biology*, 101, 619–648.

Parker GA (2000) Scramble in behaviour and ecology. *Philosophical Transactions of the Royal Society B*, 355, 1637–1645.

Parker GA (2001) Golden flies, sunlit meadows: a tribute to the yellow dung fly. In: LA Dugatkin (ed.) *Model Systems in Behavioural Ecology: Integrating Conceptual, Theoretical, and Empirical Approaches*, pp. 3–26. Princeton, NJ: Princeton University Press.

Parker GA (2006) Behavioural ecology: The science of natural history. In: JR Lucas & LW Simmons (eds.) *Essays on Animal Behaviour: Celebrating 50 Years of Animal Behaviour*, pp. 23–56. Burlington, MA: Elsevier.

Parker GA (2010) Reflections at dusk. In: L Drickamer & DA Dewsbury (eds.) *Leaders in Animal Behavior: The Second Generation*, pp. 429–464. Cambridge: Cambridge University Press.

Parker GA & Macnair MR (1979) Models of parent–offspring conflict. IV. Suppression: Evolutionary retaliation by the parent. *Animal Behaviour*, 27, 1210–1235.

Parker GA & Pizzari T (2010) Sperm competition and ejaculate economics. *Biological Reviews*, 85, 897–934.

Parker GA & Rubenstein DI (1981) Role assessment, reserve strategy, and acquisition of information in asymmetric animal conflicts. *Animal Behaviour*, 29, 221–240.

Parker GA & Stuart RA (1976) Animal behaviour as a strategy optimizer: Evolution of resource assessment strategies and optimal emigration thresholds. *American Naturalist*, 110, 1055–1076.

Perrins CD (1964) Survival of young swifts in relation to brood size. *Nature*, 201, 1147–1148.

Richards OW (1927) Sexual selection and related problems in the insects. *Biological Reviews*, 2, 298–364.

Schwartz J (2000) Death of an altruist: Was the man who found the selfless gene too good for this world? *Lingua Franca*, 10, 51–61.

Selten R (1975) A re-examination of the perfectness concept for equilibrium points in extensive games. *International Journal of Game Theory*, 4, 25–55.

Stutt AD & Willmer P (1998) Territorial defence in speckled wood butterflies: Do the hottest males always win? *Animal Behaviour*, 55, 1341–1347.

Treisman M (1977) The evolutionary restriction of aggression within a species: A game theory analysis. *Journal of Mathematical Psychology*, 16, 167–203.

Williams GC (1966) *Adaptation and Natural Selection*. Princeton, NJ: Princeton University Press.

Wynne-Edwards VC (1962) *Animal Dispersion in Relation to Social Behaviour*. Edinburgh: Oliver & Boyd.

Wynne-Edwards VC (1963) Group selection and kin selection. *Nature*, 201, 1147.

Preface

Mark Briffa & Ian C.W. Hardy

This book is about the evolution of contest behaviour in animals. It covers both predictive theories for contest evolution and empirical evidence. There are several potential strategies for organising an edited book that collects together a diverse range of study systems and a rich body of theory. One would have been to invite authors to each write a chapter about their favourite concept. For example, there are several alternative theoretical explanations (models) for contestant assessment during agonistic interactions that appear frequently in the contest literature (these are introduced in Chapter 1 and detailed in Chapter 2) and MB, along with several other contributors to this volume, has been especially interested in using a particular study-species to test the key features of these models in order to investigate assessment rules and the possible functions of repeated agonistic signals. In other words, we are all interested in how the loser makes the decision to give up, and could each have contributed a chapter along similar lines, covering the relevant theory as well as detailing our own experiments. It soon became apparent, however, that there was a potential cost associated with this layout: as a result of the tight links between theory and experimental work described above, many authors would have wanted to write about the same concepts, albeit applied to different animals, leading to much conceptual repetition between chapters.

Our alternative, and adopted, strategy for organising this book has been to divide it into two main sections, the first dealing with general theory and the second comprising a series of chapters arranged by taxon (in the somewhat uncomfortably traditional 'invertebrates to humans' sequence). The link between the theoretical and empirical sections is a chapter on analysis of contest behaviour data. This includes recent advances in our understanding of the appropriate experimental design and analytical approaches for testing hypotheses about contest behaviour, with the aim of providing practical advice to those engaged in empirical contest research. As we see it, this scheme has two main advantages.

First, all of the theory concerning contests, dyadic (pairwise) and then multi-party, is present in a contiguous narrative. This is something that we feel has been missing from recent expositions of contest behaviour, as illustrated by two recently updated textbooks. In the 4th edition of *An Introduction to Behavioural Ecology* (Davies *et al.* 2012), the foundation stones of contest theory, the Hawk–Dove game (Maynard Smith & Price 1973) and the war of attrition (Maynard Smith & Price 1973, Maynard Smith & Parker 1976) are discussed as part of a chapter on competing for resources, while the sequential assessment model ('SAM': Enquist & Leimar 1983) is dealt with in a later chapter on communication. Thus, these important components of contest behaviour appear in different parts of the book. Furthermore, the alternative explanations for repeated agonistic displays comprising the energetic war of attrition ('EWOA': Payne & Pagel 1997) and the cumulative assessment model ('CAM': Payne 1998), which have gained a great deal of recent support from empirical studies, do not fit easily into this way of organising things. In the second edition of *Principles of Animal Communication* (Bradbury & Vehrencamp 2012), these alternative assessment models are discussed in detail, but the theory surrounding other types of contest – for example, fights that result in fatalities and multi-party 'battles' – are outside the book's scope. In both of these books the choices made for dealing with contest theory suit their purposes (to introduce students to the concept of the evolutionarily stable strategy (ESS), and to discuss the role of agonistic signals, respectively). Here, we have aimed to provide an explanation of the totality of contest theory. The manner in which contest behaviour is often discussed, as an interesting addendum to, or subset of, other areas of behavioural ecology research, is possibly a result of the fact that, as noted above, contest theory intersects with so many other questions. Therefore, we believe that what has been lacking and what will be beneficial is having all of the conceptual issues surrounding contest behaviour dealt with in the same place. Recent attempts have been made to do this for assessment models at least (Arnott & Elwood 2009), sometimes within the settings of empirical papers (Kelly 2005, Stuart-Fox 2006, Briffa 2008), but Chapters 2 and 3 of this volume extend this approach to cover basic game-theory models as well as recent eco-genetic models and attrition laws for multi-party contests.

The second advantage is that this way of organising the book frees up the authors of empirical chapters to concentrate on the research findings of their own study systems. Our aim in this respect was not to in any way divorce the empirical work from its theoretical underpinnings; rather, we wanted to avoid unnecessary repetition of theoretical details that can be best exposited elsewhere, in a contiguous fashion. As far as we are aware this is the first collection of taxon-specific reviews of contest behaviour, although this structure has been implemented before in an edited book on that other 'original' topic in evolutionary game theory, sex ratios (Hardy 2002). In editing this volume we have both been struck by the fact that the choices about which sets of questions to study have very clearly been driven by the biological quirks of each study system. Hermit crabs are excellent for investigating agonistic signals (Chapter 5), beetles are ideal for looking at the development of weaponry (Chapter 9) and encounters between rival ant colonies provide insights into multi-party 'battles' (Chapter 8) and different study systems connect naturally to different areas of biological research beyond contest behaviour *sensu stricto*. Moreover, on reading these taxon-specific chapters we formed the distinct impression that the authors are often motivated as much by a passionate interest in (and detailed knowledge of) their study organisms, as by a desire to solve intriguing behavioural ecology questions. We think that the study of animal contests is all the better for this; we hope that our way of organising the book has allowed the different interconnections to be emphasised and the passions of the authors to shine through and, of course, that readers enjoy these chapters as much as we did.

The most recent previous volume to treat contest behaviour as a subject in its own right was *Animal Conflict* by Huntingford and Turner (1987). In spite (or perhaps because) of the fact that the study of animal contests has initiated so many strands within the field of behavioural ecology, contest behaviour has perhaps come to be viewed in recent years as a slightly niche topic. In particular, contests may have been superseded by, or perhaps subsumed into, the vast amount of interest in sexual selection as a potential driver of animal behaviour. We see this largely as evidence for the importance of contests, rather than as indication of their topical demise (although we were concerned to notice that a recent and large textbook explicitly on *Behavioural Ecology* (Danchin *et al.* 2008) only mentioned animal contests *en passant*, as a subset of sexual selection!). A concise version of an integrated approach to contest behaviour is provided by a dedicated chapter (Briffa & Sneddon 2010) in the recent multi-author volume *Evolutionary Behavioral Ecology* (Westneat & Fox 2010). However, in the present book, we hope to provide a more detailed account of both the historical development of, and the most recent advances in, animal contest research itself. We find a historical perspective to any research field to be immensely valuable; as Terry Pratchett has noted: 'It is important that we know where we come from, because if you do not know where you come from, then you do not know where you are, and if you do not know where you are, then you don't know where you are going. And if you don't know where you're going, you're probably going wrong' (Pratchett 2010, p. 423). The overarching aim of this book is, however, to emphasise that contests are an important aspect of the lives of diverse animals, from sea anemones competing for space on a rocky shore to fallow deer bucks competing for access to females, and are therefore a fascinating and important topic of study in their own right. As our respected colleague, and contributor to this book, Bob Elwood, has put it, 'Contests determine the unequal division of resources and thus drive natural selection to a huge extent' (R.W. Elwood, pers. comm.).

References

Arnott G & Elwood RW (2009) Assessment of fighting ability in animal contests. *Animal Behaviour*, 77, 991–1004.

Bradbury JW & Vehrencamp SL (2012) *Principles of Animal Communication*, 2nd edn. Sunderland, MA: Sinauer.

Briffa M (2008) Decisions during fights in the house cricket, *Acheta domesticus*: Mutual or self assessment of energy, weapons and size? *Animal Behaviour*, 75, 1053–1062.

Briffa M & Sneddon LU (2010) Contest behaviour. In: DF Westneat & CW Fox (eds.) *Evolutionary Behavioral Ecology*, pp. 246–265. Oxford: Oxford University Press.

Danchin E, Giraldeau L-A & Cézilly F (eds.) (2008) *Behavioural Ecology*. Oxford: Oxford University Press.

Davies NB, Krebs JR & West SA (2012) *An Introduction to Behavioural Ecology*, 4th edn. Chichester: Wiley-Blackwell.

Enquist M & Leimar O (1983) Evolution of fighting behaviour; Decision rules and assessment of relative strength. *Journal of Theoretical Biology*, 102, 387–410.

Hardy ICW (ed.) (2002) *Sex Ratios: Concepts and Research Methods*. Cambridge: Cambridge University Press.

Huntingford FA & Turner A (1987) *Animal Conflict*. London: Chapman & Hall.

Kelly CD (2005) Fighting for harems: Assessment strategies during male–male contests in the sexually dimorphic Wellington tree weta. *Animal Behaviour*, 72, 727–736.

Maynard Smith J & Parker GR (1976) The logic of asymmetric contests. *Animal Behavior*, 24, 159–175.

Maynard Smith J & Price GR (1973) The logic of animal conflict. *Nature*, 246, 15–18.

Payne RJH (1998) Gradually escalating fights and displays: The cumulative assessment model. *Animal Behaviour*, 56, 651–662.

Payne RJH & Pagel M (1997) Why do animals repeat displays? *Animal Behaviour*, 54, 109–119.

Pratchett T (2010) *I Shall Wear Midnight*. London: Corgi books.

Stuart-Fox D (2006) Testing game theory models: Fighting ability and decision rules in chameleon contests. *Proceedings of the Royal Society of London B*, 273, 1555–1561.

Westneat DF & Fox CW (eds.) (2010) *Evolutionary Behavioral Ecology*. Oxford: Oxford University Press.

Acknowledgements

It is naturally a pleasurable duty to acknowledge the many people who have contributed towards this book. First and foremost we thank the 22 chapter authors who turned in work that exceeded our expectations and, of course, Geoff Parker for the Foreword. We hope that the results of our combined efforts will help readers to develop both a deeper understanding of the theory and provide an insight of just how fundamental contests are to the lives of diverse animals.

Nearly every chapter was peer-reviewed by at least one internal referee (an author of another chapter) and at least one external referee: although this book has two editors, from the very start we regarded the refereeing process as a crucial element in its ontogeny for the same reasons that journal papers are peer-reviewed. The exceptions to this reviewing process were our brief 'intros and outros' (Chapters 1 and 16). Both were read through by Geoff Parker and the Introduction was also read by Sophie Mowles. The data analysis chapter (Chapter 4) has been reviewed internally by Sophie Mowles and subjected to the scrutiny of two external referees. For internal refereeing of the remaining core chapters we thank: Tim Batchelor, Ryan Earley, Bob Elwood, Tim Halliday, Darrell Kemp (who reviewed two chapters), Hanna Kokko, Dómhnall Jennings, Martin Gammell, Mike Mesterton-Gibbons, Armin Moczek, Sarah Pryke and Emilie Snell-Rood. For external refereeing of these chapters we thank Gareth Arnott, Thomas Breithaupt, James Cook, Sarah Collins, Duncan Irschick, Robert Jackson, Geoff Parker, Sean Rands, Claudia Rauter, Nick Royle and Christer Wiklund. In addition, Michael Reichert and Sophie Mowles initially provided external reviews of chapters but then subsequently became chapter co-authors. We thank Megan Waddington and Dominic Lewis at Cambridge University Press for patiently letting us get on with things and Pat Backwell, Teresa D. Baird, Ludek Bartoš, Patrick Bergeron, William Clarke, Tanya Detto, Sonia Dourlot, Sophie Mowles, Anthony O'Toole, Joyce Poole, Fabian Rudin, Franck Simonnet, Maria Thacker and Martin Wikelski for supplying photographic material and permissions.

Closer to home, we thank our respective families for putting up with our editorial commitments while all sorts of 'bourgeois' events in family life were occurring (such as moving houses and jobs, having children and *not* having enough holidays with them). Finally, we would like to thank each other. We have worked approximately equally on this book. The book's early conception was ICWH's but it was immediately obvious that it would benefit greatly from being co-edited with MB due to our complementary contest behaviour research experiences: in caricature, ICWH tends to focus on contest *outcomes* and also to link contests to other areas of behavioural ecology and applied biology, often using parasitoid wasps; MB tends to work on behaviours occurring *within* contests and to link these to physiological measures, mainly using crustaceans. This book was a fully joint project from well before the stage of being commissioned. To reflect this approximate equality of input, ICWH is first-named editor of the book as a whole while MB is first author of those chapters contributed to by us both: one might say that one of us has won the bouts while the other has won the contest but, actually, the thing we would most like to thank each other for is that we have managed to co-edit a book largely about agonistic dyadic behaviour without actually engaging in any between ourselves.

I.C.W.H., Sutton Bonington
M.B., Plymouth
July 2012

Abbreviations

ADI	average dominance index
AIC	Akaike Information Criterion
ART	alternative reproductive tactics
ATR	'all-trunk raised'
AVT	arginine vasotocin
BIC	Bayesian information criterion
BSA	behavioural sequence analysis
BSA	bovine serum albumin
CAM	cumulative assessment model
DS	David's score
EFOT	expected future ownership time
ESDA	exploratory sequential data analysis
ESS	evolutionarily stable strategy
EWOA	energetic war of attrition
GEE	generalised estimating equation
GLM	general linear model
GLMM	generalised linear mixed model
GLZ	generalised linear model
I&SI	inconsistencies and strength of inconsistencies
LMM	linear mixed models
MID	militarised interstate dispute
MVT	marginal value theorem
OSR	operational sex ratio
PBT	preferred body temperature
PCD	programmed cell death
PCR	polymerase chain reaction
RDNL	relative difference in nematocyst length
RHP	resource holding power/potential
RPH	relative plasticity hypothesis
RWD	relative weight difference
SAM	sequential assessment model
V	resource value
WOA	wars of attrition

Introduction to animal contests

Mark Briffa & Ian C.W. Hardy

1.1 Animal contests in nature

Next time you stand on a seashore and look carefully with your 'zoologist's eyes', you may be surprised at the high diversity of animal phyla that are present, even within a single intertidal rock pool. If you are patient and can stay still for a few minutes, another surprise in store is the preponderance of aggressive behaviour demonstrated by the intertidal fauna. Depending on which part of the world your rocky shore is in, you might observe some of the following: male Azorean blennies fighting over the nests that they need in order to attract females; pre-copula pairs of shore crabs with inter-male aggression over the ownership of recently moulted females, as these females are only receptive to sperm during a brief post-moult period; common European hermit crabs *rapping* in an attempt to evict an opponent from its gastropod shell; and, if you really have a lot of time on your hands, you might notice slow-moving sea anemones striking one another with special tentacles called *acrorhagi*, during disputes over space. Of course, aggressive behaviour is not restricted to intertidal marine animals. Take a walk in the woods and you could witness aggression over the ownership of territory; this is one of the reasons why male birds sing, why male butterflies perform many of their aerial displays and why armies of female worker wood ants try to kill individuals from a different colony. These examples illustrate two important points about aggression: first, animals will fight over a range of resources, when the ability to access those resources is a major constraint on fitness. In many cases this involves conflict over access to mates, as in the case of shore crabs. However, other resources such as territory, food and shelter are also contested, and influence the fitness of females as well as males. The second point is that aggressive behaviour is extremely widespread among animal taxa: these examples alone are drawn from three different phyla: chordates, arthropods and cnidarians.

1.2 Defining animal contests

How should we best describe the diverse behaviours that we have so far called 'aggressive'? As with many aspects of behaviour there are 'everyday' words that we might use in a fairly loose way. 'Aggression' works reasonably well for most of the examples given above, denoting a particular type of conflict of interests that is resolved through a direct and discrete interaction between the opposing parties, but aggression is a broad term that seems less appropriate for displays in butterflies (Chapter 7) than for examples that involve some form of escalated encounter, perhaps involving injuries, as in hymenopterans (Chapter 8). Huntingford and Turner (1987) discussed this issue in the preface to *Animal Conflict*, the seminal textbook dealing with aggressive behaviour. They also pointed out that a direct conflict between opponents might involve both *defensive* and *submissive* behaviours as well as *offensive* behaviours, and 'aggression' in its everyday sense does not seem to fit well with defence and submission; for example, in human interstate conflict (Chapter 15), the term 'aggressor' is usually applied to the state that is deemed to have initiated the conflict. Therefore, the term 'agonistic' might be more appropriate: agonistic behaviour is 'a system of behaviour patterns having the common function of adaptation to situations involving physical conflict' (Scott & Fredericson 1951). Agonistic is a useful term but, again, while it encompasses both defence and submission, it seems best applied to examples involving attempts to inflict damage on the opponent, or at least where there is some sort of escalated phase involving physical contact. Huntingford

Animal Contests, ed. I.C.W. Hardy and M. Briffa. Published by Cambridge University Press.

and Turner (1987) settled on the term '*Animal Conflict*' as one which is broad enough to encompass various forms of direct interaction that result from a conflict of interests. This broad term, however, might also include other types of conflict, such as conflict over survival of the prey item in a direct interaction between predator and prey. In this book we focus on a subset of interactions where the conflict occurs over the ownership of discrete resource units such as mates, food or shelter, as well as territories or positions in a social hierarchy that determine access to these items. Thus, while the form of these interactions is incredibly diverse, the underlying reason is always the same; essential resources cannot be shared and are unlikely to be available in an unlimited supply, so the best way to secure access to the resource is often to take it (or defend it) from another party. Therefore, in this volume we have chosen to use the term '*Animal Contests*'. We define a contest as *a direct and discrete behavioural interaction that determines the ownership of an indivisible resource unit*. This definition seems broad enough to include the aerial displays of butterflies (Chapter 7), the rutting of red deer (Chapter 14) and fatal fighting in ants (Chapter 8), while being specific enough to exclude other types of conflict of interest, such as scramble competition (over resources that are divisible), conflict between parents and offspring (over parental investment) or predators and prey (over dinner). This is not to say that the other terms should not be used to describe specific types of contest; it might be very appropriate to talk about aggression in shore crabs, agonistic displays in fiddler crabs or even 'fighting' in butterflies, but these terms can be thought of as elements of contest behaviour.

1.3 Animal contests and behavioural ecology

Contest behaviour appears to be a significant feature of the lives of diverse animals but, apart from involving a conflict of interests over the ownership of a resource unit, what do the different examples of contest behaviour all have in common? In terms of underlying mechanisms ('how?'), identifying unifying themes is possible but less than straightforward. Variation in testosterone, for example, influences aggressiveness in fish, birds and mammals, but this steroid is absent in non-vertebrate animals. Similarly, while the energetic demands of aggression can be a limiting factor across taxa, the metabolic pathways involved can be quite different; in vertebrates the energy storage molecule that buffers against ATP (adenosine triphosphate) depletion is creatine phosphate whereas in other animals it is arginine phosphate. It is perhaps easier, then, to identify commonalities in the 'whys' of aggression. Behavioural ecologists will recognise this approach, of focussing on questions about the function of behaviour, which is one of Tinbergen's 'four questions' in biology; the others being causation, development and evolution (Tinbergen 1963, Bolhuis & Verhulst 2009). Indeed, the analysis of contest behaviour has been of consistent interest for behavioural ecologists and the field has been characterised by a continuous and productive interplay between theoretical developments and insights gained from empirical studies. An early pioneer, Geoff Parker (see the Foreword) worked on both aspects, his insights from observations of fighting dung flies (Parker 1970) leading to the development of the study of asymmetric contests and the influence of 'resource holding power' (RHP, also termed 'resource holding potential') and 'resource value' on contest outcomes (Parker 1974, Maynard Smith & Parker 1976). RHP can be defined as 'an individual's ability to obtain or retain a resource during a contest', and may comprise several components, both intrinsic and extrinsic to the individual. Resource value (denoted by V, Chapter 2) can be defined as the value that the individual places on obtaining or retaining the resource. V may be influenced by a number of attributes of the contested resource and also the same contested resource may have different value to different contestants.

This 'RHP and V' tradition has enhanced the study of animal contests and has led to some major landmarks in our understanding of why contest behaviour has evolved. Magnus Enquist and Olaf Leimar derived the hugely influential 'sequential assessment model' (SAM: Enquist & Leimar 1983) and later used cichlid fish as an experimental system for its testing (Enquist *et al.* 1990, Chapter 10). Currently, the SAM (Enquist & Leimar 1983) and two later models, the energetic war of attrition (EWOA: Payne & Pagel 1997) and the cumulative assessment model (CAM: Payne 1998), form a triad of theories about the evolution of contest behaviour that have been the subjects of intense empirical research in recent years (e.g. Stuart-Fox 2006). The reason why these three models have been particularly influential is that (a) they are clearly differentiated by their main assumptions about the functions of agonistic behaviour and (b) they make clear predictions that

Table 1.1 Key features of three influential models that have permeated empirical studies of contest behaviour, particularly since the late 1990s. Each assumes a different 'reason' for the use of repeated agonistic behaviours, which in many cases are non-injurious. These assumptions lead to different predictions about the changes in intensity expected during pairwise contests and about the way the contests are structured, discussed in detail in Chapter 2. They also lead to inferences about the expected correlations between RHP and contest duration, discussed in Chapter 4. Chapter 4 also explores the options for appropriate statistical testing of hypotheses about contest duration, dynamics and structure.

Model	Reason for repeated actions	Giving up decision based on	Duration should correlate primarily with RHP of	Duration could also correlate with RHP of	Contest dynamics and structure
Sequential assessment model, SAM	To reduce sampling error in the opponent's estimate of the performer's RHP	Information about the opponent's RHP	Loser (+) Winner (−)		Contests structured into a series of phases characterised by increasingly intense agonistic behaviours; but within phases agonistic behaviour should be performed at a constant rate
Energetic war of attrition, EWOA	To demonstrate the performer's endurance	A threshold of costs that accrue as a result of the loser's actions	Loser (+)	Winner (+)	Constant, escalating or de-escalating within phases
Cumulative assessment model, CAM	To inflict costs upon the opponent	A threshold of costs that accrue as a result of the loser's actions plus costs that are inflicted on the loser by the actions of the opponent	Loser (+)	Winner (+) or Winner (−)	Constant, escalating or de-escalating within non-injurious phases; escalates within injurious phases

are amenable to experimental testing. (Table 1.1 provides a summary of these assumptions and predictions; Chapter 2 considers these models in more depth and Chapter 4 gives details of statistical and experimental design approaches that can be used to distinguish between them.) Thus the tradition of interplay between theory and empirical work continues in the contest literature.

The three models discussed above continue another tradition in animal contest research in that they may also provide insights that can be applied to other areas of behavioural ecology. The body of theory surrounding contest behaviour has always proved remarkably adaptable and it is worth reflecting here on the influence that this body of theory has had on the wider field of behavioural ecology as a whole. Models of contest behaviour were among the very first to utilise 'game-theoretic' reasoning to understand the evolution of animal function. Following hot on the heels of Hamilton's (1967) 'unbeatable strategy' model of sex ratio evolution, Maynard Smith and Price (1973) applied, for the first time, game theory to animal contests in their Hawk–Dove (or 'Hawk–Mouse') model, which explains the evolution of non-injurious fighting. In classical game theory, one asks 'which strategy from a set of strategies should an individual play against an opponent who may also choose from the set of strategies' (think about a game of rock–paper–scissors between two people, but also think about the cold war and nuclear standoffs between super-powers; the application of contest theory to interstate conflict is discussed in Chapter 15 and an evolutionary 'rock–paper–scissors' game occurs in lizard contests, Chapter 12). In evolutionary game theory, the underlying logic is the same but the emphasis has shifted from attempting to anticipate (using rational forethought) the decisions made by individuals (or other entities such as companies or countries) about what strategy to play, to predicting what strategies should be 'chosen' (or 'favoured' or 'selected') over evolutionary time, with natural selection acting as the optimising agent. Game theory shows that the best thing for a focal individual to do will be determined by what others are doing (and vice versa). Game-theoretical reasoning explains the evolution of behaviours that, if analysed naïvely and superficially, might seem to be counter-intuitive results of natural selection (such as using signals to resolve a contest). Rather, evolutionarily stable strategies (ESS) arise because individuals act to maximise their own fitness, and these strategies therefore tend to be at variance with what would benefit the group as a whole. Another key insight was that frequency-dependent selection can lead to the evolution of stable mixes of alternative behavioural strategies. The publication of Hamilton's (1967) and Maynard Smith and Price's (1973) game-theoretic models spawned large and successful literatures within evolutionary and behavioural ecology: for sex ratios, for instance, see Hardy (2002) or West (2009) and for contest behaviour, continue to read this book.

Acknowledgements

We thank Geoff Parker and Sophie Mowles for comments.

References

Bolhuis JJ & Verhulst S (eds.) (2009) *Tinbergen's Legacy: Function and Mechanism in Behavioral Biology*. Cambridge: Cambridge University Press.

Enquist M & Leimar O (1983) Evolution of fighting behaviour; Decision rules and assessment of relative strength. *Journal of Theoretical Biology*, 102, 387–410.

Enquist M, Leimar O, Ljunberg T, *et al.* (1990) A test of the sequential assessment game: Fighting in the cichlid fish *Nannacara anomala*. *Animal Behaviour*, 40, 1–14.

Hamilton WD (1967) Extraordinary sex ratios. *Science*, 156, 477–488.

Hardy ICW (ed.) (2002) *Sex Ratios: Concepts and Research Methods*. Cambridge: Cambridge University Press.

Huntingford FA & Turner A (1987) *Animal Conflict*. London: Chapman & Hall.

Maynard Smith J & Parker GR (1976) The logic of asymmetric contests. *Animal Behaviour*, 24, 159–175.

Maynard Smith J & Price GR (1973) The logic of animal conflict. *Nature*, 246, 15–18.

Parker GA (1970) The reproductive behaviour and the nature of sexual selection in *Scatophaga stercoraria* L. (Diptera: Scatophagidae). IV. Epigamic recognition and competition between males for the possession of females. *Behaviour*, 37, 113–139.

Parker GA (1974) Assessment strategy and the evolution of fighting behaviour. *Journal of Theoretical Biology*, 47, 223–243.

Payne RJH (1998) Gradually escalating fights and displays: The cumulative assessment model. *Animal Behaviour*, 56, 651–662.

Payne RJH & Pagel M (1997) Why do animals repeat displays? *Animal Behaviour* 54, 109–119.

Scott JP & Fredericson E (1951) The causes of fighting in mice and rats. *Physiological Zoology*, 24, 273–309.

Stuart-Fox D (2006) Testing game theory models: Fighting ability and decision rules in chameleon contests. *Proceedings of the Royal Society of London B*, 273, 1555–1561.

Tinbergen N (1963) On aims and methods of ethology. *Zeitschrift für Tierpsychologie*, 20, 410–433.

West SA (2009) *Sex Allocation*. Princeton, NJ: Princeton University Press.

Chapter 2

Dyadic contests: modelling fights between two individuals

Hanna Kokko

2.1 Summary

Animal contests were the focal topic that brought game theory to the attention of behavioural ecologists, giving rise to evolutionary game theory. Game theory has remained by far the most popular method of deriving theoretical predictions ever since, although it nowadays coexists with other methods of analysis. Here I review the developments to date and highlight similarities and differences between models. There is a clear progression from simple two-player models with fixed payoffs to explicit tracking of fitness consequences in a population context. In many cases this development has helped to discover that some of the early predictions may have been misleading. Despite the large number of current models, there are still gaps in the theoretical literature: sometimes simplifying assumptions have been relaxed in one context but not another. I hope that by highlighting these gaps theoreticians will be provided with new research ideas, and empiricists will be encouraged not only to distinguish between existing models but to be able to point out assumptions that are essential for deriving a result yet may be violated in existing systems, thus directing new modelling in the most useful direction.

2.2 Introduction

Until the mid 1960s, animal contests were viewed using group selectionist thinking. Julian Huxley (1966) thought that ritualised fights evolved to limit intraspecific damage, partly based on Konrad Lorenz's (1964, 1965) ideas that species need to evolve mechanisms that limit aggression in species that possess dangerous weapons for other reasons, e.g. as adaptations for capturing prey. Following George C. Williams' (1966) book *Adaptation and Natural Selection: A Critique of Some Current Thought*, however, biologists became aware of the need to distinguish between explanations that are based on benefits to the individual versus those that rely on benefits accruing to a group (or a species). This immediately raises the question of what limits aggression in animal populations. There are many instances where individuals interact peacefully, even cooperatively. While the evolution of cooperation is beyond the scope of this chapter, refraining from maximal aggression appears similarly puzzling if one expects selfish genes to be as ruthless as possible. Animal fights can be lethal, but very often they are not. Encounters between neighbouring territory owners, for example, often involve a lot of display and only rarely escalate to physical contact. A 'mutant' that always strikes first and kills a neighbour might be expected to spread, until the orderly territorial system is destroyed in the population. Why does it not? Likewise, male snakes wrestle for mating opportunities, but do not generally use their fangs against each other (Maynard Smith & Price 1973). Why not use all weapons available, given that the loser's genes will not be propagated into the next generation?

These kinds of questions prompted a new wave of animal behaviour studies in the 1970s. These studies made use of an approach derived from mathematical economics: John von Neumann, Oskar Morgenstern and John Nash had developed *game theory* in the early half of the twentieth century. Game-theory models seek best responses of individuals given what other individuals in the group or population are doing; in other words, game theory is about making decisions in the presence of other decision-makers. In the 1970s it was realised that game theory is the ideal tool to study the evolution of behaviour in non-human

Animal Contests, ed. I.C.W. Hardy and M. Briffa. Published by Cambridge University Press.

animals as well. Economists had made the assumption that their players are rationally calculating what is best for them, but as natural selection rewards best strategies, no conscious thought is required in evolutionary game theory (even plant growth can be modelled using game-theoretic tools, see Falster & Westoby 2003 for a review).

Many of the early papers that contain game-theoretic treatments of animal contests explicitly mention that this approach offers a novel alternative to earlier group-selectionist thinking (Maynard Smith & Price 1973, Maynard Smith 1974, Parker 1974). In papers published today, this contrast no longer needs to be made. Game theory has established itself as the most important tool for developing theories of dyadic contests, and it peacefully coexists with supporting alternatives such as individual-based simulations (e.g. Kemp 2006, Just *et al.* 2007) and genetic algorithms (Hamblin & Hurd 2007). This chapter will follow this development to date.

The focus of this chapter is on interactions between two individuals. Multi-contestant games are developed in Chapter 3, together with associated topics such as dominance hierarchies and winner–loser effects. Although in principle a dominance hierarchy can form between just two individuals, in practice these topics usually require attention to interactions that happen in larger groups (Crowley 2001). It is also a central theme of the current chapter that the history of dyadic fighting models shows an increasing appreciation of the importance of population-level phenomena, as these can have a strong influence on the traits of the two individuals in question.

2.3 Notation

Contest models have very often differed in notation. While it does not matter whether a non-aggressive individual is said to play 'Dove' (Maynard Smith 1982), 'Careful' (Crowley 2001, Kokko *et al.* 2006) or 'Mouse' (Maynard Smith & Price 1973) if the mathematical essence of the strategy remains the same, wildly varying notation makes it obviously harder to grasp similarities between models. It is fortunate that many models have now converged on roughly similar terminology. In this chapter I freely 'mistreat' published models by recasting them using the notation summarised in Table 2.1.

The most important variables, present in many models, are resource value V, cost of fighting C, time t, and the probability p that a specific individual will win a fight. The individuals in question will be labelled as A and B, where A is in the 'favoured' role (if there is one; sometimes $p_A = p_B = p$). Often 'favoured' means that $p_A > p_B$ if the situation escalates to a fight but, depending on context, the 'favoured' role has been used to mean a discoverer of a resource (such as the prior owner of a territory) or an individual with higher Resource Holding Power, or one who would benefit more from possessing a resource. Strategies will be called *commonsense* if they lead to the favoured individual being the more likely one to win, and *paradoxical* in the opposite case. Strategies are called *conventional* if one of the contestants immediately retreats after some asymmetry is observed (although in the literature one also finds 'conventional' used roughly synonymously with 'limited war', i.e. no immediate escalation to maximal use of weapons, e.g. Maynard Smith & Price 1973). Contrasting with the probability of winning a contest, p, the context-dependent variable P will be used for all other probabilities, such as the relative frequencies of different individuals, or the probability that an individual uses a specific tactic. Resource Holding Power, also called Resource Holding Potential (RHP: Parker 1974), typically does not have a variable name assigned to it, but it is often implicitly present in models such that the cost of fighting is low for an individual with high RHP (e.g. $C_A < C_B$ if A is physically larger, but see section 2.6.2.1. for why this might not always hold). In fights that consist of several bouts, or that have an explicit duration expressed in continuous time, C denotes the total cost accumulated while c

Table 2.1 Commonly used notation. The consensus notation is used in this chapter even if notation in original models differs from this. This list does not include notation that only occurs in specific models (see text).

V	Value of the resource
V_0	Value of future life
c	Rate of cost accumulation
C	Total cost; C^* if denoting the ESS
A	Favoured individual (often because of higher V/C than that of opponent)
B	Disfavoured individual (often because of lower V/C than that of opponent)
p	Probability of winning a fight
P	Generic probability (e.g. frequency of specific type of individuals)
t	Time

denotes the per-bout cost or rate of cost accumulation in the continuous-time case, respectively.

2.4 A categorisation of models

Dyadic contest models differ in several respects. Obviously, the precise *question* to be asked can differ between models: a model of lethal combat (Enquist & Leimar 1990) will take a different form than an analysis of whether the location of a territorial boundary should be settled based on a landmark (Mesterton-Gibbons & Adams 2003). The majority of models are of a generic 'winner takes all' type (the contested resource is indivisible), where the winner gains a resource of value V and the loser gains nothing (although some models additionally ask whether V is necessarily the same for both players). Of course, 'winner takes all' does not mean that the winner's fitness is large and the loser's is zero. As pointed out by Maynard Smith (1982, p. 11), the loser simply keeps what it otherwise had in life, and if V is small the interpretation may be that the winner got one food item that the loser did not. It is intuitively clear, as well being a prediction of many models, that if V is small compared to the alternative options (which are sometimes denoted by V_0), individuals are not prepared to fight to death (Enquist & Leimar 1990).

There are other important differences between the models. One can distinguish between *black box* and *open box* models. In a black box model, the link from the two strategies to the outcome of the fight is a single step and all further detail is hidden from view. For example, consider the 'Hawk–Dove' game, a classical game with somewhat unfortunate strategy labels (in reality hawks are predators of doves, but the game refers to fights among conspecifics; neither are doves especially benign species). An aggressive 'Hawk' is assumed to beat a non-aggressive 'Dove' and gain V units of fitness, while two 'Hawks' gain the expected value $(V - C)/2$ each (section 2.5.1). In neither case, nor in encounters between two Doves, does the model specify the sequence of events during a fight. All we know is that if two Hawks enter the 'black box', they emerge after a fight and one of them has some form of injury that cost it C fitness units. Mathematically it does not even matter if it is the loser or the winner who suffers the cost C, or if both suffer equally $(C/2)$: both interpretations imply that expected fitness, calculated before the winner is known, is $p(V - C)$ where $p = 1/2$.

Open box models, e.g. the sequential assessment game (Enquist & Leimar 1983), make much more detailed predictions about fight durations, the distribution of costs accumulated before the fight ends, the level of escalation reached by the end of the fight and other similar measures. The level of detail is not a goal in modelling as such: different models exist for different purposes and very simple models are often best for explaining general logical structures, such as the nature of frequency-dependence in contests.

Finally, models differ in whether individuals participating in a contest make use of *information*. The simplest Hawk–Dove games investigated frequency-dependent selection with an implicit assumption that individuals use genetically predetermined levels of aggressiveness and do not modify their behaviour according to the type of opponent encountered. The outcome of the fight, on the other hand, *does* depend on the type of opponent encountered in these models, which is why game theory is used in the first place. Best responses can evolve despite no explicit information-gathering, because evolution itself equips individuals with information about *average* frequencies of Hawks and Doves in the population. Often, however, individuals can do better than to base their decisions on the likely distribution of opponents: they may be able to estimate what kind of opponent they are facing in a given contest. Indeed, an important reason for the very large number of models of dyadic contests is that animals vary in their cognitive capabilities as well as in the situations they encounter. Depending on the situation, information gathering over behavioural time may be absent or present, and can happen before or during the contest.

Models can therefore be categorised regarding the degree of assessment that influences an animal's decision-making. For example, a war of attrition (section 2.5.2) where an individual's persistence time is picked from an exponential distribution may involve no assessment of how this particular fight might differ from any other. A second set of models is based on self-assessment: for example, an individual may be aware that it is relatively strong, or that it is guarding a resource that it knows to be more valuable than average (this information might not be available for an intruder attempting to acquire this item). Finally, models may also include 'mutual assessment'. Information transfer may occur in the form of damage accumulation during fights (as in the sequential assessment game, section 2.6.3.1), or information may be given intentionally before the fight escalates (see section 2.6.4 on signalling). For details of this type of categorisation of models see Arnott and Elwood's (2009) review.

It is not surprising that models in which information transfer occurs over behavioural time tend to be of an 'open box' type. There are exceptions; for instance, a theoretical study of 'badges of status', i.e. the idea that individuals do not engage in a fight before they have assessed each other's RHP based on a relatively cost-free 'badge' trait, might model the entire interaction as a single step (e.g. Johnstone & Norris 1993).

2.5 The beginning: ESS and the conundrum of limited war

The dawn of game-theory models of animal contests occurred simultaneously, indeed in the same publications, with the development of the concept of evolutionary stability. The precursors of the evolutionarily stable strategy arose in the different context of sex ratios (see Fisher 1930 for classical sex ratio theory, and Hamilton 1967 for an 'unbeatable strategy' for sex ratios under local mate competition; for more on history see Maynard Smith 1976, 1982, p. 174). The very next step was to develop these concepts to present a formal definition of the evolutionarily stable strategy, or ESS, in the context of animal contests (Maynard Smith & Price 1973, Maynard Smith 1974). By definition, if a population adopts the ESS, then it cannot be invaded by any other strategy that is initially rare (here the ESS differs slightly from the concept of the Nash equilibrium: it is possible that a strategy is a Nash equilibrium but not an ESS, which happens if competing strategies achieve equally high payoffs). Mutants or immigrants using a different strategy therefore cannot spread in the population (formal definitions are provided by Maynard Smith 1982, Mesterton-Gibbons 2000, McGill & Brown 2007).

What this means in practice was illustrated by Maynard Smith and Price (1973), who developed two models, both inspired by animal contests and their puzzlingly constrained form ('limited war'). This paper inspired two different mathematical routes to analysing conflict: the Hawk–Dove framework and the war of attrition.

2.5.1 Hawks and Doves

In their first model, Maynard Smith and Price (1973) presented what became a precursor of the later Hawk–Dove game. Unlike later, simplified Hawk–Dove models, Maynard Smith and Price's (1973) was an 'open box' model, where individuals played repeated moves until one of them retreated, either by choice or due to serious injury. Because of the great variety of responses one can imagine to a long sequence of events (e.g. 'I will always play Hawk except I will retreat if the fight has taken more than 122 steps and I have accumulated a damage level of at least 0.23'), there is an infinite number of potential strategies of which Maynard Smith and Price (1973) analysed only five. This highlights a feature of ESS models that one should be aware of: stability is analysed by asking whether invaders (mutants or immigrants) using a deviating rule can spread, and one has to choose the set of potential deviations in a biologically meaningful way. To take an extreme example, if a creature was able to evolve a machine gun this would change the rules of the game and easily destroy earlier stability (Davies 1979). The general point is that stability is assessed only with respect to alternatives that are judged to be realistic, not with respect to every conceivable improvement. The set of possible strategies is decided by the modeller and this decision should be subject to biological scrutiny.

In Maynard Smith and Price's (1973) model, individuals played a sequence of moves that could belong to the category of conventional moves (which in this context means not escalating), dangerous escalation, or retreat. Among the five analysed strategies some were quite complex, which meant that the authors had to resort to computer simulations to estimate the direction of evolution. As an example of complexity, their 'Prober-Retaliator' occasionally tries escalating dangerously but de-escalates if the opponent escalates in response and the response, when provoked by a fellow prober, is escalation with a high probability.

Maynard Smith and Price (1973) showed why limited aggression can evolve: if very aggressive types ('Hawks') are common, they will almost always encounter other Hawks and injury is a common result. In many situations, gaining access to a particular food item or a particular mate is less important than avoiding injury. In other words, if the value of the resource is V and an injured individual's payoff is $-C$, then $V < C$ predicts that 'Mouse' (which corresponds to 'Dove' in later models) can fare well against a 'Hawk', and 'Prober-Retaliators' can do extremely well. Of course, one can ask how a 'Mouse' can persist if it never acquires any resources. In other words, how exactly

Table 2.2 The Hawk–Dove–Bourgeois game with uncorrelated asymmetries. The Hawk–Dove game can be extracted on its own by excluding the Bourgeois row and column. Individual fitness is calculated assuming that individuals find themselves in the 'owner' and 'intruder' roles with equal frequency. Often a separate column is added for the paradoxical anti-Bourgeois but its success can be equally viewed from the current table, by noting that the 'Bourgeois' is a simple conventional solution that can refer to any conceivable and easily perceivable asymmetry.

		Opponent		
		Hawk	**Dove**	**Bourgeois**
Self	Hawk	$\frac{V-C}{2}$	V	$\frac{3V}{2}-\frac{C}{4}$
	Dove	0	$\frac{V}{2}$	$\frac{V}{4}$
	Bourgeois	$\frac{V-C}{4}$	$\frac{3V}{4}$	$\frac{V}{2}$

does reproduction take place in a population where a rare Mouse's payoff is close to zero, yet is much better than the population-wide Hawk rule? This is possible because $V < C$ predicts that Hawks have negative fitness. The proper interpretation of payoffs became a recurrent theme in later game-theory models, but the more urgent task was to simplify the Maynard Smith and Price (1973) model (which had many parameters, including fitness consequences of minor injuries such as 'scratches'). Developing a simpler model was desirable in order to capture the minimal, and thus essential, features that could explain why 'limited war' can exist in animal populations.

The model was soon simplified to the classic Hawk–Dove game (Maynard Smith & Parker 1976, Maynard Smith 1982). This game is a 'black box' type model, with individuals playing either 'Hawk' or 'Dove' (Table 2.2). It is easy to see that in a pure population of Doves an incoming Hawk will do very well. It is assumed that Doves retreat without fighting, so in these encounters Hawks take the resource V and Doves gain 0. (If two Doves meet they share resources peacefully, gaining $V/2$, although this can also be interpreted as them 'flipping a coin' to determine which contestant gets the whole resource.) Because $V > 0$, Hawks are favoured by natural selection, but as they now increase in frequency, a focal Hawk will increasingly often meet another Hawk. Hawk–Hawk encounters are not as profitable for Hawks as Hawk–Dove encounters: not only is winning no longer guaranteed, there is now also a risk of injury. The expected payoff in a fight against another Hawk is $(V - C)/2$, less than half of what it was against Doves, V: the more Hawks in the population, the smaller the success of Hawks. To be precise, if the proportion of Hawks in the population is P, the Hawk's expected payoff will be $P(V - C)/2 + (1 - P)V$. This has to be compared against the Dove's expected payoff which is 0 against Hawks and $V/2$ against other Doves. Thus, Hawks will increase in frequency as long as

$$P(V - C)/2 + (1 - P)V > (1 - P)V/2 \quad (2.1)$$

This is true up to the point where $P = V/C$, which is the ESS: a polymorphism where a proportion P of individuals play Hawk and the rest play Dove (or, as an alternative interpretation, each individual chooses to play Hawk with a probability P when participating in a contest: Maynard Smith & Parker 1976). On average, Hawks and Doves will have equal fitness, otherwise natural selection would increase the frequency of one type. If $V/C > 1$, then no such mixed strategy is possible, instead (2.1) is true for all values of P and every individual will play Hawk. This simplified model shows much more clearly than the original (Maynard Smith & Price 1973) why limited war can be stable, and it also predicts when 'total war' (*sensu* Maynard Smith & Price 1973) should take place: the higher the stakes (high V), the more likely it is that aggression evolves, even if its costs, C, are high.

2.5.2 War of attrition

Imagine a game in which you and one other player each make separate but simultaneous telephone calls costing £10 per minute (billed on a per-second basis). As soon as one player hangs up they stop accumulating further charges and the other receives a prize of £1000. How long would your call last?

This is a 'war of attrition'; an accumulating process of unpleasant damage in a two-player context. Both individuals pay costs and the individual who persists longer wins the resource. The telephone contest thought experiment was not used when the model was first presented, but it is very useful to think how one would behave in it. It might appear logical, for example, that no player should ever play beyond running up a bill of £1000. Now imagine that you have accumulated that much damage: is it not tempting to carry on for just a few seconds longer, for you might win it back? Further, would your behaviour depend on whether you believe your ability to withstand losses is lower or higher than that of your opponent?

Maynard Smith and Price (1973) introduced this type of game as their second example of 'limited war', considering a version in which there were no asymmetries between individuals. The players must individually decide on a value, C^*, which is the threshold cost (or damage) they are prepared to accumulate before they retreat. The winner gains the resource of value V. Strategies are of the form 'carry on if $C < C^*$, otherwise retreat'. A common interpretation of the war of attrition is that C increases linearly with time, t, spent in the fight, i.e. $C(t) = ct$ where c indicates the rate of damage accumulation. The analysis can also be made more general by replacing the consideration of time concepts with direct consideration of damage levels (Maynard Smith 1974, Bishop & Cannings 1978).

The interesting feature of this game is that there can be no *pure ESS* (a pure ESS would mean that a player always chooses the same value of C^* without randomness). In this game individuals do not know each other's values of C^*; if they did, it would be easy to win simply by persisting longer by an infinitesimally small amount of time. However, one way for individuals to 'know' the opponent's C^*, despite no actual information transfer upon encounter, occurs if populations converge on a single C^*. Natural selection will then favour individuals who behave as if they knew that all opponents will use C^*, and persist for slightly longer. One might then predict runaway evolution towards ever-increasing C^*, but this cannot carry on forever because if C^* exceeds $V/2$ then a mutant who does not fight at all ($C^* = 0$) will have higher expected fitness (zero) than a player who pays a larger cost than the average gain (cost: C^* each, gain: $V/2$). The statement *there is no pure ESS* means that no value of C^* exists that could prevent some deviating individuals from doing better than the rest of the population. Predictable opponents can be exploited.

Instead, Maynard Smith (1974) showed that there is a *mixed ESS* solution, in which individuals evolve to be maximally unpredictable. In a *mixed strategy*, there is some degree of randomness in individual's actions. For example, a coin-tossing individual 'persists until $C = 1$ with probability 0.5, otherwise persists until $C = 1.2$' gives an opponent a bigger challenge than does a pure strategist. If, however, all individuals in a population use such a strategy, a mutant could exploit the knowledge that its opponent will use very specific persistence times; assuming that V makes it worthwhile, one could play according to the rule 'first persist until $C = 1.001$, then if opponent hasn't given up persist until $C = 1.2001$'. Overall such a strategy is simply captured by $C^* = 1.2001$ and it can easily beat the coin-tossers (assuming $V > 1.1$ because fights will cost $(1 + 1.2)/2$ on average); but if $C^* = 1.2001$ spreads, it could again be beaten using the arguments above.

There is only one strategy that cannot be exploited by mutants who evolve to persist just slightly beyond those values of C that the opponent uses with a high probability. This strategy has a *fixed rate of retreating* if costs accumulate linearly over time (for the more general case of a fixed rate with respect to damage accumulation see Maynard Smith 1974). If all population members use such a rate, then the duration of the contest so far cannot be used to predict how long an opponent is still prepared to carry on, and thus it cannot be exploited. The ESS rate equals c/V in the case of linearly accumulating costs: higher costs predict shorter fights (higher rate of retreating per unit time), higher resource value predicts longer fights (lower rate of retreating per unit time).

Note that the formulation of a 'rate' applies whether $c > V$ or the reverse: it is a continuous-time parameter that can exceed 1 without causing mathematical problems (while retreat *probabilities* of course cannot exceed 1). To see how this works, use the above telephone contest game with $c = 200$ and $V = 100$, i.e. an outrageous charge of £200 per minute and a prize of £100. Intuition tells us that since V is relatively low compared to c, sensible individuals will only make short calls, and this is indeed the case. If populations evolved to play this game we would expect individual call durations to stop at a rate $200/100 = 2$ per unit time (minute). This means that an individual should, on average, be prepared to hold the telephone for 30 s (1/2 min). This rule will not evolve as a fixed, exploitable threshold: the ESS is to pick a duration from an exponential distribution that has the mean 0.5 min. The exponential distribution is the mathematical solution for the expected 'lifespan' of an entity (a telephone call, or a contest) that 'dies' at a constant rate.

Thus, the ESS in the symmetric war of attrition with linearly increasing costs can be formulated as 'at any point during the fight retreat at a rate c/V' or, equivalently, 'pick the persistence time t^* from an exponential distribution that has the mean V/c, i.e. $P(t^*) = c/V \exp(-ct^*/V)$'. The cost an individual is prepared to accept before retreating is ct^* and this too is distributed exponentially, with a mean V. This latter result can be shown to hold generally, without having to make an assumption of linear damage accumulation over time (Maynard Smith 1974; for an

showed that serious enough consequences of fighting can select for multiple alternative ESSs: one stable equilibrium predicts that individuals respect the size asymmetry and ignore the role asymmetry, except when sizes are equal (the 'Assessor' strategy), but another predicts that they ignore the size asymmetry and respect the role asymmetry ('Bourgeois' strategy, section 2.6.5), although paradoxical rules that force large individuals or owners to retreat immediately are also an ESS! The general conclusion is that a large enough fight cost can stabilise any arbitrary cue that allows settling the fight without having to pay the cost.

All such models are plagued by a difficulty of explaining what keeps the fight cost so large. Specifically, if B (or A) always retreats then the tendency of A (or B) to persist up to very high costs is never exposed to selection. If real fights never occur, the so-called *reserve strategy* (what an individual would do if its opponent broke the established rule) is subject to drift (Parker & Rubenstein 1981). This problem applies to Hawk–Dove games as well as to the asymmetric war of attrition, but it is more severe in Hawk–Dove games because in these the modeller simply assumes a fixed value of C instead of letting individuals evolve towards an intensity of fights that they can bear before retreating. For both types of models, relaxing the assumption of idealised situations with perfect information and discrete categories (e.g. cost can only be either 0 or C) can produce results that give qualitatively new insight (Parker & Rubenstein 1981, Hammerstein & Parker 1982, McNamara *et al.* 1997).

2.6.2.1 The Napoleon complex

One way of relaxing the assumption of perfect information is to assume that individuals occasionally perceive their roles mistakenly. This exposes the 'reserve strategy' regularly to selection and helps to maintain the 'common sense' strategy while removing the 'paradoxical' solutions. Hammerstein and Parker (1982) showed this in the context of an asymmetric war of attrition: if individuals at least sometimes makes mistakes about roles, and fights are a continuous-time process, paradoxes disappear. The predicted outcome is that an individual with a higher V/c persists for longer and thus wins. Hammerstein and Parker (1982) also discussed that paradoxical solutions might remain more plausible if fights consist of discrete bouts: this tends to increase the discrete 'zero-or-costly' nature of injuries that is necessary for paradoxical strategies to exist. Whichever role (A or B) happens to evolve to impact this level of injury on its opponent first, the receiving end will now be selected to avoid it. Therefore, it is perhaps not surprising that recent models, that have resurrected paradoxical strategies in the form of low-RHP individuals picking more fights, use single- or discrete-bout fights. The predicted ferocity of low-RHP individuals is termed the 'Napoleon complex' (Just & Morris 2003, Just *et al.* 2007), or 'Napoleon ESS' which coexists with the common-sense ESS (Morrell *et al.* 2005).

To assess how likely 'Napoleons' are in reality, future work should evaluate how realistic discrete stages of fights are, and pay more attention to the exact route to fitness after a fight has occurred. Given that a low-RHP individual is likely to stay a low-RHP individual in future fights too, one might consider making this asymmetry visible in the costs of fighting: Grafen (1987) gave a forceful argument for why C should be low for such 'desperados' (section 2.6.5.2). Models, however, rarely take this into consideration. Enquist and Leimar (1990) note that researchers tend to interpret fight costs as high when there is a high risk of injury, irrespective of how much future reproductive success was lost (but see e.g. Fromhage & Schneider 2005). Even in models that do not consider the future, their interpretation of C differs in surprisingly varied ways. Often high RHP is interpreted to cause a low C, which is then used to derive predictions that high RHP individuals should win (winning probability p is derived in these models rather than assumed a priori, e.g. Hammerstein & Parker 1982). Morrell *et al.* (2005) make the opposite assumption that RHP influences p but C is identical for all individuals, while it depends on the level of escalation the fight reaches. Just and Morris (2003) in turn assume that C differs between losers and winners, which creates an implicit correlation between RHP and cost C if RHP influences winning probability.

However, none of these models consider that the future prospects of low-RHP individuals can be fundamentally different from those of large ones; for example, Morrell *et al.* (2005) explicitly assumed that all losers have identical prospects of finding alternative resources. Contrasting the outcome of a fight with outside options (Morrell *et al.* 2005) certainly represents an improvement over earlier models that focus on one fight in isolation. It helps to make assumptions such as identical future prospects explicit, and Morrell *et al.* (2005) indeed specify that the alternative resources are uncontested. This also raises a more general question: is it realistic to contrast *one* fight with the alternative of foraging for uncontested resources? If fights are the usual route to resources then low-RHP losers will always have difficulty gaining resources, which means that the future should be modelled differently for individuals differing in RHP. If, on the other hand, resources are only rarely contested as they are abundant, then a one-fight approximation appears appropriate (see also Eshel & Sansone 1995), but this also makes it puzzling why an individual with low RHP would bother to engage in fighting.

2.6.2.2 Explicit modelling of future options

How can different futures for different individuals be dealt with? One would intuitively predict that individuals are more fierce in any one fight if their prospects otherwise remain poor (Korona 1991, Enquist & Leimar 1990, Kemp 2006), which is echoed in the verbal argument that C must be low for 'desperados' (Grafen 1987). Taking the future into account could therefore, at least in some cases, reverse the idea that high RHP must be reflected in a low C. This can indeed predict high risk-taking by individuals with low future prospects (a result well known from signalling contexts: Adams & Mesterton-Gibbons 1995, Kokko 1997).

This argument has been used to explain why small males of the butterfly *Heliconius nana* win fights over territory ownership (Hernández & Benson 1998). This butterfly has two different mating systems: males compete directly over virgin females as these emerge from their pupal stage, and in this contest large males are likely to have an advantage. Territoriality brings about new (but generally less important) mating opportunities, and ownership contests resemble a war of attrition in which low-RHP males are more persistent than large ones. Hernández and Benson (1998) conclude that C is larger for large males because they have more to lose: they are the successful males in competition over virgins. Large C predicts a smaller V/C, so it makes sense that large males are predicted to be less persistent. Note that Hernández and Benson (1998) phrase this as a 'paradoxical' outcome, but the standard phrase in game theory is 'contradictory'. The individual in the favoured role wins, so there is no paradox, while 'contradictory' captures the surprise that the low-RHP individual is the favoured one (section 2.6.1.2; Field & Hardy 2000).

How can the value of C be measured? In cases more complicated than that of *H. nana*, it might not make sense to lump together all possible effects of RHP asymmetry on the current fight with the entire future prospects of an individual and attempt to express it all as a single value, C. Enquist and Leimar (1990) proposed an alternative formulation: if the value of the current resource is V and the value of the future is V_0, then expected lifetime fitness is $pV + (1 - q)V_0$. Here p is the probability of victory and q is a factor that controls how fighting diminishes the value of the future for this individual. Fighting represents a trade-off: achieving high p probably requires doing something risky (or at the very least time is wasted), which increases q and hence diminishes $(1 - q)V_0$. If the future, V_0, is of little value compared with V (as it is for the smaller *H. nana* males) then individuals should do all they can to increase p.

Ideally we would like to know the tradeoff precisely: does q (or C) represent a risk of death (Enquist & Leimar 1990), or an opportunity cost because of time wasted in the contest, or perhaps lower lifespan or reduced RHP in future contests? While the simplistic approach of varying parameters V, V_0, C or q can clarify conceptual issues, a more thorough scrutiny of the importance of future requires, rather unsurprisingly, that the future is modelled explicitly (Houston & McNamara 1991, Korona 1991, Kemp 2006). The downside is that models get more complicated and perhaps more system-specific.

Even so, recent examples offer an intriguing contrast. If individuals can assess ownership but not RHP, yet ownership correlates with RHP because strong individuals tend to accumulate as owners, then low-RHP intruders will, to some extent at least, avoid challenging high RHP owners (Kokko *et al.* 2006). This model features no 'Napoleons', which is no real surprise: their evolution requires that fighting becomes conditional on RHP, an option not allowed in Kokko *et al.* (2006). A model that allows for such phenotypic plasticity predicts that low-RHP males do sometimes fight more, yet they are not predicted to challenge the very strongest individuals (Eshel & Sansone 2001; a similar result is also derived by Korona 1991). Both models predict that ownership that covaries with RHP is respected by at least some individuals but they are not formulated in a way that would allow easy across-model comparisons of population-wide levels of aggression. A common theme in many models is that if individuals have more information on RHP prior to the fight, the level of aggression as a whole diminishes (Crowley 2000, McNamara &

Houston 2005). However, this result might change if some individuals who know their RHP to be low correlate this with poor future prospects, as this might shift their current behaviour towards more risk-taking.

2.6.3 True open-box models

2.6.3.1 Sequential assessment game

Most models discussed so far are 'black box' models, although a war of attrition is a grey area, somewhere between 'black' and 'open'. For example, in a war of attrition with variation in resource value (Bishop *et al.* 1978), an individual with an intermediate value of V might be observed to retreat after the fight has reached this individual's persistence threshold t^*. It is simply a matter of perspective whether we state that the focal individual, once faced with a more-persistent opponent, gained information about this persistence during a fight which then made it retreat (open box interpretation), or the fight had to end at t^* because the two contestants each used their strategy and the focal individual's t^* was smaller than that of its opponent (black box interpretation where the entire fight is analysed in one step). Subsequent models have begun to treat information accumulation during a fight in much greater detail. They consider fights as a sequence of discrete bouts, making them truly open box in character.

The precursor of the sequential assessment game was a series of 'assessor' models by Parker (1974) and Parker and Rubenstein (1981). The assessor strategy attempts to estimate if it is currently player A or B, where $V_A/c_A > V_B/c_B$ (here c again refers to the rate at which cost accumulates for the two players in an asymmetric war of attrition). In the first model of Parker and Rubenstein (1981), the commonsense assessor retreats when it perceives itself to be player B, while the paradoxical assessor retreats when in the favoured role A. The paradoxical assessor is not evolutionarily stable if the reserve strategy is exposed to selection through recurrent mutation or mistakes in role perception. More intriguingly, Parker and Rubenstein (1981) produced a second model to show that when information on roles is incomplete, individuals who currently perceive themselves to be in role B should no longer immediately withdraw. They might profitably carry on fighting because each bout during a contest brings new and increasingly accurate information about RHP and/or V asymmetries, and they should only retreat once the estimated prospects of winning have become convincingly low. This model derived the prediction that contests will be settled quickly when the RHP disparity is large and/or contest costs are high relative to resource value.

The fights by Parker and Rubenstein (1981) only contained a small number of steps, and their approach was taken to a much more general level in the subsequently developed sequential assessment framework (Enquist & Leimar 1983, 1987, Leimar & Enquist 1984). In the sequential assessment model of Enquist and Leimar (1983), individuals are assumed to have identical V but they differ in the costs suffered in each bout of the game ($c_A \neq c_B$). Here, $\theta_{AB} = \ln(c_B/c_A)$ measures the relative fighting ability of A in a contest with B: high θ_{AB} indicates an ability to inflict costs on the opponent while avoiding own costs. $\theta_{AB} = \theta_{BA} = 0$ if there is no RHP difference, and in general $\theta_{AB} = -\theta_{BA}$. Individuals improve their estimate of their relative fighting ability in each bout through a sampling process: each bout gives one sample of θ but this is observed with error, thus the longer the fight lasts, the more accurate each individual's estimate of θ. Repeated sampling reduces the error but obtaining a large sample is costly. For this reason we expect individuals to retreat once they are reasonably certain that they are in role B, leaving their opponent to win the resource.

Can we quantify what 'reasonably certain' means? Enquist and Leimar (1983) show this by deriving the expected utility for two alternative behaviours of an individual who, after n steps, perceives the relative fighting ability to equal $\hat{\theta}$: at each combination of $\{n, \hat{\theta}\}$ it can either retreat or carry on for at least another bout. Both contestants are initially likely to carry on: the estimated $\hat{\theta}$ should deviate greatly from zero for an individual to retreat, given that the estimate is likely to deviate greatly from the true value of θ. After a large number of steps the diminishing uncertainty makes individuals' estimates approach the true value of θ, and it becomes more likely that one of them has crossed the threshold where the expected utility of retreating exceeds that of continuing to fight. Utility is calculated using the additive framework: it essentially equals $pV - C$ (expected benefits minus costs) but now the probability of eventually winning the fight, p, and the expected cost, C, are based on the current estimate of θ. The model predicts that B (for which fighting is costlier, $c_B > c_A$) is more likely to be this individual, but sampling error can occasionally make animal A retreat, leaving B as the winner. This is particularly likely when opponents are roughly matched ($c_B \approx c_A$), which is also when fights are expected to be particularly long. The popularity of the sequential assessment game as a target of empirical studies is no doubt due to its clearly stated predictions that are amenable to testing.

Leimar and Enquist (1984) combine the sequential assessment game with an analysis of the process by which strong individuals accumulate as owners. The great improvement over earlier models is their avoidance of the unrealistic assumption that individuals find themselves in owner and intruder roles with equal frequency. Strong individuals are more likely to win fights that assess strength in a fashion dictated by the sequential ownership game. If fights determine ownership, then at equilibrium owners are stronger than intruders. They are simultaneously likely to be more persistent than intruders, where 'persistence' is now measured as the threshold value of q_n that makes individuals retreat in a fight that has lasted for n steps. Intriguingly, numerical analysis shows that this situation allows paradoxical strategies to be stable (Leimar & Enquist 1984): the less-favoured role (intruder) chooses the more persistent strategy but the larger the variance in RHP, the less likely it is that a population will evolve towards paradoxical solutions. These results are very much in line with later results in which fights are simple 'black box' events but fitness in turn is calculated in a more realistic fashion (Kokko *et al.* 2006).

Enquist and Leimar (1987) extend the set of testable predictions to variations in resource value, V, in an owner–intruder asymmetry. They again use the sequential assessment formulation but they now consider that the owners may know V more accurately than an intruder, and the duration of the fight can therefore be used by intruders to gain some knowledge about how large V is. Even though this prolongs intruders' persistence, p_A (owner's probability of winning) will increase with resource value because owners too are more persistent when they know V to be high. Fights are therefore predicted to be particularly long if V is high; durations are particularly sensitive to V for fights that end with a takeover. Enquist and Leimar (1987) also report empirical support for these predictions.

2.6.3.2 Fatal fights

The next step in the Enquist–Leimar series of papers was a consideration of fatal fighting (Enquist & Leimar 1990). They first produced a general 'black box' argument that is similar to Grafen's (1987) verbal desperado argument. They take expected lifetime fitness, $pV + (1 - q)V_0$ (section 2.6.2.2) as a starting point and ask under what conditions fatal fights will be stable in a Hawk–Dove framework. Their derivation is simplified even further, as they only consider a setting where two Hawks always fight until one dies, and the winner cannot be known a priori (thus $p = q = 0.5$). Doves have fitness V_0 (gained through searching for uncontested resources), while a Hawk has fitness $(V + V_0)/2$. Here the winner Hawk has fitness $V + V_0$, but one might add that the winner Hawk's fitness could also be modelled as V, to represent cases where the winner will improve fitness from V_0 to V instead of keeping both fitness components V_0 and V. This minor change to the model may be necessary if, for example, V_0 refers to the expected success when searching for vacant territories and V to obtaining a territory through a contest. The loser Hawk has fitness 0, and as a Hawk cannot determine beforehand whether it will win or lose, Hawk fitness will equal, depending on the interpretation, $(V + V_0)/2$ or $V/2$. We obtain two different conditions for Hawk stability:

if the winner keeps resources V as well as V_0:

$$(V + V_0)/2 > V_0 \Leftrightarrow V > V_0 \qquad (2.2a)$$

if the winner keeps resource V only:

$$V/2 > V_0 \Leftrightarrow V > 2V_0 \qquad (2.2b)$$

Despite the apparent numerical difference, both cases lead to the same interpretation: one should be prepared to participate in fights where the probability of being killed is 0.5, if a victory more than doubles one's reproductive success (in the first case from V_0 to $V_0 + V$ where $V > V_0$, in the second case from V_0 to V where $V > 2V_0$). Enquist and Leimar (1990) then proceed to analyse the problem in a sequential assessment framework. Their model confirms the initial intuition: individuals are not expected to give up until the fight results in a fatality, if a major part of the two contestants' lifetime reproductive success is at stake.

Of course, one can ask if the winner is expected to be intact enough to make full use of the resource after such a vigorous struggle (see Mesterton-Gibbons *et al.* 1996 for a model that relaxes this assumption, although in a different context). Future work might therefore profitably explore state-dependent future fitness of individuals, perhaps combining assessments that occur in single contests with explicit modelling of the population-level process that determines the future tenure of an individual that emerges victorious but perhaps injured from a fight. One could argue that when V_0 is low, it might be indicative of an abundance of individuals who are competing over scarce resources. A weakened winner might therefore not be able to exploit its acquired resources without them being contested by new intruders who have not yet suffered damage. This could limit the evolution of fatal fighting because it creates an asymmetry in RHP that, unusually, favours intruders over owners; although this must also be contrasted against any potential accumulation of strong individuals as owners of resources (Leimar & Enquist 1984).

The future danger posed by new intruders could both limit the expected value of holding the resource (which, by diminishing V, makes fatal fighting less likely to evolve) and make individuals respond adaptively during a fight, by making them retreat before injuries reduce their ability to utilise much of the resource, V (see Mesterton-Gibbons *et al.* 1996). There is the potential for interesting feedbacks because the danger posed by future intruders clearly depends on their numbers, and their numbers will depend partially on how often fights are fatal in the population as a whole. Such feedbacks could be profitably explored in a manner explained below (section 2.6.5.3; see also Chapter 8).

2.6.4 Signalling

I have so far discussed models in which information is available directly or can be assessed by observing the duration of a fight: I have not discussed *signals*. Both signals and *cues* (e.g. body size) can transfer information; signals are traits that evolve *primarily* because of this information transfer and the effect it has on the receiver. Signals are often ephemeral (as soon as a bird stops singing, its song can no longer be heard), but they may also be morphological characters that develop over longer timescales. For either, models incorporating signalling must allow individuals to adjust their signal size, over either evolutionary or behavioural time. For example, fighting effort in the sequential assessment model is not a model of signalling because the fight yields information on traits c_A and c_B but these are fixed and do not evolve. The evolving trait in the sequential assessment model is the retreating threshold, and this is not a signal: it evolves to optimise reproductive success directly, not to make the receiver respond in a particular way (instead, once this point is reached it is 'game over').

2.6.4.1 Cumulative assessment game

Payne (1998) presents a *cumulative assessment game* as an alternative to the sequential assessment game. The difference between sequential and cumulative assessment is usually phrased as a difference in whether fights continue because individuals gain more accurate information on the same quantity (relative fighting ability) through a sampling process, as in the sequential assessment game, or whether individuals base their decision to carry on or retreat on the cumulative costs suffered so far. However, this is a relatively superficial difference: the larger the cumulative costs up to a given bout n, the more likely it is that an individual retreats in the sequential game too. The much more fundamental difference is that Payne (1998) explicitly considers that the rates of escalation are an evolving trait, and derives several empirically testable predictions regarding this characteristic. The model additionally includes a rule for retreating (retreat when cost to self exceeds a tolerance threshold), yet the evolutionary stability of this rule is not addressed.

I discuss Payne's (1998) model in a section on signalling because rates of escalation evolve to elicit a specific response from opponents. Future work could extend this model to ask why losers, who are predicted in the early stages of the fight to exhibit different behaviour than the eventual winners, do not use this observable behavioural difference to evaluate that they are in the losing role and retreat immediately to prevent further damage. This would combine the evolution of the rate of escalation with the evolution of retreat strategies. Although some models exist that approximate this task in discrete steps (e.g. Maynard Smith & Price 1973, Morrell *et al.* 2005, see also 'threat displays' section 2.6.4.3), no model so far appears to have given individuals the freedom to signal at an intensity of their choice, as well as attempt to inflict a freely chosen level of damage on an opponent at any stage during a fight, together with a decision rule that determines when retreating should occur. War of attrition games in their more general form (Maynard Smith 1974, Bishop & Cannings 1978, Haccou & Glaizot 2002) appear promisingly flexible in this respect because the cost tolerance threshold, C^*, is derived without assuming that costs accrue linearly over time; however, these models do not consider the cost inflicted as an evolving trait.

2.6.4.2 Honesty, bluff and badges of status

The above-mentioned models, in their current form, are not very helpful for analysing *displays*. These are behaviours that transfer information between contestants about future costs without yet inflicting any of this cost on the opponent. Thus, unlike the injuries modelled in a sequential assessment game, gaining display information can be cost-free (while giving this information might have costs that differ from zero, section 2.6.4.4). Despite the cost-free nature of display observation, displays are effective in changing the opponent's behaviour. This is because displays correlate with future costs received if the fight escalates. A central question is why such a positive correlation should persist. This is known as the problem of *honesty*: if displays are so effective in modifying the opponent's behaviour, why not give a dishonestly large display? Such *bluff* could, if it spread, destroy the reliability of the signal, after which no contestant is expected to pay attention to it.

A major model of display is the Badge of Status model. A badge is a 'label' carried by an individual that signals, honestly or dishonestly, its persistence (usually referred to as 'aggressiveness' in these models) should a fight occur. Carrying the label itself does not have to be costly; so what prevents bluff from spreading? Maynard Smith and Harper (1988) state that badges can be stable because dishonest cheaters, who possess too large a badge for their persistence compared to the population norm, will be selected against as they perform poorly whenever they have to participate in fights against individuals who signal at a similar level. This model assumes that fight costs must be higher for large-badged individuals in the population, but this is stated without specifying in any detail what happens in fights, or how fitness differs between honestly large-badged and dishonestly large-badged individuals. While there are many black box models, this one might be said to be truly pitch black. However, it appears to have been developed as food for further thought rather than being a fully analysed model. For example, it raises the intriguing question of whether, if cheaters suffer high costs whenever they participate in a real fight, it is individually profitable for the opponent to impose these costs (which can require escalation of a costly nature to both participants).

Johnstone and Norris (1993), consequently, state that the question of honesty becomes much more challenging when badge size and persistence ('aggressiveness') are modelled as separate variables that can become dissociated in suitable mutants. Under this formulation a surprising type of dishonest cheater can invade: this cheater does not have a larger badge than its aggressiveness would predict (such an attempt would be selected against in encounters with similar-badged individuals), but instead it is more aggressive than expected based on its badge size. Johnstone and Norris (1993) show that this mutant can spread: it otherwise obeys all conventions, but makes a quick profit by being more aggressive when similar-badged individuals expect a milder confrontation. This is the opposite assumption from that of Maynard Smith and Harper (1988) because now aggressiveness that exceeds that of the opponent's is assumed to be selected for, without any associated cost. In other words, Maynard Smith and Harper (1988) assumed there will be punishment of overly aggressive individuals, Johnstone and Norris (1993) take the absence of punishment as given.

Badge honesty requires that the spread of more-aggressive-than-expected mutants becomes halted. Therefore, aggressiveness must be costly, and because of the assumption that there is no punishment during fights, it is concluded that these costs should arise (and be condition-dependent) in some other context than actual fights (Johnstone & Norris 1993). Impaired immunocompetence is an oft-quoted example. Again, the fight is modelled as a black box, and the lack of punishment is not an outcome proven to be the opponents' optimal choice (it is simply assumed). It is therefore somewhat surprising that no further work seems to exist to assess the generality of conclusions of current Badge of Status models. If aggressiveness were modelled as a plastic trait that quickly responded to escalation, instead of a property fixed over a whole lifetime, then normally mildly behaving opponents might have a reason to escalate when they encounter a small-badged but surprisingly aggressive opponent. As stated by Maynard Smith and Harper (1988), however, it is not self-evident that they will profit by so doing. It appears that there is much work to be done on the fate of mutants who use a deviating signalling strategy in agonistic contests.

2.6.4.3 Threat displays: some of them can be dishonest

If bluff is profitable to everyone, it will spread: this is very likely the reason why threatening mammals such as cats raise their fur to appear bulkier. When every cat is bluffing, the real size of a cat may still correlate with the apparent size of a cat, so size continues to be an honest signal of RHP. Bluff becomes more interesting in the context of displays where individuals can, regardless of their true RHP, freely give a dishonestly large signal. Do threats still evolve to reflect the likely consequences of a fight? These consequences depend on the threatening individual's true motivation to escalate (strategy) as well as its true RHP, both of which are initially unknown to the receiver of the threat signal. Spread wings, exposed teeth, or a specific direction of pointed ears are typical examples of threat displays. These signals differ from badges of status in that they can be switched on or off in a rapid behavioural sequence, and the question of honesty then becomes ever more pressing. If everyone is able to adopt an equally threatening posture, and threats sometimes lead to opponents giving up without a fight, why does this behaviour not spread until it is no longer to anyone's value to pay attention to this signal?

Several models show that signals will be paid attention to even if they are not completely reliable (Adams & Mesterton-Gibbons 1995, Hurd & Ydenberg 1996). Adams and Mesterton-Gibbons (1995) provide a particularly intriguing example that also highlights why insight can sometimes depend on modelling RHP on a continuous scale rather than assuming distinct size or strength classes. In their model, an individual (the 'signaller') can, upon meeting another, make a threat or choose not to signal. A threatened individual (the 'receiver') will then either attack, in which case there will be a fight that the higher RHP individual will win, or it can flee, in which case the signaller wins the resource without costs. Fights are more costly for weaker individuals. The outcome is that with suitable parameter values, threats will be made by the strongest and the weakest members of the population while intermediates do not threaten. Contests can escalate in either case, because it may be in the receiver's best interest to attack. It is only optimal for strong individuals to attack, as a bluffing weakling would suffer severe injuries. But individuals in this model are not either 'strong', 'weak' or 'intermediate'; instead, they come in all strength varieties, which allows calculation thresholds above or below which individuals attack or threat (or both, Figure 2.1).

These thresholds are different for each behaviour (Figure 2.1). This is important because it creates a subclass of intermediate individuals who do not always attack, but do not always flee either. They instead estimate the risk associated with attacking as too high when they encounter a threat, but not as too high when they encounter an opponent who has not given the threat display (Figure 2.1). The existence of this size class of individuals ensures that a bluffer can sometimes get away with the bluff. Individuals in this receiver size class are not behaving maladaptively as their decision to flee when threatened can be correct, despite a proportion of bluffers in the population, because many of the threats are given by very high RHP individuals.

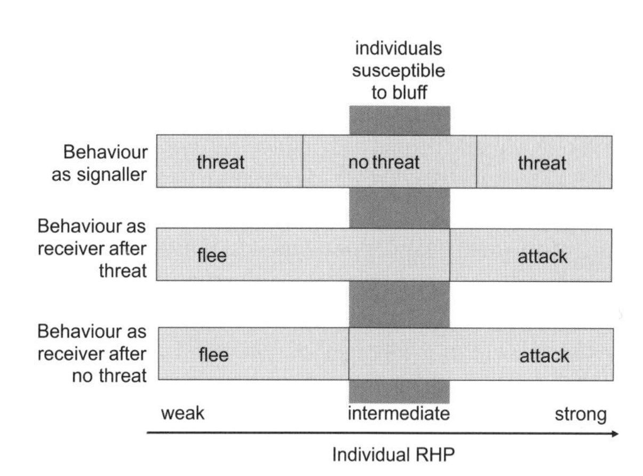

Figure 2.1 Bluff can exist as a stable part of a signalling system. Individuals give threat displays if they are either very weak or very strong; individuals always attack if they are strong, respect threat displays if they are intermediate, and never attack if they are weak. It follows that there is a class of individuals who are susceptible to bluff, and this can stabilise deceptive communication by weak individuals with no intention to attack. Stability requires that individuals cannot distinguish between bluff and real threats by strong individuals, and the bluffing class may also be absent. (Modified after Adams & Mesterton-Gibbons 1995.)

What, then, prevents bluff from spreading to all individuals if it is so effective? A threatening animal is taking a high risk because the individual-specific thresholds predict that some receivers will not believe the bluff; instead, they will attack. This risk is only worth taking when other means of acquiring resources are unlikely, which holds true for the lowest RHP individuals, or when the individual is strong and protected against large fight costs, which is true for highest RHP individuals. Other choices for parameter values in this model lead to complete honesty, where only strong individuals give threat displays. In general, a signalling system can be stable even if the majority of individuals are bluffers (Számadó 2000; see also Backwell *et al.* 2000).

2.6.4.4 Are threat signals handicaps?

The stability of signalling when participants' interests differ is usually explained using the *handicap principle* (Zahavi 1975, 1977). According to this principle, giving a dishonestly large signal must be, on average, more costly for a poor-quality signaller than for a high-quality one. Such signals are called 'handicaps' because they decrease the ability of the animal to perform well: in a threat display context they should impair the signaller's ability to attack. However, Hurd and Enquist (1998) noted that threat displays do not appear to put individuals in positions of particular vulnerability. Számadó (2003) took this approach further by asking whether threat displays might instead involve positions that confer an advantage if the conflict escalates to physical contact. A series of models (Enquist 1985, Hurd 1997, Hurd & Enquist 1998, Számadó 2003) has now shown that threat displays need not be handicaps. When the interaction consists of several steps of escalation, a handicapping signalling system can be invaded by cost-free signals that have no effect on fighting ability (conventional signals *sensu* Hurd & Enquist 1998, minimal signals *sensu* Számadó 2003). Számadó (2003) additionally considers 'negative handicaps'. These are signals that improve the individual's fighting ability should a fight occur, e.g. a posture that prepares an individual to attack. Use of negative handicaps can invade handicapping as well as cost-free signalling systems, and in his brief review of various types of threat display Számadó (2003) concludes that threatening signallers simply tend to use the initial position in the particular fighting technique of their species.

Of course, stability in all these models requires that bluff is kept at bay: much cheating might occur (Számadó 2000) but not to the extent that all signallers evolve the same signal. The fact that bluff can remain limited even if signals are 'negative handicaps' indicates that the issue of the 'handicappedness' of a signal is separate from whether giving a dishonestly large signal is costly when the entire fight sequence is taken into account. This is because the handicapping nature of a signal is evaluated by whether, assuming a given response by a receiver, the signal has detrimental consequences for the signaller. The total costs associated with giving a signal will show additional differences because receivers will react differently to different signals. This is why temporal structure is important (Hurd & Enquist 1998): it means that the stage during which the handicapping nature of a signal is evaluated does not fully capture all costs that a signal will eventually inflict on the signaller.

2.6.5 Are uncorrelated asymmetries important?

Asymmetries in RHP, and perhaps also resource value V, appear common (Maynard Smith & Parker 1976, Parker & Rubenstein 1981, Leimar & Enquist 1984). There are at least three reasons to study uncorrelated asymmetries, the idealised situation where an owner of a resource does not differ from intruders in any other way than that the owner arrived first.

First, as pointed out by Maynard Smith and Parker (1976), it is important to know if there is an a-priori reason that biases solutions towards owners winning even if owners are not stronger or have a higher V than intruders. If there is, then RHP or V differences may contribute to such ownership respect but they are not a necessary condition for it to exist.

Second, ownership respect appears to be a hardwired trait in humans (Stake 2004) which is precisely the reason we call respectful strategies 'commonsense' and disrespectful ones 'paradoxical'. Populations in which owners simply accept that they have to leave when intruders arrive contrast starkly with our intuition. We consequently have difficulty believing that owners and intruders are mathematically equivalent in models of uncorrelated asymmetries. It is not easy to accept the conclusion that for every population in which intruders respect ownership there should be another in which owners meekly retreat. While any level of ownership respect in humans is not, as such, a reason to expect that all species follow the same rules, this probably increases our intellectual curiosity about conventional rules.

Third, the debate around uncorrelated asymmetries has been useful for making two more general points. Simple Hawk–Dove or war of attrition games with asymmetries have continually utilised three important assumptions that were made explicit with varying frequency:

(a) To calculate the payoffs in an encounter between two individuals it is sufficient to know the strategies of these two individuals, not their frequency in the population as a whole;
(b) Individuals find themselves in one of two 'roles' with equal frequency;

Bourgeois state, this might indeed create a class of 'desperado' individuals, at least if the mortality of owners is not high (owner mortality would create vacancies and thus make the desperados less desperate: Kokko *et al.* 2006). Commonsense Bourgeois is then unstable and becomes replaced by mutants who behave more aggressively when in the intruder role. When a model includes population-level feedback, ownership respect is no longer destroyed completely. As soon as aggressive individuals spread and the resultant fights are sufficiently severe, as they are predicted to be given the desperado status of non-owners, then two population-level feedbacks start limiting the success of aggressive non-owners. First, the more fights, the higher the mortality or lower the reproductive success of owners. Second, the 'desperados' themselves will die more often if lethal injury is possible, or they perform perhaps less well in finding vacancies if injured. These factors are important because they all increase the expected reproductive success of individuals who refrain from fighting and simply wait for a vacancy to appear (Dunham *et al.* 1995). If other individuals are busy fighting, it may be best to hold back and then reap the benefits. If, however, every desperado were maladaptively respecting ownership, it would be beneficial for one of them to change its behaviour and begin challenging owners. This frequency-dependence forms the basis of partial respect or 'partial Bourgeois' (see Kokko *et al.* 2006 for a full derivation of its stability conditions).

In a sense, partial respect resembles the early mixed ESS in the Hawk–Dove game (Equation 2.1). However, the population-level feedback that produces partial respect is fundamentally different because it predicts that individuals still pay proper attention to their roles. While there is an equilibrium where intruders occasionally challenge owners who in turn always try to defend, there is no paradoxical mirror-image equilibrium where owners sometimes surrender and intruders always challenge them. Various feedbacks make waiting a better option for putative 'intruders' if their fellow intruders are very aggressive, while it is difficult to visualise any corresponding feedback for owners: surrendering does not automatically become a more attractive option for owners if their neighbouring owners defend against intruders more intensely (Kokko *et al.* 2006). This qualitative difference between how feedback treats owners and intruders can remove the arbitrariness of these two roles, make partial respect stable despite Selten's (1980) result, and make anti-Bourgeois unlikely to evolve, at least from ancestral populations in which respect for ownership has not yet evolved (Kokko *et al.* 2006).

Two questions remain. First, is Bourgeois truly resurrected through the inclusion of ecological feedback? It is worth remembering that RHP asymmetries probably are very common and resource value asymmetries perhaps equally so (Parker 1974, Kemp & Wiklund 2004). Nevertheless, a process that creates even a slight bias against paradoxical outcomes in the absence of other asymmetries might play a role in explaining why nature does not show a tendency for the owner–intruder asymmetry to erode, even in cases where other asymmetries must be slight (e.g. in cases where the prior residence effect is established very quickly). An RHP asymmetry can make ownership respect more robust, and its correlation with owner status has also been argued to protect the owner–intruder asymmetry against invasion by other, more arbitrary cues (such as toe length: Eshel 2005). If new results show that *all* asymmetries (uncorrelated, RHP and V) tend to operate in the direction of owner respect, it is perhaps less surprising than it otherwise would appear that 51 out of 70 studies in a list compiled by Kokko *et al.* (2006) showed evidence for a prior residence effect (some of them in multiple species). As in much of behavioural ecology, it might be pointless to treat ideas as mutually exclusive explanations that we should falsify until we are left with only one, if in reality they work best in concert. It would be more fruitful to ask why ownership is more easily overridden by RHP difference in some species than in others, and try to see if the quantified 'strength' of respect (see Hardy & Field 1998 and the supplementary material of Kokko *et al.* 2006) covaries with conditions that models predict to be particularly strongly associated with conventional solutions.

The second remaining question concerns the biological interpretation of partial respect. Ever since the classic butterfly experiments of Davies (1978) and the immediately following debate (Austad *et al.* 1979, Davies 1979), a focal topic in the owner–intruder discussion has been why an intruder should intrude in the first place if it respects ownership. Grafen's (1987) view was that individuals who appear to lose may not be playing Bourgeois but instead simply checking for the presence of an owner or evaluating its strength. On the other hand, he conceded that Bourgeois could apply when competing over resources that appear relatively abundantly and one can, by searching instead of fighting, become an owner (his example of one oestrous female illustrates this).

The new models explicitly contrast the benefits of fighting against the benefits of carrying on searching for other resources. How should we interpret the new findings that evolutionary equilibria often feature a large proportion of intruders quickly retreating and carrying on their search? Such behaviour can be an important part of a very sensible partial respect strategy with definite Bourgeois features. It follows that there is no logical reason to distinguish between an intruder who retreats because it was merely checking if an owner was there, and an individual that 'plays Bourgeois'. In both cases the causality is precisely the same: for this individual, challenging the owner does not improve fitness over what carrying on searching is expected to yield. Respect for ownership appears considerably less puzzling when outside options are taken into account in the appropriate population-level context. An empirically relevant point is that one should not underestimate respect by discarding all encounters that did not lead to physical fights (Morrell *et al.* 2005, Kokko *et al.* 2006).

2.7 Social structure

The processes that organise animal populations from simple collections of scramble competitors towards more hierarchical structures often involve fights. Many situations are best treated as multi-player games (Chapter 3). However, many conflicts in territorial or social animals involve dyadic fights, and the central theme of this chapter, that fights should not be considered in isolation from population structure, is very relevant to the analysis of such conflict.

2.7.1 Neighbours versus strangers

In territorial animals, individuals often appear to negotiate friendly relations with neighbours. The resource value V is now considerably smaller than that assumed, for example, in fatal fighting models (Enquist & Leimar 1990). V now refers to having a somewhat larger territory than one already has, thus the increase in reproductive value can be marginal and it becomes much easier to settle on conventional solutions. A clear example is the common tendency of territorial neighbours to use landmarks to mark territory boundaries (analysed mathematically by Mesterton-Gibbons & Adams 2003).

The number of different types of agonistic interactions offered by territoriality is large: individuals may have to defend not only against intruders who attempt a takeover (the focus of most models) and against neighbours who might attempt to enlarge their area (Mesterton-Gibbons & Adams 2003, Morrell & Kokko 2003, 2005), there may also be intruders who try to squeeze into the space between two or more territories (Stamps & Krishnan 2001, López-Sepulcre & Kokko 2005). In this chapter it has been argued that the V is not necessarily different between an intruder and an owner, but it is certainly different between an intruder, who owns nothing, and a neighbouring owner who might seek some additional fitness through intruding. The challenges posed by individuals of these two classes might appear similar, but the neighbour's V must be much lower, so fights between neighbours are far more likely to be settled by a convention (Korona 1991). The outcome is the 'dear enemy' effect (Temeles 1994): neighbours are less agonistic towards each other than towards strangers. In addition to a brief mention by Korona (1991) there appears to be practically no theoretical work on this phenomenon, although the model of Switzer *et al.* (2001) of nomadic versus persistent intruders on feeding territories has a flavour of this contrast. In nature, intruder pressure can even create a scope for defensive coalitions between owners (as observed in fiddler crabs, Backwell & Jennions 2004).

With increasing social organisation the question of aggression *within* groups becomes relevant. Analysing fights in social organisms will have to take into account potential relatedness between players (Grafen 1979). Relatedness forms a reason why opponents might treat each other more gently than one otherwise would expect, although the complexity of social behaviour gives rise to other possibilities. *Reproductive skew theory* is a framework that can be used to ask what keeps aggression limited in social groups despite commonly found inequalities in the division of reproduction: why isn't there relentless struggle for power (Cant & Johnstone 2000)? A full-length review of reproductive skew theory is beyond the scope of this chapter, but it is interesting to note that it portrays social groups as a complicated web of interactions, some of them arising because they prevent aggression. For example, it may be beneficial for a dominant to accept a situation in which skew is not complete (subordinates use colony resources for their own reproduction) because this makes them less likely to challenge the dominant position (Cant *et al.* 2006).

2.8 Conclusions and future directions

Throughout this chapter, I have highlighted current gaps in our theoretical understanding. There are certain issues that I have not touched upon. Fitness is typically higher for individuals in the 'favoured' role, which often means higher RHP. Very little work has been done on the *evolution* of RHP: why does selection not increase RHP indefinitely? Or, what maintains variation in RHP? RHP is a trait that may have a strong environmentally determined component, or it may be age-dependent; nevertheless, natural selection should act to improve it. Such selection is of an intriguing nature since benefits are frequency-dependent while the costs are typically absolute. Maynard Smith (1982) discusses a 'size game' in the context of life history evolution in which larger than average individuals win fights but growing large is costly, and Härdling (1999) investigates the effects of different cost functions of armaments, predicting cyclic dynamics of RHP under some situations.

A related question is the maintenance of variability at the level of strategies. Models with distinct 'Hawks' and 'Doves' are certainly simplistic, but they show that discrete strategies can form a stable polymorphism, at least in principle. An emerging trend is to examine *behavioural syndromes*: for example, aggressive behaviour selected for in one context might 'spill over' to other contexts as well, and lifetime fitness cannot be evaluated without taking such correlations into account (Sih *et al.* 2004, Rudin & Briffa 2012). A behavioural syndrome (an aspect of 'animal personality') might maintain variability in, for example, individual risk-taking (see Wilson *et al.* 2011), but its ability to do so is still debated (Wolf *et al.* 2007, McElreath *et al.* 2007). A general question is whether disruptive selection truly exists or whether, instead, distinct types only evolve when the modeller only permits discrete behavioural categories (or ignores population feedback: McElreath *et al.* 2007). Throughout this chapter the necessity of quantifying lifetime fitness in a population context approach has been emphasised. The evolution of RHP, and the evolution of personalities, are good examples of problems that will benefit from such an approach.

This chapter has concentrated on theoretical developments to date, but the majority of readers of this book will be empiricists. I have intentionally focused on exposing models' assumptions rather than extracting a simplistic list of testable predictions. It is tempting to read a theoretical paper scanning for predictions that could be tested. The clearer the list originally provided, the more successful a theory paper is in attracting empirical attention (the large number of tests distinguishing between sequential and cumulative assessment testifies to this). An oft-repeated instruction to test the models' *assumptions*, not just predictions, is less often followed. However, this would be an extremely fruitful approach, not only for distinguishing between current models, but for highlighting where assumptions may not reflect reality in an appropriate way.

For example, numerous early models simply assumed that individuals find themselves in the role of intruder and owner with equal frequency, and many (in my view, too many) models to date assume that the value of the future does not differ between contestants. In many cases models contain assumptions hidden from immediate view, most likely because the modellers themselves did not realise the importance of an assumption of, say, identical future prospects for all individuals, and consequently this is not emphasised in the paper. I have here tried to highlight *why* models differ in their predictions: they cannot be based on different versions of mathematics, thus different predictions must be based on different underlying biological assumptions. The more familiar empiricists are with the hidden as well as the visible assumptions that the various models make, the more constructively critical their view of the theory will become and theoretical developments will benefit reciprocally.

Acknowledgements

Extensive comments by Mark Briffa, Ian Hardy, Mike Mesterton-Gibbons, Geoff Parker and Ruben Requejo improved the chapter. Funding was provided by the Academy of Finland and the Australian Research Council.

References

Adams ES & Mesterton-Gibbons M (1995) The cost of threat displays and the stability of deceptive communication. *Journal of Theoretical Biology*, 175, 405–421.

Arnott G & Elwood RW (2009) Assessment of fighting ability in animal contests. *Animal Behaviour*, 77, 991–1004.

Austad SN, Jones WT & Waser PM (1979) Territorial defence in speckled wood butterflies: Why does the resident always win? *Animal Behaviour*, 27, 960–961.

Backwell PRY & Jennions MD (2004) Coalition among male fiddler crabs. *Nature*, 430, 417.

Backwell PRY, Christy JH, Telford SR, *et al.* (2000) Dishonest signalling in a fiddler crab. *Proceedings of the Royal Society of London B*, 267, 719–724.

Belsky G & Gilovich T (1999) *Why smart People Make Big Money Mistakes – And How to Correct Them: Lessons from the New Science of Behavioral Economics.* New York, NY: Simon & Schuster.

Bishop DT & Cannings C (1978) A generalized war of attrition. *Journal of Theoretical Biology*, 70, 85–124.

Bishop DR, Cannings C & Maynard Smith J (1978) The war of attrition with random rewards. *Journal of Theoretical Biology*, 74, 377–388.

Cant MA & Johnstone RA (2000) Power struggles, dominance testing and reproductive skew. *American Naturalist*, 155, 406–417.

Cant MA, English S, Reeve HK, *et al.* (2006). Escalated conflict in a social hierarchy. *Proceedings of the Royal Society of London B*, 273, 2977–2984.

Clark CW (1994) Antipredator behavior and the asset-protection principle. *Behavioral Ecology*, 5, 159–170.

Crowley PH (2000) Hawks, doves, and mixed-symmetry games. *Journal of Theoretical Biology*, 204, 543–563.

Crowley PH (2001) Dangerous games and the emergence of social structure: Evolving memory-based strategies for the generalized Hawk–Dove game. *Behavioral Ecology*, 12, 753–760.

Curio E (1987) Animal decision-making and the Concorde fallacy. *Trends in Ecology and Evolution*, 2, 148–152.

Davies NB (1978) Territorial defence in the speckled wood butterfly (*Pararge aegeria*): The resident always wins. *Animal Behaviour*, 26, 138–147.

Davies NB (1979) Game theory and territorial behaviour in speckled wood butterflies. *Animal Behaviour*, 27, 961–962.

Dunham ML, Warner RR & Lawson JW (1995) The dynamics of territory acquisition: A model of two coexisting strategies. *Theoretical Population Biology*, 47, 347–364.

Elwood RW & Briffa M (2001) Information gathering and communication during agonistic encounters: A case study of hermit crabs. *Advances in the Study of Behaviour*, 30, 53–97.

Enquist M (1985) Communication during aggressive interactions with particular reference to variation in choice of behaviour. *Animal Behaviour*, 33, 1152–1161.

Enquist M & Leimar O (1983) Evolution of fighting behaviour: Decision rules and assessment of relative strength. *Journal of Theoretical Biology*, 102, 387–410.

Enquist M & Leimar O (1987) Evolution of fighting behaviour: The effect of variation in resource value. *Journal of Theoretical Biology*, 127, 187–205.

Enquist M & Leimar O (1990) The evolution of fatal fighting. *Animal Behaviour*, 39, 1–9.

Eshel I (2005) Asymmetric population games and the legacy of Maynard Smith: From evolution to game theory and back? *Theoretical Population Biology*, 68, 11–17.

Eshel I & Sansone E (1995) Owner–intruder conflict, Grafen effect and self-assessment. The Bourgeois principle re-examined. *Journal of Theoretical Biology*, 177, 341–356.

Eshel I & Sansone E (2001) Multiple asymmetry and concord resolutions of a conflict. *Journal of Theoretical Biology*, 213, 209–222.

Falster DS & Westoby M (2003) Plant height and evolutionary games. *Trends in Ecology and Evolution*, 18, 337–343.

Field SA & Hardy ICW (2000) Butterfly contests: Contradictory but not paradoxical. *Animal Behaviour*, 59, F1–F3.

Fisher RA (1930) *The Genetical Theory of Natural Selection.* Oxford: Oxford University Press.

Fromhage L & Schneider JM (2005) Virgin doves and mated hawks: Contest behaviour in a spider. *Animal Behaviour*, 70, 1099–1104.

Grafen A (1979) The Hawk–Dove game played between relatives. *Animal Behaviour*, 27, 905–907.

Grafen A (1987) The logic of divisively asymmetric contests: Respect for ownership and the desperado effect. *Animal Behaviour*, 35, 462–467.

Haccou P & Glaizot O (2002) The ESS in an asymmetric generalized war of attrition with mistakes in role perception. *Journal of Theoretical Biology*, 214, 329–349.

Hamblin S & Hurd PL (2007) Genetic algorithms and non-ESS solutions to game theory models. *Animal Behaviour*, 74, 1005–1018.

Hamilton WD (1967) Extraordinary sex ratios. *Science*, 156, 477–488.

Hammerstein P (1981) The role of asymmetries in animal contests. *Animal Behaviour*, 29, 193–205.

Hammerstein P & Parker GA (1982) The asymmetric war of attrition. *Journal of Theoretical Biology*, 96, 647–682.

Hammerstein P & Riechert SE (1988) Payoffs and strategies in territorial contests: ESS analyses of two ecotypes of the spider *Agelenopsis aperta*. *Evolutionary Ecology*, 2, 115–138.

Maynard Smith J (1983) Game theory and the evolution of cooperation. In: DS Bendall (ed.) *Evolution from Molecules to Men*, pp. 445–456. Cambridge: Cambridge University Press.

McGregor PK & Peake TM (2000) Communication networks, social environments for receiving and signalling behaviour. *Acta Ethologica*, 2, 71–81.

McNamara JM & Houston AI (2005) If animals know their own fighting ability the evolutionarily stable level of fighting is reduced. *Journal of Theoretical Biology*, 232, 1–6.

Mesterton-Gibbons M (1999) On the evolution of pure winner and loser effects, a game-theoretic model. *Bulletin of Mathematical Biology*, 61, 1151–1186.

Mesterton-Gibbons M & Sherratt TN (2006) Victory displays: A game-theoretic analysis. *Behavioral Ecology*, 17, 597–605.

Mesterton-Gibbons M & Sherratt TN (2007*a*) Social eavesdropping: A game-theoretic analysis. *Bulletin of Mathematical Biology*, 69, 1255–1276.

Mesterton-Gibbons M & Sherratt TN (2007*b*) Coalition formation: A game-theoretic analysis. *Behavioral Ecology*, 18, 277–286.

Mesterton-Gibbons M & Sherratt TN (2009*a*) Animal network phenomena: Insights from triadic games. *Complexity*, 14, 44–50.

Mesterton-Gibbons M & Sherratt TN (2009*b*) Neighbor intervention: A game-theoretic model. *Journal of Theoretical Biology*, 256, 263–275.

Mesterton-Gibbons M & Sherratt TN (2011) Information, variance and cooperation: Minimal models. *Dynamic Games and Applications*, 1, 419–439.

Mesterton-Gibbons M & Sherratt TN (2012) Signalling victory to ensure dominance: a continuous model. *Annals of the International Society of Dynamic Games*, 12, 25–38.

Mesterton-Gibbons M, Hardy ICW & Field J (2006) The effect of differential survivorship on the stability of reproductive queueing. *Journal of Theoretical Biology*, 242, 699–712.

Mesterton-Gibbons M, Gavrilets S, Gravner J, *et al.* (2011) Models of coalition or alliance formation. *Journal of Theoretical Biology*, 274, 187–204.

Nakamaru M & Sasaki A (2003) Can transitive inference evolve in animals playing the Hawk–Dove game? *Journal of Theoretical Biology*, 222, 461–470.

Noë R (1994) A model of coalition formation among male baboons with fighting ability as the crucial parameter. *Animal Behaviour*, 47, 211–213.

Noë R & Sluijter AA (1995) Which adult male savanna baboons form coalitions? *International Journal of Primatology*, 16, 77–105.

Oliveira RF, McGregor PK & Latruffe C (1998) Know thine enemy, fighting fish gather information from observing conspecific interactions. *Proceedings of the Royal Society of London B*, 265, 1045–1049.

Packer C (1977) Reciprocal altruism in *Papio anubis*. *Nature*, 265, 441–443.

Packer C, Gilbert DA, Pusey AE, *et al.* (1991) A molecular genetic analysis of kinship and co-operation in African lions. *Nature*, 351, 562–565.

Panchanathan K & Boyd R (2004) Indirect reciprocity can stabilize cooperation without the second-order free rider problem. *Nature*, 432, 499–502.

Pandit SA & van Schaik CP (2003) A model for leveling coalitions among primate males, toward a theory of egalitarianism. *Behavioral Ecology and Sociobiology*, 55, 161–168.

Peake TM (2005) Eavesdropping in communication networks. In: P McGregor (ed.) *Animal Communication Networks*, pp. 13–37. Cambridge: Cambridge University Press.

Plowes NJR & Adams ES (2005) An empirical test of Lanchester's square law, mortality during battles of the fire ant *Solenopsis invicta*. *Proceedings of the Royal Society of London B*, 272, 1809–1814.

Poundstone W (1993) *Prisoner's Dilemma, John Von Neumann, Game Theory and the Puzzle of the Bomb.* New York, NY: Anchor Books Doubleday.

Radford AN & du Plessis MA (2004) Territorial vocal rallying in the green woodhoopoe: Factors affecting contest length and outcome. *Animal Behaviour*, 68, 803–810.

Rubin PH (2002) *Darwinian Politics: The Evolutionary Origin of Freedom*. New Brunswick, NJ: Rutgers University Press.

Rutte C, Taborsky M & Brinkhof MWG (2006) What sets the odds of winning and losing? *Trends in Ecology and Evolution*, 21, 16–21.

Scheffran J (2006) The formation of adaptive coalitions. In: A Haurie, S Muto, LA Petrosjan, *et al.* (eds.) *Advances in Dynamic Games. Annals of the International Society of Dynamic Games*, 8, 163–178. Boston, MA: Birkhäuser.

Semyonova A (2003) *The Social Organization of the Domestic Dog; A Longitudinal Study of Domestic Canine Behavior and the Ontogeny of Domestic Canine Social Systems*. The Hague: The Carriage House Foundation (www.nonlineardogs.com, version 2006).

Silk JB, Alberts SC & Altmann J (2004) Patterns of coalition formation by adult female baboons in Amboseli Kenya. *Animal Behaviour*, 67, 573–582.

Strashny A (2007) *Dynamic Paired Comparison Models.* PhD Thesis, University of California, Irvine.

Thomas ML, Payne-Makkrisâ CM, Suarez AV, *et al.* (2006) When supercolonies collide: Territorial aggression in an invasive and unicolonial social insect. *Molecular Ecology*, 15, 4303–4315.

Tsuji K & Tsuji N (2005) Why is dominance hierarchy age-related in social insects? The relative longevity hypothesis. *Behavioral Ecology and Sociobiology*, 58, 517–526.

Waal FBM de & Harcourt AH (1992) Coalitions and alliances, a history of ethological research In: AH Harcourt & FBM de Waal (eds.) *Coalitions and Alliances in Humans and Other Animals*, pp. 1–19. Oxford: Oxford University Press.

Walters JR (1980) Interventions and the development of dominance relationships in female baboons. *Folia Primatologia*, 34, 61–89.

Whitehead H & Connor R (2005) Alliances I. How large should alliances be? *Animal Behaviour*, 69, 117–126.

Wilson EO (1998) *Consilience, the Unity of Knowledge*. New York, NY: Random House.

Wilson MI, Britton NF & Franks NR (2002) Chimpanzees and the mathematics of battle. *Proceedings of the Royal Society of London B*, 269, 1107–1112.

Zabel CJ, Glickman SE, Frank LG, *et al.* (1992) Coalition formation in a colony of prepubertal spotted hyaenas. In: AH Harcourt & FBM de Waal (eds.) *Coalitions and Alliances in Humans and Other Animals*, pp. 113–135. Oxford: Oxford University Press.

Analysis of animal contest data

Mark Briffa, Ian C.W. Hardy, Martin P. Gammell, Dómhnall J. Jennings, David D. Clarke & Marlène Goubault

4.1 Summary

In this chapter we outline and discuss statistical approaches to the analysis of contest data, with an emphasis on testing key predictions and assumptions of the theoretical models described in Chapters 2 and 3. We use examples from an array of animal taxa, including cnidarians, arthropods and chordates, to illustrate these approaches and also the commonality of many key aspects of contest interactions despite the differing life histories and morphologies (including weaponry) of these organisms. We first deal with the analysis of contest outcomes, a useful approach for determining which traits contribute to an individual's resource holding potential (RHP). Here we outline alternative statistical approaches that treat the outcome as either an explanatory (independent) variable or as the response (dependent) variable. In both cases, we treat a single contest as one 'experimental unit' and consider ways in which multiple measures taken from the same experimental unit should be accounted for in the analysis. Thus, we introduce paired and repeated measures approaches for contest data and also the calculation of composite measures. We then discuss more complex mixed models, which are particularly useful for dealing with multiparty contests when multiple individuals from the same group occur in more than one observation. Having established what factors influence RHP, one might then ask questions about the roles of information-gathering and decision-making during contests. These questions are prompted by the theoretical models of dyadic contests discussed in Chapters 1 and 2, and we consider the advantages and limitations of using analysis of contest duration to distinguish between 'mutual-' and 'self-assessment' type contests. An additional tool that we can use to address this question is the analysis of escalation and de-escalation patterns, and we thus shift the focus to within-contest behavioural changes. As well as discussing changes in rates of agonistic behaviour we introduce 'sequence analysis', which is a tool for analysing transitions between different behaviours. Thus, it is possible to investigate changes in agonistic behaviour in both quantitative and qualitative terms. We then turn to the wider social setting within which many contests take place and consider the range of methods currently used for analysis of dominance hierarchies. We then return to the idea of information and describe an approach for analysing the potential for bluffing during a contest. This question of signal honesty is one that pertains not only to contest behaviour but also to the wide range of contexts where signals are used to advertise some aspect of individual quality. Finally, we suggest some key points that should be considered when designing an experiment on contest behaviour.

4.2 Introduction

Understanding contests from an evolutionary perspective is complicated by the fact that they involve at least two individuals and the benefits that accrue to one contestant will be influenced both by its own behaviour and by that of its opponent. When contests take place between opposing groups of individuals, the situation is complicated further because outcomes will be determined not only by the size of each rival group, but also by the behavioural interactions of the individuals within each group. For these reasons simple optimality models are often insufficient to account for the full complexity of animal contests. Therefore, approaches based on game-theoretic modelling, which deals with decisions made in the presence

Animal Contests, ed. I.C.W. Hardy and M. Briffa. Published by Cambridge University Press.

of other decision-makers, have been used to develop the conceptual understanding of contest interactions (Chapters 2 and 3).

The fact that contests are interactions between two or more individuals also has implications for how data from empirical studies should be analysed statistically. From a statistical point of view, the types of question that experiments on animal contest behaviour are often designed to address can be grouped as follows:

- Differences between winners and losers, which we term 'analysis of outcomes' (section 4.3);
- Relationships between key correlates of resource holding potential (also termed resource holding power, RHP, Chapter 2) and/or resource value, *V*, as identified in analyses of outcomes, and the duration of the contest. We term this 'analysis of duration' (section 4.4); and
- Changes in rates of agonistic behaviour as the contest progresses, which we term 'analysis of contest dynamics' (section 4.5.1). Associated with this, there may be a need to use a robust statistical approach to identify transitions between distinct phases of contests, or to identify the effects of particular events within a contest, which we term 'analysis of contest structure' (section 4.5.2).

For dyadic contests in particular, these three approaches in combination are very useful in distinguishing between the different assessment models proposed by theory. By examining RHP, *V*, contest duration and contest structure we can attempt to test the assumptions and predictions of the mainstream theory of dyadic contests. For instance, the different modelling assumptions concerning whether giving-up decisions are made by 'mutual' or 'self' assessment (Chapter 2) should yield different relationships between RHP and contest duration (section 4.4). These models also vary in their predictions about changes in the rate of agonistic behaviours as the contest progresses (Chapter 2, section 4.5).

In addition to testing hypotheses about dyadic contests, we may be interested in analyses of social structures (section 4.6), such as dominance hierarchies, which arise or are maintained as a result of contest behaviour, or in the analyses of 'attrition' rates between rival groups that engage in fatal 'battles'. Finally, we consider approaches to analyse the role of honesty and the potential for bluffing in the performance of agonistic signals (section 4.7).

These approaches have been chosen on the assumption that many readers will be interested in designing experiments that aim to test predictions stemming from the body of game theory and related areas of contest theory that is at the core of this book (e.g. Chapters 2 and 3). Studies of aggressive behaviour and contests may, however, be conducted with different emphases: to better understand other research areas within behavioural ecology (e.g. clutch size, Chapter 8) or in other areas of basic biological research, such as the neuronal basis of aggression (e.g. in crustaceans, Chapter 5) and developmental biology (e.g. developmental plasticity in the development of horns and mandibles in beetles, Chapter 9). Contest studies may also have ultimately applied aims, such as improving biological pest control (Chapter 8) or animal welfare (Chapter 16). Thus, although focussed on answering questions about contest behaviour from a functional perspective, the issues surrounding the analysis of data on fights and other agonistic encounters pertain to studies in a far broader range of disciplines than behavioural ecology *sensu stricto*.

To illustrate the key statistical techniques, we draw on both recent and 'classic' studies and use data from very different animals, such as sea anemones, parasitoid wasps, swordtail fish and African elephants. For instance, Cnidarians such as the beadlet anemone, *Actinia equina*, are small marine animals, lack a centralised nervous system, have no specialised organs for excretion or gas exchange and their sensory cells are the simplest among metazoans (Rudin & Briffa 2011, Figure 4.1). In contrast, elephants, *Loxodonta africana*, are the largest extant land animals, have central nervous systems and possess very complex respiratory and social structures (Poole 1982, 1989, Figure 4.2). These zoologically dissimilar species nevertheless engage in conceptually and analytically similar contests.

4.3 Analysis of contest outcome

The simplest question that one could ask about a contest, in terms of its outcome, is 'Who won?' If we observed a number of such contests we would then probably want to know 'What is different about winners and losers?' Many studies have shown that winners tend to be larger than losers, but there may also be more subtle differences such as the size of weapons, prior experiences of winning or losing,

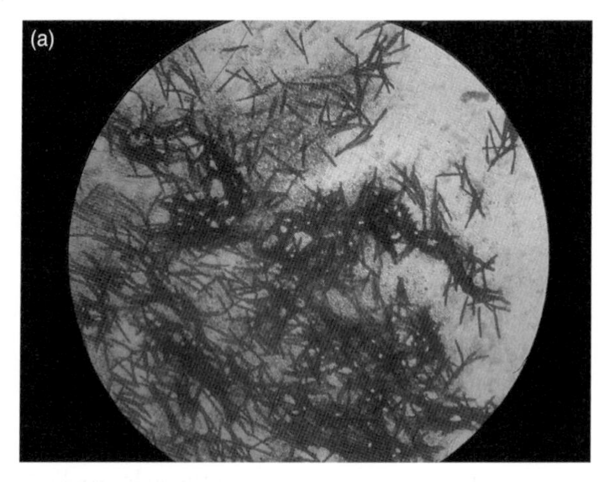

Figure 4.1 (**a**) Nematocysts, the stinging harpoon-like organelles of cnidarians, stained and viewed under light magnification. These nematocysts were from a sample of approximately 1 mm^2 of acrorhagial epidermis. (**b**) *Actinia equina* in combat: the acrorhagi are the blue bud-like tentacles visible on the individual on the left. The individual on the right has received a series of acrorhagial contacts and blue acrorhagial peels are visible on its epidermis. (Photograph credit (a) Mark Briffa, (b) Fabian Rudin.) For colour version, see colour plate section.

Figure 4.2 Agonistic interaction between two male African elephants, *Loxodonta africana*. These individuals, known to researchers as Dionysus (left) and Iain (right) are both in musth. (Photograph reproduced with permission © Joyce Poole, ElephantVoices.) For colour version, see colour plate section.

aggressiveness or how much each individual values the resource. Regardless of which factor we are interested in, if it is possible to quantify it we can (given sufficient replication; Taborsky 2010, Smith *et al.* 2011) determine whether winners and losers differ significantly within our sample of contests. In a formal statistical sense we are treating the outcome of each contest as an explanatory ('predictor' or 'independent') variable with two levels, 'winner' or 'loser'. The measure of interest (e.g. body size) is then treated as the response (dependent) variable. Although this proposition seems very straightforward, there is a second way to ask the same question: 'What made the winner win and the loser lose?' or, using different words, 'What factors determine the outcome of contests?' In many cases, asking the question this way round might be more intuitive; after all, if large individuals are better at fighting it is the large body size of the winner (relative to the loser) that made it win, not winning that caused it to be big (at least not in the timescale of a single agonistic encounter). In the second formulation of the question, it is the measured variable of interest (e.g. body size) that is treated as the explanatory variable, and the categorical outcome (win or lose) that is treated as the response variable.

Treating the outcome as the response variable is therefore a truer reflection of the direction of causality in the relationship between body size and outcome and some readers may well find this solution more satisfactory from a statistical view-point. However, there are several advantages to the approach of using the outcome as an explanatory variable, which we detail below (section 4.3.1). It has recently been argued that both of the above approaches are valid and that a key distinction between them is largely semantic: 'Are winners larger than losers?' or 'Does being big increase the chance of winning?' both represent appropriate ways of determining what traits may contribute to RHP (Briffa & Elwood 2010). Below we discuss the merits of each approach, and describe another key way in which they differ; how data from two or more individuals should be handled to avoid the potential for pseudoreplication (Hurlbert 1984).

Pseudoreplication is the treatment of non-independent observations as if they were independent with the concomitant, and incorrect, inflation of sample size and thus statistical power. Pseudoreplication promotes Type I error, which is concluding that a fundamentally absent effect is present. The potential for pseudoreplication to lead to a Type I error is well known, but we also describe below a less well-known phenomenon in which treating non-independent data as independent could also lead to a Type II error of a 'false negative' result, i.e. failure to detect an effect that is actually present.

4.3.1 Contest outcome as an explanatory variable

4.3.1.1 Paired analyses

As noted above, a contest involves at least two opponents and when measures are taken from each opponent, these measures could be correlated and would therefore not be independent. It is the contest that is the *experimental unit* while the contestants themselves are *experimental subunits*. The reasoning can be illustrated by considering cases where the variable of interest is a physiological parameter that is expected to change during the contest. For example, numerous studies have shown that lactic acid accumulates during contests and winners tend to have lower lactic acid than do losers (e.g. crustaceans, Chapter 5; fish, Chapter 10; snakes, Chapter 12). Measures of lactic acid in winners and losers are not independent: the two opponents have been in the same contest, competing for the same amount of time for the same resource and are likely to have experienced the same level of escalation, even if they have been employing different contest tactics. Thus, winner and loser lactate levels may be positively correlated within contests. While these individuals share a common experience, across a set of data, obtained from observing many contests, there will be a range of different experiences of contest durations and levels of escalation. Thus, compared to within-contest differences in the variable of interest, there might be a greater amount of between-contest variation between winners and losers. Such 'background variation' (Heath 1995) has the potential to obscure the presence of relative differences between winners and losers within fights (Briffa & Elwood 2010), and 'paired' analyses are required to avoid this problem.

The same argument can be applied to non-physiological variables which could influence RHP but which will not vary during the encounter, e.g. body size or weapon size. This is because winner or loser status is determined by the contestants rather than being experimentally assigned (Briffa & Elwood 2010). Furthermore, contestants may be size-matched if size-assortative fighting naturally occurs or if there is a size-dependent spatial distribution of individuals in their habitat (Morrell *et al.* 2005). Within-fight correlations between opponents may also occur if the value of each resource is higher for one size class of potential contestants, as could occur for fights over burrows of different sizes. In these cases there will be a correlation between the size of winners and losers but across all observed contests the difference between winners and losers may be obscured if the identities of the rivals that fought each other is not accounted for. In other words, there are several reasons why a failure to account for the non-independence of data might lead to Type II errors (failing to detect an effect that is underlyingly present).

In our example of the study of sea anemones, an important agonistic behaviour is 'stinging' the opponent, which involves contacting it with specialised fighting tentacles called acrorhagi, which contain high densities of nematocytes (cells that contain the harpoon-like organelles called nematocysts, Figure 4.1a). When triggered by contact, the nematocysts pierce the opponent's dermis and inject a powerful toxin. Such attacks result in visible 'peels' of the acrohagial ectoderm being left on the epidermis of the recipient (Figure 4.1b). In a small proportion of fights (17/82 staged encounters, see Rudin & Briffa 2011 for full experimental details), both individuals inflict peels on their opponent. Imagine we want to ask whether there is a difference in the number of peels inflicted by winners and losers. Clearly, data on one opponent in an encounter will not be independent of data on the other; an individual's ability to inflict peels is likely to be influenced by the number of peels (and hence the amount of toxin) that it has itself received during the encounter; moreover, an individual might attempt to match the number of peels inflicted by the opponent. Interestingly, there is a significant positive correlation between the number of peels inflicted by each opponent (Pearson correlation coefficient: $r_{15} = 0.53$, $p < 0.05$), which suggests the latter. Regardless of the cause of the correlation, the data are clearly non-independent and a paired t-test is the correct

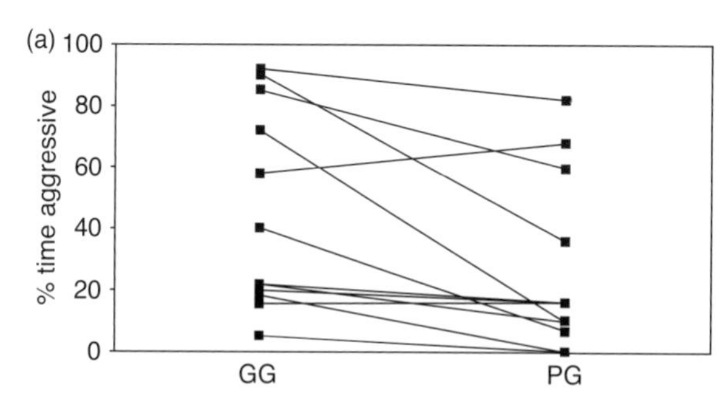

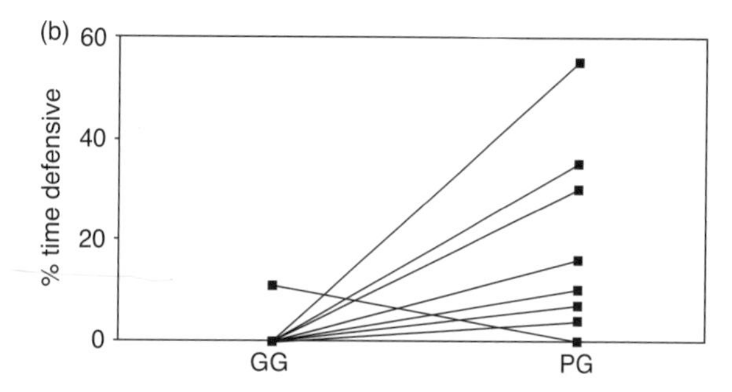

Figure 4.3 The paired analyses of the effects of treatment (GG or PG) of male swordtails (fish) during development on their subsequent agonistic behaviour. The squares represent individual male aggressive and defensive scores (proportion of observation time) and the lines join pairs of males that fought one another, and therefore comprise experimental units. The appropriate analyses reveal significant differences between GG and PG individuals for both defensive and offensive behaviours. If the data are analysed using an unpaired analysis, however, only the difference in defensive behaviour remains significant. (Figure adapted from Royle *et al.* 2005.)

analysis to use. This yields $t_{16} = 7.1, p < 0.0001$, indicating that winners inflicted more peels on losers than did losers on winners. In this example an unpaired *t*-test on the same data results in $t_{32} = 4.9, p < 0.0001$, which would also strongly indicate the presence of a difference between winners and losers, but the analysis is nevertheless incorrect (as indicated by the inflated degrees of freedom).

In this case the difference between the paired and unpaired analyses may not appear critical because both analyses yield highly significant *p*-values. However, the lower *t*-value for greater degrees of freedom in the case of the unpaired test shows how it is behaving in a more conservative way than the paired test in this example. An example of a study where incorrect use of an unpaired test would have changed the final result is given by Royle *et al.* (2005), who examined the effect of 'catch-up' growth on agonistic behaviour in a fish, the green swordtail *Xiphophorus helleri*. The treatment applied to each opponent before the fight is the explanatory variable, rather than the explanatory variable being the outcome, but the reason for using a paired analysis is the same. The two treatments were as follows: during the first year of development, males were provided with either a 'good' high-protein diet for both the first 6 months and the second 6 months of development ('GG' individuals) or a 'poor' low-protein diet during the first 6 months followed by the good high-protein diet during the second 6 months ('PG' individuals). After 6 months the GG individuals were larger than the PG individuals, but after 1 year there was no difference in body size between the two groups due to the PG individuals having undergone catch-up growth during the second 6 months. Encounters were then staged between 1-year-old GG and PG individuals to determine whether the process of catch-up growth influenced fighting behaviour, quantified as the proportion of time spent on aggressive and defensive behaviours. Here, a paired analysis is appropriate to determine the effect of 'developmental experience' (GG or PG) on agonistic behaviour, because each fight contains a GG and a PG individual. Therefore, a fight is the experimental unit and PG and GG individuals are experimental subunits. In this case the response variable was proportionate, so the non-parametric equivalent of a paired *t*-test, the Wilcoxon signed rank test, was used. The analysis indicated that GG individuals were more aggressive ($Z = 2.67$, $n = 13$, $p = 0.008$) than PG individuals (Figure 4.3). Had an equivalent non-paired Mann–Whitney *U* test been used, the result would have been that, despite the artificially elevated sample size, the difference in aggressive behaviour between the two groups was not significant ($U = 47.8$, $n = 26$, $p = 0.060$). This example illustrates the possibility of obtaining 'false negative' results (Type II errors) by using an unpaired analysis to make comparisons between opponents that took part in the same fights. Although a false negative is less likely if the effect is very strong, unnecessarily large Type II errors can be avoided by using the appropriate paired analysis. For another illustration of this point see the hypothetical example discussed in Briffa and Elwood (2010).

4.3.1.2 Repeated measures analysis

Paired designs are common in biology and are appropriate whenever two measures are taken from the same experimental unit or 'when the treatment is given to two individuals sharing a common experience' (Sokal & Rohlf 1995, pp. 352–356). In many studies on contest behaviour there might be further comparisons that we are interested in, in addition to the paired comparison between winners and losers. For example, we might want to know whether differences between winners and losers are consistent for different levels of resource value, V. Lindström and Pampoulie (2005) investigated contests between male sand gobies, *Pomatoschistus minutus*, over small (low V) and large (high V) nests. By investigating male size and nest size simultaneously they found that the amount of time that males spent guarding a nest was influenced by both factors, with defence time increasing with both the owner's RHP and the quality of the nest. Similarly, in some types of contest, such as shell fighting in hermit crabs (Chapter 5), the opponents adopt two distinct roles, which fight in different ways, and we might want to test for differences between roles as well as between outcomes. While a straightforward paired t-test is sufficient to analyse data where the only explanatory variable is the outcome status of the contestants, this approach cannot be used to analyse data sets where there are additional factors such as resource value (e.g. high or low V) and role (e.g. resident or intruder; attacker or defender).

In these situations, a 'repeated measures analysis of variance' (repeated measures ANOVA) can be used to determine whether there are significant differences (1) between outcomes, regardless of the condition under which fights were staged, or (2) between conditions under which fights were staged, regardless of outcome status, and (3) by examining the 'outcome × treatment' interaction, whether any differences between winners and losers are consistent across the two sets of conditions (or 'treatments') under which the fights were staged. Typically, the 'repeated measure' would be analogous to the paired comparisons of non-independent data on winners and losers from the same fights, discussed above. Another way to put this is to call the factor denoting outcome status (winner or loser) the 'within-subject' factor (the term '*subject*' being equivalent to '*experimental unit*'), as this comparison is being made 'within' experimental units. The 'treatment' is a 'between-groups' factor, because different fights occur under different conditions; the data from fights with high or low resource value are independent of one another. The analysis can be described as a 'one within, one between repeated measures ANOVA' or sometimes as 'repeated measures ANOVA with between-subjects factors'. This analysis offers a statistically powerful way to make comparisons between winners and losers in fights that have been staged under different sets of experimentally manipulated conditions.

A common misconception about the validity of repeated measures analysis when applied to animal contests arises from the term 'repeated measures'. In an analysis of independent data points, an unpaired t-test or an ANOVA is appropriate if there are two groups for comparison and, because t-tests can only compare two groups, an ANOVA is used if there are data from more than two categorical groups to be compared (e.g. A, B and C: one-way ANOVA) or if there is more than one type of group (factor) in the experiment (e.g. a comparison between A and B under conditions of Y and Z: two-way ANOVA or, more generally, 'factorial ANOVA'). Similarly, if there are more than two groups of non-independent data, or if two groups of non-independent data need to be analysed under different treatment conditions, a repeated measures ANOVA accounts for non-independence in the same way that a paired t-test (or a Wilcoxon signed ranks test for non-parametric data) would do so in a simpler experiment. Here, the term '*repeated measures*' is used instead of '*paired*' simply because it prescribes no upper limit on the number (> 1) of non-independent groups that can be compared. Here we have described the use of repeated measures analysis for dealing with data collected from the same experimental unit, that were collected at the same time (e.g. measures from the winner and the loser at the end of the fight). Such measures are sometimes described as '*simultaneous*' data from non-independent groups. Note that repeated measures analyses can also be used to investigate measures taken from the same experimental units but at different time points. This type of '*longitudinal*' data might be collected when we are interested in analysing changes in agonistic behaviour as contests progress, and this is discussed in section 4.5.

4.3.1.3 Contest outcomes as influences on future contests

Contest outcomes can also affect the resolution of subsequent contests, with prior victories usually increasing, and prior losses decreasing, the probability of winning a future contest (e.g. Chapters 2, 3, 8 and 10). To test the existence of these so-called 'winner and loser effects', one might be tempted to use binomial tests to determine whether the probability of winning the second contest is significantly above or below 0.5 for prior winners and prior losers, respectively. However, the correct null hypothesis depends on the procedure used to give fighting experiences to focal individuals. The two types of protocols are referred to as 'random-selection' or 'self-selection procedures' (Chase *et al.* 1994, Beaugrand & Goulet 2000). In the self-selection methods, two individuals (usually matched for size, as a proxy for matching RHP) are allowed to fight. The resulting winner and loser are then each separately paired against a (different) naïve opponent and their respective winning success is measured. The problem with this apparently 'natural' way of obtaining prior winners and losers is that the selected individuals do not only differ in terms of their previous fighting experience but most likely also in their intrinsic RHP (e.g. individuals winning the first contest will tend to be larger than losers; Chase *et al.* 1994, Bégin *et al.* 1996). As a result, the effect of prior fighting experience may be confounded with that of fighting ability.

An alternative is to use the random-selection procedure: individuals are randomly chosen and receive a predetermined experience (victory or defeat). To generate predetermined winners, focal individuals are paired against opponents that are expected to have a low success rate (i.e. an individual with inferior RHP) during a first (training) contest. In contrast, predetermined losers are obtained from training contests by pitting focal individuals against individuals expected to win (i.e. individuals with a large RHP). The probability of winning the second contest can then be compared to the null hypothesis of 0.5. Kasumovic *et al.* (2010) suggested that one-sided binomial tests can be used owing to the assumptions that, if winner and loser effects exist, prior winner's success probability should be above 0.5, and prior losers' under 0.5. The advantage of the random-selection procedure over self-selection is that it randomises intrinsic differences between competitors and thus focuses on the effects of experience.

While training in the random-selection procedure has used asymmetries in RHP (smaller/larger or lighter/heavier trainers, e.g. Hsu *et al.* 2009, Kasumovic *et al.* 2010), the approach can be readily adapted for species where contest outcomes are strongly influenced by the difference in the value that contestants place on the resource, V. For instance, in *Eupelmus vuilleti*, a parasitoid wasp fighting for host access (Chapter 8), contests are generally won by the female with higher egg load (V correlates with egg load because females with more mature eggs in their ovaries are more ready to lay and hosts are worth more to them than to females with fewer eggs). To explore how fighting experience influences future contests, designated winners and losers were, respectively, paired against low and high egg-load trainers. In the second contest, they were then paired against naïve individuals matched for V. Surprisingly, this showed the existence of a winner effect in the absence of evident loser effects (Goubault & Decuignière 2012, Chapter 8).

When contests escalate, however, the value of information on previous fighting experience has been found to diminish compared to that acquired during the current contest, with contests being mainly settled according to asymmetries in RHP and V between current contestants (Hsu *et al.* 2006). Other factors that can affect the detection of winner–loser effects are the frequency and duration of experience training, the interval between the end of training and the subsequent contest and the isolation of focal individuals which, in some species, affects individuals' aggression levels compared to those reared together (Hsu *et al.* 2006). Thus, when designing experiments, all of these factors would need careful consideration in order, for example, to determine what sample size would be adequate to detect the effect of previous fighting experience in subsequent fights.

4.3.2 Contest outcome as the response variable

4.3.2.1 Generalised linear models

By asking 'What effect does a particular trait have on the probability of winning a fight?' we are dealing with a binary (win = 1 or lose = 0; out of a maximum of 1, see below) response variable, rather than the normally distributed responses discussed above (section 4.3.1). While normally distributed response variables can be readily analysed within the 'general linear model' (which can be termed GLM, e.g. Grafen & Hails 2002) family of statistical tests (including standard regression and standard ANOVA), the appropriate (quasi-parametric) approach for non-normally distributed data can be to utilise one of the wider class of 'general*ised* linear models', often now referred to as GLZs. GLZs assume that outcomes conform to one of the 'exponential family' of distributions and significance is assessed using 'change in deviance', G (a likelihood ratio). This is in contrast to the more familiar F-ratio tests used by GLMs that assume normally distributed outcomes. The test statistic G is usually very similar to χ^2 so, if the statistical package does not generate a p-value for you (some do not!), χ^2 tables can be used to look up p for the appropriate degrees of freedom. Note that while G and χ^2 values are similar they are not identical (usually the differences are larger when sample sizes are smaller), which means that hypothesis testing using GLZs involves greater uncertainty when estimates of p are close to the conventional 0.05 threshold (Crawley 1993, Warton & Hui 2011). While Wald tests are the default for logistic analysis in many statistical packages, likelihood ratio tests, G, are often preferable, because they are based on fewer approximations and because Wald tests have increasing Type II error rates when probability is close to 0.0 or 1.0 (Warton & Hui 2011): however, the biological interpretation of the two tests will usually be the same.

GLZs include logistic regression and logistic ANOVA (for binary responses) as well as log-linear equivalents for small integer responses (e.g. Crawley 1993, Wilson & Hardy 2002, O'Hara & Kotze 2010, Warton & Hui 2011). Binary and count data are integer data: count data have a theoretical maximum of infinity (e.g. the number of fights won by an elephant during its entire lifetime), while proportional data are counts out of a given total (e.g. the number of fights the elephant won out of the total of, say, five it entered into on a given day). If the given total is 1 then the data are binary (e.g. the elephant won a particular fight, 1/1, or lost this fight, 0/1). For small count data the Poisson distribution is a suitable initial assumption for the distribution of outcomes and for binary data the equivalent assumption is the binomial distribution.

The GLZ approach can also be used for continuous response data that are not normally distributed. For example, with skewed data on contest duration we could implement a GLZ adopting a gamma distribution of outcomes, rather than following the more traditional method of normalising data by a mathematical transformation (e.g. taking logarithms) prior to GLM analysis (Crawley 1993). A problem, however, is that currently this type of functionality is not available in many statistical packages. Even in the relevant packages of highly flexible open source software, such as *R*, the ability to perform some of these GLZs seems to be poorly supported at present and even the validity of these tests is the subject of ongoing debate. Opting for this approach can make for a frustrating time if you are attempting to analyse your latest contest duration data. Fortunately, one approach that is well supported across commonly used statistical software is the type of GLZ analysis that is appropriate for understanding contest outcome (as the response variable): logistic analysis.

4.3.2.2 Logistic analysis

Logistic analyses are GLZs in which outcomes are assumed to be binomially distributed and a logit link function, relating the model to the response variable, is specified (e.g. Crawley 1993, Hardy & Field 1998, Warton & Hui 2011). Note that when binary data are 'grouped' (there are several binary outcomes within a replicate, e.g. the elephant had five fights, each with a binary outcome, during an observation period) the assumption of binomiality can be modified (quasi-binomiality) to fit better the actual distribution of outcomes (discussed by Crawley 1993, Wilson & Hardy 2002), but contest outcome data are usually 'ungrouped' (one binary response per replicate, e.g. the outcome of just one elephant fight is recorded) and in these cases logistic analyses are constrained to assume exactly binomially distributed outcomes.

Logistic analysis should by now have replaced the 'traditional' and inferior practice of handling binomial data by first applying an arcsine square root transformation to attempt to normalise and stabilise the variance, followed by GLM analysis assuming constant and normally distributed outcomes (Wilson & Hardy 2002, Warton & Hui 2011): we hope this chapter speeds the completion of this change in practice.

In most statistics software logistic analysis is relatively straightforward to implement; *SPSS* (version 18), for example, takes the user through a series of fairly intuitive screens where these options are chosen. *GenStat* (11th edition) allows the user to construct, very flexibly, a GLZ from a general menu (our preference) or to opt for a pre-structured menu. In less user-friendly but nonetheless flexible packages, such as *R* or *GLIM*, a degree of programming is required. In all of these packages performing logistic analyses is only very marginally more complex than using their standard (GLM) equivalents.

In its simplest form, 'binary logistic regression' tests for a relationship between a continuous

may recur in the data set up to four times due to the four possible combinations of musth statuses. This does not mean that either the field observations or the data set are in any way 'incorrect', indeed we find them rather impressive, but it does mean that this particular analysis is pseudoreplicated (section 4.3) and we must thus be wary of our conclusions. We return to this example and offer a solution in section 4.3.3.

4.3.2.3 More complex logistic models

There is, of course, great flexibility to the form that logistic analysis can take, such as exploring the effects of several continuous explanatory variables (logistic multiple regression), one categorical explanatory variable with two or more 'factor levels' (logistic ANOVA), or several categorical variables (logistic 'two-way' or 'factorial' ANOVA), or exploring a mix of continuous and categorical variables (logistic ANCOVA) as mentioned in both the elephant and sea anemone examples (section 4.3.2.2). It is also possible to explore effects on a response variable with more than two types of response. In a contest study we might, for instance, have a situation where pairs of individuals either did not behave agonistically, engaged in non-injurious chases or engaged in full-on fighting, generating a 3-level response variable describing aggression. The effect of categorical explanatory variables on the probability of each level of such trinomial, or more generally multinomial, responses occurring can be examined by adopting the assumption of Poisson distributed outcomes: we refer the interested reader to Crawley (1993, p. 297).

One issue that will often pertain to more complex logistic models (as well as complex GLMs and GLZs in general) in which there are many explanatory variables, is that of how best to include these variables in the model. There may be a large number of potential interactions between the 'main effect' explanatory variables, and decisions will need to be made as to how many, and which, should be considered in the analysis. As it is seen as generally better (e.g. Crawley 1993, 2002) to begin with a complex model and then find which terms within it are significant, rather than progressively adding terms into an initially simple model, the above decisions mainly pertain to which terms (main effects and their various interactions) should be retained in an initially complex statistical model. Two pragmatic considerations surrounding the inclusion of interaction terms are first that it is probably not sensible to try to include so many interaction terms that their number exceeds the number of replications in the data set and, second, that higher-order interactions may be very difficult to interpret if significant. For decisions about main effects, a concern is that correlation between explanatory variables (known as collinearity or multi-collinearity) can lead to untrustworthy parameter estimates and p-values. Such interpretational problems can be reduced by not including in the initial model explanatory variables that are effectively redundant (Quinn & Keough 2002): for instance, include either weight *or* length as the measure of contestant body size but not both. Of course, including only a judicious subset of potential main effects in the initial model also reduces the number of possible interaction terms and helps manage the problem of 'too many and too complex interactions' mentioned above.

Once terms of interest have been included in an initial 'maximal' model there are numerous possible ways to proceed towards a more parsimonious model (Crawley 1993, Quinn & Keough 2002, Bolker *et al.* 2008). One currently commonly used method is to remove terms sequentially from the maximal model, starting with the highest-order interactions and testing the significance of each term on its deletion. Non-significant terms are then left out of the model whereas significant terms are re-entered. Significant categorical main effects (factors) with more than two levels can be simplified by aggregating some of the levels, and the simplified version retained if the change does not produce a significant change in the explanatory power of the model. This 'backward elimination' approach leads to a 'minimal adequate model' containing only those terms that are themselves significant or form part of a significant interaction term. For contest data, this might, for instance, provide the parsimonious (i.e. simplest adequate) description of which measured properties of contestants influence contest outcome. This was the approach to statistical model selection used to generate the results in Table 4.1, and is described in further detail by Crawley (1993, 2002), Hardy and Field (1998), Quinn and Keough (2002) and Wilson and Hardy (2002). Some further recent examples of such logistic model simplification in the contest literature are provided by Bentley *et al.* (2009), Innocent *et al.* (2011) and Lizé *et al.* (2012).

An alternative method of simplifying from complex initial maximal models is to choose as the 'best fit' one of the less complex possible models (i.e. the possible simpler versions containing subsets of the initially included terms) on the basis of its having the lowest value of the Akaike information criterion (AIC). The AIC value will be lower for a model that is simpler (has fewer parameters) if it fits the data equally well, and lower for a model that fits the data better if it has the same number of parameters; thus, the value of the AIC is a compromise between model simplicity and model fit. Note that, unlike performing a series of significance tests during the backwards elimination procedure outlined above, employing AIC to choose a parsimonious model does not in itself involve tests of null hypotheses, although significance tests can be obtained for given model parameters, for example by performing *G*-tests (e.g. Kemp *et al.* 2006). While this may take some getting used to for many behavioural ecologists, there are compelling arguments for reducing the current emphasis on null hypothesis significance testing (Nakagawa & Cuthill 2007). Examples of using the information-theoretic approach to logistic model selection for contest outcome data are given by Kemp *et al.* (2006) and Bamford *et al.* (2010) for pairwise contests, and by Batchelor and Briffa (2010, 2011) for multi-party contests. In addition, in the study by Royle *et al.* (2005) on *Xiphophorus helleri* (section 4.3.1.1), comparison of AIC values was used in the initial analysis of growth rates, prior to the paired analysis of contest outcomes.

The choice of which of the many possible approaches to model simplification to adopt may be influenced by pragmatic considerations, such as AIC not being readily available in a particular statistical package, but we do note that there is concern in the statistical literature that the backwards elimination procedures we outline above are not always able to select the most influential variables from an initial set. These issues are the subject of ongoing statistical debate and interested readers (or readers who have collected data on a lot of explanatory variables and find themselves in the position of needing to know) are referred to Crawley (2002), Quinn and Keough (2002), Bolker *et al.* (2008) and Mundry and Nunn (2009) for further details. Currently, the use of information criteria (often AIC, but also alternatives such as the Bayesian information criterion, BIC) seems well established when mixed models, which can account for 'random' factors (section 4.3.3), are used. The popularity of this approach in mixed models might be due to the fact that there are disagreements about the validity of generating *p*-values for parameter estimates in mixed models. Note, however, that in many cases, one is not interested in the random effect per se; rather, one merely wants to account for it in the analysis (although for an exception in the contest literature see the study by Wilson *et al.* (2011), on fighting and personality in green swordtail fish).

4.3.2.4 Further advantages and disadvantages of logistic analysis

While the paired and repeated measures analyses described in section 4.2 are well developed, easy to illustrate and are probably very familiar to most researchers, there are a number of advantages of using the more complex logistic approach for contest data analyses. First, logistic analysis, which treats contest outcome as the response variable, is often the more intuitive formulation of explanatory and response variables. Being larger than an opponent might be a cause of victory over a smaller rival, but defeating a rival is unlikely to be the cause of larger size in the victor, at least within the timescale of a single encounter, as mentioned above. Furthermore, the plots from logistic analyses are easy to interpret visually and are often comparable to probability plots predicted by game theory (see figure 2 in Hammerstein 1981, figure 4 in Enquist & Leimar 1983, figure 1 in Leimar & Enquist 1984, figure 4 in Enquist & Leimar 1990, figure 1 in Mesterton-Gibbons 1994). Such plots can also facilitate comparison among empirical studies, allowing for example a ready comparison of the effect of weapon size across studies of diverse species where the 'weapons' may be very different in morphological terms (e.g. Figure 4.3 here can be readily compared with figure 2 in Sneddon *et al.* 1997, showing the effect of relative difference in claw size in the shore crab *Carcinus maenas*).

As with other types of GLZ, logistic analysis is an extremely flexible approach for the exploration of the effects of multiple explanatory variables and their interactions. This is useful for studies conducted in the field where a range of variables beyond the control of the observer can nevertheless be recorded and included in the analysis, and for studies where a range of potential correlates of RHP and/or *V* have been obtained. As the accessibility of logistic regression increases (it is now widely available in programmes with user-friendly interfaces such as *SPSS* and *GenStat* and in the perhaps less user-friendly, but free, package *R*) it has become an increasingly popular approach for the analysis of contest behaviour (e.g. Sneddon *et al.* 1997, Neat *et al.* 1998, Humphries *et al.* 2006, Kemp *et al.* 2006, Peixoto & Benson 2008, Mohamad *et al.* 2010, Lizé *et al.* 2012). Using current software, GLZ is not much more difficult to perform and illustrate than the more traditional paired analyses (section 4.3.1) and these advantages of GLZ can be gained with just a little more statistical work than is required for the 'outcome as explanatory variable' approach.

However, some aspects of logistic analyses do not compare favourably with paired analyses, where the data from both individuals (if available) can be analysed rather than using a composite measure of their differences. Various approaches for utilising all available information have been suggested, and building marginal models through the use of a generalised estimating equation (GEE) appears to have undergone the greatest level of development. GEEs fit marginal models to repeated measurements and can be used when the response has a distribution in the exponential family, including binomially distributed contest outcome data. Likelihood ratio tests are not valid for hypothesis-testing when using GEEs but Wald tests can be used, and the information-theoretic approach (section 4.3.2.3) can also be used for selection of GEE models. However, there are still uncertainties about the conditions under which this semi-parametric approach is valid (Davis 2002 section 9.5.9 provides a discussion; see also Quinn & Keough 2002) and it is not widely available in all commonly used statistical software. Therefore, in logistic regression the problem of non-independence of contest data is usually avoided by using a *composite* measure of intra-contest inter-contestant differences (section 4.3.2.3) and the consequently reduced amount of information in the analysis lowers its statistical power. This may not be a major concern if sample sizes are large. For instance, in a typical-sized study involving around 60–80 staged encounters, an effect that cannot be detected by logistic regression might not be biologically important, but not all studies will be able to acquire such large sample sizes. Field studies will typically be most limited in terms of sample size but even in laboratory studies the effective sample size may be limited by the behaviour of the animals under study. For example, in different studies conducted in the same laboratory using the same species of parasitoid wasp (*Goniozus legneri*), the percentage of staged contests having a clear winner can vary considerably (65% in Bentley *et al.* 2009; 80–86% in Lizé *et al.* 2012, with 80–128 replicates per study): in these cases the conditions that influenced the probability of contest resolution could be informatively explored using logistic regression (with 'contest resolved or not?' as the response variable) and then the subset of replicates in which contests were resolved could be subjected to logistic analysis of factors influencing the outcomes of these contests (with the identity of the winner as the response variable). In the anemone study of Rudin and Briffa (2011), out of 82 contests staged, only 17 (20%) involved stinging by both opponents. While logistic regression was a good option for analysing the data set as a whole, and produced figures readily comparable with other studies (and results that were comparable with the paired version of the analysis described in section 4.3.1.1), for the subset of data where 'mutual stinging' occurred, a paired *t*-test was preferable to a logistic analysis due to the reduction in statistical power when a composite measure was used.

Moreover, there are situations where a test of difference may be fundamentally more appropriate regardless of concerns over sample size and statistical power. In some cases the a priori assumption might be that the variable of interest is in fact influenced by the outcome, for example in an analysis of a hormone or metabolite that is expected to vary as a result of the experience of winning or losing (e.g. Sneddon *et al.* 2000). A final consideration is that if a data set is obtained from a laboratory experiment, rather than from the field, there may be several factors that have been included in the experimental design but relatively few 'nuisance variables' that need to be included in the analysis. In such cases, the flexibility offered by GLZ may be redundant and a test of difference may fit better with the experimental design, enhancing the ability to illustrate (through the use of interaction bar charts) and interpret the results. In sum, both approaches are statistically valid, and the choice of which approach to use should therefore be informed by considerations such as the structure of the data, the number of independent observations (which may be constrained by welfare considerations), and the ease with which the results can be presented.

4.3.3 Mixed-effect models

We have thus far assumed that although more than one measure is taken from each fight (i.e. at least one measure from each contestant), each individual appears in the analysis only once. By using the repeated measures approach or by calculating composite measures of difference between contestants we can use information on both opponents while avoiding pseudoreplication (sections 4.3.1.2 and 4.3.2.2). Pseudoreplication can also arise when the same individual appears in more than one experimental unit. This might occur when data are collected from observations of contests occurring naturally between wild animals and individuals are not easily distinguishable and thus particular individuals may have been observed in several contests, as in a field study on Lappet-faced vulture contests by Bamford *et al.* (2010) or when multiple interactions between known individuals are recorded during long-term field studies (Poole 1982, Figure 4.2). A similar problem is encountered when replication is achieved by staging encounters between genetically similar individuals (e.g. social insects, Chapter 8), a situation that is likely to occur when multi-party contests are being analysed to test the predictions of Lanchester's attrition laws (Chapters 3 and 8).

One solution is to use a mixed-model procedure that incorporates 'individual identity' (or 'colony identity') (if known) into the analysis as a 'random-effect' explanatory variable, in contrast to the 'fixed-effect' explanatory variables we have so far discussed. A random effect may be loosely thought of as something operating at the population level and beyond direct experimental control. An example that will be familiar to many readers is 'blocking' which is an experimental design method that attempts to minimise, by using blocks, the influence of uncontrollable variation in experimental conditions on the exploration of the fixed effect of interest (blocks are thus the random effect). However, exact definitions of fixed and random effects are still being debated and also whether a particular effect should be regarded as fixed or random will depend on context and experimental design (Bolker *et al.* 2008).

For those interested in contests, dealing with individuals appearing non-independently in several experimental units (i.e. a given individual appearing in multiple contests or genetically similar individuals appearing separately in a series of contests) might be thought of as a different issue from that of how to deal with two measures taken from the same experimental unit (i.e. a winner and a loser from each contest). In a statistical sense, however, both problems are consequences of having data that are structured hierarchically. In the case of individuals appearing in multiple contests, mixed effect models allow us to account for this source of pseudoreplication by including individual or colony identity as a random factor. Similarly, in an 'outcome as independent' type analysis, winners and losers are nested within the same fight. Therefore, mixed models could also be used to achieve a paired or repeated measures analysis to account for this. In this case 'fight' would be specified as a random factor. For those used to a repeated measures approach there are many aspects of mixed models that might seem unfamiliar, such as the use of model simplification techniques (section 4.3.2.3). An initial hurdle to appreciate is the difference in how the data are stored. In a traditional repeated measures ANOVA data are usually required in 'wide format', whereas an equivalent mixed model would need the data to be arranged in 'long' format. Table 4.2 provides a comparison of the two formats for a set of hypothetical contest data.

Mixed-model analyses are possible for both 'outcome as explanatory variable' analyses (i.e. linear mixed models, LMM) and 'outcome as response variable' analyses (i.e. generalised linear mixed models, GLMMs). The conditions under which some of these more advanced mixed models (i.e. GLMM) may be implemented, and the best options for setting up the analyses, have been the subject of considerable recent debate. We direct interested readers to some excellent papers which provide guidance that is aimed at evolutionary ecologists: Krackow and Tkadlec (2001), Bolker *et al.* (2008) and Warton and Hui (2011).

Returning to the analysis of elephant contest data conducted in section 4.3.2.2, the logistic analysis is pseudoreplicated because the number of contests analysed is far larger than the number of elephants observed. We can address this by running a logistic GLMM with the identities of the two males within each dyad fitted as random effects and the same 'fixed-effects' explanatory variables as in section 4.3.2.2. GLMM analysis leads us to conclude that there is no interaction between difference in musth status and difference in age (Wald $\chi^2_2 = 0.2$, $p = 0.903$) and that both of these main effects affect contest outcome significantly (Age: Wald $\chi^2_1 = 69.93$, $p < 0.001$; Musth: Wald $\chi^2_2 = 16.25$, $p < 0.001$), i.e. the same biological conclusions as in the simpler analysis but now we can be more confident of our interpretation of these data.

Table 4.2 Contest data in (a) 'wide' and (b) 'long' formats. For a traditional repeated measures ANOVA or paired analysis, data are usually arranged in wide format. If a mixed model is used to cope with repeated measures taken from the same contest, long format is usually used. An important difference is that in a repeated-measures ANOVA or a paired *t*-test, the information on observation number (i.e. a single 'data point' or 'experimental unit') is only for data management purposes and is not used in the analysis. In the case of a mixed model, information on the observation number would be treated as a random factor. Note that in some statistical software, specifying 'subjects' might also be required if a mixed model is to be used for this type of analysis. In large data sets converting between the formats manually is laborious but some software packages (e.g. *SPSS* and *R*) provide tools to do this.

(a) 'Wide' format for paired *t*-test or repeated-measures ANOVA			(b) 'Long' format for a mixed model analysis		
Observation number (informative, not a factor in the analysis)	Weight (repeated measure)		Observation number (a random factor in the analysis)	Outcome (a fixed factor in the analysis)	Weight
	Winner	Loser			
1	110	50	1	Winner	110
2	75	70	1	Loser	50
3	85	81	2	Winner	75
4	90	72	2	Loser	70
5	60	40	3	Winner	85
			3	Loser	81
			4	Winner	90
			4	Loser	72
			5	Winner	60
			5	Loser	40

For further examples and discussion of how to implement GLMMs in contest data analysis we refer readers to Bamford *et al.* (2010), Batchelor and Briffa (2010, 2011), Furrer *et al.* (2011) and Wilson *et al.* (2011).

4.4 Analysis of contest duration

Having established what traits influence RHP and/or *V* (sometimes called 'correlates' of RHP and *V*), the aim of many researchers is to determine how these traits influence the decision of the loser to give up. As noted in section 4.2, this question is often prompted by the theory discussed in Chapter 2. 'Mutual' and 'self-' assessment are key assumptions that distinguish different models of dyadic fighting and distinguishing between mutual and self-assessment is thus an important part of understanding agonistic behaviour. Individuals of low RHP are those that reach their giving-up threshold quickly (either because it is set at a low level, because they accrue costs rapidly or a combination of these factors). This means that if the giving up decision is based on self-assessment only, contest duration should increase as a function of increasing absolute RHP of the weaker opponent. If contests are settled on the basis of mutual assessment, however, it should be easier for the weaker opponent to discover that it is weaker when the RHP asymmetry is large. Thus, there should be a negative relationship between increasing RHP difference and contest duration. This relationship had been used in many studies as a marker for the presence of mutual assessment (reviewed by Taylor & Elwood 2003), but there is a problem with this approach: RHP difference may show a negative relationship with the absolute RHP of losers. Taylor and Elwood (2003) used simulation models to show that if this relationship is strong enough, the apparent negative relationship between RHP difference and contest duration could be an artefact of a stronger positive relationship between loser absolute RHP and contest duration. In this case it is much more likely that the decision is made through self-assessment than by mutual assessment. Thus, analysis of the relationship between RHP difference and contest duration alone does not unequivocally demonstrate the presence of mutual assessment.

Taylor and Elwood (2003) suggested that in order to show that mutual assessment is used it is necessary to examine the relationships between contest duration and both loser and winner RHP. As the difference in RHP will be easier for the eventual loser to establish when the asymmetry is large, contests should still increase in duration with increasing loser RHP: as losers increase in contest ability it will be progressively more difficult for them to ascertain that they are weaker than the opponent because the RHP asymmetry will be smaller. By the same logic, there should also be a negative relationship between winner absolute RHP and contest duration: as winners decrease in fighting ability it will also be progressively more difficult for the eventual loser to ascertain that they are the weaker opponent because the RHP disparity will be smaller. Thus, it has been suggested that a positive relationship between loser RHP and contest duration plus a negative relationship between winner RHP and contest duration together indicate the presence of mutual assessment (Figure 4.6).

While the identification of these diagnostic relationships for the different modes of assessment (Taylor & Elwood 2003) was a great advance in understanding how to analyse contests, it was noted that even this approach may have its limitations (Taylor & Elwood 2003, Briffa & Elwood 2009). Essentially, the relationships described above occur because opponents are influenced directly by each others' activities during contests. In the case of mutual assessment this involves an exchange of information, such that by performing agonistic behaviours each contestant directly influences the information state of its rival. The decision to give up is thus dependent on the RHP of both opponents, and this is clearly different from wars of attrition (WOA, Chapter 2) where the loser gives up solely on the basis of its own RHP. However, there are other ways in which opponents could influence each others' giving-up decisions without (in theory at least) the exchange of any information. If the opponents can directly influence each others' rate of cost accrual (for example, by inflicting injuries) they could increase the rate at which the loser reaches its individual maximum cost threshold. Under this scenario, contest duration should still increase with the absolute RHP of the loser: losers with greater RHP should have higher cost thresholds and the rate at which they approach these should decrease, in both cases prolonging the time before the giving-up decision is made. Furthermore, contest duration could also decrease with increasing RHP in the winner: winners with large RHP will have greater ability to inflict costs on their opponents, leading to the losers reaching their giving-up threshold more quickly (Figure 4.6). In terms of relationships between winner and loser absolute RHP and contest

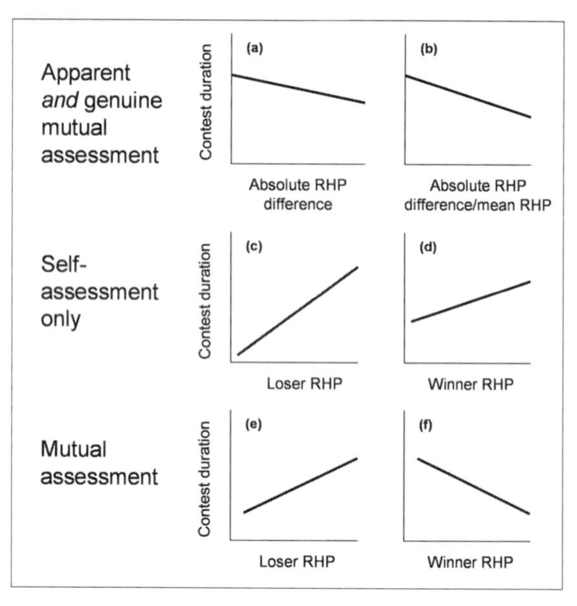

Figure 4.6 Expected relationships between contest duration and RHP. Contest duration has commonly been found to decrease as the RHP asymmetry between contestants increases (**a**, **b**), in accord with models that assume mutual assessment. When contests are determined by Loser RHP only, contest duration will increase with an increase in Loser RHP (**c**) and also with Winner RHP, although less steeply (**d**): analysis of these contests employing composite measures of winner and loser RHPs generates negative relationships between duration and RHP asymmetry (**a**, **b**) even though there is no mutual assessment. When contests are genuinely determined by mutual assessment, relationships between contest duration and Loser RHP should be approximately equal, but opposite in sign, to relationships between contest duration and Winner RHP (**e**, **f**). Thus, the key marker for mutual assessment would be a negative relationship between duration and winner RHP, rather than the weak positive relationship expected under self-assessment. Expected relationships are based on figures derived from analyses of simulated data in Taylor and Elwood (2003; see also Gammell & Hardy 2003).

duration, this is the same result as would be obtained if the contest was settled by mutual assessment, but here there is no assumption of an exchange of information about RHP.

Contests of this form are described by the cumulative assessment model (CAM; Payne 1998, Chapters 1 and 2). Note that while it seems logical that an individual receiving costs as a direct result of the actions of the opponent might benefit from using the severity of these costs to assess its opponent's RHP, this model makes no such assumption that information may be inferred in this way. Although we might expect, from the perspective of minimising costs, selection to favour mutual assessment, constraints on cognitive abilities may preclude its evolution in some taxa. As in the energetic war of attrition (EWOA; Mesterton-Gibbons *et al.* 1996, Chapter 2), giving up in the CAM occurs when the accumulated costs of persisting exceed a maximum cost threshold. The only difference is that in the EWOA costs accrue only as a result of performing agonistic behaviours whereas under the CAM the costs inflicted by the opponent are also added. Although contestants directly influence each others' costs, they do not influence each others' information state in terms of providing extraneous information about opponent RHP. Although CAM- and EWOA-type contests may yield the same pair of relationships between the absolute RHP of winners and losers and contest duration as seen under mutual assessment, the giving-up decision is nevertheless made on the basis of what the loser knows about itself. The presence of these relationships is thus not always indicative of the presence of mutual assessment. While they do indicate that contests are settled by RHP difference rather than on the basis of only loser RHP, this is possible under mutual assessment and, during CAM-type contests, in the absence of mutual assessment.

4.4.1 Statistical analysis of contest duration

Analysis of RHP and contest duration is not enough on its own to rule out the possibility of 'self-assessment' (due to the possibility of CAM-type fights) but it is capable of ruling out 'mutual assessment'. Thus, it is a worthwhile analysis to perform on contest data and the techniques involved are straightforward. The simplest methods would be to conduct a series of correlation or simple regression analyses (which should be 'Model II' regression if neither variable is controlled for; the extent to which one can control for variation in RHP is debateable and likely to vary among study systems) between contest duration and each of the following; loser RHP, winner RHP and a composite measure of intra-contest inter-contestant differences in RHP, similar to those used in logistic analyses of contest outcomes (section 4.3.2). RHP difference has been calculated in several ways, ranging from a simple 'absolute' difference in RHP ('winner *minus* loser' or 'larger *minus* smaller') to various measures of 'relative' difference in RHP, where the RHP of the weaker individual is treated as a proportion of that of the stronger individual (see Taylor & Elwood 2003 for examples). If individuals do not vary a great deal within the experiment in the chosen correlate of RHP (e.g. length, weight) then absolute RHP difference will probably suffice. If there is a large amount of variation between observations then a relative measure would be better. For example, in a hermit crab shell fight where the attacker weighed 0.3 g and the defender weighed 0.25 g, the absolute difference in weight would be the same as in a fight where the attacker weighed 1.0 g and the defender weighed 0.95 g; in relative terms, however, the size differential would be quite disparate between the two fights. Having conducted a series of three regression or correlation analyses (duration against loser RHP, winner RHP and RHP difference) one could compare the results to the predictions described by Taylor and Elwood (2003) and illustrated in Figure 4.6. Clearly, the key test that allows us to discount the possibility of mutual assessment is for a negative relationship between contest duration and winner RHP. If a significant negative relationship is absent we can discount the possibility of mutual assessment, if it is present we conclude that either (a) a mutual-assessment type decision is being used by the loser or (b) another type of decision, that is influenced by the ability of the attacker (e.g. as in CAM-type contests), is being used by the loser.

Rudin and Briffa (2011) tested for correlations between contest duration and RHP measures in their study of fighting in *Actinia equina*. In fights that did not involve stings (when one opponent retreats after tentacular contact has been observed, but in the absence of contact with the fighting tentacles and acrorhagial peels) they found a negative relationship between contest duration and relative difference in dry weight but no relationship in either direction between duration and winner or loser absolute dry weight. In fights that did involve stings there was a weakly negative relationship between duration and relative difference in nematocyst length, a weakly positive relationship between the length of loser's nematocysts and contest duration, and no relationship between winner nematocyst length and contest duration. Thus, in this empirical example it was not possible to map the results of these analyses (Figure 4.7) directly onto those predicted by Taylor and Elwood's (2003) simulation exercise. On the basis of these analyses, Rudin and Briffa (2011) concluded (very tentatively) that mutual assessment was unlikely because there were no negative correlations between duration and winner dry weight or between duration and winner nematocyst length.

Sets of correlation analyses such as these are one way to test for associations between RHP and contest

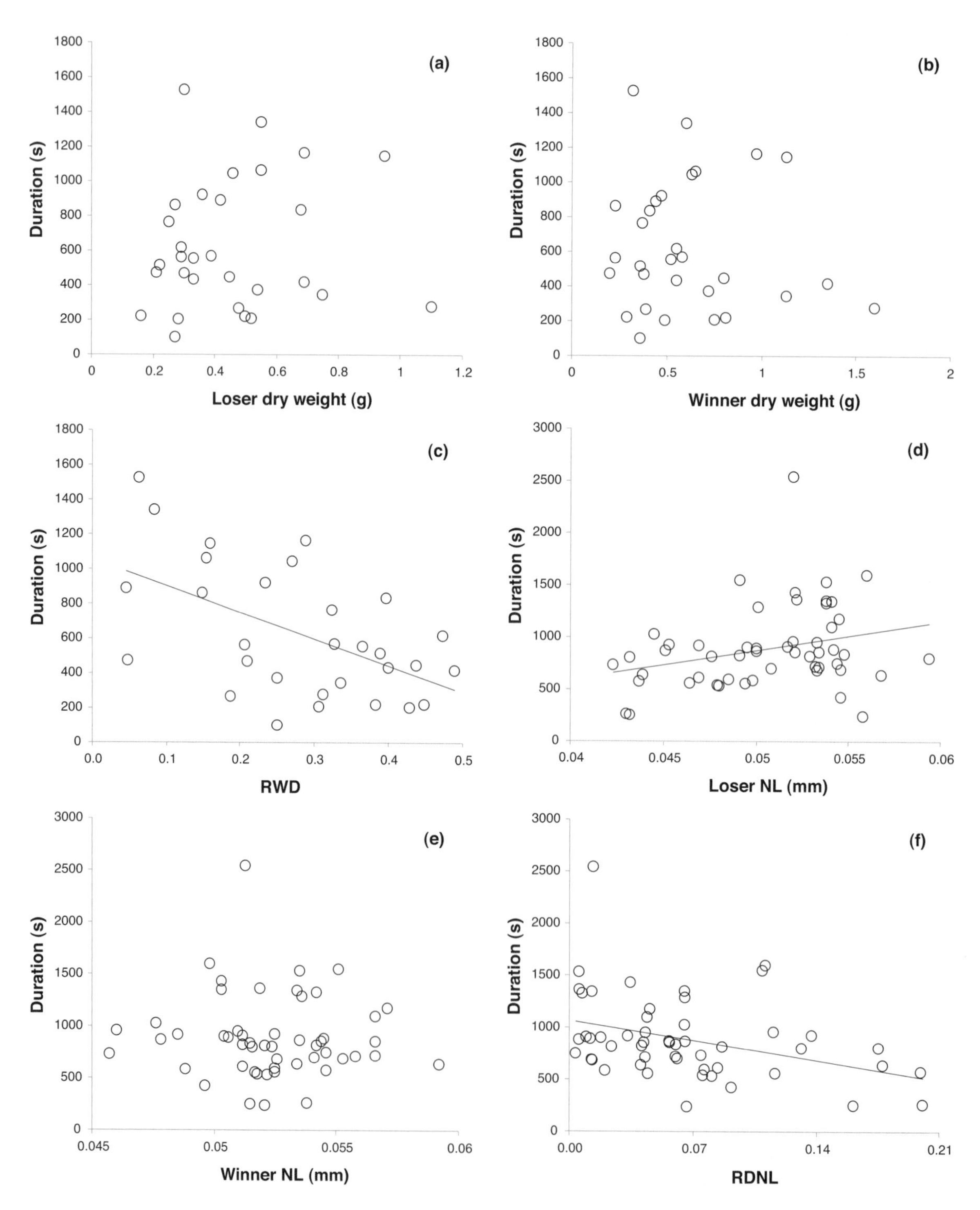

Figure 4.7 Ambiguous results obtained from analysing correlations between contest duration and measures of RHP in *Actinia equina*, from Rudin and Briffa (2011). Tests for correlations between absolute and relative measures of RHP and contest duration for contests without (**a**–**c**) and with (**d**–**f**) nematocyst stings. (**a**) Loser dry weight, (**b**) winner dry weight and (**c**) relative dry weight difference (RWD), (**d**) loser nematocyst length (NL), (**e**) winner nematocyst length (NL), (**f**) relative nematocyst length difference (RDNL). Regression lines were used to illustrate the presence of significant correlations. Compare these two sets of results with the predicted results for mutual and self-assessment illustrated in Figure 4.6.

duration. Taylor and Elwood (2003) also suggested a slightly different approach that would be more suitable if the real data from a study can be easily mapped onto the predictions obtained from their analyses of the simulated data. Because the relationship between winner RHP and contest duration is the only one that will differ between contests that *might* be settled through mutual assessment and contests that definitely *are not* settled by mutual assessment, they suggested that the following steps represent the most parsimonious solution. First, a multiple regression where the explanatory variables are winner and loser RHP and the response variable is duration would allow one to determine how the absolute measures of RHP vary with contest duration. If this analysis revealed a positive association with loser RHP and a negative association with winner RHP, then one might replace the multiple regression with a simple regression where the single explanatory variable is a measure of RHP difference. From such a result we would conclude that duration is being driven by RHP difference rather than a threshold that the loser reaches before the winner but, as noted above, this is compatible with both mutual assessment and self-assessment in a CAM-type contest.

4.5 Analysis of contest dynamics and structures

As well as considering the durations and outcomes of contests it can be useful to examine behavioural changes within contests as they progress. Here we discuss the analysis of changes in rates of agonistic behaviour as the contest progresses and analysis of the sequence of behavioural events within contests.

4.5.1 Analysis of contest dynamics

Given the limitations of using correlations between RHP and contest duration (section 4.4), what else can we do to distinguish between mutual- and self-assessment? Commonly tested models, such as the sequential assessment model (SAM; Chapter 2), EWOA and CAM, have a range of assumptions and predictions, not only concerning contest duration but also contest outcomes and dynamics (Table 1.1). For example, CAM and SAM make different assumptions about assessment rules but both assume duration to be determined by RHP difference and both could yield similar relationships between absolute measures of RHP and contest duration (section 4.4). However, CAM and SAM make different predictions about fight structure: SAM predicts only escalation between phases and constant intensity within a phase, whereas CAM allows escalation within phases and, in phases that do not involve injuries, the possibility of de-escalation. Further, SAM predicts that opponents should match each other's performances while CAM does not.

Examining fight structure alone may yield limited information: EWOA and CAM make similar predictions about fight structure (both under certain circumstances allowing within-phase escalation and de-escalation), but they are based on different assumptions about the nature of assessment rules, which means that they should make different predictions about relationships between RHP and contest duration, and these models also make different predictions about matching between opponents. Thus, in order to distinguish between mutual- and self-assessment it is necessary to establish whether contest duration is based on RHP difference or loser absolute RHP and to investigate the ethological details of contest behaviour.

Empirical studies on a range of systems illustrate the types of ethological detail that can be readily obtained and complement analyses of relationships between contest duration and RHP. In hermit crabs, *Pagurus berhardus*, one opponent, the 'attacker', performs bouts of 'shell rapping' by striking its gastropod shell against that of a 'defender'. The bout structure of this activity has been extensively analysed (Chapter 5) revealing that the vigour of shell rapping can both escalate and de-escalate as the contest progresses. Interestingly, the pattern varies between outcomes. As there are two different roles in these encounters there is no matching between opponents and this, in combination with escalation and de-escalation, indicates that a CAM-type model may describe these contests reasonably well. On the other hand, analysis of the energetic status of the opponents (Chapter 5) indicates that they may use different decision rules, a possibility that has yet to be modelled. In house crickets, *Acheta domesticus*, contests are divided into a series of distinct phases of increasing intensity in terms of energy expenditure (Hack 1997*a*) and losers appear to pay higher energetic costs than winners (Briffa 2008). Furthermore, careful analysis of the agonistic behaviour reveals the presence of escalation and de-escalation in the choice of agonistic tactics used (Hack 1997*b*).

Because duration shows negative relationships with measures of loser RHP and positive relationships with measures of winner RHP (Briffa 2008), these analyses taken together again indicate the presence of a cumulative assessment rule. Contest dynamics can also be investigated during field observations of large animals such as fallow deer, *Dama dama*: Jennings *et al.* (2005, Chapter 14) found that the rate of head-on jump-clashing declined as fights progress, a pattern compatible with self- but not mutual assessment-based models. Here, each contest was divided into four quarters and the rate of each activity was compared between quarters.

In the study on sea anemones, Rudin and Briffa (2011) followed the same approach as Jennings *et al.* (2005) and divided the contest into four quarters, to look at changes in the duration of acrorhagial contacts as the fight progressed. These data were analysed using a two-factor repeated-measures ANOVA. The first repeated measure was 'fight quarter' (one, two, three or four) and the second repeated measure, as described in section 4.3, was 'outcome' (winner or loser). Thus, they were able to determine whether the duration of contacts changed as the fight progressed and also whether the duration of contacts performed by winners and losers was different. Moreover, by examining the interaction (fight quarter × outcome) effect between these two factors, they were able to explore whether the temporal pattern of changes in behaviour differed between winners and losers. In this case, the number of contacts varied significantly between fight quarters ($F_{3,153} = 9.3, p < 0.0001$). 'Post-hoc' Scheffé's tests indicated that the duration of contact in the first quarter was lower than that in the third ($p < 0.0001$) and fourth ($p < 0.003$) quarters and that the duration in the second quarter was lower than in the third quarter ($p < 0.02$). Winners spent significantly more time performing acrorhagial contact than did losers ($F_{1,51} = 42.5, p < 0.0001$). Finally, a significant interaction effect ($F_{3,153} = 5.1, p < 0.01$), indicates that the pattern of change between fight quarters was different for winners and losers, with the pattern of change being more marked for winners than for losers (Figure 4.8).

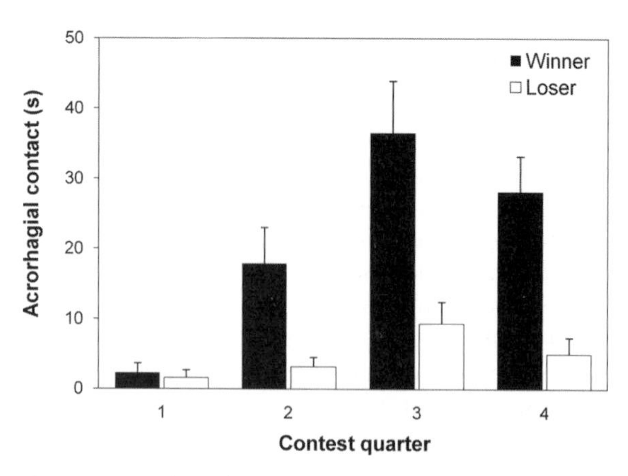

Figure 4.8 Analysis of contest dynamics in *Actinia equina* (Rudin & Briffa 2011). Both winners and losers showed escalation (quarters 1 to 3) followed by de-escalation (quarters 3 to 4), but the pattern was more marked in winners.

4.5.2 Analysis of contest structure

The sequence of behaviours performed by animals during contests will often be of interest in a proximate sense, but the proximate understanding of behavioural sequences may also shed light on the ultimate function of a particular behaviour. Here we illustrate this possibility via a re-analysis of data on contest behaviour in a parasitoid wasp, *Goniozus legneri* (all data derive from the study of Goubault *et al.* 2006, where full experimental details are given). There is a well-developed functional understanding of many aspects of female–female contests in this species (Chapter 8). It is also known that females may emit a volatile chemical, a spiroacetal, upon losing a contest and that the probability of emission is correlated with the level of aggression within the contest (Box 8.3). While aggression is reduced after spiroacetal release, the function of release is not clear: two candidates are that the chemical acts as a weapon of rearguard action, creating an opportunity for the releaser to retreat from a contest, but it may also function as a signal of submission. While we would a priori favour the former possibility because the spiroacetal involved is known to be toxic to other insect species, distinguishing between these candidate functions is nonetheless challenging because no negative effects of exposure have as yet been observed in *G. legneri* or its close relatives (possibly due to practical constraints with working with this particular chemical rather than its lack of toxicity to parasitoids). Examination of behavioural sequence data may, in principle, be informative in this regard. Before carrying out an illustrative exploration of contest sequence data we provide a brief overview of the statistical approaches and techniques involved.

4.5.2.1 Sequential and time-interval analysis

Sequence analysis goes under a variety of names, including 'behavioural sequence analysis' (BSA) and 'exploratory sequential data analysis' (ESDA). It is used in ethology, social psychology, developmental psychology and ergonomics. Sequence analysis can provide answers to a number of questions in a single coherent representation; typically a kind of 'path map'. We can ask whether there is a typical sequence (or several inter-related sequences) in which events tend to occur, and whether certain events typically trigger other events of interest. We can ask whether there are key switch points at which a single choice may determine the future course of events, or possibly the outcome of a strategic situation. We can also examine whether a behavioural system has 'absorbing states' (areas where the entry probability is much higher than the exit probability). If it does, the animals will appear to be drawn towards those behaviours and to maintain them, even though no identifiable motivation may exist. For instance, if one animal lays a trap for another (even without 'bait') there is a certain probability that the prey will stray into the trap, if only by chance, and no probability they will leave; so the illusion is created that they are drawn to the trap. As time goes by, the probability increases that the prey will be caught, and as a matter of mathematical principle, as time 'tends towards infinity', the likelihood tends towards certainty that the all the prey animals will be in the trap, whether it exerts any actual attraction on them or not.

In the simplest case, a record of behaviour over time can be treated as a succession of discrete events, each belonging to one of a modest number of types or classes. So *abc* might represent a behaviour (or other event) of type *a*, followed by one of type *b*, and so on. This version of the data is created by coding the raw observations or films in three stages. First the record is 'parsed' or 'unitised', identifying the break-points between behaviours, or points where one behaviour is succeeded by another. In the psychological case, this has been found to raise issues about the structure of the behaviour stream, and the way it is perceived (Dickman 1963, Newtson 1973). Next, the discrete events or behaviours are categorised, replacing each one in the data string by the type to which it belongs. A great deal depends on what types are used and how each is defined. Ideally a type should not consist merely of behaviours that are similar in form or function but of behaviours that are similar in the ways they form sequences with others (van Hooff 1973). This is a recurring concept in the analysis of structures in various fields, including linguistics, where it is called 'test-frame analysis' or 'the principle of commutation'. Successive events may be quite diverse in form and duration but they cannot blend, overlap or be simultaneous (analogous to beads of assorted shapes and colours on a single string). Alternative approaches may define events by duration, for example sampling the behaviour stream every 10 seconds. Finally, the relative prevalence of particular strings of types is examined. The elementary building blocks of this technique are pairs of successive event-types called 'transitions', often tallied in tables called transitional frequency matrices, or converted to probabilities in transitional probability matrices. These tables can then be examined to see whether they show non-random properties of sequential organisation overall and whether particular transitions, such as *ab* or *bc*, have made a singular contribution to the pattern that warrants their inclusion in a flow diagram or transition map of the behaviour.

While a given transition, such as *ab*, could be by far the most common transition in the set of data, this may simply be due to *a* and *b* both being very common events. The standard solution is to evaluate the matrix using a test (e.g. χ^2) which corrects for such effects of chance by comparing the observed frequency of occurrence of each transition with an expected frequency ([row total × column total]/grand total). Individual cells (transitions) are of interest when their standard normal residual (in effect their contribution to the χ^2 value of the table) reaches a criterion value (Colgan & Smith 1978). In effect this is to ask does *ab* occur so often because *a* is raising the probability of *b*, and likewise for all other transitions? This is appropriate when the question is a causal one: what controls the occurrence of *b*? However, the analysis is probably over-used and a different logic is needed for prediction. If the question is 'After *a*, what will happen next?', the important measure is the conditional probability of *b* given *a*, regardless of whether the presence of *a* has raised that above the baseline or unconditional probability of *b*. In that case, the use of χ^2 testing will produce Type II errors.

Transitions of special interest, typically those that are statistically over-represented, are often summarised in a transition map, such as Figure 4.9, constructed for the illustrative example of *G. legneri*

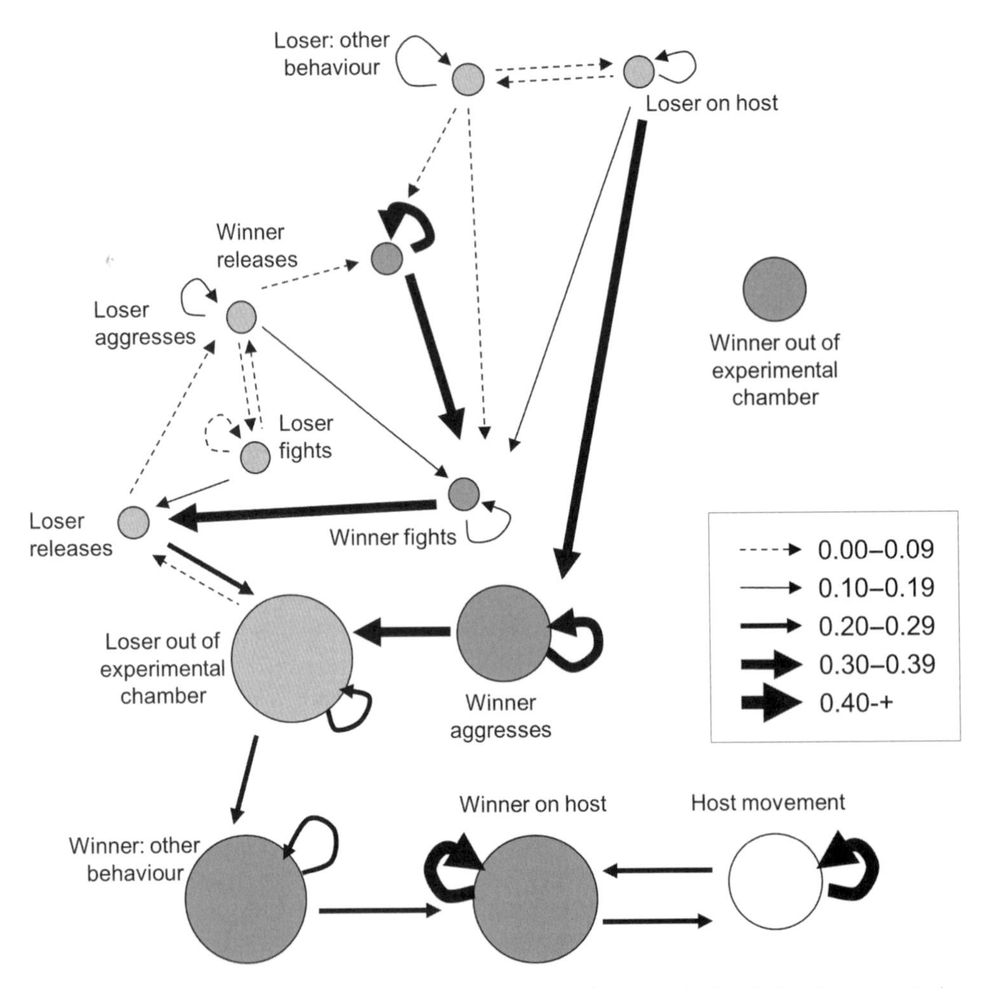

Figure 4.9 Transition map of behaviours before and after release of spiroacetal in female–female contests in the parasitoid *Goniozus legneri* at lag 2 (that is, relating each behaviour to the *next-but-one* in the sequence). The radius of each circle reflects the frequency of each type of behaviour as a percentage of the total. The widths of the arrows indicate the conditional probability of the event at the head of the arrow being next but one to occur, given the event at the tail of the arrow.

contest interactions. However, transitions picked out in this way are not transitive: *ab* could be more common than chance, and so could *bc*, without the chain *abc* being remarkable (or even possible). It is hard to draw inferences about longer and more complex sequential patterns without subscribing to this fallacy, but Dawkins (1976) provided an ingenious solution, the essence of which is to search for long chains of behaviour selectively around the shorter chains that are known to exist. The problem of non-transitive transitions arises because some sequences are 'higher-order' and events are affected by specific combinations of *several* previous events. So *dbc* might be common while *abc* is not. The 'brute force' way to deal with this is to have a three-, four- or higher-dimensional matrix to capture patterns that occur in strings of three, or four or more events. This can soon lead to a 'combinational explosion', in that there need to be enough data to populate N^m cells where N is the number of event categories, and m is the 'order' of the analysis. For instance, if this approach were used to characterise all the words of 3 letters or fewer in a language with 26 letters, the transition matrix would have 26^3 or 17,576 cells. This makes it important to restrict the coding to a 'modest' number of behaviour-types, typically around 10–15 (this problem is analogous to the 'number of interaction terms' issue discussed in section 4.3.2.3).

It may be that the most interesting contingencies occur between events some distance apart in the

Ellis L (1995) Dominance and reproductive success among nonhuman animals: A cross-species comparison. *Ethology and Sociobiology*, 16, 257–333.

Elo AE (1961) The new U.S.C.F. rating system. *Chess Life*, 16, 160–161.

Elo AE (1978) *The Rating of Chess Players, Past and Present*. New York, NY: Arco.

Enquist M & Leimar O (1983) Evolution of fighting behaviour: Decision rules and assessment of relative strength. *Journal of Theoretical Biology*, 102, 387–410.

Enquist M & Leimar O (1987) Evolution of fighting behaviour: The effect of variation in resource value. *Journal of Theoretical Biology*, 127, 187–205.

Enquist M & Leimar O (1990) The evolution of fatal fighting. *Animal Behaviour*, 39, 1–9.

Field SA & Calbert G (1998) Patch defence in the parasitoid wasp *Trissolcus basalis*: When to begin fighting? *Behaviour*, 135, 629–642.

Field SA, Calbert G & Keller MA (1998) Patch defence in the parasitoid wasp *Trissolcus basalis* (Insecta: Scelionidae): The time structure of pairwise contests, and the 'waiting game'. *Ethology*, 104, 821–840.

Fushing H, McAssey MP, Beisner B, *et al.* (2011) Ranking network of a captive rhesus macaque society: A sophisticated corporative kingdom. *PLoS One*, 6, e17817.

Furrer RD, Kyabulima S, Willems EP, *et al.* (2011) Location and group size influence decisions in simulated intergroup encounters in banded mongooses. *Behavioral Ecology*, 22, 493–500.

Gammell MP & Hardy ICW (2003) Contest duration: Sizing up the opposition? *Trends in Ecology and Evolution*, 18, 491–493.

Gammell MP, Vries H de, Jennings DJ, *et al.* (2003) David's score: A more appropriate dominance ranking method than Clutton-Brock *et al.*'s index. *Animal Behaviour*, 66, 601–605.

Gottman, JM & Roy AK (1990) *Sequential Analysis: A Guide for Behavioral Researchers*. Cambridge: Cambridge University Press.

Goubault M & Decuignière M (2012) Prior experience and contest outcome: winner effects persist in absence of evident loser effects in a parasitoid wasp. *American Naturalist*, 180, 364–371.

Goubault M, Batchelor TP, Linforth RST, *et al.* (2006) Volatile emission by contest losers revealed by real-time chemical analysis. *Proceedings of the Royal Society of London B*, 273, 2853–2859.

Grafen A (1990) Biological signals as handicaps. *Journal of Theoretical Biology*, 144, 517–546.

Grafen A & Hails R (2002) *Modern Statistics for the Life Sciences*. Oxford: Oxford University Press.

Hammerstein P (1981) The role of asymmetries in animal contests. *Animal Behaviour*, 29, 193–205.

Hack MA (1997*a*) Assessment strategies in the contests of male crickets, *Acheta domesticus* (L.). *Animal Behaviour*, 53, 733–747.

Hack MA (1997*b*) The energetic costs of fighting in the house cricket, *Acheta domesticus* L. *Behavioral Ecology*, 8, 28–36.

Hardy ICW & Field SA (1998) Logistic analysis of animal contests. *Animal Behaviour*, 56, 787–792.

Heath D (1995) *An Introduction to Experimental Design and Statistics for Biology*. London: UCL Press.

Hemelrijk CK, Wantia J & Gygax L (2005) The construction of dominance order: Comparing performance of five methods using an individual-based model. *Behaviour*, 142, 1037–1058.

Hooff JARAM van (1973) A structural analysis of the social behaviour of a semi-captive group of chimpanzees. In: M von Cranach & I Vine (eds.) *Social Communication and Movement*, pp. 75–162. London: Academic Press.

Hsu Y, Earley RL & Wolf LL (2006) Modulation of aggressive behaviour by fighting experience: Mechanisms and contest outcomes. *Biological Reviews*, 81, 33–74.

Hsu Y, Lee IH & Lu CK (2009) Prior contest information: Mechanisms underlying winner and loser effects. *Behavioral Ecology and Sociobiology*, 63, 1247–1257.

Hughes M (1996) The function of concurrent signals: Visual and chemical communication in snapping shrimp. *Animal Behaviour*, 52, 247–257.

Hughes M (2000) Deception with honest signals: Signal residuals and signal function in snapping shrimp. *Behavioral Ecology*, 11, 614–623.

Humphries EL, Hebblethwaite AJ, Batchelor TP, *et al.* (2006) The importance of valuing resources: Host weight and contender age as determinants of parasitoid wasp contest outcomes. *Animal Behaviour*, 72, 891–898.

Hurlbert SH (1984) Pseudoreplication and the design of ecological field experiments. *Ecological Monographs*, 54, 187–211.

Innocent TM, West SA, Sanderson JL, *et al.* (2011) Lethal combat over limited resources: Testing the importance of competitors and kin. *Behavioral Ecology*, 22, 923–931.

Jameson KA, Appleby MC & Freeman LC (1999) Finding an appropriate order for a hierarchy based on probabilistic dominance. *Animal Behaviour*, 57, 991–998.

Jennings DJ, Gammell MP, Payne RJH, *et al.* (2005) An investigation of assessment games during fallow deer fights. *Ethology*, 111, 511–525.

Jennings DJ, Carlin CM, Hayden TJ, *et al.* (2011) Third-party intervention behaviour during fallow deer fights: The role of dominance, age, fighting and body size. *Animal Behaviour*, 81, 1217–1222.

Kasumovic MM, Elias DO, Sivalinghem S, *et al.* (2010) Examination of prior contest experience and the retention of winner and loser effects. *Behavioral Ecology*, 21, 404–409.

Kemp DJ, Alcock J & Allen G R (2006) Sequential size assessment and multicomponent decision rules mediate aerial wasp contests. *Animal Behaviour*, 71, 279–287.

Klass K & Cords M (2011) Effect of unknown relationships on linearity, steepness and rank ordering of dominance hierarchies: Simulation studies based on data from wild monkeys. *Behavioural Processes*, 88, 168–176.

Krackow S & Tkadlec E (2001) Analysis of brood sex ratios: Implications of offspring clustering. *Behavioral Ecology and Sociobiology*, 50, 293–301.

Langbein J & Puppe B (2004) Analysing dominance relationships by sociometric methods – A plea for a more standardised and precise approach in farm animals. *Applied Animal Behaviour Science*, 87, 293–315.

Leimar O & Enquist M (1984) Effects of asymmetries in owner–intruder conflicts. *Journal of Theoretical Biology*, 111, 475–491.

Lindström K & Pampoulie C (2005) Effects of resource holding potential and resource value on tenure at nest sites in sand gobies. *Behavioral Ecology*, 16, 70–74.

Lindquist WB & Chase ID (2009) Data-based analysis of winner–loser models of hierarchy formation in animals. *Bulletin of Mathematical Biology*, 71, 556–584.

Lizé A, Khidr SK & Hardy ICW (2012) Two components of kin recognition influence parasitoid aggression in resource competition. *Animal Behaviour*, 83, 793–799.

Mesterton-Gibbons M (1994) The Hawk–Dove game revisited: Effects of continuous variation in resource holding potential on the frequency of escalation. *Evolutionary Ecology*, 8, 230–247.

Mesterton-Gibbons M, Marden JH & Dugatkin LA (1996) On wars of attrition without assessment. *Journal of Theoretical Biology*, 181, 65–83.

Mohamad R, Monge J-P & Goubault M (2010) Can subjective resource value affect aggressiveness and contest outcome in parasitoid wasps? *Animal Behaviour*, 80, 629–636.

Morrell LJ, Backwell PRY & Metcalfe NB (2005) Fighting in fiddler crabs *Uca mjoebergi*: What determines duration? *Animal Behaviour*, 70, 653–662.

Mundry R & Nunn CL (200) Stepwise model fitting and statistical inference: Turning noise into signal pollution. *American Naturalist*, 173, 119–123.

Nakagawa S & Cuthill IC (2007) Effect size, confidence interval and statistical significance: A practical guide for biologists. *Biological Reviews*, 82, 591–605.

Neat FC, Taylor AC & Huntingford FA (1998) Proximate costs of fighting in male cichlid fish: The role of injuries and energy metabolism. *Animal Behaviour*, 55, 875–882.

Neumann C, Duboscq J, Dubuc C, *et al.* (2011) Assessing dominance hierarchies: Validation and advantages of progressive evaluation with Elo-rating. *Animal Behaviour*, 82, 911–921.

Newtson D (1973) Attribution and the unit of perception of ongoing behaviour. *Journal of Personality and Social Psychology*, 28, 28–38.

O'Hara RB & Kotze DJ (2010) Do not log transform count data. *Methods in Ecology and Evolution*, 1, 118–122.

Paoli T, Palagi E & Tarli SMB (2006) Reevaluation of dominance hierarchy in bonobos (*Pan paniscus*). *American Journal of Physical Anthropology*, 130, 116–122.

Payne RJH (1998) Gradually escalating fights and displays: The cumulative assessment model. *Animal Behaviour*, 56, 651–662.

Peixoto PEC & Benson WW (2008) Body mass and not wing length predicts territorial success in a tropical Satyrine butterfly. *Ethology*, 114, 1069–1077.

Petersen G & Hardy ICW (1996) The importance of being larger: Parasitoid intruder–owner contests and their implications for clutch size. *Animal Behaviour*, 51, 1363–1373.

Poisbleau M, Jenouvrier S & Fritz H (2006) Assessing the reliability of dominance scores for assigning individual ranks in a hierarchy. *Animal Behaviour*, 72, 835–842.

Poole JH (1982) Musth and male–male competition in the African elephant. PhD thesis, University of Cambridge.

Poole JH (1989) Announcing intent: The aggressive status of musth in African elephants. *Animal Behaviour*, 37, 140–152.

Quinn GP & Keough MJ (2002) *Experimental Design and Data Analysis for Biologists*. Cambridge: Cambridge University Press.

Rohwer S & Rohwer FC (1978) Status signalling in Harris sparrows: Experimental deceptions achieved. *Animal Behaviour*, 26, 1012–1022.

Royle NJ, Lindström J & Metcalfe NB (2005) A poor start in life negatively affects dominance status in adulthood independent of body size in green swordtails *Xiphophorus helleri*. *Proceedings of the Royal Society of London B*, 272, 1917–1922.

Rudin FS & Briffa M (2011) The logical polyp: Assessments and decisions during contests in the beadlet anemone *Actinia equina*. *Behavioral Ecology*, 22, 1278–1285.

Sanvito S, Galimberti F & Miller EH (2007) Vocal signalling of male southern elephant seals is honest but imprecise. *Animal Behaviour*, 73, 287–299.

Sapolsky RM (2005) The influence of social hierarchy on primate health. *Science*, 308, 648–652.

Schjelderup-Ebbe T (1922) Beiträge zur sozialpsychologie des haushuhns. *Zeitschrift für Psychologie*, 88, 226–252.

Sharpe T & Koperwas J (2003) *Behavior and Sequential Analysis: Principles and Practice*. London: Sage.

Shizuka D & McDonald DB (2012) A social network perspective on measurements of dominance hierarchies. *Animal Behaviour*, 83, 925–934.

Smith DR, Hardy ICW & Gammell MP (2011) Power rangers: No improvement in the statistical power of analyses published in *Animal Behaviour. Animal Behaviour*, 81, 347–352.

Sneddon LU, Huntingford FA & Taylor AC (1997) Weapon size versus body size as a predictor of winning in fights between shore crabs, *Carcinus maenas* (L.). *Behavioral Ecology and Sociobiology*, 41, 237–242.

Sneddon LU, Taylor AC, Huntingford FA, *et al.* (2000) Agonistic behaviour and biogenic amines in shore crabs *Carcinus maenas*. *Journal of Experimental Biology*, 203, 537–545.

Sokal RR & Rohlf FJ (1995) *Biometry*, 3rd edn. New York, NY: WH Freeman & Co.

Taborsky M (2010) Sample size in the study of behaviour. *Ethology*, 116, 185–202.

Taylor PW & Elwood RW (2003) The mismeasure of animal contests. *Animal Behaviour*, 65, 1195–1202.

Vries H de (1995) An improved test of linearity in dominance hierarchies containing unknown or tied relationships. *Animal Behaviour*, 50, 1375–1389.

Vries H de (1998) Finding a dominance order most consistent with a linear hierarchy: A new procedure and review. *Animal Behaviour*, 55, 827–843.

Vries H de (2009) On using the DomWorld model to evaluate dominance ranking methods. *Behaviour*, 146, 843–869.

Vries H de & Appleby MC (2000) Finding an appropriate order for a hierarchy: A comparison of the I&SI and the BBS methods. *Animal Behaviour*, 59, 239–245.

Vries H de, Stevens JMG & Vervaecke H (2006) Measuring and testing the steepness of dominance hierarchies. *Animal Behaviour*, 71, 585–592.

Warton DI & Hui FKC (2011) The arcsine is asinine: The analysis of proportions in ecology. *Ecology*, 92, 3–10.

Whitehead H (2008) *Analyzing Animal Societies: Quantitative Methods for Vertebrate Social Analysis*. Chicago, IL: University of Chicago Press.

Whitehead H (2009) SOCPROG programs: Analyzing animal social structures. *Behavioral Ecology and Sociobiology*, 63, 765–778.

Wilson AJ, Boer M de, Arnott G, *et al.* (2011). Integrating personality research and animal contest theory: Aggressiveness in the green swordtail *Xiphophorus helleri*. *PLoS ONE*, 6, e28024.

Wilson K & Hardy ICW (2002) Statistical analysis of sex ratios: An introduction. In: Hardy ICW (ed.) *Sex Ratios: Concepts and Research Methods*, Pp. 48–92 Cambridge: Cambridge University Press.

Wittemyer G & Getz WM (2006) A likely ranking interpolation for resolving dominance orders in systems with unknown relationships. *Behaviour*, 143, 909–930.

Zahavi A (1975) Mate selection – A selection for a handicap. *Journal of Theoretical Biology*, 53, 205–214.

Chapter 5

Contests in crustaceans: assessments, decisions and their underlying mechanisms

Mark Briffa

5.1 Summary

Crustaceans have been used extensively to study aggression from a variety of perspectives. Traits that make them good models for studies of aggression include the possession of prominent weapons and a ready willingness to fight, the ease with which they may be obtained and their ease of maintenance in the laboratory. Furthermore, hard exoskeletons mean that equipment such as heartbeat sensors can be easily attached directly to the animals and they are very amenable to physiological investigation. Therefore a feature of studies into the contest behaviour of crustaceans is a strong link between ultimate 'functions' and proximate 'mechanisms'. Given the very wide range of studies on crustacean aggression any review of this group could potentially be extremely broad in focus. Rather than attempt such a broad review, I focus on studies that have combined ethological data with data on underlying mechanisms in order to address questions that have arisen from the body of evolutionary contest theory described in Chapter 2. This has meant that some areas of research on aggression in the Crustacea, such as the neuroendocrine control of aggression and studies of social aspects of aggression such as dominance hierarchies, are outlined only briefly. I consider evidence for information-gathering and decision-making in respect of resource holding potential and resource value, and describe how studies of the underlying motivation to fight can provide useful insights into what information fighting animals might use when making strategic decisions. I also consider studies of agonistic signals, which in crustaceans include visual, tactile and chemical modalities, with a particular consideration of the extent to which 'dishonesty' might play a role in crustacean agonistic dealings. I then review physiological aspects of fighting in crustaceans, including studies based on post-fight assays of metabolites and by-products, studies based on 'real-time' non-invasive techniques and studies based on a functional morphology approach of investigating whole organism performance capacities. Finally, I suggest ways in which studies on crustacean contests could inform new theoretical models of contest behaviour and discuss the potential for applying approaches used to study crustaceans to the study of contests in other taxa.

5.2 Introduction

Crustaceans have been used extensively as models for the study of aggressive behaviour, both from a behavioural ecology perspective, dealing with the resolution of contests, and from the perspective of investigating the proximate mechanisms involved in aggression. Many studies have focused on taxa from the order Decapoda ('ten legged' crustaceans) such as the Brachyura ('true' crabs including swimming crabs and fiddler crabs), the Anomura (including hermit crabs and squat lobsters), the Astacidea (lobsters and crayfish) and the Caridea ('prawns' or 'shrimps') (Figure 5.1). In addition to studies on decapods there have also been studies on stomatopods (mantis shrimps) and peracarids (amphipods and isopods). In general, crustaceans are readily available and in the case of intertidal and freshwater species, they can often be collected from the field at little cost. Crustaceans have a global distribution. Hermit crabs, for example, are found intertidally in northwestern Europe, North America, the tropics and the Arctic. Freshwater crustaceans can be equally easy to obtain across a global range of locations. Aside from deep-sea species, crustaceans are usually easy to maintain in the laboratory. As many study species are relatively small in size, yet

Animal Contests, ed. I.C.W. Hardy and M. Briffa. Published by Cambridge University Press.

Figure 5.1 Examples of decapod crustaceans from groups that have been the focus of studies on aggression. (**a**) A common European hermit crab, *Pagurus bernhardus* (Anomura) occupying a *Littorina littorea* shell. Note the asymmetrically sized chelipeds; (**b**) a male fiddler crab, *Uca mjoebergi* (Brachyura: Ocypodidae), waving its enlarged and conspicuously coloured major cheliped, an adaptation that is used to attract females and during fights against rival males; (**c**) a velvet swimming crab *Necora puber* (Brachyura: Portunidae), an example of a 'true crab' that engages in agonistic encounters; (**d**) a pair of Australian freshwater crayfish or 'yabbies', *Cherax dispar* (Astacidae), grappling with their chelipeds. (Photo credits: (**a**) Sophie Mowles, (**b**) Tanya Detto, (**c**) Thomas Guest, (**d**) Anthony O'Toole.) For colour version, see colour plate section.

large enough for accurate direct observation (Huntingford *et al.* 1995), they can be housed in moderately sized aquaria, which can also be used as fight arenas. As a result of their suitability for laboratory work the study of crustacean physiology is a well-developed field and this has influenced the direction of research into contest behaviour in this group.

A theme permeating much work on crustaceans has been to link the study of adaptive functions with that of proximate mechanisms. Proximate mechanisms have been studied from three perspectives. First, the overarching idea of 'motivation' or 'willingness' to fight, although not formally identified with specific physiological pathways, has been studied to gain insights into the use of information during contests. For example, studies on hermit crabs (e.g. Elwood 1995, Briffa & Elwood 2001*a*), have used 'motivational probing' techniques to determine the extent to which an opponent has access to information on its rival's resource holding potential (RHP) and on the value of the resource (V). Second, energy metabolism (e.g. Smith & Taylor 1993, Huntingford *et al.* 1995, Thorpe *et al.* 1995, Sneddon *et al.* 1999, 2000*a*, Briffa & Elwood 2001*b*, 2002, 2004, 2005, Rovero *et al.* 2000, Brown *et al.* 2003) and whole organism performance capacities (e.g. Lailvaux *et al.* 2008, Bywater *et al.* 2008, Wilson *et al.* 2009, Mowles *et al.* 2009, 2010) have been investigated to gain insights into the physiological constraints on agonistic behaviour. Third, the neuroendocrine mechanisms of aggression have been investigated (e.g. Sneddon *et al.* 2000*b*, Briffa & Elwood 2007). Although such proximate mechanisms are often thought of as being outside of the usual remit of behavioural ecology, studies of crustacean contests show how investigating mechanisms can be extremely relevant to understanding behaviour from a functional viewpoint. The core theory of contest behaviour makes key assumptions about the use of information and the role of costs (in some models, specifically energetic costs) during contests. Neuroethological studies of mechanisms involved in 'aggression' and dominance have run somewhat in parallel to the behavioural ecology work on 'contests', but both areas have a shared interest in decision-making during agonistic encounters, and there is evidence of increasing communication of ideas between the two fields (e.g. Huber *et al.* 1997*a,b*, Sneddon *et al.* 2000*b*, Huber 2005). Although studies of hormonal and neurological mechanisms in crustaceans are potentially highly relevant to the question of decision-making in contests, relatively few attempts (e.g. Sneddon *et al.* 2000b, Briffa & Elwood 2007) have been made to relate these mechanisms directly to the evolutionary decision rules proposed by contest theory (Chapter 2). On the other hand, links between metabolic pathways and the functional aspects of contest behaviour have been extensively studied in crustaceans and I therefore discuss these studies in detail in section 5.4.2.

Even for behavioural ecologists *sensu stricto*, who do not stray into the 'black box' of proximate mechanisms, the Crustacea offer several advantages as a group to study. Due to their hard calcareous cuticle it is very easy to gain accurate measures of size, often considered to be effectively synonymous with RHP, in addition to overall body mass. Further, most species possess weapons, usually modified claw (= 'chela', plural 'chelae') bearing appendages ('chelipeds') that are used to perform agonistic signals as well as direct fight activities such as grappling, striking or piercing

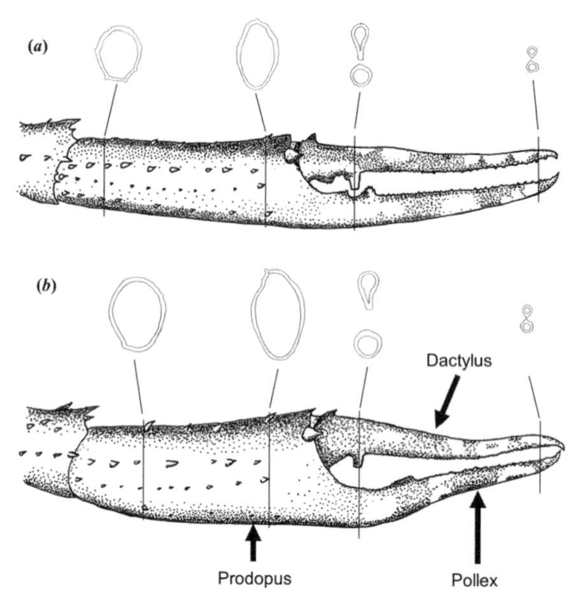

Figure 5.2 Chelar dimorphism and RHP in the squat lobster *M. rugosa*. In flat claws (**a**) the marked tooth near the base of the dactylus appears to be less effective at inflicting puncture wounds on opponents than in the case of arched claws (**b**). Analysis of cross-sectional measurements, taken along the claw, indicates that arched claws are also 'bulkier' than straight claws, allowing greater leverage of the moveable dactylus and higher closing force against the fixed pollex (the non-moving 'finger' of the claw that grows out from the prodopus). (Reproduced with permission from Claverie & Smith 2007.)

the opponent to cause injuries. In addition to metrics of body size, the strength and appearance of these appendages can also be quantified (Figure 5.2). Thus, crustaceans offer a range of potential correlates of RHP for investigation. They will also fight readily in the laboratory, over a range of resources such as food (Huntingford *et al.* 1995), shelters (Figler *et al.* 2005) and mates (Sneddon *et al.* 2003). These resources can be manipulated directly by placing a contested resource in an arena or indirectly by use of odour cues associated with these resources (e.g. food odour or pheromones) because aquatic organisms rely strongly on chemoreception and are very responsive to odour cues. In addition to their suitability for laboratory study, in intertidal and semi-terrestrial species agonistic behaviour can also be studied in the field, allowing insights into how agonistic behaviour is influenced by environmental factors such as the oxygen concentration of seawater (Sneddon *et al.* 2000*a*) or the risk of predation (Reaney & Backwell 2007). Again, such factors relate to the direct and circumstantial costs of fighting that feature in theoretical models.

The Crustacea is a diverse group. It is currently classified as a subphylum within the Arthropoda, although molecular evidence suggests that the crustaceans and hexapods should be grouped together in a single subphylum, the 'Pancrustacea' (Regier *et al.* 2005). To put crustacean diversity, as currently defined, into perspective, the vertebrates also comprise a single subphylum and the two taxa contain approximately the same number of known species. Due to the high diversity of species that are amenable to study there is the clear potential for comparative studies, which could provide insights into the evolution of agonistic behaviour. This is an approach that has been applied to animal communication but has yet to be applied to contests, even though the use of agonistic signals is common. Nevertheless, for the reasons outlined above, crustaceans have been used as model species for the range of questions about contest behaviour that are interesting to behavioural ecologists. In the first half of this chapter (section 5.3), I review studies where crustaceans have been used as models to investigate the role of information, assessments and decision-making during contests. I focus primarily on investigations of 'correlated' asymmetries, that is differences between opponents in RHP and their subjective resource value V (Chapter 2, section 5.1), and deal with uncorrelated asymmetries (e.g. the effects of prior experiences and ownership status) as they arise. Special attention is given to the question of what factors might influence RHP, the use of agonistic signals and to studies of motivation for fighting. In the second half of the chapter (section 5.4), I focus on work linking strategic decisions and fight tactics to the costs of non-injurious fighting in crustaceans.

5.3 Information, assessments and decisions

Contest outcomes are ultimately determined by a range of asymmetries, i.e. key differences between the opponents that will influence their chance of winning. Information about these asymmetries is potentially available to the opponents such that they may perform assessment activities and base strategic and tactical decisions on the information they have gained during the contest. Indeed, the widespread use of agonistic signals is one of the most striking features of contest behaviour in the crustaceans as a whole. Despite the fact that agonistic signals are common, the question of what information is available in the fight is currently the focus of much interest, in the crustaceans as in other groups. In this section I discuss the influence on contest outcomes and structure of the two key 'correlated' asymmetries that influence payoffs, fighting ability (RHP) and the value of the resource (V).

5.3.1 Strategic decisions and resource holding potential

Swimming crabs of the family Portunidae such as *Necora puber*, *Liocarcinus depurator* and *Carcinus maenas* have long been subjects for the study of contest behaviour (Huntingford *et al.* 1995) and remain an important group of model species. Agonistic behaviour in these crabs involves a combination of postural displays using the chelipeds and walking legs and physical contact in the form of grappling. The claws may also be used to inflict injuries on the opponent.

A slightly unusual feature of contests in these crabs is that the fights are rather unstructured with no clear level of organisation into bouts and no clear pattern of escalation in intensity between phases. This lack of structure in the contest behaviour of portunid crabs places a constraint on our ability to distinguish between the different ways in which they might make the strategic decision to withdraw from a contest. As discussed in Chapter 2, some of the key theoretical models of agonistic behaviour differ in their predictions about patterns of escalation within and between contest phases. Nevertheless, in shore crab contests, several studies have investigated assessment rules by determining how contest duration varies with RHP (Chapter 4 provides an overview of this approach). As discussed in Chapter 2, the sequential assessment model (SAM) assumes mutual assessment whereas the energetic war of attrition (EWOA) assumes decisions are made by self-assessment. The cumulative assessment model (CAM) also predicts self-assessment, but because costs in the CAM can also be caused by the actions of the opponent, it may be difficult to distinguish from the SAM on the basis of contest duration alone (Briffa & Elwood 2009).

The most basic prediction of any model of contest behaviour is that the opponent of greater RHP should be more likely to win. RHP can be reasonably assumed to increase with overall body size (the terms have been and often continue to be used interchangeably) and in addition to early work on other taxa such as spiders (Chapter 6), studies of the portunid crabs *L. depurator* (Glass & Huntingford 1988), *N. puber* (Smith *et al.* 1994) and *C. meanas* (Sneddon *et al.* 1997*a,b*) support this prediction, although the smaller crab may still win if the size difference is very small. Interestingly, although size difference is clearly linked to the decision of the loser to quit, there does not appear to be the same effect on the decision to initiate a fight as the first move is as likely to be made by the smaller individual as the larger (Huntingford *et al.* 1995). This indicates that information on size is either not available by visual assessment alone (Huntingford *et al.* 1995) or, perhaps, that mutual assessment of size is of limited importance. Examining relationships between size difference and contest duration can indicate whether mutual assessment plays a role, although recent developments (Taylor & Elwood 2003, Briffa & Elwood 2009, Chapter 4) demonstrate the importance of analysing absolute correlates of RHP as well as RHP differences.

In male–male fights in *N. puber* contest duration and intensity show a negative relationship with size difference (Smith *et al.* 1994), but surprisingly the opposite pattern is seen in fights between females where fights are short when the size difference is small (Thorpe *et al.* 1994). This raises an interesting point about contests in general: although mating opportunities are often the most 'highly prized' resource (Briffa & Sneddon 2007) over which contests occur, contests over other types of resource are also frequent, such that investigating aggression within the framework of sexual selection may not always be appropriate. Indeed, studies on crustaceans show that females and males can engage in contests over the resources such as food or shelter that are required by both sexes. These studies illustrate the general point that we should not neglect contests over resources other than mates. The study on velvet swimming crabs (Thorpe *et al.* 1994) suggests that males and females might fight according to different rules; the pattern of settling contests more quickly when opponents are evenly matched seen in females is the opposite trend for that usually reported in male–male fights. In hermit crabs there are also striking differences between the agonistic behaviour of males and females, even when the fight is over the same resource. In this example, females fight more intensely over ownership of empty gastropod shells than do males (Briffa & Dallaway 2007). In the fiddler crab, *Uca annulipes*, both males and females defend burrows on mudflats, but there appears to be a form of sexual segregation in agonistic encounters, driven by females choosing to retreat from male invaders but to put up a fight against female invaders (Milner *et al.* 2010). Finally, in the crayfish *Cherax dispar*, greater variance in claw strength in males compared to females allows males but not females to employ dishonest signals (section 5.3.4.1) of RHP (Wilson *et al.* 2009). Thus, evidence from studies of a range of crustaceans shows that not only do females frequently engage in fights, but males and females may fight in different ways. Nevertheless, assessment models are often tested using males only, and there is therefore the possibility that conclusions drawn from studies of males only may not be applicable to all cases of fights, even within the same species.

Although the negative association between contest duration and RHP difference shown in male–male fights could indicate the use of mutual assessment, these studies of swimming crab fights from the 1980s and 1990s anticipated the importance highlighted by Taylor and Elwood (2003) of analysing absolute RHP as well as relative RHP. In *L. depurator* fight duration also increased with the absolute size of the opponents (Glass & Huntingford 1988), indicating that decisions could be based on self-assessment mechanisms such as a maximum threshold of costs rather than mutual assessment. However, continued attempts to match any single assessment rule to fights in portunid crabs have produced equivocal results. In the green shore crab, *C. maenas*, support for mutual assessment is evidenced by negative relationships between contest duration and RHP difference (e.g. Sneddon *et al.* 2003, Smallegange *et al.* 2007). On the other hand, contest duration also varies with the absolute RHP of the smaller contestant (Smallegange *et al.* 2007), which, according to the statistical framework described by Taylor and Elwood (2003, Chapter 4), supports individual threshold-based models. The real problem with trying to match swimming crab fights to any assessment game is that, regardless of how contest duration varies with absolute and relative measures of RHP, the structure of the contests simply bears little resemblance to those assumed by any current models. There is no pattern of escalating phases, which rules out the SAM, and there is no behavioural matching between opponents (Smallegange *et al.* 2007), which rules out war of attrition-type models (Chapter 2). The opponents do, however, interact directly with one another, pushing, grappling and striking with the chelipeds (Sneddon *et al.* 1997, 2000c, Sneddon & Swaddle 1999). Although these activities are not necessarily injurious (Smallegange *et al.* 2007), they could allow a contestant to inflict costs on its opponent, thereby increasing the rate at which the costs will accumulate towards a giving-up threshold. Such costs could derive from the energetic consequences of fighting; for example, pushing the opponent down might force an increase in energy metabolism while it struggles to maintain its position. If fighting swimming crabs do inflict costs on one another these contests might fit the mode of fighting assumed by the CAM. Although this model does not predict mutual assessment, contest duration may still show the strongest relationships with relative RHP, due to the ability of contestants to influence the costs experienced by the opponent (Briffa & Elwood 2009).

Hermit crabs are another example of a crustacean system where the nature of the contest raises some problems in applying current assessment models but, as in the case of portunid crabs, they have nonetheless proved useful in investigating the use of information about RHP during contests. Hermit crabs are anomurans characterised in most cases by a weakly calcified abdomen, which rather than showing the typical carcinised 'crab-like' morphology, of being reduced in size and tucked flush under the cephalo-thorax, has remained elongate as in shrimp and crayfish. Instead of producing a hardened cuticle over this region, hermit crabs are adapted to take advantage of empty objects, which act as 'portable burrows', protecting the crab from attack by predators and environmental extremes. In most cases this object is an empty gastropod shell, which will ideally be large enough to adequately protect the abdomen and to provide a refuge into which the crab can withdraw its cephalothorax and walking legs when threatened. In most species the body is distinctly asymmetric with a curved abdomen and marked size asymmetry in the chelipeds. These are specific adaptations for occupying gastropod shells; the muscular abdomen wraps around the internal columnellar axis of the shell, gripping it from the inside, while the larger 'major' cheliped acts as an operculum, closing off the aperture of the shell when the crab has withdrawn. As the crab grows it will require a series of shells of increasing size to maintain the level of protection required, and as the supply of empty shells is limited this often necessitates taking a shell from another hermit crab. 'Attackers' initiate the fight by approaching a smaller defender that nevertheless occupies a shell that is of better quality than the attacker's current shell. There may be a period of agonistic displays using the chelipeds before the attacker lunges forwards and uses its walking legs to grasp the defender's shell. At this point the defender usually withdraws tightly into its shell, closing off the aperture with the major chelus. The attacker meanwhile rotates the defender's shell, and runs its chelipeds and walking legs over the external surface before inserting its chelae in through the aperture of the defender's shell and using the claws to grab the defender's chelipeds. Using the muscles in its walking legs and abdomen the attacker rapidly and repeatedly swings the two shells together such that their external surfaces strike. This 'shell rapping' occurs in a series of bouts which are separated by pauses. After a number of bouts, the fight may end in one of two ways. First, the attacker can give up, simply releasing its grasp of the defender's shell. Second, the defender may make the strategic decision to give up. In this case the defender will release its abdominal grip of its shell, allowing the attacker to evict it by pulling it out through the aperture. In this case the attacker will have won the fight and have the choice of moving into the shell vacated by the defender or remaining in its original shell. Usually the attacker will decide to take the shell that it has won, but it may take some time to make this decision and will perform further investigations of the empty shell (even swapping a few times between the two shells). During this process, the defender is left without a shell and is vulnerable to attack, but once the attacker has made its decision the defender is free to occupy the shell it has discarded. Thus, hermit crab fights are divided into a clear sequence of phases but involve two roles which perform different activities.

Because the defender, as well as the attacker, can gain (by moving from a shell that is too big to one that is closer to its optimal size) these encounters are sometimes called 'shell exchanges' rather than the more usual 'shell fights'. Indeed, in the tropical hermit crab *Clibanarius vittatus* it has been shown that the defender is more likely to allow itself to be evicted when it stands to gain in shell quality and shell exchanges have been described as a process of 'negotiation' (Hazlett 1978, 1982, 1983, 1984, 1987, 1989, 1996). In the common European hermit crab, *Pagurus bernhardus*, however, while there is still the potential for the defender to gain in shell quality, there is no evidence that this influences the outcome of the encounter and shell fights are considered to be aggressive in nature (Elwood & Briffa 2001). Indeed, naked attackers, not in possession of a gastropod shell, are able to evict the defender even when the evicted crab will have no shell immediately available (Dowds & Elwood 1985).

Rather than focusing on contest duration, as in the studies of shore crabs, early studies on hermit crabs focused on links between absolute and relative measures of RHP and the series of decisions made by attackers and defenders during shell fights. At the start of the encounter, there is a significant bias for the larger of the two crabs to initiate the fight, indicating that in contrast to fights in portunid crabs some assessment of size must take place during the early stages of the encounter (Dowds & Elwood 1985). Evidence from experiments involving crabs with autotomised (removed) chelae indicate that this information on size is derived from 'cheliped displays' during the opening phases of the encounter (Neil 1985). Such chelar displays are common during agonistic encounters in crustaceans and are clearly linked to an individual's RHP because, on average, chelar size is a good indication of body size. Once the fight is underway, the major cheliped also appears to be important to the defender for mounting a successful defence of its shell. Defenders perform bouts of cheliped flicking between bouts of shell rapping by the attacker and, although the function of cheliped flicking is not entirely clear (but see Mowles & Briffa 2012), defenders with missing chelipeds are more likely to be evicted (Neil 1985). A missing cheliped therefore appears to reduce the RHP of defenders but size difference also appears to influence RHP-based decisions. After the attacker has grabbed the defender's shell they are more likely to escalate and perform shell rapping if they are larger than the defender and large attackers are more likely to evict the defender than those which are equal in size or smaller than the defender (Dowds & Elwood 1985). Size difference continues to influence decisions even after the defender has been evicted. If the size difference between the attacker and the evicted defender is low then the attacker makes a less thorough assessment of its newly acquired shell before vacating the area where the fight occurred.

A third group of crustaceans in which size-based RHP has been studied intensively are the fiddler crabs. These are semi-terrestrial brachyurans of the family Ocypodidae, comprising a single genus, *Uca*. They are found on sandy shores and mudflats in the Pacific and western Atlantic and west African coasts. The only species found in Europe is *U. tangerii*, at the northern limit of its range in Portugal, but most work on agonistic behaviour has been conducted on eastern Pacific species such as *U. mjoebergi* and *U. perplexa*. Fiddler crabs are characterised by a vastly enlarged and conspicuously coloured major cheliped in males, a sexually selected trait used for agonistic displays and grappling in male–male competition and in 'claw-waving' courtship displays for attracting females. Different aspects of the claw appear to be important for each type of activity. In *Uca mjoebergi*, for example, females assess ultraviolet components of the male's claw colouration while these components are ignored by rival males, where claw size appears to be the key feature (Detto & Backwell 2009). Nevertheless, both males and females engage in contests over burrow ownership and the foraging territories around the burrow entrances (these territories are also used for claw-waving in males). Burrows are excavated by the crabs and provide refugia into which they retreat during low tide and fights often occur between resident burrow owners and wandering males. If fights occur at the burrow mouth there appears to be a tactical asymmetry that can give the resident a mechanical advantage (Fayed *et al.* 2008), an example of how payoff uncorrelated asymmetries can also influence fight outcomes (also see Tricarico & Gherardi 2007 and Peeke *et al.* 1995 for examples of prior ownership and resident advantage effects in crustaceans). In fights in the open, however, the use of agonistic signals and largely symmetric fight tactics means that the current assessment models may be readily applicable to much of the contest behaviour of fiddler crabs.

Morrell *et al.* (2005) investigated the possibility of mutual assessment in *U. mjoebergi* using Taylor and Elwood's (2003) analytical framework. In this field experiment, resident males were captured, translocated and then released near the burrows of other residents. Thus encounters were staged between owners and newly created intruders. This avoided a potential problem of using naturally occurring wandering males, which could have been weak individuals, already displaced from their own burrows. It also allowed contests over a range of size differences to be staged. This would not occur naturally because a feature of fighting in fiddler crabs is that contests are strongly size-assortative, where large males avoid initiating fights with males that are much smaller. This is because small males will excavate burrows that are unsuitably small for larger males, and it appears that for this reason opponent size is assessed at the start of the encounter and used in the decision to initiate a contest (Jennions & Backwell 1996). By using translocation to increase the chance of encounters between differently sized males it was possible to determine whether similar information was used in the decision to give up. When intruders were released away from their original burrows the duration and intensity of the interaction with the first resident encountered was recorded. Intruders won significantly fewer fights than residents, but the size of the intruder still influenced the outcome, with intruders more likely to win when they were larger than the resident. Contest duration (and intensity, which co-varied with duration) increased with both winner and loser size and decreased with various measures of size difference. Further analysis of these relationships through stepwise regression indicates that although loser claw size had a strong positive correlation with duration, the duration also shows a negative correlation with the winner's claw size. This suggests that information about the opponent is available to the loser, supporting the idea of mutual assessment. On the other hand, although both relationships were significant, the statistical effect size of loser claw length was greater than the effect of winner claw length. Thus, although both sources of information might be available, the stronger effect of loser claw length (compared to that of the winner) on contest duration indicates that a contestant's information about itself might be more important than information about the opponent. Morrell *et al.* (2005) concluded that the best fit to these data could be given by a modification of the CAM, which accounted for the possibility of size-assortative fighting, a specific feature of these contests as they occur naturally. Under the CAM, the decision to give up is based on individual thresholds but the opponents can influence how quickly the other approaches their thresholds by inflicting direct costs. In the case of fiddler crabs it is possible that, when there is a size disparity, costs accumulate more quickly for the smaller opponent.

Studies of fighting fiddler crabs indicate that the best way to understand assessment rules may be to modify the basic assessment models as appropriate for the specific features of the contest under consideration. This would also be a useful approach in understanding contests in other examples where agonistic behaviour deviates from that assumed by models, for example in future studies of shore crabs or hermit crabs. Furthermore, fiddler crabs are a good example of how 'real' contests, especially those which occur under field conditions rather than encounters that are staged in the laboratory, might involve many more sources of information than those that are encapsulated by current models of pairwise contests (Chapter 2). For instance, in densely populated mudflats, *U. mjoebergi* will respond more aggressively to unfamiliar individuals than to neighbours that stray into the territory surrounding their burrow (Booksmythe *et al.* 2010*a*) and may even assist established neighbours in defending their burrows against intruders (Backwell & Jennions 2004) (a courtesy that even extends to individuals of a different species, *Uca elegans*; Booksmythe *et al.* 2010*b*). This 'dear enemy' effect occurs because a new neighbour would require a costly renegotiation of territory boundaries and indicates that information on identity is being used to make tactical decisions about the level of aggression to employ. Similarly, individual recognition plays a role in maintaining dominance hierarchies in the hermit crab *Pagurus longicarpus* (Gherardi & Atema 2005).

5.3.1.1 What determines RHP?

Regardless of what is assessed, by whom and when, it is worth considering what factors determine an individual's RHP. Overall body 'size' is often used interchangeably with RHP but in fights where weapons are used features of these may have a stronger influence on fight outcomes. In Morrell *et al.*'s (2005) study on fiddler crabs, for example, the relative and absolute measures of major chelus length were used as the predictor of fight outcomes and duration, rather than measures of carapace width. Indeed, fiddler crabs bearing regenerated claws are weaker opponents than those with brachychelous (non-regenerated) claws (Backwell *et al.* 2000, Lailvaux *et al.* 2008). Furthermore, the effectiveness of a weapon may be influenced by morphological plasticity as well as overall strength or size. In the squat lobster, *Munida rugosa*, there is a dimorphic claw architecture and males that possess 'arched' claws appear to be better at inflicting puncture wounds on their opponents than males that possess 'flat' claws (Claverie & Smith 2007) (Figure 5.2). Sneddon *et al.* (1997*a*) compared the effects of carapace width and chelae length on the outcome of fights in *C. maenas* and found that asymmetry between opponents in chelar length was a more reliable predictor of contest outcome than was asymmetry in carapace width. Weapon size and strength (Gabbanini *et al.* 1995, Sneddon *et al.* 2000*c*) could therefore be as important as overall body size in determining fight outcomes and could actually be more readily assessed than overall body size by fighting opponents.

Obtaining accurate measures of weaponry is particularly feasible in the crustacea and could be an increasingly important focus of research. In a study on the giant freshwater prawn *Macrobrachium rosenbergii* by Barki *et al.* (1997), it was shown that only claw size, and not body size, differs significantly between the winners and losers of fights. In other studies it has been shown that quite subtle aspects of appendage morphology, even in appendages that do not bear claws, reliably predict fight outcomes. During size-matched fights in the hermit crab *P. longicarpus*, the third pair of non-claw-bearing walking legs (periopods) in winners are wider and flatter than in losers, which may provide an advantage in grasping the opponent's shell (Tricarico *et al.* 2008). Similarly, Sneddon and Swaddle (1999) found that shore crabs, *C. maenas*, were disadvantaged during fights by an asymmetry in the lengths of the fifth pair of periopods, possibly due to a lack of postural stability while grappling. Appendage morphology, particularly of those that bear weapons, therefore seems important, but this does not mean it necessarily plays an equally important role in all cases. In the snapping shrimp, *Alpheus heterochaelis*, body size independent of chelar size influences the chance of victory but, even though the chelae are used in agonistic displays, chelar size independent of body size has no effect on contest outcomes (Hughes 1996). Claw morphology is not the only morphological trait that might influence the outcome of a contest. In hermit crabs the power of impact achieved during shell rapping influences the chance of success (Briffa & Elwood 2002, Briffa *et al.* 2003) and attackers with greater abdominal muscle mass relative to their overall body mass are more likely to evict the defender than attackers with lower muscle mass (Mowles *et al.* 2011).

In addition to morphology, individuals may differ in their physiology. As noted above, fighting is costly so the ability to pay those costs may influence RHP. The costs of fighting have been examined extensively in crustaceans and are discussed in detail in section 5.4.2. Finally, while individuals differ from one another in physical traits (morphology and physiology), they may vary in 'psychological' terms. 'Animal personalities' (consistent between-individual variation in behaviour) have been recently demonstrated in *P. bernhardus* (Briffa *et al.* 2008). Further, a link between risk-taking and aggression has been shown in fiddler crabs (Reaney & Backwell 2007). This is suggestive of a behavioural syndrome for risk-taking and aggression, which could clearly influence an individual's agonistic behaviour. Such 'animal personalities' or 'behavioural types' derive from mechanisms involved in information-gathering and decision-making, processes which play key roles in contest behaviour. Thus RHP has traditionally been considered to vary with an individual's overall size, such that the larger opponent is assumed to be the better competitor, but work on crustaceans shows that more subtle features such as variation in physiology, weapon morphology and behavioural type can also influence RHP.

5.3.2 Strategic decisions and resource value

In addition to varying with RHP, agonistic behaviour may also vary with the value of the contested resource. Similar to RHP, resource value (V) can be discussed in absolute and relative terms, since opponents might place a different value for the same resource. The subjective nature of V means that it is likely to influence the 'willingness' or motivational state for agonistic behaviour, but it would be an oversimplification to assume a direct relationship as motivation could also vary with RHP (Hammerstein & Parker 1982, Enquist & Leimar 1987) and perhaps with uncorrelated asymmetries such as experience or ownership. Therefore, despite the fact that V should have a strong influence on motivation it is useful to maintain a distinction between the two terms. Nevertheless, motivational state is a useful factor to investigate and is discussed in detail below.

Considering the effects of V on contests is very relevant to the various assessment models. Cost threshold-based models (e.g. WOA, EWOA and CAM, Chapters 1 and 2) all assume that the loser will give up when the costs of fighting exceed its individual threshold. The theoretical maximum cost an opponent will pay should be a function of its RHP but the actual cost it is willing to pay might be somewhere below this according to the level of motivation for fighting. Hermit crabs and a range of other crustaceans have been key models for investigating the effects of resource value on agonistic behaviour and hermit crabs in particular have been used to investigate the links between information-gathering, fight tactics and motivational change during contests.

Because hermit crabs fight over empty gastropod shells it is relatively easy to manipulate the value of the resource by supplying shells of different sizes to attackers and defenders. Moreover, it is possible to quantify the value of the resource. In free shell choice experiments, the preferred weight of shell chosen by the crabs shows a highly significant positive relationship with crab weight. By using regression equations generated from such experiments it is possible to determine the weight of an optimal shell for a given weight of crab and to thus manipulate resource value by a known amount by varying either the size or species of shell supplied to attackers and defenders (Dowds & Elwood 1983, Elwood & Neil 1986, Elwood 1995, Elwood & Briffa 2001). Dowds and Elwood (1983) supplied crabs with either non-preferred *Gibbula cineraria* shells or preferred *Littorina obtusata* shells to investigate the effect of resource value on the decisions of both attackers and defenders. The experiment involved four groups in which the species of shell supplied to the attacker and defender was varied between groups but where the weight of both shells was the preferred shell weight of the larger crab. The decision of the larger crab to initiate a fight (lunging at the smaller crab and grabbing its shell, causing the small crab to withdraw into the shell), and taking on the role of attacker was more likely to occur if it was supplied with a low-quality *G. cineraria* shell, but the quality of the smaller crab's shell did not influence the chance of the larger crab initiating a fight. This indicates that the decision to initiate a fight is based on information about the value of the currently owned resource rather than on more elaborate information about the potential gain in resource value that an exchange of shells would entail. This predominance of current shell quality, over the potential gain in shell quality, in the decision to launch an attack was also found in a similar study on a different species, *Pagurus longicarpus* (Gherardi 2006). The sequence of events leading up to the point when the larger crab initiates a fight is therefore similar in *P. bernhardus* and *P. longicarpus* but the encounters progress rather differently once the attacker has grabbed the defender. As described above, in *P. bernhardus* there is a period of further investigation of the external surface of the defender's shell, prior to a period of shell rapping (Elwood & Neil 1986, Briffa *et al.* 1998, Elwood & Briffa 2001). By contrast these events are rare in *P. longicarpus* (Gherardi 2006). These differences in agonistic behaviour might explain different effects of shell quality on subsequent decisions. In *P. bernhardus*, as with the decision to initiate a fight, the decision to escalate and commence shell rapping is significantly affected by the quality of the attacker's shell but now the quality of the defender's shell also influences the behaviour of attackers with shell rapping more likely to occur when the defender is in a good quality shell (Dowds & Elwood 1983) and the persistence and success rates of attackers increase with the potential gain available (Elwood & Neil 1992). In *P. longicarpus*, attacker shell quality continues to affect agonistic behaviour, such that attackers in poor-quality shells persist for longer before giving up than those in good-quality shells. However, there is no effect of defender shell quality on the duration of contests where attackers give up (Gherardi 2006). As noted (section 5.2), cross-species comparison is a technique that has yet to be applied to contest behaviour in a formal sense, but comparing these similar studies on different species of hermit crab hints at how such an approach could provide new insights into contests.

In *P. bernhardus*, while there are opportunities for attackers to gain information about the shell of the defender once the fight has been initiated (although the attacker may still not have access to information about internal features of the defender's shell that could influence its quality: Arnott & Elwood 2007) it seems unlikely that the withdrawn defender is able to gather information about the quality of the attacker's shell (Elwood & Briffa 2001). It has been suggested that information about attacker shell quality could be revealed through shell rapping (Hazlett 1978, 1982, 1985, 1987, 1996) as the acoustic pitch of the raps might reveal the size of the attacker's shell. Although there is evidence to support the idea that the decision of defenders to give up is strongly influenced by the quality of the shell they may be left with, in some tropical species ('negotiation' hypothesis) analysis of the temporal pattern of shell rapping in *P. bernhardus* suggests that shell exchanges in this species should be considered to be aggressive interactions (Briffa *et al.* 1998, 2003, Briffa & Elwood 2000*a,b*) (Figure 5.3). This illustrates a point that could apply to a wide range of examples of contests. It is clear that asymmetries in RHP and perceived *V* should influence fight outcomes, but there can also be asymmetries in the tactics used by the opponents in contests where they adopt distinct roles, such as attacker and defender or owner and intruder. Different roles may afford different opportunities for information gathering such that there may also be asymmetries in information about RHP and *V* and therefore opponents may not necessarily use the same

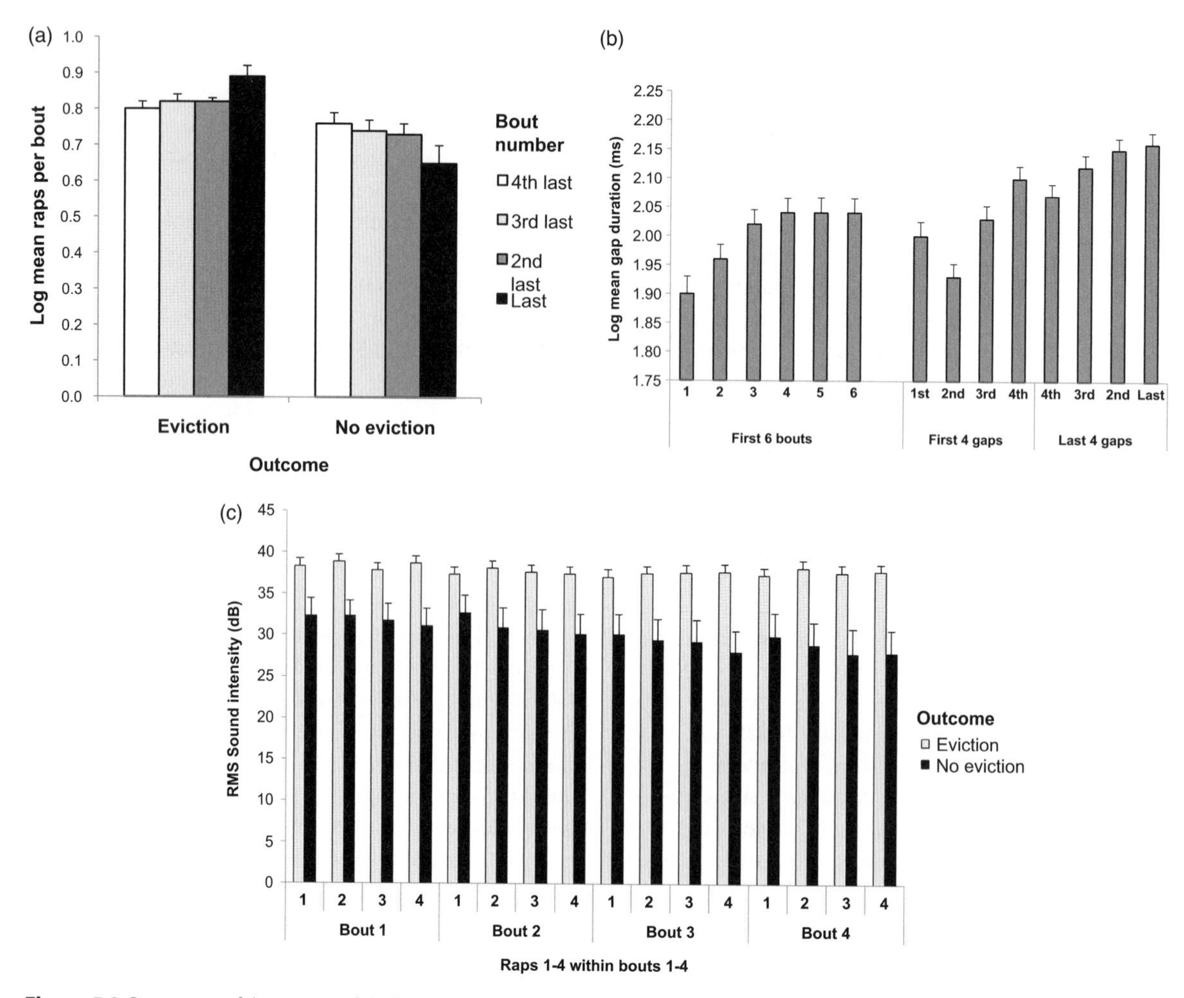

Figure 5.3 Parameters of the pattern of shell rapping analysed in hermit crab shell fights. (**a**) Successful attackers perform increasing raps in each bout up until the end of the fight whereas those that give up perform fewer raps in each bout. (**b**) For all attackers the duration of gaps increases from bout to bout and within bouts. (**c**) The sound intensity of rapping remains relatively constant in bouts for successful attackers (black bars) but declines towards the end of the bout for attackers that give up (white bars). These changes between and within bouts and differences between successful attackers and those that give up suggest that shell rapping is a demanding activity. This suggests that its primary function is to advertise the RHP of the attacker, perhaps through a demonstration of stamina. (Figures re-drawn from Briffa *et al.* 1998, Briffa & Elwood 2000*a*,*b*.)

decision rules. Many examples of contests involve two distinct roles, often occurring when one individual holds a resource which its opponent attempts to prise away such as fights over ownership of guarded females in amphipods (Plaistow *et al.* 2003) or during contests over burrows in fiddler crabs (Fayed *et al.* 2008).

Hermit crabs are a useful system for the analysis of resource value due to the way in which shell quality can be accurately quantified and manipulated. Nevertheless, the asymmetry in roles between attacker and defender, and the fact that perceived resource value and the level of difficulty of evicting the defender co-vary with size difference are clearly complicating factors. Resource value has also been investigated in species where contests are more symmetric in nature and may map more readily onto current assessment models. In one respect, however, the results are broadly similar to those obtained from studies of hermit crabs: agonistic behaviour varies with the value of the resource such that fighting animals might reveal some information on the value they place on the contested resource. In shore crabs, *C. maenas*, fights are more intense in the presence than in the absence of food, but are also resolved more quickly (Sneddon *et al.* 1997*b*). In contests where the number of food items is varied, duration increases as food becomes more scarce (Smallegange *et al.* 2007). Links between resource value and duration are perhaps not unexpected but any link between resource value and behavioural variables that could potentially reveal a contestant's persistence time are interesting because theory predicts that fighting animals should not reveal this information (Turner & Huntingford 1986). In order to fully understand this process, however, it is necessary to establish the specific ('subjective') value that each opponent places on the resource in addition to knowing the absolute value of the resource.

5.3.3 What do fighting animals know? Studies of motivation

Manipulative experiments, where asymmetries in various correlates of RHP such as size and perceived resource value are manipulated indicate the sources of information used during the decision-making process in fights. From the examples discussed above it can be seen that a range of crustaceans have been successfully used to address these questions. Nevertheless, there is another way of gaining insights into the information available to animals that has been used in a range of contexts but in the case of aggressive behaviour has been used uniquely in the crustacea. This has been to investigate the underlying 'motivational state' of fighting animals, in effect to 'ask' the opponents directly what sources of information they are using. 'Motivation' is a term often used in ethology that may be defined as 'a reversible aspect of the animal's state that plays a causal role in behaviour' (McFarland 2006). This makes it distinct from directional or permanent changes in internal state such as those caused by learning or development such that there is the potential for motivation to change in response to information gathered during the contest. For example, if information about the RHP of the weaker opponent is revealed as the contest progresses, its stronger opponent could experience an increase in motivation. Similarly, for an intruder attempting to relieve the current owner of a contested resource, proximity to the resource during the contest might allow a more accurate assessment of its value than was possible before the encounter was initiated; if the resource was revealed to be of lower value than anticipated, then the intruder might experience a decrease in motivation. In the context of aggression, motivation is sometimes referred to as an individual's 'willingness to fight' (e.g. Hoffman & Schildberger 2001). Motivation is also likely to be influenced by physiological variables such as hunger or reproductive status that can influence the subjective value of a given resource. Thus, although motivation is primarily a feature of the 'internal mechanisms' that promote particular behaviours, these interact with information gathered within the timescale of a contest and should clearly influence strategic decisions. Indeed, analyses based on differences in V, in addition to analyses of RHP, can help distinguish between the use of mutual and self assessment (Chapter 4).

As noted above, resource value appears to have a clear effect on the motivation of fighting animals as evidenced by enhanced persistence in highly motivated individuals, which in turn appears to enhance the chance of victory. While persistence time is a good measure of motivation, we can only know the persistence time of losers. In the case of winners, their persistence has been curtailed by the decision of the loser to retreat. It might be possible to gain an index of motivation by analysis of the vigour with which agonistic behaviour is performed, such that vigorous agonistic behaviour is assumed to reflect high motivation. There is, however, a potential limitation on our ability to make inferences about the internal motivational state of fighting animals from observations of the external expression of their agonistic behaviour: although the chance of winning a fight might influence motivation (Enquist & Leimar 1987, Hammerstein & Parker 1982), fighting animals might be expected to conceal their motivation, or at least not reveal it 'intentionally' (here we use the term 'intentional' to denote the primary function of an agonistic behaviour, or of 'what it has evolved for') (Turner & Huntingford 1986). In a war of attrition scenario (Hammerstein & Parker 1982), for example, revealing willingness to fight for a time t would deliver a guaranteed victory to the opponent because they could simply decide to fight for a time of $t + 1$ (RHP permitting). In other words, fighting animals should attempt to 'play their cards close to their chest' and, it has been argued (Hammerstein & Parker 1982, Turner & Huntingford 1986), agonistic behaviours that reveal motivation (at least in WoA-type contests) should be selected against.

Thus, although agonistic behaviours might covary with the underlying motivation to fight, we

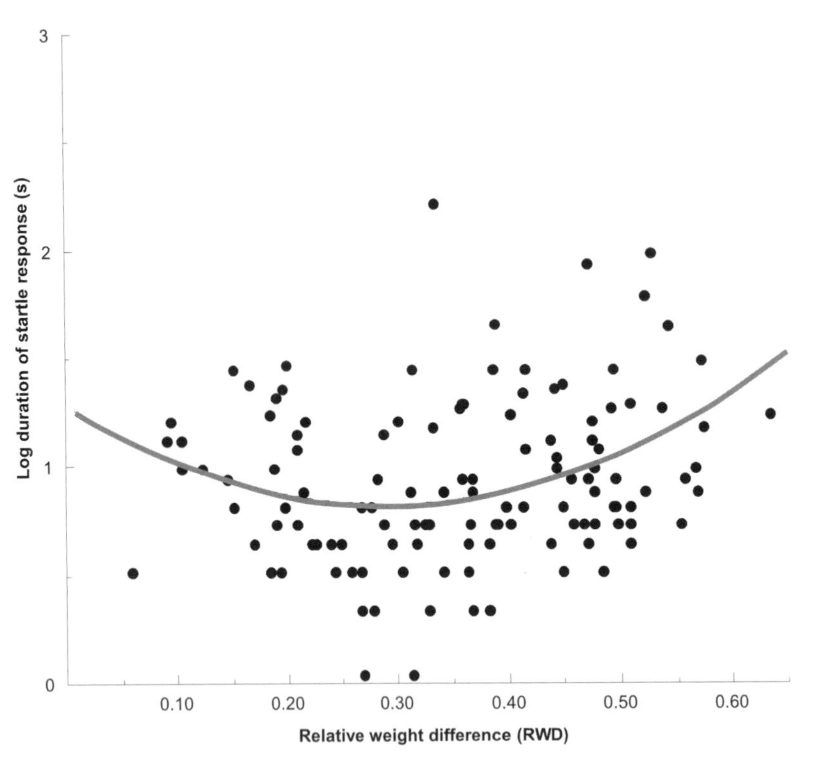

Figure 5.4 The duration of startle-responses and size asymmetry between attackers and defenders during shell fights in *P. bernhardus*. In the lower half of the range of size asymmetries the startle-response of attackers declines as their size relative to the defender increases. This indicates an increase in motivation with increasing disparity in size. When size asymmetry increases further, however, startle-responses begin to increase, indicating a decline in motivation. This is probably due to the fact that as defenders get smaller relative to the attacker, but are provided with shells of ideal size for the attacker, they can retreat further into their shells and are difficult to evict. Such studies using a motivational probe give insights into the information available to fighting animals and the decision-making process. (Figure redrawn from Briffa & Elwood 2001*a*).

cannot assume that this is necessarily the case. What is required in order to investigate motivation is some measure that is independent of the ongoing agonistic behaviour. One means of achieving this is by interrupting the fight with a novel stimulus. If this provokes a startle-response, which the subject can recover from and then recommence agonistic behaviour, the duration of the startle-response will be inversely proportional to the motivational state for resuming the fight. Hermit crabs, due to their response of withdrawing into the gastropod shell, are ideal subjects for this approach. The one limitation is that, once the fight has been initiated, the defender is already withdrawn into its shell, so only measures of attacker motivation can be gained. Nevertheless, it is possible to determine the motivation of victorious attackers as well as those that give up.

In studies of motivation in hermit crabs, the novel stimulus consists of a weighted square of card suspended above the fight arena (a crystallising dish containing seawater and a layer of sand). At the point of interest, the card is released so that it impacts upon the rim of the dish but does not enter the water. Attackers respond to the resultant shock by withdrawing into their shell. Elwood *et al.* (1998) investigated the effects of the size difference between attackers and defenders (relative RHP) and the quality of the attacker's shell (subjective *V*). When the potential gain was high, attackers rapped more vigorously and were more likely to evict the defender, but for the reasons outlined above these might not be reliable indicators of attacker motivation. Attackers in the high-gain group, however, showed shorter startle-responses than those in the low-gain group. In this study the stimulus was applied very early in the fight, after the attacker had initiated but before the onset of shell rapping. As only attacker shell quality was manipulated in this experiment, it does not provide any support for previous studies that indicate that information about the defender's shell is available to attackers (Dowds & Elwood 1983), but it does unequivocally demonstrate the use of information about current resource value during the contest. Interestingly, size difference, a measure of the attacker's relative RHP, did not influence the startle duration.

In a subsequent study (Briffa & Elwood 2001*a*) a non-linear relationship between attacker motivation and relative weight difference was revealed (Figure 5.4). Across the lower range of increasing RHP difference the startle-response of attackers declined, indicating that attackers become more motivated as the chance of winning increased. Over the upper range of RHP differences, however, the startle-responses of attackers started to increase. This is most likely due to co-variation between the size of shell supplied to the defender and the difference in weight between the two crabs; when the relative weight difference (RWD) was high the defenders were supplied with relatively large shells (the defender's shell was completely adequate for the attacker), into which they could withdraw further. This made it more difficult for the attacker to effect an eviction and effectively decreased their relative RHP. Thus, startle-responses were shorter when attacker relative RHP was high, indicating that the chance of winning does influence the motivation for fighting. Given this link between RHP and motivation, a key question about an individual's willingness to fight is whether this information is available to the opponent. Elwood *et al.* (1998) found that the average duration of the pauses between bouts of rapping showed a positive correlation with startle-response duration, indicating that attackers with low levels of motivation rapped less vigorously. It is not clear why the attackers should reveal their motivation in this way, especially when performing the shell rapping signal entails significant metabolic costs (section 5.4.2), which are presumably wasted if the attacker indicates that it will not persist for long. One possible explanation is that, as demonstrated by Briffa and Elwood (2001*a*), motivation varies as a function of the attacker's relative RHP, in terms of its ability to perform shell rapping. The link between startle duration and shell rapping could be due to low motivation in attackers of low RHP, as the chance of winning is low. A probable function of the shell-rapping signal is to advertise RHP by demonstrating 'stamina' (section 5.4.2) and attackers with low stamina that leave long pauses between bouts of rapping might be poorly motivated if the defender appears to be a strong competitor. Thus, while the pattern of shell rapping might demonstrate the attacker's stamina, information about the attacker's underlying motivation might be revealed 'unintentionally' simply because the two factors (motivation and RHP) co-vary. Whether defenders can use this information to estimate the attacker's willingness (rather than ability) to fight has yet to be ascertained. A recent study (Doake & Elwood 2011) indicates that the link between RHP and motivation in hermit crabs might not always be clear-cut. When attacking crabs were placed in very small gastropod shells, their motivation to fight was revealed to be very high. Compared to less-motivated attackers in slightly larger shells, however, their ability to fight was low, as evidenced by high levels of circulating lactic acid (section 5.4.2).

Thus, there are theoretical reasons (Hammerstein & Parker 1982, Turner & Huntingford 1986), and lines of empirical evidence (Doake & Elwood 2011), which suggest that we should not attempt to infer the motivational state of fighting animals from measures of their ongoing behaviour. Rather, we should use some form of 'motivational probe' that provides a measure that is independent of the agonistic behaviour. In the case of hermit crabs it appears that the motivation to fight does, in fact, vary with measures of ongoing agonistic behaviour (and the reasons for this link have yet to be established). Nevertheless, without the use of a motivational probe that was independent of agonistic behaviour it would not have been possible to determine whether variation in rates of agonistic behaviour was linked to variation in motivation.

These studies on hermit crabs show that motivational probing, by interrupting ongoing behaviour, is a potentially powerful tool that can reveal the sources of information available to fighting animals. Although only applied to crustaceans thus far, this technique could be equally applicable to other examples of contest; all that is required is a means of interrupting agonistic behaviour so as to produce a startle-response from which the subjects can recover and recommence fighting. Although the concept is ethological in origin and focussed on understanding internal mechanisms, studies of motivation are relevant to the key assessment models of fighting. A current issue, for example, is whether fighting animals should make decisions on the basis of relative or absolute RHP, also termed 'mutual' or 'self' assessment (see Chapters 2 and 4 for a summary of the theory and analytical approaches for distinguishing between assessment modes, and Chapter 6 for empirical examples of studies of assessment rules in spiders). Comparing relationships between measures of relative and absolute RHP and startle-response durations could be useful in determining what information about RHP is available.

5.3.4 Agonistic signals

Empirical evidence from studies of crustacean agonistic behaviour supports the idea that information should play a critical role in contests. Agonistic signals are used very widely among the Crustacea, and while they are likely to advertise RHP, motivational studies show that they might also contain information on the willingness to fight. Although the functions may be similar, there is a great variety between taxa in the actual format of the displays. In many examples the periopods and chelae are used for displays. This usually involves holding the chelae in a conspicuous posture. Portunid crabs, for example, hold the chelae laterally in a 'meral spread' when threatened by an opponent or a potential predator (Huntingford *et al.* 1995). Similar cheliped displays are used by hermit crabs during the opening phases of shell fights (Elwood *et al.* 2006, Arnott & Elwood 2007) and sometimes immediately after the shell exchange has occurred (Elwood & Neil 1992). In snapping shrimp such as *Alpheus heterochaelis* a more subtle chelar display is given. Here, the claw must be held open in a 'cocked chelar' display in order to provide information on RHP (Hughes 1996). Male fiddler crabs use claw-waving displays to attract females (Backwell *et al.* 1998), but the claws are also used for assessment and as weapons during agonistic encounters between males (Backwell *et al.* 2000). In addition to visual displays, auditory signals may be used. In the terrestrial genus of hermit crab *Coenobita*, for example, a 'chirping' sound is produced by defenders during shell fights (Hazlett 1966).

There is a key difference between relatively static 'cheliped displays' seen during agonistic encounters in portunids and the more dynamic 'claw-waving' displays used during courtship by male fiddler crabs. In static displays the magnitude of the signal is given by the size of the feature, such as a claw, being displayed. In dynamic displays, the size of the ornament or weapon used to produce it may contribute to the signal, but information will also be contained in parameters of the performance, such as the rate and range of displacement of the appendage. Indeed, in non-visual displays the signal may be given exclusively by parameters of the performance. A distinction between the two types of signal was made by Hurd (1997), who differentiated between dynamic 'performance signals' and static 'strategic signals'. The main difference between these types of signals is how the costs of using them accrue. In the case of strategic signals, the cost has been paid during development, whereas in the case of performance signals the cost arises from the demands of the performance itself.

Performance signals have been studied in detail in European hermit crabs, *Pagurus bernhardus*. The shell-rapping behaviour of attackers involves direct contact with the shell of the defender, which receives a series of percussive impacts. There is therefore the possibility that in addition to transmitting information, this behaviour could have direct detrimental effects on the recipient, which influences their decision to give up and relinquish the shell. In this sense the shell-rapping behaviour could act as a dual-role 'hybrid' behaviour (Bradbury & Vehrencamp 1998, pp. 355–356) with both signalling and direct tactical functions (this possibility would fit well with the key assumptions of the CAM). 'Damping' the force of impact by coating the external surface of the attacker's shells with latex did indeed reduce the chance of the defender being evicted, but attackers with rubberised shells still managed to evict defenders (Briffa & Elwood 2000*a*) and attackers without shells that rap using their naked abdomens may still effect an eviction (Dowds & Elwood 1985). Furthermore, detailed analysis of the temporal pattern of shell rapping and contest outcomes shows that it could also contain information about the attacker that would be available to defenders if they monitor the pattern of rapping. In a series of studies the following parameters of the 'vigour' of shell rapping were analysed: the total number of raps and bouts of rapping, the mean number of raps per bout, the mean duration of pauses between bouts (Briffa *et al.* 1998), the mean duration of intra-bout intervals ('gaps') between individual raps within bouts (Briffa & Elwood 2000*b*) and the sound intensity of the raps (Briffa *et al.* 2003) (Figure 5.3). Apart from the total number of raps and total number of bouts, these parameters were analysed over fights as a whole and with respect to position in the fight such that any changes in the vigour of shell rapping that occurred as the fights progressed could be investigated. First, as noted above, shell rapping does seem to be influenced by the potential gain in shell quality available to attackers, indicating that it could reveal information about the attacker's motivational state and hence persistence. Attackers in shells that were only 50% adequate performed more raps and bouts than those in 80% adequate shells indicating that those with higher motivation persisted for longer (Briffa *et al.* 1998). The key differences, however, were between attackers that gave up and victorious attackers that evicted the defender. Over the whole fight, successful attackers perform more raps per bout, leave shorter pauses between bouts (Briffa *et al.* 1998), shorter gaps between raps (Briffa & Elwood 2000*b*) and produce raps of greater sound intensity, indicating a greater force of impact (Briffa *et al.* 2003) than those that gave up. If rapping vigorously induces the defender to give up, it would clearly benefit attackers to rap with high vigour, yet unsuccessful attackers do not do so. This indicates that there are constraints that prevent unsuccessful attackers from rapping with high vigour and one possibility is that they are subject to greater levels of fatigue than are successful attackers, due to the demands of performing shell rapping. Comparison between successful and unsuccessful attackers in the changes in vigour that occur as the fight progresses towards the loser's giving-up decision support this possibility. In fights where the attackers win, they perform a progressively greater number of raps in each bout before evicting the defender. In fights where attackers lose, they perform progressively fewer raps in each bout prior to releasing the defender (Briffa *et al.* 1998, Figure 5.3a). Similar differences between outcomes are seen in the duration of pauses (Briffa *et al.* 1998) and intra-bout gaps (Briffa & Elwood 2000b, Figure 5.3b). Unsuccessful attackers also perform increasingly weak raps of declining sound intensity both within and between bouts, whereas successful attackers maintain a performance of powerful rapping across successive bouts (Briffa *et al.* 2003, Figure 5.3c). In *P. bernhardus* outcomes are clearly related to the pattern of repetitive shall rapping performed by attackers. In this respect, shell rapping is similar to other 'repeated signals' used in a wide range of taxa such as vertebrates and insects, during both courtship and agonistic encounters. One possible function of such repeated signals (or at least the signal value, *sensu* Bradbury & Vehrencamp 1998, in the case of activities that also play a tactical role) is to advertise RHP by demonstrating the sender's stamina.

Agonistic signals involving obvious displays of weaponry or conspicuous repetition of activities may be immediately obvious to an observer but crustaceans also utilise more subtle (to human observers) forms of agonistic signal. Hermit crabs, for example, appear to use chemical cues during fights (Briffa & Williams 2006) and in the tropical species *Calcinus tibicen* the use of directed jets of fluid (possibly produced with the scaphognathites or 'gill bailers') combined with the usual chelar displays has been documented (Baron & Hazlett 1989). Chemical communication is likely to be very important for any aquatic organism and crustaceans are known to be chemo-responsive to odour cues in a wide range of contexts. Odours released from an opponent could reveal information about its status, for example hunger levels or hormonal state (Breithaupt & Atema 1993, 2000). Trying to understand the functions of such odour release (e.g. whether it is used intentionally to transmit information) is extremely difficult without being able to visualise it. While the vast majority of studies of crustacean aggression have focused on marine species, freshwater crustaceans are especially useful in this respect. A proportion of chemical cues are likely to be borne in the urine and freshwater species tend to urinate very frequently, producing large amounts compared to marine crustaceans. This makes it far easier to investigate the use of urine-borne signals in freshwater species than in marine species. Briethaupt and Eger (2002) conducted a study on the freshwater crayfish *Astacus leptodactylus*, which confirms the use of urine in contests. By injecting a fluorescent dye into the crayfish, which is subsequently excreted in the urine, they were able to observe the use of directed urine plumes during staged agonistic encounters between pairs of males (Figure 5.5). Furthermore, urine release is more likely during fights than during other activity and winners released urine more often than losers. Indeed, visual chelar signals in the absence of urine do not cause the opponent to retreat but urine cues alone can cause retreat in blindfolded opponents. It therefore appears clear that urine-borne signals may advertise RHP and perhaps also reveal motivation to fight in a range of crustacean species. As noted in section 5.3.3, there are good theoretical arguments about why fighting animals should not reveal their motivation, but on the evidence of shell rapping in hermit crabs and urine-borne cues in crayfish, it appears that some agonistic behaviours may in fact vary with motivational state. The question of whether the function of these behaviours includes providing information on motivation, or whether the link with motivation is an 'unintended consequence' of the agonistic behaviour, remains to be resolved.

5.3.4.1 Honest and deceptive communication

Crustaceans therefore use a wide variety of signals, comprising visual, tactile and chemical modalities, during contests. If these signals advertise RHP it would clearly pay the sender if the signal could be used to exaggerate its RHP. Various versions of Zahavi's handicap principle (Grafen 1990) have considered how prohibitively high costs of cheating could maintain a system where signals are honest most of the time and selection should favour receivers who avoid being fooled by deceptive signals. Nevertheless, low levels of

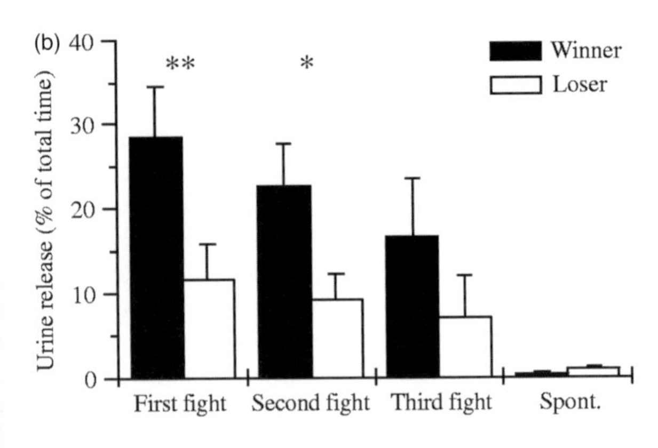

Figure 5.5 (**a**) Use of fluorescein to visualise chemical communication via urine release in blindfolded crayfish, *A. leptodactylus*. Reproduced from Berry & Breithaupt (2010). (**b**) When two opponents first meet, winners release urine more frequently than do losers, but by the third encounter between the same two opponents the difference is no longer significant (** $p < 0.001$, * $p < 0.05$). In all encounters, however, urine release occurs more frequently during a fight than spontaneously when not fighting. (Reproduced with permission from Breithaupt & Eger 2002). For colour version of 5.5(a), see colour plate section.

cheating may be evolutionarily stable (Johnstone 1994) and, although the question of signal honesty is usually considered in the context of courtship displays, the question of deception during agonistic encounters has been investigated in several studies of agonistic signalling in crustaceans. These studies have investigated both strategic and performance signals.

As we have seen, weapons are incredibly important in aggression in the Crustacea and there is evidence that certain features of the chelae may have a stronger influence on RHP than does overall body size. In general, larger claws are stronger and make better weapons but are costly in terms of growth and maintenance (e.g. Doake *et al.* 2010). Agonistic signals are often given by a display of the chelipeds and this presents opportunities for dishonest signalling that are probably peculiar to the arthropods, if not the Crustacea. First, for a period after the moult has been completed the new exoskeleton will be soft and the chelae will thus be very ineffective as weapons (Adams & Caldwell 1990). Nevertheless, they may be similar in appearance to the chelae at later stages in the moult cycle, when fully hardened and useful as weapons. Second, if an individual loses a cheliped, for example if it is autotomised following damage, it will be replaced by a new regenerated cheliped. Regenerated 'leptochelous' claws, although similar to non-regenerated 'brachychelous' claws in appearance and (due to their rapid growth) eventually in size, tend to be lighter and weaker (Backwell *et al.* 2000) and potentially less effective as weapons (e.g. Lailvaux *et al.* 2008, see below). Therefore, there is the possibility of a dishonest 'bluff' if an unhardened claw or a leptochelous claw is displayed. In mantis shrimps, *Gonodactylus bredini*, newly moulted males still defend their shelters with a 'bluffed' meral spread (Adams & Caldwell 1990). Interestingly, the use of this bluff is adjusted according to the risk of being found out. If the opponent is large, and therefore more likely than a smaller opponent to continue to probe after receiving the dishonest display, the meral spread is less likely to be used.

The question of signal dishonesty has been studied in detail in male fiddler crabs. Similar to the example of *G. bredini*, male *U. annulipes* fiddler crabs with a leptochelous major cheliped continue to use it to ward off opponents during agonistic encounters even though in a fight it would be poor as a weapon (Backwell *et al.* 2000). Males with regenerated claws also continue to perform waving displays to attract females, applying less effort for the same effect as males that wave with heavier brachychelous claws (Backwell *et al.* 2000). In fact further work utilising a 'whole-organism performance capacity' approach shows how leptochelous males may be cheating twice. Analysis of claw strength (closing force and pull resisting force) reveals that these regenerated claws are not only lighter but also weaker than a brachychelous claw of the same size (Lailvaux *et al.* 2008).

Hughes (2000) analysed signal residuals to formally test the possibility of dishonest signalling in the big clawed snapping shrimp *A. heterochaelis*. This was based on the relationship between actual RHP and signal structure within a population. Due to the imperfect relationship between actual RHP and the agonistic signal used to advertise it, individuals with the same RHP were suggested to use signals of different magnitude. For senders whose signal is smaller than the population average for their RHP, signal residuals are negative; for senders whose signal is greater than the population average for their RHP, signal residuals are positive. Individuals with a positive residual could then exaggerate their RHP. In *A. heterochaelis*, there is a very tight relationship between chelar size as revealed by the cocked chelar display and body size (the key determinant of RHP). However, there is still a degree of error in the relationship and individuals with positive residuals (chelar size larger than expected for the body size) use the cocked chelar display more often than those with negative signal residuals, and it was therefore concluded that there could be a payoff for dishonest signalling (Chapter 4 gives details of the statistical methods for analysing signal residuals). Similar results were found in *P. bernhardus*, where individuals with larger than predicted chelipeds for their size performed a greater number of 'pre-fight displays' prior to the shell rapping phase of the encounter (Arnott & Elwood 2010). While cheating appears to be possible for *A. heterochaelis*, and *P. bernhardus*, in other examples a close relationship between signal size and the component of RHP that it advertises might prevent cheating. In the crayfish *C. dispar*, males show a loose relationship between chelar force and claw size and the frequency of cheating is high (Wilson *et al.* 2009). In females, however, there is a very tight relationship between claw size and strength, and claw size thus appears to provide a reliable signal of chelar strength (Bywater *et al.* 2008).

These are perhaps special cases of dishonest signalling, where apart from the danger of being probed on their true RHP, there is no prohibitive cost incurred by individuals performing a dishonest display (if anything, the cost of the display may be lower than that of the equivalent honest signal). For performance signals, however, such as those that may advertise stamina, the energetic costs of maintaining a high rate of performance so as to give an exaggerated signal of RHP may well be prohibitively high. Shell-rapping signal residuals were examined in *P. bernhardus* (Briffa 2006) using a similar approach to Hughes' (2000) study on snapping shrimp. Here one of the measures of RHP used was energetic status, and in contrast to the finding of cheating in the use of pre-fight cheliped displays (Arnott & Elwood 2010), there was no evidence that hermit crabs could exaggerate their stamina by performing shell rapping with an unrepresentatively high level of vigour. As with the use of many other examples of performance signals, including other examples of agonistic signals, the vigour of rapping appears to be constrained by the energetic costs of performance. Theory suggests that such costs of engaging in contest behaviour, both in terms of performing signals and other types of agonistic behaviour, such as grappling or trying to injure the opponent, should accumulate during the contest. Further, according to threshold-based models, these accumulated costs may form the basis of strategic decisions, that is, once the contest is underway, the decision to give up or withdraw. In order to gain a full understanding of the use of information, the basis of strategic decisions and the functions of agonistic signals, it is therefore necessary to investigate some of the potential costs of fighting, especially when contests are non-fatal or non-injurious.

5.4 Costs of contests

The previous section dealt with the use of information during contests, a key idea from theory. The second key assumption of theoretical models is that engaging in contests, even if they are non-injurious, should entail significant costs. Indeed these costs, either anticipated or realised, are expected along with information-gathering to influence strategic decision-making.

5.4.1 Direct and circumstantial costs

It is useful first to distinguish between costs that are circumstantial to engaging in a contest and costs that accrue as a direct result of engaging (Huntingford & Turner 1986). The rate of accrual of circumstantial costs will not vary directly with the agonistic behaviours performed. Thus, time allocated to the contest may be considered to be a circumstantial cost because the rate of time expenditure cannot vary. Other examples of circumstantial costs are an increased risk of predation due to lack of vigilance or increase of visual presence during a fight. These are costs which arise from external factors rather than as a direct result of performing agonistic behaviour. As circumstantial costs are not 'specific' to contest behaviour (they could accrue equally for any other type of activity) they are less likely to have played an important role in its evolution than direct costs (apart perhaps for 'waiting game'-type encounters, although these are best considered as interference competition rather than contests per se; see Smallegange *et al.* 2006 for a study investigating the distinction between contest behaviour and interference competition in *C. maenas*). In the case of direct costs, which arise as a result of the activities performed, the rate of cost accrual will vary with the intensity of behaviours performed. Injuries, for example, are clearly a direct result of fighting and the rate (or chance) of an injury would increase with the intensity of the contest. As noted above, in the Crustacea there are examples of injurious fighting (snapping shrimp, Hughes 1996; squat lobsters, Claverie & Smith 2007, Figure 5.2) but there are no reports of fatal fighting which is presumably rare. Indeed, even though injuries may occasionally occur, non-injurious fighting appears to be far more frequent. Nevertheless, we have seen that contests in the Crustacea appear to be demanding, often involving intensive activities such as grappling and bouts of agonistic signals. These are likely to entail significant metabolic consequences, which if they represent the main direct costs of agonistic behaviour should play a role in strategic decision-making.

5.4.2 Energy and fighting

While some costs may be circumstantial, costs that accrue directly as a result of performing agonistic behaviour should be of particular functional significance. The idea of energetic costs is central to wars of attrition without assessment (Mesterton-Gibbons *et al.* 1996), the energetic war of attrition, where agonistic signals are assumed to advertise 'stamina' (Payne & Pagel 1996, 1997) and cumulative assessment models (Payne 1998). As discussed in section 5.3.3, some agonistic signals such as shell rapping in hermit crabs seem to be constrained by 'fatigue' (Briffa *et al.* 1998, Briffa & Elwood 2000*a*,*b*,*c*). 'Stamina' or 'endurance' are fairly loose phrases, roughly equivalent to 'the ability to resist fatigue' or, to borrow a phrase from functional morphology, 'maximal performance capacity'. These ideas have nevertheless been invoked as key correlates of RHP in a range of examples, and endurance capacity is an idea that appears to have strongly influenced theory.

Intuitively, violent clashes involving frenetic bouts of agonistic behaviour, sometimes over protracted periods of time, appear to be energetically demanding. The first anecdotal evidence of high metabolic costs associated with contest behaviour came from reports of apparent exhaustion during fights in red deer (Clutton-Brock & Albon 1979) and appears to support this assumption, but the first direct evidence of this came from studies on crustaceans. Briefly, energetic costs will be due primarily to the units of glucose expended during the contest. Under aerobic conditions, metabolism of one molecule of glucose will generate a theoretical maximum of 38 molecules of ATP (2 ATP from anaerobic glycolysis and 36 ATP from aerobic respiration). ATP is the molecule that interacts with muscle filaments, the final step in converting the chemical energy stored in food to the kinetic energy used in behaviour. The glucose used for ATP production may originate from molecules freely circulating in the haemolymph, but demanding activity may also involve the mobilisation of muscular glycogen stores.

The first studies that addressed the question of whether the metabolic costs of fighting could actually relate to the direct costs other than injuries assumed by models were conducted on portunid crabs. In *Necora puber*, ventilation rates correlate strongly with rates of aerobic respiration (Smith & Taylor 1993) so they can be used as an index of aerobic respiration and therefore energy expenditure. This measure was obtained by attaching small electrodes to the scaphognathites, which vibrate within the gill chamber to provide the respiratory current of water. These were attached to an impedance device via very fine wires, suspended over the body of the crab with floats. This preparation allows each individual sweep of a scaphognathite through the gill chamber to be recorded, such that the rate can be accurately calculated. Although slightly invasive, the great advantages of this technique are that an index of aerobic energy expenditure can be obtained continuously through the contest, in real time, and individually for each opponent. Smith and Taylor (1993) staged fights between size-matched males prepared in this way. Rates of scaphognathite beating were significantly elevated (compared to resting levels) during agonistic activities and there was a positive association between contest duration and the level of elevation in beating. Interestingly, elevated scaphognathite beating was found not only during phases of physical contact between the opponents but also during the use of display activities. The level of advertisement given by such displays will obviously correlate with claw size, but this association with scaphognathite beating indicates that it is also associated with an elevated metabolic rate, and thus may involve a performance signal component in addition to a strategic signal component.

Studies of aerobic respiration in contests indicate elevated rates of energy expenditure as a result of engaging in agonistic behaviour. Further, they indicate that all types of agonistic behaviour (displays, trials of strength and dangerous fighting) entail metabolic costs. An alternative approach to estimating the costs of performance during a fight is to use respirometry to estimate metabolic rates. Such techniques have the advantage of being non-invasive and do not require equipment to be attached to either contestant. The obvious drawback is that in a contest involving two animals it is not possible to distinguish between the respiration rate of each opponent during the contest, but it is possible to analyse differences in resting metabolic rate between winners and losers. Brown *et al.* (2003), for example, found that contests over dominance in the freshwater prawn *Macrobrachium rosenbergii* were won by individuals with a high resting metabolic rate, independent of the effects of body size on metabolism, indicating the possibility that some individuals may have a physiological advantage over others. Similarly, studies on the concentration of haemocyanin (the protein responsible for oxygen transport in most crustaceans, which freely circulates in the haemolymph) (Mowles *et al.* 2009) and the concentration of circulating metal ions that modulate the oxygen affinity of haemocyanin (Mowles *et al.* 2008) indicate that in *P. bernhardus*, winners appear to have a greater aerobic capacity than losers. Thus, studies on prawns and hermit crabs show that RHP might be influenced by variation in physiology as well as by variation in morphological traits such as weapons and musculature (section 5.3.1.1). Such initial differences in physiology between opponents could lead to differences in the pattern of change in energetic status as the contest proceeds. For example, an individual with low haemocyanin levels might experience the onset of anaerobic respiration sooner than an opponent with greater levels of haemocyanin. Theory predicts that the costs that accrue during fights should play a key role in strategic decisions, in particular that accumulated costs could influence the decision to give up. If energetic costs do play a significant role in these decisions we might expect to see not only differences in the initial physiological condition of eventual winners and losers at the start of the fight but also differences in their post-fight physiological status.

Rovero *et al.* (2000) used a similar approach to Smith and Taylor (1993) but measured heartbeat rate instead of scaphognathite beating rates during fights in *C. maenas*. They found that although oxygen consumption increases with heartbeat rate, this measure only explains 47% of variation in oxygen consumption. This suggests that, although such techniques are valuable for estimating energy expenditure during contests, they may not be able to detect small-scale differences in energy consumption. Nevertheless, both studies (Smith & Taylor 1993, Rovero *et al.* 2000) indicate increasing energy expenditure with increasing contest duration. Conversely, neither study detected any relationship between energetic costs and contest intensity. This indicates that the intensity of agonistic behaviour is not constrained by the limits of aerobic metabolism. Indeed, both studies concluded that fighting is less costly than forced exercise. In *C. maenas*, however, the costs appeared to be higher for losers than winners because they showed increased rates of ventilation during the contests and during a post-contest recovery phase, whereas post-contest elevation was not seen in winners. Thus, although aerobic metabolism does not appear to be related to fight intensity, the aerobic demands do appear to increase as the contest progresses, indicating an escalation in costs which could be related to the decision of losers to give up.

Although studies on rates of aerobic respiration (Smith & Taylor 1993, Rovero *et al.* 2000) and aerobic capacity (Mowles *et al.* 2008, 2009) suggest a role for energetic costs in strategic decisions, it is surprising that aerobic energy metabolism does not appear to constrain agonistic behaviour as there is no correlation between ventilation rate and intensity. However, investigation of aerobic respiration will not necessarily reveal all of the energetic costs of fighting. If the agonistic behaviour is performed with high intensity the ventilation and circulatory systems may not be able to supply oxygen in sufficient quantities to allow the energetic demands to be met through aerobic respiration. In this case the shortfall in ATP generation can be achieved in the absence of oxygen but such anaerobic respiration is more costly than aerobic respiration. First, the process is far less efficient in terms of glucose depletion, yielding only two molecules of ATP from glycolysis. Second, pyruvate, the end product of glucolysis, which under aerobic conditions is converted to acetyl CoA (thereby transferring the bulk of the energy stored in the original glucose molecule to the citric acid cycle to produce ATP aerobically), is instead fermented to produce large amounts of lactic acid, which accumulates in muscle tissue. This reaction is necessary to produce NAD^+, which is needed to maintain glycolysis under anaerobic conditions. However, while the exact mechanism is the subject of ongoing debate, accumulated lactate (the lactic acid will quickly dissociate to lactate and H^+ ions) is known to be associated with impaired muscle function and a major cause of 'fatigue' (depleted ATP levels are unlikely; low ATP would permanently damage muscle tissue and levels are buffered by a pool of intermediary high-energy phosphates such as arginine phosphate in invertebrates). The effects of prolonged anaerobic respiration appear to be particularly marked in decapod crustaceans. The 'oxygen debt' of accumulated lactate results in a post-activity recovery phase, which constrains activity rates until lactate levels have been reduced to resting levels in the presence of oxygen. Links between energetic costs of fighting and the intensity of agonistic behaviour, as well as with strategic decisions, may therefore be revealed by assay of lactate, a key detrimental by-product of anaerobic respiration, and of metabolites such as circulating glucose and muscular glycogen stores. This will allow post-fight differences in energetic status to be established, which could indicate differences in costs paid by winners and losers. Lactic acid is particularly interesting because under anaerobic respiration it will continue to accumulate for as long as intense activity is sustained. Although not an energetic cost in the sense that it directly represents depleted energy reserves, the cumulative nature of the cost fits well theoretical models of fighting. The EWOA and CAM both predict that costs other than time should accumulate as contests progress, until the loser crosses an individual cost-threshold. These costs are assumed to accrue from a variety of sources (depending on the model) but are envisaged as occurring in arbitrary and interchangeable payoff units. The detrimental effects of high lactate could thus affect the payoffs of engaging in the contest.

In the velvet swimming crab, *N. puber*, post-fight assay suggests that there is no effect of fighting on the levels of haemolymph glucose and tissue glucose and glycogen levels (Thorpe *et al.* 1995). In contrast, forced exercise appears to cause significant mobilisation of muscular glycogen reserves. This suggests that fighting is less costly in terms of depletion of energy stored than intense exercise. However, results appear to be highly variable between species, even close relatives. In *C. maenas* fighting does result in elevated haemolymph glucose levels compared to resting levels (Sneddon *et al.* 1999). Elevated glucose is also seen during fights in the common European hermit crab *Pagurus bernhardus* (Briffa & Elwood 2001*b*, 2002, 2004). Defenders that resist being evicted from their shell have higher circulating glucose levels than evicted defenders (Briffa & Elwood 2001*b*) and a greater proportion of their muscular energy reserves mobilised as glucose rather than stored as glycogen (Briffa & Elwood 2004), but there are no differences between successful and unsuccessful attackers. Thus, in these contests where there are two distinct roles, the use of energy reserves does seem to be associated with the giving-up decision but only for one of the roles.

Elevated glucose appears to be a feature of contests in several crustaceans (e.g. *C. maenas*, *N. puber*, *P. bernhardus*) and mobilisation of energy reserves may be partly responsible for depleted muscular glycogen seen in fighting decapods (Thorpe *et al.* 1995, Sneddon *et al.* 1999, Briffa & Elwood 2004). However, glycogen stores may also be depleted via anaerobic respiration and investigations of the effects of hypoxia (low oxygen) on fighting indicate that this may be the case. Lowering the oxygen content of seawater is a useful way to investigate the anaerobic costs of fighting as it should impose a period of anaerobic respiration, providing a rough means of manipulating lactate levels between opponents. This technique is also ecologically relevant (Sneddon *et al.* 1999, 2000*a*) as intertidal crustaceans, such as certain species of portunid crab, or the smaller size classes of hermit crab, will experience a daily cycle of change in oxygen tension as algal respiration in the absence of sunlight will reduce dissolved oxygen in tidepools through the night. Further, tidepools may become hypoxic when there are high daytime temperatures, due to oxygen dissolution at high temperature. In *C. maenas* under normoxic conditions, fighting results in significant elevation of circulating glucose but not to the extent seen in crabs that have undergone forced exercise. In hypoxic conditions, however, the level of elevation exceeds that seen in exercised crabs, and the fights are of shorter duration (Sneddon *et al.* 1999); decreased fight vigour under hypoxia has been shown during shell fights in *P. bernhardus* (Briffa & Elwood 2000*c*). The most direct way to establish the effect of lactate accumulation is to assay post-fight levels, either in haemolymph or muscle samples.

Again, the relationships between energetics and contest behaviour differ between species. Fighting did not cause a significant increase in circulating lactate in *N. puber* but significant elevation was seen in crabs subjected to forced exercise to exhaustion (Thorpe *et al.* 1995). In *C. maenas* elevated lactate does occur as a result of fighting, although as in the case of *N. puber*, fighting appears not to be as demanding as forced exercise (Sneddon *et al.* 1999, 2000*a*). In these portunids, where contest intensity must be quantified categorically (due to the unstructured nature of the contests), links between energy metabolism and the intensity of agonistic behaviour may be somewhat obscured. In species that perform distinct bouts of agonistic behaviour such as displays, however, such links are likely to be clearer such that there is the possibility of determining whether metabolic factors constrain agonistic behaviour in terms of limiting contest duration but also in terms of limiting the intensity of agonistic behaviour within the timescale of the encounter. In male fiddler crabs, *Uca perplexa*, claw-waving displays commence when the burrows are exposed by the retreating tide and claw waving can continue for up to 2.5 h after exposure. Matsumasa and Maurai (2005) showed that haemolymph lactate increases during this period while the rate of claw waving declines. A similar result is seen during shell fights in hermit crabs, but in this case there are also differences between the two roles and with contest outcomes. In both attackers and defenders lactate increases as a result of fighting (Briffa & Elwood 2001b, 2002), although in attackers this affect may be most marked when attackers enter the contest while occupying particularly small shells (Doake & Elwood 2011). It is not possible to take haemolymph samples without stopping the contest, but an insight into how metabolic costs may accumulate during fight can be gained by stopping contests at specific points, such as after a set number of bouts. In *P. bernhardus*, stopping fights after 0, 2, 4 or 6 bouts reveals significant increases in lactate with the number of bouts performed by attackers. This is seen only in attackers; defenders are also subject to elevated lactate but there is no clear pattern of change with the number of bouts they have received (Briffa & Elwood 2005). As noted above, attackers that give up perform progressively weaker bouts of shell rapping in the bouts preceding the decision to give up. While this behavioural evidence indicates that successful and unsuccessful attackers may have different abilities to resist 'fatigue', this analysis demonstrates a clear link between sustained and intense agonistic behaviour and the accrual of specific metabolic costs, which are key components of fatigue. Indeed, negative relations between the vigour of shell rapping and post-fight lactate concentration (Briffa & Elwood 2001*b*, 2002) indicate that the cost of accumulated lactate does constrain the performance of attackers during the fight. Because losers have higher lactate than winners there is also the possibility that the cost is either more severe or accrues more rapidly for poor-quality individuals.

Lactate accumulation thus appears to be a constraint on agonistic behaviour and seems to relate well to the cumulative costs of fighting predicted by models such as the EWOA (Payne & Pagel 1996, 1997) and the CAM (Payne 1998). A key assumption of the EWOA is that non-injurious agonistic behaviour demonstrates the performer's stamina. In the CAM, performance is also linked to stamina, but there is the additional possibility of inflicting direct harm on the opponent. As noted above, 'stamina' could be driven by a range of underlying physiological traits; understanding these traits can reveal details of strategic decision-making during contests, but studying any one trait does not investigate 'stamina' per se. In an experiment on *P. bernhardus*, Mowles *et al.* (2010) investigated the two components of stamina (endurance and rate of performance) by forcing hermit crabs to move around a circular track. After a recovery period, the crabs were then used in staged shell fights as described above. For both attacking and defending roles, successful individuals were on average capable of greater rates of locomotion than those that gave up. Thus, one component of stamina does appear to be an important facet of RHP in hermit crabs.

Although aspects of the EWOA and CAM appear to fit well with hermit crab contests both intuitively and with elements of the empirical data, there is a limitation in the applicability of these models in that they are both based on the assumption of 'matched performances'. The precise meaning of this criterion is open to interpretation. In hermit crabs, attackers and defenders clearly do not match one another's performances behaviourally. Also, although the overall measure of stamina appears to be important for both roles, the metabolic costs for each role have a different basis. On the other hand, returning to the idea that costs ultimately relate to the payoffs, in interchangeable 'fitness units', of being in the contest there is still the possibility that the total costs accrued by each opponent could be similar, albeit arising from different physiological mechanisms. Nevertheless, new models (perhaps adaptations of existing models based on cost-thresholds) that specifically cater for contests with asymmetric roles that fight in different ways and are subject to different sources of cost would clearly be a useful development. As noted above, fights that involve two distinct roles may involve asymmetries in the information available to each opponent. In *P. bernhardus*, attackers seem to have better information about the potential gain in resource value than do defenders, whereas defenders may have better information on the opponent's RHP than attackers. There is the potential for the two roles to use different sources of information in their strategic decision-making, and analysis of the physiological consequences of shell fighting indicates that the decisions are clearly associated with different metabolic costs.

It appears that in crustaceans, which are limited in terms of aerobic capacity, lactate accumulation is a significant cost of contest behaviour, but it may also play a significant role in a wide range of taxa. Although a study by Abrahams *et al.* (2005) on Siamese fighting fish, *Betta splendens*, indicates that lactate accumulation could have a significant role during contests, relatively few studies have examined this potential cost of agonistic behaviour in non-crustacean models. Neat *et al.* (1998) show that muscular glycogen reserves are depleted anaerobically during fights in the cichlid fish *Tilapia zillii*, resulting in elevated lactate. Another vertebrate example is Schuett and Grober's (2000) study of copperhead snakes, *Agkistrodon contortix*, where losers but not winners show elevated lactate compared to resting levels. Such studies, along with anecdotal evidence such as that provided for fights in ungulates, indicate that the metabolic consequences of fighting may influence strategic decisions in a range of taxa. Thus, advances that have been made through studies linking function to proximate mechanisms in crustaceans might indicate useful future directions for the study of contests in other taxa.

5.5 Conclusions

Crustaceans have been used extensively as models for investigating the assessment rules used during contests. A particular feature of studies conducted on this group is that the costs and sources of information assumed or predicted by theory have been investigated through studies of proximate mechanisms. A key difference between theoretical models is that some assume that the contest is settled by mutual assessment (e.g. the SAM), whereas others are based on the idea of self-assessment and giving-up decisions triggered by individual thresholds of cost (e.g. the EWOA and the CAM, although this model can be difficult to distinguish from mutual assessment models). Analyses of relations between contest duration and absolute and relative measures of RHP have been conducted on a range of crustacean species, all of which have their particular ways of fighting. Considering the nature of the contests under investigation can give clues as to which theoretical model might resemble the assessment rules used. Unstructured fights, with no clear pattern of escalation, for example, would be unlikely to involve sequential assessment. Similarly, in fights with asymmetric roles each opponent may have different sources of information available and it is possible that current models do not describe such encounters well. Even in examples that appear to be well-suited to current models, there still appear to be limits to how well models match the empirical data; in the case of fiddler crabs, it is not yet clear which assessment mode is used, let alone whether any specific model adequately describes the fight. While it is not expected that theoretical models will completely match empirical data, the evidence from studies of assessment of RHP in crustaceans suggests that new models are needed in the advancement of our understanding of the functions of agonistic behaviour. In particular, models that took account of (1) asymmetric roles with access to different sources of information and (2) the possibility that different sources of information are used as the contest progresses would be useful for analysing the contests in crustaceans described in this chapter.

Acknowledgements

I am grateful to Thomas Breithaupt and Bob Elwood for their constructive comments on this chapter. I am also grateful to the following for helping to source figures and allowing their images to be used in the chapter: Pat Backwell, Thomas Breithaupt, Thomas Claverie, Tanya Detto, Sophie Mowles, Anthony O'Toole and Robbie Wilson.

References

Abrahams MV, Robb TL & Hare JF (2005) Effect of hypoxia on opercular displays: evidence for an honest signal? *Animal Behaviour*, 70, 427–432.

Adams ES & Caldwell RL (1990) Deceptive communication in asymmetric fights of the stomatopod crustacean *Gonodactylus bredini*. *Animal Behaviour*, 39, 706–716.

Arnott G & Elwood RW (2007) Fighting for shells: How private information about resource value changes hermit crab pre-fight displays and escalated fight behaviour. *Proceedings of the Royal Society of London B*, 274, 3011–3017.

Arnott G & Elwood RW (2010) Signal residuals and hermit crab displays: flaunt it if you have it! *Animal Behaviour*, 79, 137–143.

Backwell P & Jennions M (2004) Coalition among male fiddler crabs. *Nature*, 430, 417.

Backwell P, Jennions M, Passmore N, *et al.* (1998) Synchronized courtship in fiddler crabs. *Nature*, 391, 31–32.

Backwell PRY, Christy JH, Telford SR, *et al.* (2000) Dishonest signalling in a fiddler crab. *Proceedings of the Royal Society of London B*, 267, 719–724.

Barki A, Harpaz S & Karplus I (1997) Contradictory asymmetries in body and weapon size, and assessment in fighting male prawns, *Macrobrachium rosenbergii*. *Aggressive Behaviour*, 23, 81–91.

Baron LC & Hazlett BA (1989) Directed currents: A hydrodynamic display in hermit crabs. *Marine Behavior and Physiology*, 15, 83–87.

Berry FC & Breithaupt T (2010) To signal or not to signal? Chemical communication by urine-borne signals mirrors sexual conflict in crayfish. *BMC Biology* 8, 25.

Booksmythe I, Jennions MD & Backwell PRY (2010*a*) Investigating the 'dear enemy' phenomenon in the territory defence of the fiddler crab, *Uca mjoebergi*. *Animal Behaviour*, 79, 419–423.

Booksmythe I, Jennions MD & Backwell PRY (2010*b*) Interspecific assistance: Fiddler crabs help heterospecific neighbours in territory defence. *Biology Letters*, 6, 748–750.

Bradbury J & Vehrencamp S (1998) *Principles of Animal Communication*. Sunderland, MA: Sinauer.

Breithaupt T & Atema J (1993) Evidence for the use of urine signals in agonistic interactions of the American lobster. *Biological Bulletin*, 185, 318–318.

Breithaupt T & Atema J (2000) The timing of chemical signaling with urine in dominance fights of male lobsters (*Homarus americanus*). *Behavioral Ecology and Sociobiology*, 49, 67–78.

Breithaupt T & Eger P (2002) Urine makes the difference: Chemical communication in fighting crayfish made visible. *Journal of Experimental Biology*, 205, 1221–1231.

Briffa M (2006) Signal residuals during shell fighting in hermit crabs: Can costly signals be used deceptively? *Behavioral Ecology*, 17, 510–514.

Briffa M & Dallaway D (2007) Inter-sexual contests in the hermit crab *Pagurus bernhardus*: Females fight harder but males win more encounters. *Behavioral Ecology and Sociobiology*, 61, 1781–1787.

Briffa M & Elwood RW (2000*a*) The power of shell rapping influences rates of eviction in hermit crabs. *Behavioral Ecology*, 11, 288–293.

Briffa M & Elwood RW (2000*b*) Analysis of the finescale timing of repeated signals: Does shell rapping in hermit crabs signal stamina? *Animal Behaviour*, 59, 159–165.

Briffa M & Elwood RW (2000*c*) Cumulative or sequential assessment during hermit crab shell fights: Effects of oxygen on decision rules. *Proceedings of the Royal Society of London B*, 267, 2445–2452.

Briffa M & Elwood RW (2001*a*) Motivational change during shell fights in the hermit crab *Pagurus bernhardus*. *Animal Behaviour*, 62, 505–510.

Briffa M & Elwood RW (2001*b*) Decision rules, energy metabolism and vigour of hermit-crab fights. *Proceedings of the Royal Society of London* B, 268, 1841–1848.

Briffa M & Elwood RW (2002) Power of shell-rapping signals influences physiological costs and subsequent decisions during hermit crab fights. *Proceedings of the Royal Society of London B*, 269, 2331–2336.

Briffa M & Elwood RW (2004) Use of energy reserves in fighting hermit crabs. *Proceedings of the Royal Society of London B*, 271, 373–379.

Briffa M & Elwood RW (2005) Rapid change in energetic status in fighting animals: Causes and effects of strategic decisions. *Animal Behaviour*, 70, 119–124.

Briffa M & Elwood RW (2007) Monoamines and decision making during contests in the hermit crab *Pagurus bernhardus*. *Animal Behaviour*, 73, 605–612.

Briffa M & Elwood RW (2009) Difficulties remain in distinguishing between mutual and self-assessment in animal contests. *Animal Behaviour*, 77, 759–762.

Briffa M & Sneddon LU (2007) Physiological constraints on contest behaviour. *Functional Ecology*, 21, 627–637

Briffa M & Williams R (2006) Use of chemical cues during shell fights in the hermit crab *Pagurus bernhardus. Behaviour*, 143, 1281–1290.

Briffa M, Elwood RW & Dick JTA (1998) Analysis of repeated signals during shell fights in the hermit crab *Pagurus bernhardus. Proceedings of the Royal Society of London B*, 265, 1467–1474.

Briffa M, Elwood RW & Russ JM (2003) Analysis of multiple aspects of a repeated signal: power and rate of rapping during shell fights in hermit crabs. *Behavioral Ecology*, 14, 74–79.

Briffa M, Rundle SD & Fryer A (2008) Comparing the strength of behavioural plasticity and consistency across situations: Animal personalities in the hermit crab *Pagurus bernhardus. Proceedings of the Royal Society of London B*, 275, 1305–1311.

Brown JH, Ross B, McCauley S, *et al.* (2003) Resting metabolic rate and social status in juvenile giant freshwater prawns, *Macrobrachium rosenbergii. Marine and Freshwater Behaviour and Physiology*, 36, 31–40.

Bywater CL, Angilletta MJ Jr & Wilson RS (2008) Weapon size is a reliable indicator of strength and social dominance in female slender crayfish (*Cherax dispar*). *Functional Ecology*, 22, 311–316.

Claverie T & Smith IP (2007) Functional significance of an unusual chela dimorphism in a marine decapod: Specialization as a weapon? *Proceedings of the Royal Society of London B*, 274, 3033–3038.

Clutton-Brock TH & Albon SD (1979) The roaring of red deer and the evolution of honest advertisement. *Behaviour* 69, 145–170.

Detto T & Backwell PRY (2009) The fiddler crab *Uca mjoebergi* uses ultraviolet cues in mate choice but not aggressive interactions. *Animal Behaviour*, 78, 407–411.

Doake S & Elwood RW (2011) How resource quality differentially affects motivation and ability to fight in hermit crabs. *Proceedings of the Royal Society of London B*, 278, 567–573.

Doake S, Scantlebury M & Elwood RW (2010) The costs of bearing arms and armour in the hermit crab *Pagurus bernhardus. Animal Behaviour*, 80, 637–642.

Dowds BM & Elwood RW (1983) Shell wars: Assessment strategies and the timing of decisions in hermit crab shell fights. *Behaviour*, 85, 1–24.

Dowds BM & Elwood RW (1985) Shell Wars 2. The influence of relative size on decisions made during hermit crab shell fights. *Animal Behaviour*, 33, 649–656.

Elwood RW (1995) Motivational change during resource assessment by hermit crabs. *Journal of Experimental Marine Biology and Ecology*, 193, 41–55.

Elwood RW & Briffa M (2001) Information gathering and communication during agonistic encounters: A case study of hermit crabs. *Advances in the Study of Behavior*, 30, 53–97.

Elwood RW & Neil SJ (1986) Asymmetric contests involving two resources. *Journal of Theoretical Biology*, 120, 237–249.

Elwood RW & Neil SJ (1992) *Assessments and Decisions: A Study of Information Gathering by Hermit Crabs.* London: Chapman & Hall.

Elwood RW, Wood KE, Gallagher MB & Dick JTA (1998) Probing motivational state during agonistic encounters in animals. *Nature*, 393, 66–68.

Elwood RW, Pothanikat E & Briffa M (2006) Honest and dishonest displays, motivational state, and subsequent decisions in hermit crab shell fights. *Animal Behaviour*, 72, 853–859.

Enquist M & Leimar O (1987) Evolution of fighting behavior – The effect of variation in resource value. *Journal of Theoretical Biology*, 127, 187–205.

Fayed SA, Jennions MD & Backwell PRY (2008) What factors contribute to an ownership advantage? *Biology Letters*, 4, 143–145.

Figler M, Blank GS & Peeke HVS (2005) Shelter competition between resident male red swamp crayfish *Procambarus clarkii* (Girard) and conspecific intruders varying by sex and reproductive status. *Marine and Freshwater Behaviour and Physiology*, 38, 237–248.

Gabbanini F, Gherardi F & Vannini M (1995) Force and dominance in the agonistic behavior of the freshwater crab *Potamon fluviatile. Aggressive Behavior*, 21, 451–462.

Gherardi F (2006) Fighting behavior in hermit crabs: The combined effect of resource-holding potential and resource value in *Pagurus longicarpus. Behavioral Ecology and Sociobiology*, 59, 500–510.

Gherardi F & Atema J (2005) Memory of social partners in hermit crab dominance. *Ethology*, 111, 271–285.

Glass C & Huntingford FA (1988) Initiation and resolution of fights between swimming crabs (*Liocarcinus depurator*). *Ethology*, 77, 237–249.

Grafen A (1990) Biological signals as handicaps. *Journal of Theoretical Biology*, 144, 517–546.

Hammerstein P & Parker GA (1982) The asymmetric war of attrition. *Journal of Theoretical Biology*, 96, 647–682.

Hazlett BA (1966) Observations on the social behavior of the land hermit crab, *Coenobita Clypeatus* (Herbst). *Ecology*, 47, 316–317.

Hazlett BA (1978) Shell exchange in hermit crabs: Aggression, negotiation or both? *Animal Behaviour*, 26, 1278–1279.

Hazlett BA (1982) Resource value and communication strategy in the hermit crab *Pagurus bernhardus* (L). *Animal Behaviour*, 30, 135–139.

Hazlett BA (1983) Interspecific negotiations – mutual gain in exchanges of a limiting resource. *Animal Behaviour*, 31, 160–163.

Hazlett BA (1984) Variations in the pattern of shell exchange among hermit crabs. *Animal Behaviour*, 32, 934–935.

Hazlett BA (1985) Communication about shell condition during hermit crab shell exchanges. *American Zoologist*, 25, A4-A4.

Hazlett BA (1987) Information-transfer during shell exchange in the hermit-crab *Clibanarius antillensis*. *Animal Behaviour*, 35, 218–226.

Hazlett BA (1989) Shell exchanges in the hermit crab *Calcinus tibicen*. *Animal Behaviour*, 37, 104–111.

Hazlett BA (1996) Assessments during shell exchanges by the hermit crab *Clibanarius vittatus*: The complete negotiator. *Animal Behaviour*, 51, 567–573.

Hoffman HA & Schildberger K (2001) Assessment of strength and willingness to fight during aggressive encounters in crickets. *Animal Behaviour*, 62, 337–348.

Huber R (2005) Amines and motivated behaviors: A simpler systems approach to complex behavioural phenomena. *Journal of Comparative Physiology* A 191, 231–239.

Huber R, Orzeszyna M, Pokorny, N, *et al.* (1997*a*) Biogenic amines and aggression: Experimental approaches in crustaceans. *Brain, Behavior and Evolution*, 50, 60–68.

Huber R, Smith K, Delago A, *et al.* (1997*b*) Serotonin and aggressive motivation in crustaceans: Altering the decision to retreat. *Proceedings of the National Academy of Sciences USA*, 94, 5939–5942.

Hughes M (1996) The function of concurrent signals: Visual and chemical communication in snapping shrimp. *Animal Behaviour*, 52, 247–257.

Hughes M (2000) Deception with honest signals: Signal residuals and signal function in snapping shrimp. *Behavioral Ecology*, 11, 614–623.

Huntingford FA & Turner A (1986) *Animal Conflict*. London: Chapman & Hall.

Huntingford FA, Taylor AC, Smith IP, *et al.* (1995) Behavioural and physiological studies of aggression in swimming crabs. *Journal of Experimental Marine Biology and Ecology*, 193, 21–39.

Hurd PL (1997) Is signalling of fighting ability costlier for weaker individuals? *Journal of Theoretical Biology*, 184, 83–88.

Jennions MD & Backwell PRY (1996) Residency and size affect fight duration and outcome in the fiddler crab, *Uca annulipes*. *Biological Journal of the Linnean Society*, 57, 293–306.

Johnstone RA (1994) Honest signalling, perceptual error and the evolution of 'all or nothing' displays. *Proceedings of the Royal Society of London B*, 256, 169–175.

Lailvaux SP, Reaney LT & Backwell PRY (2008) Dishonest signalling of fighting ability and multiple performance traits in the fiddler crab, *Uca mjoebergi*. *Functional Ecology*, 23, 359–366.

Matsumasa M & Murai M (2005) Changes in blood glucose and lactate levels of male fiddler crabs: Effects of aggression and claw waving. *Animal Behaviour*, 69, 569–577.

McFarland D (2006) Motivation. In: *A Dictionary of Animal Behaviour*, pp. 133–135. Oxford: Oxford University Press.

Mesterton-Gibbons M, Marden JH & Dugatkin LA (1996) On wars of attrition without assessment. *Journal of Theoretical Biology*, 181, 65–83.

Milner RNC, Booksmythe I, Jennions MD, *et al.* (2010) The battle of the sexes? Territory acquisition and defence in male and female fiddler crabs. *Animal Behaviour*, 79, 735–738.

Morrell LJ, Backwell PRY & Metcalfe NB (2005) Fighting in fiddler crabs *Uca mjoebergi*: What determines duration? *Animal Behaviour*, 70, 653–662.

Mowles SL & Briffa M (2012) Forewarned is forearmed: early signals of RHP predict opponent fatigue in hermit crab shell fights. *Behavioral Ecology*, 23, 1324–1329.

Mowles SL, Briffa M, Cotton PA, *et al.* (2008) The role of circulating metal ions during shell fights in the hermit crab *Pagurus bernhardus*. *Ethology*, 114, 1014–1022.

Mowles SL, Cotton PA & Briffa M (2009) Aerobic capacity influences giving up decisions in fighting hermit crabs: Does stamina constrain contests? *Animal Behaviour*, 78, 735–740.

Mowles SL, Cotton PA & Briffa M (2010) Whole-organism performance capacity predicts resource holding potential in the hermit crab *Pagurus bernhardus*. *Animal Behaviour*, 80, 277–282.

Mowles SL, Cotton PA & Briffa M (2011) Flexing the abdominals: Do bigger muscles make better fighters? *Biology Letters*, 7, 358–360.

Neat FC, Taylor AC & Huntingford FA (1998) Proximate costs of fighting in male cichlid fish: The role of injuries and energy metabolism. *Animal Behaviour*, 55, 875–882.

Neil SJ (1985) Size assessment and cues: Studies of hermit crab contests. *Behaviour*, 92, 22–38.

Payne RJH (1998) Gradually escalating fights and displays: The cumulative assessment model. *Animal Behaviour*, 56, 651–662.

Payne RJH & Pagel M (1996) Escalation and time costs in displays of endurance. *Journal of Theoretical Biology*, 183, 185–193.

Payne RJH & Pagel M (1997) Why do animals repeat displays? *Animal Behaviour*, 54, 109–119.

Peeke HVS, Sippel J & Figler MH (1995) Prior residence effects in shelter defense in adult signal crayfish (*Pacifastacus leniusculus* (Dana)): Results in same- and mixed-sex dyads. *Crustaceana*, 68, 873–881.

Plaistow S, Bollanche L & Cezilly F (2003) Energetically costly precopulatory mate guarding in the amphipod *Gammarus pulex*: Causes and consequences. *Animal Behaviour*, 65, 683–691.

Reaney LT & Backwell PRY (2007) Risk-taking behavior predicts aggression and mating success in a fiddler crab. *Behavioral Ecology*, 18, 521–525.

Regier JC, Shultz JW & Kambic RE (2005) Pancrustacean phylogeny: Hexapods are terrestrial crustaceans and maxillopods are not monophyletic. *Proceedings of the Royal Society of London B*, 272, 395–401.

Rovero F, Hughes RN, Whiteley NM, *et al.* (2000) Estimating the energetic cost of fighting in shore crabs by noninvasive monitoring of heartbeat rate. *Animal Behaviour*, 59, 705–713.

Schuett GW & Grober MS (2000) Post-fight levels of plasma lactate and corticosterone in male copperheads, *Agkistrodon contortrix* (Serpentes, Viperidae): Differences between winners and losers. *Physiology and Behavior*, 71, 335–341.

Smallegange IM, van der Meer J & Kurvers RHJM (2006) Disentangling interference competition from exploitative competition in a crab-bivalve system using a novel experimental approach. *Oikos*, 113, 157–167.

Smallegange IM, Sabelis MW & Meer J van der (2007) Assessment games in shore crab fights. *Journal of Experimental Marine Biology and Ecology*, 351, 255–266.

Smith IP & Taylor AC (1993) The energetic cost of agonistic behaviour in the velvet swimming crab, *Necora* (= *Liocarcinus*) *puber* (L). *Animal Behaviour*, 45, 375–391.

Smith IP, Huntingford FA, Atkinson RJA, *et al.* (1994) Strategic decisions during agonistic behaviour in the velvet swimming crab, *Necora puber* (L). *Animal Behaviour*, 47, 885–894.

Sneddon LU & Swaddle JP (1999) Asymmetry and fighting performance in the shore crab *Carcinus maenas*. *Animal Behaviour*, 58, 431–435.

Sneddon LU, Huntingford FA & Taylor AC (1997*a*) Weapon size versus body size as a predictor of winning in fights between shore crabs, *Carcinus maenas* (L.). *Behavioral Ecology and Sociobiology*, 41, 237–242.

Sneddon LU, Huntingford FA & Taylor AC (1997*b*) The influence of resource value on the agonistic behaviour of the shore crab, *Carcinus maenas* (L.). *Marine and Freshwater Behaviour and Physiology*, 30, 252–237.

Sneddon LU, Taylor AC & Huntingford FA (1999) Metabolic consequences of agonistic behaviour: Crab fights in declining oxygen tensions. *Animal Behaviour*, 57, 353–363.

Sneddon LU, Taylor AC & Huntingford FA (2000*a*) Combined field and laboratory studies on agonistic behavior in shore crabs, *Carcinus maenas*: Metabolic consequences of variable oxygen tensions. *Biodiversity Crisis and Crustacea*, 12, 201–210.

Sneddon LU, Taylor AC, Huntingford FA, *et al.* (2000*b*) Agonistic behaviour and biogenic amines in shore crabs *Carcinus maenas*. *Journal of Experimental Biology*, 203, 537–545.

Sneddon LU, Huntingford FA, Taylor AC, *et al.* (2000*c*) Weapon strength and competitive success in the fights of shore crabs (*Carcinus maenas*). *Journal of Zoology*, 250, 397–403.

Sneddon LU, Huntingford FA, Taylor AC, *et al.* (2003) Female sex pheromone-mediated effects on behavior and consequences of male competition in the shore crab (*Carcinus maenas*). *Journal of Chemical Ecology*, 29, 55–70.

Taylor PW & Elwood RW (2003) The mismeasure of animal contests. *Animal Behaviour*, 65, 1195–1202.

Thorpe KE, Huntingford FA & Taylor AC (1994) Relative size and agonistic behaviour in female velvet swimming crabs, *Necora puber*. *Behavioural Processes*, 32, 234–246.

Thorpe KE, Taylor AC & Huntingford FA (1995) How costly is fighting? Physiological effects of sustained exercise and fighting in swimming crabs, *Necora puber* (L) (Brachyura, Portunidae). *Animal Behaviour*, 50, 1657–1666.

Tricarico E & Gherardi F (2007) The past ownership of a resource effects the agonistic behavior of hermit crabs. *Behavioral Ecology and Sociobiology*, 61, 1945–1953.

Tricarico E, Benvenuto C, Buccianti A, *et al.* (2008) Morphological traits determine the winner of 'symmetric' fights in hermit crabs. *Journal of Experimental Marine Biology and Ecology*, 354, 150–159.

Turner GF & Huntingford FA (1986) A problem for game-theory analysis – Assessment and intention in male mouthbrooder contests. *Animal Behaviour*, 34, 961–970.

Wilson RS, James RS, Byewater C, *et al.* (2009) Costs and benefits of increased weapon size differ between sexes of the slender crayfish *Cherax dispar*. *Journal of Experimental Biology*, 212, 853–858.

Chapter 6

Aggression in spiders

Robert W. Elwood & John Prenter

6.1 Summary

Both genders of spiders compete for a variety of resources. They typically use non-contact display, followed by increasingly escalated contact phases comprising touching and sparring and then escalated grappling and biting. Studies of spiders have been central to the understanding of assessment strategies and, for the most part, the data support self-assessment rather than mutual-assessment models. There is good evidence for effects of resource value and ownership on the conduct and outcome of these contests. Fights may have short-term consequences with respect to fatigue, but can have longer-term effects such as loss of appendages or death. Specific experience of winning or losing contests may also influence future encounters. Spiders have also been used in studies of the underlying genetic basis for variation in contest behaviour. Spiders have been the inspiration for motivational models of aggression and we propose a new two-dimensional model that uses cost and resource value as the major factors influencing motivational state and hence choice and duration of activities.

6.2 Introduction

Spiders present eminently tractable systems for the study of animal contests, both in field and laboratory settings, and have been the subjects in several seminal studies of aggression. While some taxonomic groups build webs and use them to capture prey and others do not, aggressive behaviour spans the taxonomic and foraging strategy distinction. It occurs in sedentary web-builders (e.g. orb-web spiders, money spiders, comb-footed spiders), sedentary ambushing species (e.g. crab spiders) and more cursorial hunting spiders (e.g. wolf and jumping spiders).

6.2.1 Contests between males

Male spiders often guard females prior to mating and other males may attempt to take over the guarding position. The high value of reproductive opportunities and the ease with which males may be induced experimentally to engage in contests have made these male–male interactions particularly amenable for research. Contests typically start with non-contact displays and proceed to touching with legs followed by grappling and biting. The biting involves locking the chelicerae (jaws) or biting the body or appendages. Non-contact displays may be visual, vibratory or seismic (Huber 2005), the form of which principally depends on whether or not a web is used. For example, male jumping spiders (Salticidae) use visual displays (Jackson 1982*a*, 1988, Faber & Baylis 1993, Taylor *et al.* 2001, Taylor & Jackson 1999) and/or a substrate-borne vibratory signal (Elias *et al.* 2008). In the ant-mimicking jumping spider, *Myrmarachnae lupata*, males have extremely elongated chelicerae that they spread and display visually to rivals at a distance before escalating to 'embracing' with the chelicerae in pushing contests (Jackson 1982*b*). Male wolf spiders (Lycosidae) start their encounters with various displays of the front legs and body postures and use their palps (Fernandez-Montraveta & Ortega 1993, Aspey 1994), abdomen (Kotiaho *et al.* 1999, Delaney *et al.* 2007) or forelegs (Delaney *et al.* 2007) to tap on the substrate.

Males of orb-web species may start a contest when a wandering male encroaches on a female's web that is already guarded by a resident male (Christenson & Goist 1979, Hack *et al.* 1997, Bridge *et al.* 2000). The resident typically orients and waves his front legs towards the intruder and each spider may then tap and beat their legs upon the other with varying degrees

Animal Contests, ed. I.C.W. Hardy and M. Briffa. Published by Cambridge University Press.

of intensity, leading to grappling and biting at each other's chelicerae (Hack *et al.* 1997, Bridge *et al.* 2000). In *Nephila clavipes*, a resident male responds by shaking his body laterally and by vigorous plucking on strands of the web before moving forward with the front legs extended and attempting to bite the opponent (Christenson & Goist 1979). Sparring with the first two pairs of legs, grappling and then locking of the chelicerae occur in the bowl and doily spider, *Frontinella pyramitela* (Linyphiidae) (Austad 1983), and in *Argyrodes antipodiana* (Theriididae) the contests escalate from moving and 'shuddering' of the abdomen that vibrates the web, to touching the opponent and grappling and then to wrestling and biting (Whitehouse 1997). Thus, in most male–male contests there is a progression from non-contact display to low-cost touching to grappling and then to biting that might result in injury and ultimately the death of a contestant.

6.2.2 Contests between females

In marked contrast to males, female spiders do not typically compete over access to mates. However, contests for other resources still occur; for example, in web builders there may be competition for space or for the use of specific webs (Leborgne & Pasquet 1987). Contests start with signalling with legs or vibratory activities leading to contact with cheliceral locking, biting and tumbling around (Riechert 1978*a,b*). Contests may also occur between wandering females that may fight to avoid being cannibalised or to cannibalise another female. For example, in wolf spiders, *Lycosa* spp., interactions may involve a large repertoire of activities with many variants of leg display plus touching and grappling (Nossek & Rovner 1984). However, the complexity of these fights seems to be comparable with male contests in wolf spiders (Aspey 1994). Thus, despite the difference in the contested resource, the activities seen in contests between females are very similar to those in contests between males (Buskirk 1975, Jackson 1982*b*, Schuck-Paim 2000). However, the frequency and duration of those activities may differ between the genders as shown in *Phidippus clarus* (Elias *et al.* 2010).

6.2.3 Contests between the sexes

There are various reasons that males and females might fight. For example, courtship interactions can be tinged with the threat of sexual cannibalism (cannibalism of the male by the female sometime before, during or shortly after mating, reviewed by Elgar 1992, Prenter *et al.* 2006*a*) and males may attempt to fight off the female. Aggression and cannibalism is especially serious to the male when he is 'dwarfed' by the female, as with the golden orb-weaving spider *Nephila plumipes* (e.g. Schneider *et al.* 2000) and the Australian redback spider and other members of the widow spider genus *Latrodectus* (Forster 1992, Andrade 1996, Sergoli *et al.* 2008). In the orb-web spider *Araneus diadematus*, hungry females are more likely to cannibalise a courting male, especially if the size difference between the pair is large (Roggenbuck *et al.* 2011). Sexual cannibalism, however, may also occur in moderately dimorphic wolf spiders in which food-deprived females are more likely to consume males but more so if the male is small relative to the female (Wilder & Rypstra 2008). Thus, there is evidence of sexual cannibalism being a feeding strategy (Andrade 1998), but it might also be used as means of mate choice by eliminating inferior males (Prenter *et al.* 2006*a*, Stoltz *et al.* 2009). Reversed sexual cannibalism may occur when males are the larger sex, for example in the water spider *Argyroneta aquaticus* (Schutz & Taborsky 2005). In cases of sexual cannibalism it is not the cannibalism that is of interest here but the attempts to defend against cannibalistic attack. Another setting that might result in inter-sexual aggression is when a female attempts to defend her eggs from sexually selected infanticide, as occurs in *Stegodyphus lineatus* (Schneider & Lubin 1997). However, studies on cannibalism and infanticide have focussed on the outcome of the interactions and give little detail on the manner of the contests.

6.3 Assessment strategies in spider contests

For each species contests are variable in terms of duration, degree of escalation or intensity and sometimes in the choice of activity within the main fight phases. That is, some contests are short, and resolved during the non-contact display phase, whereas others escalate to fierce fighting. The typical cost of the contest will thus vary (Riechert 1978*a,b*, Austad 1983, Bridge *et al.* 2000, Taylor *et al.* 2001, Decarvalho *et al.* 2004). The understanding of this variation was boosted by the application of game theory, which examined how different strategies might be employed by each contestant and how the winner is determined (Maynard Smith & Price 1973, Parker 1974, Maynard Smith & Parker 1976, Chapter 2). Some initial game-theory models considered symmetric contests (in terms of no size difference between contestants and with each contestant valuing the resource equally) in which the contestants differ only in their choice of behaviour, e.g. choice between 'Hawk' and 'Dove' strategies or between different durations of activities (e.g. Maynard Smith 1982, Chapter 2). A property of these models is that information is not gathered during the contest and thus cannot be used in decision-making. These models, however, led to the development of more biologically realistic models in which contestants were expected to gather information about asymmetries and to use that information in deciding the behaviour to be exhibited (Parker & Stuart 1976, Parker & Rubenstein 1981, Chapter 2).

Studies of spiders have been at the forefront of attempts to test these models. They have been influential in modifying how the models are viewed and used in the wider literature on animal contests. In particular, they have been central to challenging the dogma that mutual assessment pervades animal contests and informs decision-making processes during aggressive interactions. They have contributed significantly to a critical re-evaluation of evidence, and influenced how such evidence is analysed and interpreted, and have sparked a resurgence of interest in this area of animal behaviour. In the remainder of this section we review the major factors that affect contestant behaviour in aggressive interactions in spiders and how they fit predictions of game-theory models.

6.3.1 Resource holding potential

Contestants are likely to differ in size, development of weaponry, energetic state and other aspects of physiology, experience or position. Any of these factors might affect the ability of the animal to win a contest and together are termed resource holding potential or resource holding power (RHP) (Parker & Stuart 1976, Chapters 1 and 2). The contestant with the higher RHP is likely to incur lower costs in the fight because it will accumulate wounds at a lower rate and may use a lower proportion of available energy in fighting. If both contestants could reliably assess their opponent's RHP relative to their own, then the one with the lower RHP could terminate the contest immediately, thus reducing the costs of time, energy and risk of injury from engaging in a contest that it would inevitably lose. This mutual assessment is central to the asymmetric war of attrition (Parker & Rubenstein 1981, Hammerstein & Parker 1982, Chapter 2) and sequential assessment models (SAM; Enquist & Leimar 1987, Chapter 2). The latter have been particularly influential and assume that successive parts of the interaction become increasingly costly but provide increasingly accurate information about the opponent. SAMs also assume that each contestant compares that information with information it has about its own RHP. As soon as an asymmetry is apparent, with sufficient reliability, the weaker animal will concede defeat and give up. However, an alternative scenario is that each contestant only has information about its own abilities or state and incurs costs up to a particular threshold, at which point it gives up (Bishop *et al.* 1978, Parker & Rubenstein 1981). This is the feature of several models (Chapter 2): war of attrition without assessment (Mesterton-Gibbons *et al.* 1996), energetic war of attrition (Payne & Pagel 1996, 1997) and the cumulative assessment game (Payne 1998). These self-assessment models are inferior to mutual assessment because the loser will always incur costs up to the threshold and, perhaps for that reason, until recently, were rarely given serious consideration.

As spiders typically show displays that escalate to contact, grappling and biting they seem intuitively to follow the pattern of increasing costs from phase to phase suggested for the mutual-assessment models, and spiders were used in early tests of those models. For example, Austad (1983) examined fights between male bowl and doily spiders, *Frontinella pyramitela*, to test the predictions of the war of attrition with perfect assessment (Parker & Rubenstein 1981). In common with studies on other taxa, Austad (1983) found that fight duration was negatively correlated with size difference. As size is typically a strong determinant of RHP, this finding was taken as convincing evidence that both contestants assessed the RHP of their opponent and of themselves. Large differences in RHP between opponents would be more easily determined, so a relatively smaller individual would give up earlier rather than persisting in a contest it would certainly lose. Furthermore, Austad (1983) noted that the probability of injury increased with fight duration. He was able to calculate the cost of the contest in terms of reduced reproductive success and found that the rate of cost accrual was greatest for the smaller male when there was a substantial size difference between opponents. The conclusion was that the males were 'unequivocally' exhibiting an assessment strategy and used information on the relative fighting ability of their opponent to make decisions about persistence in contests. A negative relationship between duration of jaw lock and mass difference between opponents was also reported for this species, with the identical conclusion regarding support for mutual assessment of RHP (Suter & Keiley 1984). This species was also used specifically to test the predictions of the SAM (Leimar *et al.* 1991) and again broad support was claimed for that model.

In a jumping spider, *Hypoblemum albovittatum* (formerly *Euophrys parvula*), there was no significant negative relationship between duration and size difference but there was between level of escalation and size difference (Wells 1988, note that this study had modest sample size). For another jumping spider, *Zygoballus rufipes*, a relationship was found between size difference and both duration and level of escalation (Faber & Baylis 1993). Pre-contact abdominal drumming during agonistic encounters was also examined with respect to weight difference of opponents in wolf spiders, *Hygrolycosa rubrofasciata* (Kotiaho *et al.* 1999). When the sizes of the contestants were similar, the drumming rate was also similar; but with a size difference of greater than 30%, there was a clear disparity in drumming rate, with the heavier male showing the higher rate. However, weight difference did not predict the occurrence of an escalated fight and support for the SAM was not clear (Kotiaho *et al.* 1999). Perhaps the clearest evidence for weight asymmetry influencing fight duration was found for the orb-web spider *Metellina segmentata* (Hack *et al.* 1997). Fights were longer when size difference was small, irrespective of whether the intruder or resident won the contest, and thus the findings were considered to be strong support for the predictions of mutual assessment of RHP (Parker & Rubenstein 1981, Hammerstein & Parker 1982, Enquist & Leimar 1987).

In the studies outlined above, the analyses concentrated on the relative size of the opponents because that was the focus of the mutual-assessment models. Very little attention was given to the absolute size of the opponents other than sometimes noting that the average size of opponents was not related to duration (Leimar *et al.* 1991, Hack *et al.* 1997). However, when males were matched for size, large pairs of *Argyrodes antipodiana* were more likely to escalate than were small pairs, suggesting that some factor other than relative size influenced contest escalation (Whitehouse 1997). Further, when the absolute size, as well as relative size, of opponents was investigated in the orb-web spider *Metellina mengei* (Bridge *et al.* 2000), the results could not easily be reconciled with those of earlier studies: as with previous studies there was a negative relationship between size difference and contest duration, but this was not as strong as the positive relationship between the size of the loser and fight duration. It was concluded that fight duration was determined primarily by the absolute size of the loser. The lack of a relationship between winner size and fight duration indicated that assessment of relative size was not occurring in these contests. Because other studies had not reported separate effects of winners and losers on contest duration, it was suggested that assessment of relative size may not be 'a universal feature in animal contests, even though it forms a central part of the theory' (Bridge *et al.* 2000).

Using the salticid, *Plexippus paykulli*, Taylor *et al.* (2001) examined effects of absolute size in contests and found a strong positive effect of the smaller contestant's size but no effect of the larger contestant on duration. Further, in size-matched contests, fights were longer if the contestants were large, which is not predicted under mutual assessment. It was concluded that individuals fought for a duration (or cost) that was determined by their own size, as suggested by game-theory models such as the war of attrition without assessment or energetic war of attrition (Mesterton-Gibbons *et al.* 1996, Chapter 2), and not by relative size. It was also noted that such an effect of loser's size would create a misleading relationship between relative size and cost: if there is a small size difference it is possible that some of the size-disadvantaged opponents are quite large whereas if the difference is large then, of logical necessity, the size-disadvantaged animal will be very small. These relationships were explored further and means of discriminating the various models of assessment were provided by Taylor and Elwood (2003, see also Chapter 4). Self-assessment and mutual-assessment models both predict a strong positive relationship between the size of the smaller contestant (or the loser) and duration but the relationship between larger (or winner) contestant's size and contest duration differs between the models. If mutual assessment occurs, the larger (winner) contestant's size should be strongly negatively related to duration, whereas for self-assessment it should be weakly positive. Further, large matched pairs should fight for longer than small matched pairs in self-assessment of size, but in mutual assessment there should be no difference.

Elias *et al.* (2008, 2010) tested those predictions on *P. clarus*, a salticid, and the findings implied that self-assessment of size is used rather than mutual assessment (also see Kasumovic *et al.* 2011). The results of Whitehouse's (1997) study on *A. antipodiana* are also consistent with this interpretation. We should note, however, that the cumulative assessment model (CAM) also predicts that animals should incur costs up to a size-dependent threshold but that the actions of the opponent might influence how quickly that threshold is reached. Although it is not a model of mutual assessment, CAM is difficult to distinguish from mutual assessment models (Briffa & Elwood 2009, Arnott & Elwood 2009, Chapters 2 and 5).

Rather than examining how contestant size affected contest duration, Constant *et al.* (2011) examined how size affected the probability of escalation in *N. clavipes*. They showed that escalation usually occurred when both contestants were large, but if one or both were small, escalation did not occur, a finding consistent with self-assessment. Further, a study of *Neriene litigosa* (Linyphiidae) showed that the only significant factor to influence escalation was the weight of the lighter male, again indicating self-assessment (Keil & Watson 2010). In a linear regression analysis that included both the mass of the lighter animal and mass difference of the opponents, only the former had a significant effect, again indicative of self-assessment. However, further analyses on contest duration showed that both had a significant effect, thus suggestive of mutual assessment. Unfortunately separate effects of the weights of both the lighter and heavier animals on duration were not included in the statistical model. A simple comparison of these variables typically allows discrimination between pure self-assessment and either cumulative assessment or mutual assessment (Taylor & Elwood 2003, Arnott & Elwood 2009, Briffa & Elwood 2009). Nevertheless, the authors concluded that the data indicated 'opponent assessment'. However, even if we accept that the analyses showed that this was a possibility, the possibility of cumulative assessment was not excluded. The CAM does not require any assessment of the opponent but just an ability of the opponent to impose costs (Chapter 2).

When suitable methods have been used to test alternative models of assessment in contests, the general conclusion is that spiders do not gather information about their opponents (but see Keil & Watson 2010). However, one example of opponent assessment, albeit in an indirect manner, comes from a study of contests between female 'hermit' spiders, *Nephilengys cruentata* (Schuck-Paim 2000). These are web-builders and the size of the web mesh is positively correlated with the resident builder's size. The ability of an intruder to judge the size of the resident by web parameters was tested by replacing either small or large residents with a large new resident. Thus, in the former case, the web indicated a small resident but the actual resident was large. In the latter case, there was a match between web structure and the large new resident. Intruders were then introduced to those webs. The behaviour of the intruders depended on the web structure: fewer intruders left the web immediately if the web had been built by a small female and intruders onto 'small webs' were less likely to retreat in response to behaviour of the new resident. Thus, in this species, the web provides cues as to the size of the web-builder and these were used by intruders (Schuck-Paim 2000). However, the opponent size cue provided by web size is available only to intruders and whether the resident can assess the size of the intruder by other means has not been determined. It has also been demonstrated that female *Portia labiata* can utilise information from draglines (single lines of silk used as safety lines) to assess the fighting abilities of conspecific females (Clark *et al.* 1999).

Taylor and Elwood (2003) discussed self-assessment and mutual assessment in terms of distinct models but, in practice, they are two ends of a continuum (Prenter *et al.* 2006*b*). Animals are likely to have good information about their own ability or state but the reliability of information about their opponent may vary in quality. If no information is gathered, then pure self-assessment will, by definition, pertain. If accurate information is available about the opponent we would expect pure mutual assessment. However, it seems likely that the accuracy of the information about the opponent will be somewhere between very poor and very good (Prenter *et al.* 2006*b*) and this is supported by the work of Elias *et al.* (2008) on jumping spiders. This opens up a promising avenue for future research: we need to tease out how much information is gathered about the opponent's RHP relative to that of a focal animal. Furthermore, different levels of information about the opponent may be available at different stages of the contest. Intuitively, we would predict that contestants would start the contest with good information about their own abilities and gather increasing amounts of information about the opponent as the contest progresses (although perhaps naïve young spiders may not have a good estimate of their own ability and this is why they are prone to escalate contests, e.g. Whitehouse 1997). To determine the information gained at different contest stages, low-intensity fights may be analysed separately from longer, high-intensity fights. In contradiction of the expectation of increasing information being available as the contest progresses, two studies (on fish and fig wasps) have indicated mutual assessment at early stages of the contest, apparently when deciding whether to escalate (from mutual display to attack) and self-assessment thereafter (Hsu *et al.* 2008, Moore *et al.* 2008, Chapters 8 and 10). This suggests that more information about the opponent is available from 'displays' but that the key determinant of persistence in an escalated fight is a contestant's own condition. Perhaps spiders (and other animals) also take note of morphological attributes during pre-contact displays but not during escalated phases because that could detract from their ability to defend themselves. To use an analogy from boxing, it might be possible to judge the bicep size of the opponent during the introductions but it might be better to concentrate on the process of attack and defence once the fight actually starts. Further, in spiders, there may be positional effects that favour one opponent in gathering information about the other, as with the case of an intruder using cues from the owner's web (Schuck-Paim 2000). Alternatively, a stationary owner might detect and gather visual cues about a moving intruder before that intruder detects the owner (Taylor *et al.* 2001, section 6.3.3). Thus, determining the degree to which spiders can gather information about opponents is likely to be complex.

6.3.2 Resource value

The value of the resource (V), has a strong effect on the likely payoff from a contest and thus is predicted to influence how contests proceed (Parker 1974, Maynard Smith & Parker 1976, Parker & Stuart 1976, Parker & Rubenstein 1981, Enquist & Leimar 1987, Chapter 2). In general, fight costs should vary with the resource value but, of course, if competing individuals are to adjust behaviour with regard to V they must be able to gather and use relevant information (Arnott & Elwood 2008). The higher the value the animal places on the resource the greater will be the motivation to fight and the greater the cost (C) the animal is prepared to pay to gain the resource. Resource assessment is typically inferred from the costs in terms of contest duration (e.g. Hack *et al.* 1997), vigour or degree of escalation (Cross *et al.* 2006, Jackson *et al.* 2006), probability of injury (Austad 1983) or probability of winning (Fromhage & Schneider 2005) being positively related to V.

There are several studies showing that exposure to a resource increases the motivation of spiders to fight. For example, Cross and Jackson (2011) noted that exposure to odour of prey items enhances escalated contests between female *Portia fimbriata*. Further, Kotiaho *et al.* (1999) investigated the pre-contact drumming phase of contests between males and found that previous introduction to a receptive female increased the drumming rate in *H. rubrofasciata*. In addition, previous losers that had been exposed to the female subsequently won contests with previous winners that had not been exposed, especially if the size difference between the males was small. Wells (1988) showed that fights between male jumping spiders were more intense when males fought in the presence of a dead female glued to a moveable cork (males readily court such models). These results have been confirmed in the same, and a number of additional, jumping spider species (Cross *et al.* 2006). Contests staged in the presence of a freeze-dried conspecific female were more escalated than in the presence of a female of a different species (Cross *et al.* 2006; also see Jackson *et al.* 2006). Also, contests between male *N. litigosa* for females showed that the most intense contests occurred over virgin females rather than females that had a lower reproductive value (Keil & Watson 2010) and in the wolf spider *Pardosa milvina* both mating status and nutritional condition of females affected how males fought for those females (Hoefler *et al.* 2009).

Riechert (1979) staged owner–intruder contests over web sites in the funnel-web spider, *Agelenopsis aperta*, and found a positive relationship between contest cost and site quality. Site quality can vary naturally in terms of food resources or sites can be experimentally supplemented with food (Riechert 1984). The same trend was found when owners and intruders were analysed separately suggesting that both individuals had accurately assessed web-site quality. Riechert (1979) sought to distinguish if mutual resource assessment occurred or if only the owner had made the assessment, with the intruder merely taking cues from the behaviour of the owner. Contests were staged between two intruders that had not previously been exposed to the contested web. In this case, the relationship between contest cost and web-site quality was lost, indicating that the actual assessment of web-site quality appeared to be made only by the owner (also see Riechert 1984). Importantly, this suggests that other cases of owners and intruders both apparently assessing the resource value need to be re-examined to determine if the intruder used cues from the owner rather than from the resource.

There are several examples of an asymmetry in the ability of spiders to gather information about the resource value. In contests between males for a female, information on female quality may not be available to an intruder. For example, in the orb-web spiders *Metellina segmentata* (Hack *et al.* 1997) and *M. mengei* (Bridge *et al.* 2000) a male guards a female by waiting at the web periphery for her to catch a large fly, after which he initiates courtship (Prenter *et al.* 1994*a,b,c*). During the wait he may gather information about the size, and hence reproductive value, of the female. Male *M. segmentata* appear to use a chemical cue on the web to assess some aspects of resident females (Prenter *et al.* 1994*d*) but an intruding male does not appear to do this prior to a contest. In both species owner persistence was positively correlated with female value when determined from fights in which the owner gave up. However, when the intruder gave up there was no relationship between intruder persistence and female size. Thus it was concluded that intruders had no knowledge of the value of the female whereas the resident did. Furthermore, in the jumping spider *P. clarus* owners had information about the female size but this was not available to intruders (Kasumovic *et al.* 2011). Intruders in these cases did not cue into the responses of the owners as found by Riechert (1984) for *A. aperta*.

A particularly detailed and informative study of asymmetry in information about resource value is that of the bowl and doily spider, *Frontinella pyramitela* (Austad 1983). In this species there is first-male priority in sperm competition (Austad 1982), so if a male lost a fight halfway through mating, he would lose only future reproductive success, i.e. the takeover would not affect his past success (Austad 1982). Thus a copulating male has information about the remaining value of the female although there is no evidence that he can assess her size and hence egg number (female size and reproductive output are positively related in spiders, e.g. Prenter *et al.* 1999). On first encounter with a female the male performs an intromission without sperm transfer and this is thought to be a means of assessing her mating history (Austad 1982). The experimental procedure involved males being allowed to mate with females for different durations and hence those females would represent different future reproductive values that should be known to the resident male. Intruder males were then introduced but these had no information about the previous copulation duration and could not assess this until they had taken over the females by winning fights against the resident males. Only at that time could an intruder conduct the pre-insemination intromission. Austad (1983) found that the fight durations of the resident male varied in a way predicted by female value. The resident fought for longer and was more likely to win when the female was of high value. However, there was no indication that the intruder adjusted its fight behaviour in response to female value (see also Leimar *et al.* 1991).

Enquist and Leimar (1990) suggest an extension to the theory on resource value, specifically that the value of the current resource should be balanced against the potential cost of a contest in terms of lost future reproductive success. In the orb-web spider *Nephila fenestrata*, the male's pedipalps suffer damage during mating and cannot be used again. Thus a mated male has zero future reproductive potential whereas a virgin intruder risks losing his future opportunity of mating if severely wounded. A resident mated male benefits from victory because he maintains his paternity of about half of the potential brood that would otherwise be taken by a successful intruding virgin (Fromhage & Schneider 2005). In this type of contest, resident mated males were more likely to win against intruding virgins and fights staged between two mated males were particularly fierce. This is despite the loss of legs that can occur in males during mating and, when all else is controlled, leg loss reduces success in contests (see also Taylor & Jackson 2003). The conclusion was that competing males make decisions based on the ratio of the value of the resource and the cost of lost future fertilisations should death or serious injuries occur (Fromhage & Schneider 2005). Similarly, males that had lost their pedipalps during copulation were more aggressive than intact males in contests in the orb-web spider *Nephilengys malabarensis* (Kralj-Fišer *et al.* 2011).

6.3.3 Owner–intruder asymmetries

Owners of resources or 'prior residents' are often reported as more likely to win contests than are intruders, and this is the case in spiders, for example, in the crab spider *Misumenoides formosipes* (Hoefler 2002). Similar resident advantages were noted by Riechert (1978*b*) in her studies of funnel-web spiders, by Leimar *et al.* (1991) for the bowl and doily spider and by Kasumovic *et al.* (2011) for a jumping spider, *P. clarus*. In the early literature on game theory it was suggested that this general trend might be due to a convention by which intruders gave way to residents (e.g. Maynard Smith & Parker 1976, Maynard Smith 1982); however, numerous other factors may account for such effects (reviewed by Kokko *et al.* 2006). Some effects may be due specifically to the experimental method. For example, in the studies on orb-web spiders noted above (Hack *et al.* 1997, Bridge *et al.* 2000), natural guarding males were either left on the webs of their females or the males were removed and then used as intruders. This involved capture, holding and placement of only the 'intruders' and any of these procedures might have adversely affected their performance in fights. Another reason that intruders may have reduced success may be taken from a study on jumping spiders that was not specifically using owner–intruder asymmetries (Taylor *et al.* 2001). It demonstrated that whichever spider that first detected and oriented to the other had a distinct advantage. It is likely that a stationary resident could observe the movement of the intruder before the latter is aware of the presence of the owner. In these cases, ownership or residency confers an increased success due to increasing RHP simply because owners get the first indication that a contest might occur. Alternatively, there may be an accumulation of good-quality owners when the resources are frequently contested (Rubenstein 1987, Prenter *et al.* 2003) and losers may be energetically drained and/or wounded (section 6.4), so that they may be less likely to win a subsequent encounter as an intruder.

In other cases, ownership may increase the value of the resource, possibly because that spider has become familiar with the immediate surroundings and, in the case of mate-guarding, may be familiar with the female. This may take time and effort that an intruder will have to pay in the future. Furthermore, owners may 'know' the true value of the resource and thus be more likely to win important contests (Bridge *et al.* 2000). It should be noted, however, that not all studies have reported an owner advantage (e.g. Austad 1983, Fromhage & Schneider 2005). No unambiguous case of a convention settling contests in favour of owners has been demonstrated in spiders.

6.4 Consequences of aggressive encounters

The most obvious consequence of a contest is that one contestant gets the resource and the other does not. Other consequences include the loser incurring a higher cost in terms of depletion of energy, temporary fatigue or, more seriously, wounding or death. These outcomes may have far-reaching implications at the individual level with respect to fitness (Austad 1983), for the evolution of particular traits (Prenter *et al.* 1995, Bridge *et al.* 2000, but see Foellmer & Fairbairn 2004) and for population-level phenomena such as size-assortative pairing (because large males preferentially compete for and win access to large females).

6.4.1 Physiological consequences

Contests are costly in terms of physiological demand (Briffa & Sneddon 2007). Energetic demand may be particularly important during spider contests because of the use of relatively inefficient booklungs, the tracheae not being connected directly to the muscles and an open circulatory system for oxygen and metabolite transport (Anderson & Prestwich 1982, 1985, Schmitz 2005). Most spiders are incapable of sustained activity and fatigue rapidly after short periods (10–20 s) of intense activity. Many of their actions are maintained anaerobically, resulting in the accumulation of costly metabolites such as lactic acid (Prestwich 1988*a*,*b*), and they require considerable recovery time before physiological state is restored to resting levels. For example, Prestwich (1983*a*,*b*) suggests 30 min for full recovery from a mere 2 min of continuous activity. It is likely that persistence in contests will be limited by anaerobic metabolism and consequent accumulation of anaerobic metabolites. Further, one interaction may markedly influence performance in a subsequent interaction.

Energetic costs of aggression have been examined in male sierra dome spiders, *N. litigosa*, with respect to different phases of the contest (Decarvalho *et al.*

2004). Phase 1 involved non-contact displays that raised energy use by three times above resting rate. Phase 2 involved 'pedipalp wrestling' and this raised energy use to seven times resting level. Phase 3 involved the animals locking chelicerae and wrestling and raised energy use to nine times resting value. However, these estimates of energetic costs suffer from an assumption of aerobic metabolism and an inability accurately to account for the anaerobic metabolism that seems important in supporting movement and activity in spiders. Furthermore, estimates of energetic expenditure were derived from the CO_2 output for paired males in the respirometry apparatus used to obtain these data. A combination of the study of the anaerobic condition of contestants after fights and respirometry studies, measuring both oxygen consumption and CO_2 production, would help determine when interacting spiders switch from aerobic to anaerobic respiration and provide the causal link between physiological status and contest behaviour and decisions.

6.4.2 Physical consequences

Because escalated fights often involve biting, wounds may be common. The loss of a leg may occur, especially one of the front legs (Austad 1983, Johnson & Jakob 1999, Taylor & Jackson 2003, Fromhage & Schneider 2005) and this may have long-term consequences for future contests. The loss of a single leg impairs the ability to win subsequent contests for territories in the funnel-web spider, *Agelenopis aperta* (Riechert 1988), for access to females in *Nephila fenestrata* (Fromhage & Schneider 2005) and in contests for females in *Misumenoides formosipes* (Dodson & Beck 1993, although the study of Dodson & Schwaab 2001 cast doubt on the finding for *M. formosipes*). Leg loss in *Holocnemus pluchei* (Pholcidae) did not reduce ability to win contests over prey items. However, the intensity of the fight had an effect on which contests were lost (Johnson & Jakob 1999): those with a missing limb were more likely to lose intense, escalated contests than they were to lose contests settled at earlier stages of interaction. As noted in section 6.3.2, escalated contests tend to occur over high-value resources and thus it is possible that the consequences of leg-loss are greater than suggested because the fights lost may be those with important fitness consequences. In the jumping spider, *Trite planiceps*, the loss of a leg offsets a size advantage in contests: the larger of two spiders is generally favoured but limb loss reduces the probability of victory (Taylor & Jackson 2003). Further, in size-matched contests, intact spiders beat those with one missing limb and the latter beat those with two missing limbs (Taylor & Jackson 2003). Thus, non-lethal injury in contests may have major subsequent fitness effects.

6.4.3 Fatalities

Deaths have been observed in numerous studies in fights between males (e.g. jumping spiders, *P. johnsoni* and *J. queenslandica*, Jackson 1980, 1988; web spiders, bowl and doily spiders, *F. pyramitela*, Suter & Keiley 1984) and between females (e.g. the jumping spider *Portia* in which female–female interactions are described as ferocious, Jackson 1982*a*, Jackson & Pollard 1997, and *Jacksonoides queenslandica*, Jackson 1988, Mediterranean tarantula, *Lycosa tarantula*, Moya-Laraño *et al.* 2002*a*). Such deadly fights tend to occur when the value of the resource is high and future reproductive potential low (Enquist & Leimar 1990). Furthermore, spiders are notorious, but not necessarily prolific, sexual cannibals, and the consequences of aggression between the sexes may be the death of a losing male (Elgar 1992). As a consequence, the 'threat' of sexual cannibalism appears to have shaped mating behaviour, with male behaviour being characterised by caution (Elgar 1992, Foellmer & Fairbairn 2004), even to the extent of feigning death (thanatosis) in the nuptial gift-giving *Pisaura mirabilis* (Bilde *et al.* 2006, Hansen *et al.* 2008). A less dramatic example is of male *M. segmentata* that guard at the webs of females and wait until the female has captured a fly of size sufficient to occupy her for long enough to allow the male to approach and begin courtship; this still contains aggressive elements and the male protects himself from attack (Prenter *et al.* 1994*a*,*b*). Males of other web-building species are also known to employ a similar strategy (Austin & Anderson 1978, Robinson & Robinson 1980). In *M. segmentata*, the male's waiting strategy has the knock-on effect of heightening male–male competition, increasing the probability of aggressive interactions between males for access to females and mating (Prenter *et al.* 2003). To date, however, intersexual aggressive interactions have been typically examined from the viewpoint of intersexual conflict over mating (Schneider & Lubin 1998, Schneider *et al.* 2006), male sacrifice, female feeding strategy (e.g. Andrade 1996, 1998, 2003) and sperm competition (e.g. Elgar 1998, Schneider & Elgar 2001), but not from the context of contests (but see Wilder & Rypstra 2008). Fights leading to cannibalism are also seen in same-sex encounters and large size and hunger of a focal animal both enhance the probability of the opponent being consumed (Petersen *et al.* 2010). However, because these interactions involve attack and defence they should also be considered in the context of aggression.

In many animal species a male may kill young sired by another male and consequently gain in reproduction because females produce the next batch of young sooner than if infanticide had not occurred and the killer may sire those young (Hrdy 1979). This infanticide is typically disadvantageous to the female but advantageous to the male and is viewed as a prime example of sexual conflict (reviewed in Arnqvist & Rowe 2005). Schneider and Lubin (1997, 1998) reported infanticide in the spider *Stegodyphus lineatus* (Eresidae) and observed that the female fights to avoid attacks on her egg sac. Males were found to be responsible for a third of all egg losses and lost broods were replaced by females. In this species males are under particularly strong selection to destroy eggs fertilised by other males because the offspring consume the mother after emergence. When males were released near brood-guarding females, the majority tried to enter the female's tube-like retreat and gain access to the eggs. Females that succeeded in keeping the male out of the retreat were larger than the male and, as the sexes are generally similar in size, about half of the females were successful. If the male won the contest, however, the female withdrew to the far end of the retreat. As the egg sac hangs nearer to the front of the retreat, victorious males could gain access to the brood and typically removed the threads holding the sac and pushed the sac out of the tube. Thus, a major consequence for a *S. lineatus* female of losing a contest is the loss of her current brood.

6.4.4 Future behaviour

Another consequence of contests is that experience of fighting may alter the outcome and behaviour in future contests (Hsu *et al.* 2006), as in the crab spider *M. formosipes* (Dodson & Schwaab 2001, Hoefler 2002). In the jumping spider, *P. clarus*, previous winners show an increased probability of winning contests shortly thereafter but a decreased probability of winning after about 5 h (Kasumovic *et al.* 2009, 2010). Losers showed a decreased probability of winning for 2 h after the initial contest (Kasumovic *et al.* 2010). Whitehouse (1997) noted that naïve males were more prone than experienced males to escalate contests in *Argyrodes antipodiana* (Theriidae). When matched for size, large naïve males were more likely to escalate contest behaviour and experienced winners were more likely to win contests than were experienced losers. To some extent these observations of more highly escalated activities being performed by younger and inexperienced spiders suggest that spiders need to learn fighting strategies (Whitehouse 1997, Hoefler 2002). Naïve spiders appear to over-estimate their own RHP (Whitehouse 1997), which could be modified by experience of aggressive interactions and especially by losing. Experienced winners seem to retain high estimates of their abilities whereas losers seem to have lower estimates (Hoefler 2002). In the crab spider *Misumena vatia*, younger males generally won contests against older males and engaged in more escalated activities involving contact between contestants than did older males (Hu & Morse 2004). The fact that younger males engaged in more highly escalated potentially injurious activity is contrary to game-theoretical predictions in that young males have more to lose in potential future reproductive success than do older males (Parker 1974), thus potentially dangerous contests should be delayed and confined to older males.

6.4.5 Accumulation of high quality male residents

As noted in section 6.3.3, a possible explanation for owners winning in naturally observed contests is that repeated contests lead to an accumulation of resource holders with high RHP. This possibility was examined in the orb-web spider, *M. segmentata*: as is found for all species studied, there is a strong size advantage in fights (Hack *et al.* 1997), which if sufficiently large can overcome a residency advantage. Further, males may guard females for several days prior to mating and there may be numerous interactions between males in that time (Rubenstein 1987, Prenter *et al.* 1994*c*). During the reproductive season not all females are mature on any one day so many do not have males guarding their webs and this increases competition for the mature females. Males defeated at one web typically wander to locate another possibly unguarded web, but these males are difficult to locate in the dense vegetation. To determine if these wandering males were of lower quality (RHP) than guarding males, all guarding males were removed in some locations but left on the webs in other locations. New males then appeared at the webs that were not guarded. When all males were subsequently collected, those that had arrived on the webs from which previous male owners had been removed were significantly smaller than those from undisturbed sites. This demonstrates that winners had excluded inferior males from reproductive opportunity (Prenter *et al.* 2003). In the orb-web spider *Zygiella x-notata* there was a clear large-male advantage in competition for females in the time prior to the female moult (Bel-Venner & Venner 2006). Takeovers were frequently noted, with small males being ousted by large males and large males competed specifically for females close to moult and hence close to mating (Bel-Venner & Venner 2006). Large males, therefore, reduced the time spent guarding and thus increased the rate at which they mated. This is in keeping with a general model of mate-guarding (Hardling *et al.* 2004). In *N. clavipes*, although the males are much smaller than the females, there is an advantage for males in being large relative to other males (Christenson & Goist 1979, Vollrath 1980). Males at the central hub area are larger than males restricted to the web periphery and those at the hub get more frequent and longer copulations.

A consequence of an accumulation of good-quality (large size, high RHP) males at webs of females, coupled with an ability to assess female quality in terms of her size and hence fecundity, is that the best-quality males may be found guarding the best-quality females. This may be the reason for the often observed size-assortative pairing found in spiders (Rubenstein 1987, Prenter *et al.* 1994*b*, Hoefler 2007). However, other mechanisms may also drive size-assortative pairing (Crespi 1989, Elwood & Dick 1990).

6.4.6 Sexual size dimorphism

Many spider species show extreme size dimorphism, with very small males compared to females (e.g. Vollrath & Parker 1992, Coddington *et al.* 1997, Vollrath 1998, Hormiga *et al.* 2000, Foellmer & Moya-Laraño 2007). There is typically selection for large female size as this correlates with fecundity (Prenter *et al.* 1994*b*, 1999). Small male size may be selected for by mating advantages of early male maturation, but when there is strong competition for females, selection may favour large males. For example, in *M. segmentata*, females weigh more than males but this is primarily due to their large egg-carrying opisthosoma (Prenter *et al.* 1995). If just the prosoma and legs are compared, the male is larger than the female, having a large prosoma and long powerful legs that are used in male–male contests (Prenter *et al.* 1995). Differing selection forces may influence male shape and size in ways that maximise success in fights but also enable early maturation (Prenter *et al.* 1995, Bel-Venner & Venner 2006, Maklakov *et al.* 2004). Male *Mecolaesthus longissimus* (Pholcidae) do not show the allometric leg growth frequently found in spiders and do not use the legs in any particular way during contests. However, they do use their abdomens and these are more than twice as long in males compared to females, probably due to long abdomens being more effective in contests (Huber 2005). In *S. lineatus* the size of the cephalothorax is important in deciding winners in male–female contests (Schneider & Lubin 1996) and investment in this body part varies according to population density. In low densities the value of winning these contests is high for males and they invest more in the cephalothorax at the expense of investment in the abdomen (Maklakov *et al.* 2006).

The trade-off between early maturation and large size may depend on the likelihood of contests; if contests are not common, due to low population density, for instance, then scramble competition will be more important, resulting in selection for extreme dwarfism in males (Vollrath & Parker 1992, Legrand & Morse 2000, Moya-Laraño *et al.* 2002*b*, Maklakov *et al.* 2004, Foellmer & Moya-Laraño 2007, De Mas *et al.* 2009). The large body size variation seen in male Australian golden orb-weavers, *N. plumipes*, has been attributed to the conflicting effects of selection for smaller males to avoid sexual cannibalism (females responded aggressively less often to small compared to large males crossing their web) and selection for larger body size through male–male contests (in which larger males can exclude smaller males from the central hub of the web) (Elgar & Fahey 1996). However, evidence for a general advantage of small male size in avoiding sexual cannibalism is lacking (Foellmer & Fairbairn 2005, Prenter *et al.* 2006*a*). It has been demonstrated that larger male *N. clavipes* have a much greater reproductive success than smaller males when density of males on the web is high but not when it is low (Rittschof 2010). In the latter condition male–male contests are rare and variation in density and its effects on success are thought to maintain the large variation in male size in *Nephila* species. In a different, extremely dimorphic, orb-web spider, *Argiope aurantia*, both scramble competition and interference competition seem to operate but, contrary to the earlier predictions with regard to scramble competition and sexual size dimorphism, both these mechanisms appear to select for larger male body size (Foellmer & Fairbairn 2004, 2005).

Kasumovic *et al.* (2011) suggest that the advantage to resource holders in contests (section 6.3.3) may directly influence selection for early maturation and small size in male spiders. They advocate that this owner advantage along with an experience advantage that would result from owners winning contests could amount to a large enough advantage to promote selection for early maturation and small size in males. Males that mature early would have a greater probability of securing first access to females and availing of the ownership advantage. If this mechanism promotes sexual size dimorphism in spiders, then we might expect a correlation between the strength of the owner advantage and the degree of sexual size dimorphism in spiders.

6.4.7 Eavesdropping

Aggressive contests do not occur in isolation and the signals involved in interactions and the information they provide about the contestants may be available to other individuals through eavesdropping. The concept of information being available in the wider social domain (McGregor 1993, 2005) has been an area of active research in other taxa in recent years. In web-spiders, information about male RHP may be available to the female (and to any additional males present), when male–male contests take place. Further, in spiders that aggregate in groups of webs that share attachment points, like *Nephila* spp., vibratory cues may be available to other males and females in connected webs. Visual and vibrational (seismic) information about contestants and their displays could be available to jumping spiders in the vicinity of two displaying or physically interacting spiders and vibrational information (seismic and sonic) from drumming displays is potentially available to eavesdropping wolf spiders 'overhearing' the display of interacting spiders. Information about rivals may also be available when males 'line-up' in order of dominance at the webs of females, e.g. *Nephila* spp. (but see Foellmer & Fairbairn 2005).

Chan *et al.* (2008) concluded that females eavesdropped on male–male contests in the jumping spider *Thiania bhamoensis*. In contests between size-matched males, females that observed showed no subsequent preference for winners or losers; however, when females had not observed the contest, they showed a significant preference for the losers of contests. Winners of contests subsequently courted females for longer than losers and achieved higher mating success due to increased courtship persistence. However, Chan *et al.* (2008) also found that the outcome of male–male contests was not a reliable predictor of mating success for males. The visual eavesdropping proposed by Chan *et al.* (2008) could enable the female to obtain information about relative aspects of fighting ability, male quality and condition. However, either the information available to females appears insufficient to allow a preference between the two males to be formed, or females do not utilise information gained from eavesdropping as a basis for mate choice.

6.5 Motivational change and decision-making in spider contests

The use of motivational models to help us understand the behaviour in contests has a long history. The 'conflict hypothesis' suggests that behaviour in contests is determined by two major opposing drives, one being 'attack', the other 'flee'. Because these are mutually incompatible, displacement activities were thought to occur and become 'ritualised' for use in agonistic contexts (reviewed by Baerends 1975). The concept of 'drives' and their interaction has given way to a state space approach in which combinations of causal factors are thought to determine particular activities uniquely (McFarland & Sibly 1975). This approach has been developed specifically with respect to spider contests using a two-dimensional state space (Maynard Smith & Riechert 1984). The two axes of this space were envisaged as 'scalar variables' or sets of causal factors for 'aggression' and 'fear', and used to examine how these might explain decision-making during contests in the desert spider, *A. aperta* (Maynard Smith & Riechert 1984). It was noted that the terms could have been 'estimate of one's own fighting ability and willingness to continue' and 'estimate of opponent's ability and willingness to continue', respectively, and thus specifically modelled a process of mutual assessment,

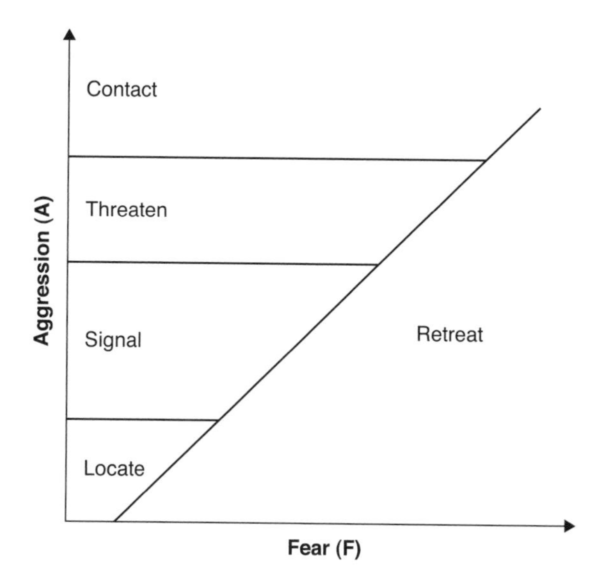

Figure 6.1 Motivational change and decision-making in contests was presented graphically as a causal factor space by Huntingford and Turner (1987) as an interpretation of Maynard Smith and Riechert's (1984) motivational model with aggression and fear as the key variables. (Reproduced with permission from Huntingford & Turner 1987.)

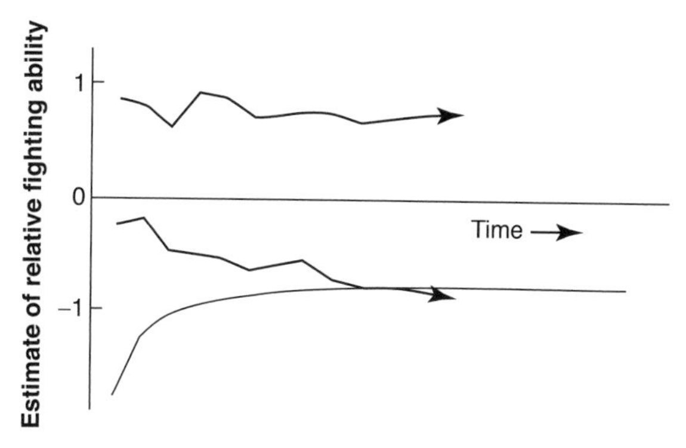

Figure 6.2 The sequential assessment model is presented graphically showing a switching line (the lower curve) and the trajectories of the motivation of the two contestants as they gather information about relative fighting ability. The contestant that estimates that it has the lower fighting ability will eventually cross the switching line and give up the contest at that point. With high-value resources the switching line will be lower and motivation of the eventual loser will take longer to reach that line. (Reproduced with permission from Enquist & Leimar (1987). See also Figure 15.1 for more detail).

which was assumed at the time to be virtually ubiquitous in contests. A pictorial representation of the two-dimensional state space enhanced the understanding of the model (Huntingford & Turner 1987, Figure 6.1); it showed that the key decisions in the fights were to persist or quit but while remaining in the contest various activities could occur. Contests started with zero aggression and zero fear producing low-intensity 'locate' behaviour. When aggression increased, 'signal', 'threat' and escalated 'contact' activities were seen. Should an animal perceive that it was smaller than the opponent then fear levels would increase and if these exceeded aggression levels the animal would quit the contest, and this could occur at any stage in the contest. The lines in the two-dimensional state space are termed 'switching lines', i.e. lines between combinations of causal factors that give rise to different activities (McFarland & Sibly 1975). Aspects of the resource value were also assumed to influence aggression and territory owners were thought to have higher 'aggression' because they knew the value of the resource (V). However, the full role of resource value is not immediately clear in this model.

An attempt to produce a more general model was made with the sequential assessment model (Enquist & Leimar 1983, 1987, Chapter 2), which, although primarily couched in terms of game theory, also had key aspects of a motivational model. The model is again one of mutual assessment, thus allowing an estimate of likely cost. Information is gathered during each bout of the contest and the estimate of cost becomes more reliable as the contest progresses. Eventually, the estimate of likely cost for one contestant will cross a switching line (between continue and quit) and that individual will quit the contest (Enquist & Leimar 1987, Figure 6.2). The value of the resource (V) typically affects persistence or accepted cost (C) in contests and this is visualised in the model by changing the position of the switching line (Enquist & Leimar 1987). With a high resource value the switching line shifts so that more bouts of sampling of relative resource holding potential (RHP) are required to achieve a statistically reliable estimate, i.e. the contestant becomes more persistent.

6.5.1 A new two-dimensional motivational model

Here we propose a two-dimensional conceptual model to provide a simple visualisation of the two major factors that typically influence the course and outcome of fights, these being cost (C) and resource value (V) (Chapter 2). Both C and V have been repeatedly demonstrated to influence the motivational state of contestants (section 6.3, Chapter 2) and we suggest these two factors should be explicitly used in motivational models. We seek a general model that will allow for either mutual- or self- (own-size) assessment to be considered in a single approach. With mutual assessment, cost will be an estimate of future cost and with own-size assessment an estimate of accrued cost at that time. In either case, a single scalar variable may be employed that integrates information from a variety of sources. It can also be used for cases in which the animal has accurate information about its own state and ability but only partial information about the opponent (Prenter *et al.* 2006*b*, Elias *et al.* 2008). Thus a single variable may be used to visualise the level of causal factors for cost. A second set of causal factors, viewed as a separate scalar variable, will relate to the animal's estimate of the value of the resource. This may be determined by direct information about the resource or by indirect information by reference to the opponent's actions or by some average value of resource when no specific information is available. Such a model would fit well with the terms normally used in game theoretic approaches to assessment strategies. Here we tailor the model for spider fights in which three increasing levels are seen in the contest, e.g. pre-contact displays, low-intensity contact using legs and more highly escalated (potentially injurious) behaviour such as grappling, cheliceral locking and/or biting (Figure 6.3).

We suggest that at the beginning of the contest actual or estimated costs start at zero because no energy has been expended or, with mutual assessment, no information has been gathered. However, the estimate of *V* will typically be greater than zero because contestants usually have information about the presence of a resource even if not the absolute value of that resource. Thus the point in state space initially is somewhere on the vertical axis. As estimated or actual costs increase, the point moves to the right, and how long it takes to reach the switching line for quitting (Figure 6.3) depends, in part, on the speed with which actual or estimated costs change. For example, with self-assessment the smaller animal is likely to accrue costs at a faster rate than the larger and thus reach its switching line first. With mutual assessment, when the smaller animal gathers information about its likely inferiority the point in state space will move to the switching line and the smaller spider will quit. In contrast, the larger spider will perceive its superiority, and hence likely low costs, and remain willing to fight. However, in this case, if the smaller animal does not

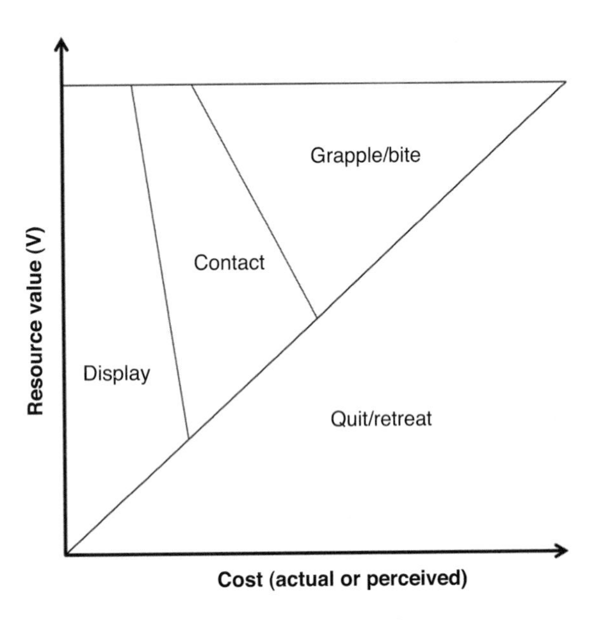

Figure 6.3 Generalised causal factor space for aggression in spiders (and other organisms) showing how information about resource value and cost (actual or predicted) influences the pattern of agonistic encounters.

quit, the larger animal will accrue costs of remaining in the contest and thus a particularly persistent smaller animal may force the larger opponent to quit.

Information about resource value moves the point in state space along the vertical axis in the pictorial representation (Figure 6.3). If a contestant estimates a low value it may only engage in pre-contact displays before quitting. If the estimated value is higher the point in state space starts further from the switching line and thus will persist for longer, accruing higher costs, before quitting. In this case, a spider may move through all three fight phases before it quits. When the two spiders differ in estimates of resource value it could be the smaller animal that places the higher value and this will account for the persistence of some small competitors.

Thus varying the balance of causal factors accounts for a progression of activities and quitting at any stage of the fight and seems to be suitable as a general model for spider fights and for contests in general (Elwood & Neil 1992). It also accounts for what happens if information about resource value is gathered during the contest. In that case, a revised estimate of a higher value will push the point in state space up into a region away from the main switching line for quit. A revised estimate of resource value that it is lower than the first estimate will push the point in state space down, nearer to the quitting line and into an area of low-intensity activities. The crucial innovation of the model is that the two major sets of causal factors relate to the two major variables that have consistently been shown to influence contest behaviour, namely resource value and cost.

6.6 Selection and contest behaviour

The key point of our model (section 6.5.1) is that spiders should gather information about the actual or likely cost and also about the value of the resource. How these values integrate and match up with particular activities will depend on selection pressures that are population- or species-specific. Thus we would expect to find differences between closely related species or even populations within species that are dependent on the ecological settings in which their contests occur. The key ecological factors include the availability and hence value of resources, population density and variability of competitors' RHP, severity of wounds, propensity of predators to prey on contestants either during or after contests and the prevalence of disease in wounded animals. These and more will determine the fitness consequences of contest behaviour. Thus all those factors, and undoubtedly more, will have an influence on natural selection as relates to contest behaviour.

To see how different selection processes may shape aggression, Riechert (1978*b*, 1986) observed funnel-web spiders, *Agelenopsis aperta*, from riparian and grassland populations. The riparian spiders had less-intense, shorter contests and were typically found within shorter distances from conspecifics than were grassland spiders. Good web sites in the grassland area were less common (had higher value) and thus the higher aggression in this population seemed to be adaptive. Curiously, the F_1 hybrids from matings of riparian and grassland individuals had more costly fights, i.e. they were more likely to end in injury or death than either of the parental populations. Following the logic of the motivational model of Maynard Smith and Riechert (1984, Figure 6.1) it was proposed that two genes or gene complexes were each responsible for two major traits, aggression and fear. The grassland spiders were presumed to be homozygous for both genes but to have a double dominant for high aggression and double recessive for low fear. The riparian spiders were presumed to have a double recessive for low aggression and double dominant for high fear. Thus the F_1 would have a dominant for both high aggression and low fear resulting in fights that are more costly than seen in either natural population. A later study examined the F_2 generation and their backcrosses, and demonstrated that aggression was influenced by both autosomal and sex-linked genes (Riechert & Maynard Smith 1989). Further, because there was no increase in the overall variation in contest behaviour or territory size in the F_2 generation, it was suggested that many loci were involved in these traits. There is, therefore, sufficient genetic complexity to provide for change of the competitive equilibrium, dictated by local costs and benefits and frequency-dependent criteria (Hammerstein & Riechert 1988).

Hodge and Uetz (1995) compared two closely related species of colonial web-builders, *Metepeira atascadero* and *M. incrassata*. The former is a desert species with low food levels and the latter has a more productive habitat with high food levels. The rare but defendable resources of the desert spiders were regarded as more valuable and these spiders showed higher levels of escalated fighting. It was suggested that selection has shaped aggressive behaviour of the two species because of the differing habitats.

6.7 Conclusion and future directions

Clearly, studies of the aggressive behaviour of spiders have had a major influence in elucidating factors that affect contests, in the testing of emergent theoretical approaches and, specifically, the predictions of various different game-theoretic models. They have also been at the vanguard in the recent challenge to the assumption of ubiquitous mutual assessment and the ongoing evaluation of assessment behaviour and strategies. We would expect this pattern to continue and for studies of spider contests to remain central in advancing our understanding, especially given the tractability of spider systems to observation and manipulation in empirical studies.

Gaps in our knowledge also remain. It is still not clear how prevalent self-assessment is, both in spiders, where there is evidence that it is used in contests, and in other taxa. Whether individuals employ more than one assessment strategy is also in question and can only be answered by detailed investigation of different aspects or phases of contests. The mechanisms underlying self-assessment are poorly understood and require further investigation. Little is known about the physiological costs of contests in spiders and these are likely to play an important role in decisions in contests, especially in the internal thresholds envisioned to operate under self-assessment. Thus an integrative approach, combining mechanistic and evolutionary approaches to the study of aggression in spiders, is advocated in order to better understand how these different aspects combine to produce the intriguingly complex and diverse contest behaviour exhibited by this group.

Acknowledgements

We thank Gareth Arnott, Rowan McGinley and two referees for comments and the BBSRC (UK) for funding aspects of our work.

References

Anderson JF & Prestwich KN (1982) Respiratory gas exchange in spiders. *Physiological Zoology*, 55, 72–90.

Anderson JF & Prestwich KN (1985) The physiology of exercise at and above maximal aerobic capacity in a theraphosid (tarantula) spider, *Brachypelma smthi* (F.O. Pickard-Cambridge). *Journal of Comparative Physiology*, 155, 529–539.

Andrade MCB (1996) Sexual selection and male sacrifice in the Australian redback spider. *Science*, 271, 70–72.

Andrade MCB (1998) Female hunger can explain variation in cannibalistic behaviour despite male sacrifice in redback spiders. *Behavioral Ecology*, 9, 33–42.

Andrade MCB (2003) Risky mate search and male self-sacrifice in redback spiders. *Behavioral Ecology*, 14, 531–538.

Arnott G & Elwood RW (2008) Information-gathering and decision-making about resource value in animal contests. *Animal Behaviour*, 76, 529–542.

Arnott G & Elwood RW (2009) Assessment of fighting ability in animal contests. *Animal Behaviour*, 77, 991–1004.

Arnqvist G & Rowe L (2005) *Sexual Conflict*. Princeton, NJ: Princeton University Press.

Aspey WP (1994) Agonistic behaviour and dominance–subordinance relationships in the wolf spider *Schizocosa crassipes*. *Proceedings of the 6th International Arachnological Congress*, p. 102.

Austad SN (1982) First male sperm priority in the bowl and doily spider, *Frontinella pyramitela* (Walckenaer). *Evolution*, 36, 777–785.

Austad SN (1983) A game theoretical interpretation of male combat in the bowl and doily spider (*Frontinella pyramitela*). *Animal Behaviour*, 31, 59–73.

Austin AD & Anderson DT (1978) Reproduction and development of the spider *Nephila edulis* (Koch) (Araneidae: Araneae). *Australian Journal of Zoology*, 26, 501–518.

Baerends GP (1975) An evaluation of the conflict hypothesis as an explanatory principle for the evolution of displays. In: GP Baerends, C Beer & A Manning (eds.) *Function and Evolution of Behaviour*, pp. 187–228. Oxford: Clarendon Press.

Bel-Venner MC & Venner S (2006) Mate-guarding strategies and male competitive ability in an orb-weaving spider: Results from a field study. *Animal Behaviour*, 71, 1315–1322.

Bilde T, Tuni C, Elsayed R, *et al.* (2006) Death feigning in the face of sexual cannibalism. *Biology Letters*, 2, 23–35.

Bishop DT, Cannings C & Maynard Smith J (1978) The war of attrition with random rewards. *Journal of Theoretical Biology*, 74, 377–388.

Bridge AP, Elwood RW & Dick JTA (2000) Imperfect assessment and limited information preclude optimal strategies in male–male fights in the orb-weaving spider *Metellina mengei*. *Proceedings of the Royal Society of London B*, 267, 273–279.

Briffa M & Elwood RW (2009) Difficulties remain in distinguishing between mutual and self assessment in animal contests. *Animal Behaviour*, 77, 759–762.

Briffa M & Sneddon LU (2007) Physiological constraints on contest behaviour. *Functional Ecology*, 21, 627–637.

Buskirk RE (1975) Aggressive display and orb defence in a colonial spider, *Metabus gravidus*. *Animal Behaviour*, 23, 560–567.

Chan JPY, Lau PR, Tham AJ & Li D (2008) The effects of male–male contests and female eavesdropping on female mate choice and male mating success in the jumping spider, *Thiania bhamoensis* (Araneae: Salticidae). *Behavioral Ecology and Sociobiology*, 62, 639–646.

Christenson TE & Goist KC (1979) Costs and benefits of male–male competition in the orb weaving spider, *Nephila clavipes*. *Behavioural Ecology and Sociobiology*, 5, 87–92.

Clark RJ, Jackson RR & Waas JR (1999) Draglines and assessment of fighting ability in cannibalistic jumping spiders. *Journal of Insect Behavior*, 12, 753–766.

Coddington JA, Hormiga G & Scharff N (1997) Giant female or dwarf males? *Nature*, 385, 687–688.

Constant N, Valbuena D & Rittscof CC (2011) Male contest investment changes with male body size but not female quality in the spider *Nephila clavipes*. *Behavioural Processes*, 87, 218–223.

Crespi BJ (1989) Causes of assortative mating in arthropods. *Animal Behaviour*, 38, 980–1000.

Cross FR & Jackson RR (2011) Effects of prey-spider odour on intraspecific interactions of areneophagic jumping spiders. *Journal of Ethology*, 29, 321–327.

Cross FR, Jackson RR, Pollard SD, *et al.* (2006) Influence of optical cues from conspecific females on escalation decisions during male–male interactions of jumping spiders. *Behavioural Processes*, 73, 136–141.

Decarvalho TN, Watson PJ & Field SA (2004) Costs increase as ritualized fighting progresses within and between phases in the sierra dome spider, *Neriene litigiosa*. *Animal Behaviour*, 68, 473–482.

Delaney KJ, Roberts A & Uetz GW (2007) Male signalling behaviour and sexual selection in a wolf spider (Areneae: Lycosidae): A test for dual functions. *Behavioral Ecology and Sociobiology*, 62, 67–75.

De Mas E, Ribera C & Moya-Laraño J (2009) Resurrecting the differential mortality model of sexual size dimorphism. *Journal of Evolutionary Biology*, 22, 1739–1749.

Dodson GN & Beck NW (1993) Pre-copulatory guarding of penultimate females by male crab spiders, *Misumoides formosipes*. *Animal Behaviour*, 46, 951–959.

Dodson GN & Schwaab AT (2001) Body size, leg autotomy, and prior experience as factors in the fighting success of male crab spiders, *Misumenoides formosipes*. *Journal of Insect Behavior*, 14, 841–855.

Elgar MA (1992) Sexual cannibalism in spiders and other invertebrates. In: MA Elgar & BJ Crespi (eds.) *Cannibalism: Ecology and Evolution among Diverse Taxa*, pp. 128–155. Oxford: Oxford Scientific.

Elgar MA (1998) Sperm competition and sexual selection in spiders and other arachnids. In: TR Birkhead & AP Møller (eds.) *Sperm Competition and Sexual Selection*, pp. 307–339. London: Academic Press.

Elgar MA & Fahey BF (1996) Sexual cannibalism, competition, and size dimorphism in the orbweaving spider *Nephila plumipes* Latreille (Araneae: Araneoidea). *Behavioral Ecology*, 7, 195–198.

Elias DO, Kasumovic MM, Punzalan D, *et al.* (2008) Assessment during aggressive contests in male jumping spiders. *Animal Behaviour*, 76, 901–910.

Elias DO, Botero CA, Andrade MCB, *et al.* (2010) High resource valuation fuels 'desperado' fighting tactics in female jumping spiders. *Behavioral Ecology*, 21, 868–875.

Elwood RW & Dick JTA (1990) The amorous *Gammarus*: The relationship between precopula duration and size assortative mating in *Gammarus pulex*. *Animal Behaviour*, 39, 828–833.

Elwood RW & Neil SJ (1992) *Assessment and Decisions: A Study of Information Gathering by Hermit Crabs*. London: Chapman and Hall.

Enquist M & Leimar O (1983) Evolution of fighting behaviour: Decision rules and assessment of relative strength. *Journal of Theoretical Biology*, 102, 387–410.

Enquist M & Leimar O (1987) Evolution of fighting behavior – The effect of variation in resource value. *Journal of Theoretical Biology*, 127, 187–205.

Enquist M & Leimar O (1990) The evolution of fatal fighting. *Animal Behaviour*, 39, 1–9.

Faber DB & Baylis JR (1993) Effects of body size on agonistic encounters between male jumping spiders (Arenae: Salticidae). *Animal Behaviour*, 45, 289–299.

Fernández-Montraveta C & Ortega J (1993) Sex-differences in the agonistic behaviour of a lycosid spider (Araneae, Lycosidae). *Ethology, Ecology and Evolution*, 5, 293–301.

Foellmer MW & Fairbairn DJ (2004) Males under attack: Sexual cannibalism and its consequences for male morphology and behaviour in an orb-weaving spider. *Evolutionary Ecology Research*, 6, 163–181.

Foellmer MW & Fairbairn DJ (2005) Competing dwarf males: Sexual selection in an orb-weaving spider. *Journal of Evolutionary Biology*, 18, 629–641.

Foellmer MW & Moya-Laraño J (2007) Sexual size dimorphism in spiders: patterns and processes. In: DJ Fairbairn, WU Blanckenhorn & T Székely (eds.) *Sex, Size and Gender*, pp. 71–81. Oxford: Oxford University Press.

Forster LM (1992) The stereotyped behaviour of sexual cannibalism in *Latrodectus hasselti* Thorell (Araneae: Theridiidae), the Australian redback spider. *Australian Journal of Zoology*, 40, 1–11.

Fromhage L & Schneider JM (2005) Virgin doves and mated hawks: Contest behaviour in a spider. *Animal Behaviour*, 70, 1099–1104.

Hack MA, Thompson DJ & Fernandes DM (1997) Fighting in males of the autumn spider, *Metellina segmentata*: Effects of relative body size, prior residency and female value on contest outcome and duration. *Ethology*, 103, 488–498.

Hammerstein P & Parker GA (1982) The asymmetric war of attrition. *Journal of Theoretical Biology*, 96, 647–682.

Hammerstein P & Riechert SE (1988) Pay-offs and strategies in territorial contests: ESS analyses of two ecotypes of the spider *Agelenopsis aperta*. *Evolutionary Ecology*, 2, 115–138.

Hansen LS, González SF, Toft S, *et al.* (2008) Thanatosis as an adaptive male mating strategy in the nuptial gift-giving spider *Pisaura mirabilis*. *Behavioral Ecology*, 19, 546–551.

Hardling R, Kokko H & Elwood RW (2004) Priority versus brute force: When should males begin guarding resources? *American Naturalist*, 163, 240–252.

Hodge MA & Uetz GW (1995) A comparison of agonistic behaviour of colonial web-building spiders from desert and tropical habitats. *Animal Behaviour*, 50, 963–972.

Hoefler CD (2002) Is contest experience a trump card? The interaction of residency status, experience, and body size on fighting success in *Misumenoides formosipes* (Araneae: Thomisidae). *Journal of Insect Behavior*, 15, 779–790.

Hoefler CD (2007) Male mate choice and size-assortative pairing in a jumping spider, *Phidippus clarus*. *Animal Behaviour*, 73, 943–954.

Hoefler CD, Guhanarayan G, Persons MH, *et al.* (2009) The interaction of female condition and mating status on male–male aggression in a wolf spider. *Ethology*, 115, 331–338.

Hormiga G, Scharff N & Coddington JA (2000) The phylogenetic basis of sexual size dimorphism in orb-weaving spiders (Araneae, Orbiculariae). *Systematic Biology*, 49, 435–462.

Hrdy SB (1979) Infanticide among animals: A review, classification, and examination of the implications of the reproductive strategies of females. *Ethology and Sociobiology*, 1, 13–40.

Hsu Y, Earley RL & Wolf LL (2006) Modulation of aggressive behaviour by fighting experience: Mechanism and contest outcomes. *Biological Reviews*, 81, 33–74.

Hsu Y, Lee S-P, Chen M-H, *et al.* (2008) Switching assessment strategy during a contest: Fighting in killifish *Kryptolebias marmoratus*. *Animal Behaviour*, 75, 1641–1649.

Hu HH & Morse DH (2004) The effect of age on encounters between male crab spiders. *Behavioral Ecology*, 15, 883–888.

Huber BA (2005) Sexual selection research on spiders: Progress and bias. *Biological Reviews*, 80, 363–385.

Huntingford F & Turner A (1987) *Animal Conflict*. London: Chapman and Hall.

Jackson RR (1980) The mating strategy of *Phidippus johnsoni* (Araneae, Salticidae): III. Intermale aggression and a cost–benefit analysis. *Journal Arachnology*, 8, 241–249.

Jackson RR (1982*a*) The biology of *Portia fimbriata*, a web-building jumping spider (Araneae, Salticidae) from Queensland: Intraspecific interactions. *Journal of Zoology, London*, 196, 295–305.

Jackson RR (1982*b*) The biology of ant-like jumping spiders: Intraspecific interactions of *Myrmarachne lupata* (Araneae, Salticidae). *Zoological Journal of the Linnean Society*, 76, 293–319.

Jackson RR (1988) The biology of *Jacksonoides queenslandica*, a jumping spider (Araneae: Salticidae) from Queensland: Intraspecific interactions, web-invasion, predators and prey. *New Zealand Journal of Zoology*, 15, 1–37.

Jackson RR & Pollard SD (1997) Jumping spider mating strategies: Sex among cannibals in and out of webs. In: JC Choe & BJ Crespi (eds.) *The Evolution of Mating Systems in Insects and Arachnids*, pp. 340–351. Cambridge: Cambridge University Press.

Jackson RR, Walker MW, Pollard SD, *et al.* (2006) Influence of seeing a female on the male–male interactions of a jumping spider, *Hypoblemum albovittatum*. *Journal of Ethology*, 24, 231–238.

Johnson SA & Jakob EM (1999) Leg autotomy in a spider has minimal costs in competitive ability and development. *Animal Behaviour*, 57, 957–965.

Kasumovic MM, Elias DO, Punzalan D, *et al.* (2009) Experience affects the outcome of agonistic contests without affecting the selective advantage of size. *Animal Behaviour*, 77, 1533–1538.

Kasumovic MM, Elias DO, Sivalinghem S, *et al.* (2010) Examination of prior contest experience and the retention of winner and loser effects. *Behavioral Ecology*, 21, 404–409.

Kasumovic MM, Mason AC, Andrade MCB, *et al.* (2011) The relative importance of RHP and resource quality in contests with ownership asymmetries. *Behavioral Ecology*, 22, 39–45.

Keil PL & Watson PJ (2010) Assessment of self, opponent and resource during male–male contests in the sierra dome spider, *Neriene litigosa*: Linyphiidae. *Animal Behaviour*, 80, 809–820.

Kokko H, Lopez-Sepulcre A & Morrell LJ (2006) From hawks and doves to self-consistent games of territorial behaviour. *American Naturalist*, 167, 901–912.

Kotiaho JS, Alatalo RV, Mappes J, *et al.* (1999) Honesty of agonistic signalling and effects of size and motivation asymmetry in contests. *Acta Ethologica*, 2, 13–21.

Kralj-Fiser S, Gregoric M, Zhang S, *et al.* (2011) Eunuchs are better fighters. *Animal Behaviour*, 81, 933–939.

Leborgne R & Pasquet A (1987) Influences of aggregative behaviour on space occupation in the spider *Zygiella x-notata* (Clerck). *Behavioural Ecology and Sociobiology*, 20, 203–208.

Legrand RS & Morse DH (2000) Factors driving extreme sexual dimorphism of a sit-and-wait predator under low density. *Biological Journal of the Linnean Society*, 71, 643–664.

Leimar O, Austad SN & Enquist M (1991) A test of the sequential assessment game – Fighting in the bowl and doily spider, *Frontinella pyramitela*. *Evolution*, 45, 862–874.

Maklakov AA, Bilde T & Lubin Y (2004) Sexual selection for increased male body size and protandry in a spider. *Animal Behaviour*, 68, 1041–1048.

Maklakov AA, Bilde T & Lubin Y (2006) Intersexual combat and resource allocation into body parts in the spider, *Stegodyphus lineatus*. *Ecological Entomology*, 31, 564–567.

Maynard Smith J (1976) Evolution and theory of games. *American Scientist*, 64, 41–45.

Maynard Smith J (1982) *Evolution and the Theory of Games*. Cambridge: Cambridge University Press.

Maynard Smith J & Parker GA (1976) The logic of asymmetric contests. *Animal Behaviour*, 24, 159–175.

Maynard Smith J & Price GR (1973) Logic of animal conflict. *Nature*, 246, 15–18.

Maynard Smith J & Riechert SE (1984) A conflicting tendency model of spider agonistic behaviour: Hybrid–pure population line comparisons. *Animal Behaviour*, 32, 564–578.

McFarland DJ & Sibly R (1975) The behavioural final common path. *Philosophical Transactions of the Royal Society B*, 270, 265–293.

McGregor PK (1993) Signalling in territorial systems: A context for individual identification, ranging and eavesdropping. *Philosophical Transactions of the Royal Society B*, 340, 237–244.

McGregor PK (2005) *Animal Communication Networks*. Cambridge: Cambridge University Press.

Mesterton-Gibbons M, Marden JH & Dugatkin LA (1996) On wars of attrition without assessment. *Journal of Theoretical Biology*, 181, 65–83.

Moore JC, Obbard DJ, Teuter C, *et al.* (2008) Fighting strategies in two species of fig wasp. *Animal Behaviour*, 76, 315–322.

Moya-Laraño J, Orta-Ocana JM, Barrientos JA, *et al.* (2002*a*) Climbing to reach females: Romeo must be small. *Evolution*, 56, 420–425.

Moya-Laraño J, Halaj J & Wise DH (2002*b*) Territoriality in a cannibalistic burrowing wolf spider. *Ecology*, 83, 356–361.

Nossek ME & Rovner JS (1984) Agonistic behaviour in female wolf spiders (Araneae, Lycosidae). *Journal of Arachnology*, 11, 407–422.

Parker GA (1974) Assessment strategy and evolution of fighting behavior. *Journal of Theoretical Biology*, 47, 223–243.

Parker GA & Rubenstein DI (1981) Role assessment, reserve strategy, and acquisition of information in asymmetric animal conflicts. *Animal Behaviour*, 29, 221–240.

Parker GA & Stuart RA (1976) Animal behavior as a strategy optimizer: Evolution of resource assessment strategies and optimal emigration thresholds. *American Naturalist*, 110, 1055–1076.

Payne RJH (1998) Gradually escalating fights and displays: The cumulative assessment model. *Animal Behaviour*, 56, 651–662.

Payne RJH & Pagel M (1996) Escalation and time costs in displays of endurance. *Journal of Theoretical Biology*, 183, 185–193.

Payne RJH & Pagel M (1997) Why do animals repeat displays? *Animal Behaviour*, 54, 109–119.

Petersen A, Nielsen KT, Christensen CB, *et al.* (2010) The advantage of starving: Success in cannibalistic encounters among wolf spiders. *Behavioral Ecology*, 21, 1112–1117.

Prenter J, Elwood RW & Montgomery WI (1994*a*) Male exploitation of female predatory behaviour: Cannibalism reduction in male autumn spiders, *Metellina segmentata*. *Animal Behaviour*, 47, 235–236.

Prenter J, Elwood R & Colgan S (1994*b*) The influence of prey size and female reproductive state on the courtship of the autumn spider, *Metellina segmentata*: A field experiment. *Animal Behaviour*, 47, 449–456.

Prenter J, Montgomery WI & Elwood RW (1994*c*) Patterns of mate guarding in *Metellina segmentata* (Araneae: Metidae). *Bulletin of the British Arachnological Society*, 9, 241–245.

Prenter J, Elwood R & Montgomery WI (1994*d*) Assessments and decisions in *Metellina segmentata* (Araneae: Metidae): Evidence of a pheromone involved in mate guarding. *Behavioral Ecology and Sociobiology*, 35, 39–43.

Prenter J, Montgomery WI & Elwood RW (1995) Multivariate morphometrics and sexual dimorphism in the orb-web spider *Metellina segmentata* (Clerck 1757) (Araneae, Metidae). *Biological Journal of the Linnean Society*, 55, 345–354.

Prenter J, Elwood RW & Montgomery WI (1999) Sexual size dimorphism and reproductive investment by female spiders: A comparative analysis. *Evolution*, 53, 1987–1994.

Prenter J, Montgomery WI & Elwood RW (2003) Mate guarding, competition and variation in size in male orb-web spiders, *Metellina segmentata*: A field experiment. *Animal Behaviour*, 66, 1053–1058.

Prenter J, MacNeil C & Elwood RW (2006*a*) Sexual cannibalism and mate choice. *Animal Behaviour*, 71, 481–490.

Prenter J, Elwood RW & Taylor PW (2006*b*) Self-assessment by males during energetically costly contests over precopula females in amphipods. *Animal Behaviour*, 72, 861–868.

Prestwich KN (1983*a*) Anaerobic metabolism in spiders. *Physiological Zoology*, 56, 112–121.

Prestwich KN (1983*b*) The roles of aerobic and anaerobic metabolism in active spiders. *Physiological Zoology*, 56, 122–132.

Prestwich KN (1988*a*) The constraints on maximal activity in spiders: Evidence against the fluid insufficiency hypothesis. *Journal of Comparative Physiology B*, 158, 437–447.

Prestwich KN (1988*b*) The constraints on maximal activity in spiders: Limitations imposed by phosphogen depletion and anaerobic metabolism. *Journal of Comparative Physiology*, 158, 449–456.

Riechert SE (1978*a*) Energy-based territoriality in populations of the desert spider, *Agelenopsis aperta* (Gertsch). *Symposia of the Zoological Society of London*, 42, 211–222.

Riechert SE (1978*b*) Games spiders play I: Behavioural variability in territorial disputes. *Behavioural and Ecological Sociobiology*, 3, 135–162.

Riechert SE (1979) Games spiders play II: Resource assessment strategies. *Behavioural Ecology and Sociobiology*, 6, 121–128.

Riechert SE (1984) Games spiders play III: Cues underlying context-associated changes in agonistic behaviour. *Animal Behaviour*, 32, 1–15.

Riechert SE (1986) Spider fights: A test of evolutionary game theory. *American Science*, 74, 604–610.

Riechert SE (1988) The energetic costs of fighting. *American Zoologist*, 28, 877–884.

Riechert SE & Maynard Smith J (1989) Genetic analyses of two behavioural traits linked to individual fitness in the desert spider, *Agelenopsis aperta*. *Animal Behaviour*, 37, 624–637.

Rittschof CC (2010) Male density affects large-male advantage in the golden silk spider, *Nephila clavipes*. *Behavioral Ecology*, 21, 979–985.

Robinson MH & Robinson B (1980) Comparative studies of courtship and mating behaviour of tropical araneis spiders. *Pacific Insects Monographs*, 36, 1–218.

Roggenbuck H, Pekár S & Schneider JM (2011) Sexual cannibalism in the European garden spider *Areneus diadematus*: The roles of female hunger and mate size dimorphism. *Animal Behaviour*, 81, 749–755.

Rubenstein DI (1987) Alternative reproductive tactics in the spider *Meta segmentata*. *Behavioural Ecology and Sociobiology*, 20, 229–237.

Schmitz A (2005) Spiders on a treadmill: Influence of running activity on metabolic activity in *Pardosa lugubris* (Araneae, Lycosidae) and *Marpisa mucosa* (Araneae, Salticidae). *Journal of Experimental Biology*, 208, 1401–1411.

Schneider JM & Elgar MA (2001) Sexual cannibalism and sperm competition in the golden orb-web spider *Nephila plumipes* (Araneoidea): Female and male perspectives. *Behavioral Ecology*, 12, 547–552.

Schneider JM & Lubin Y (1996) Infanticidal male eresid spiders. *Nature*, 381, 655–656.

Schneider JM & Lubin Y (1997) Infanticide by males in a spider with suicidal maternal care *Stegodyphus lineatus* (Eresidae). *Animal Behaviour* 54, 305–312.

Schneider JM & Lubin Y (1998) Intersexual conflict in spiders. *Oikos*, 83, 496–506.

Schneider JM, Herberstein ME, de Crespigny FEC, *et al.* (2000) Sperm competition and small size advantage for males of the golden orb-web spider *Nephila edulis*. *Journal of Evolutionary Biology*, 13, 939–946.

Schneider JM, Gilberg S, Fromhage L, *et al.* (2006) Sexual conflict over copulation duration in a cannibalistic spider. *Animal Behaviour*, 71, 781–788.

Schuck-Paim C (2000) Orb-webs as extended-phenotypes: Web design and size assessment in contests between *Nephilengys cruentata* females (Arenae, Tetragnathidae). *Behaviour*, 137, 1331–1347.

Schütz D & Taborsky M (2005) Mate choice and sexual conflict in the size dimorphic water spider *Argyroneta aquatica* (Araneae: Argyronetidae). *Journal of Arachnology*, 33, 767–775.

Sergoli M, Arieli R, Sierwald P, *et al.* (2008) Sexual cannibalism in the brown widow spiders (*Latrodectus geometricus*). *Ethology*, 114, 279–286.

Stoltz JA, Elias DO & Andrade MCB (2009) Male courtship effort determines female response to competing rivals in redback spiders. *Animal Behaviour*, 77, 79–85.

Suter RB & Keiley M (1984) Agonistic interactions between male *Frontinella pyramitela* (Araneae, Linyphiidae). *Behavioural Ecology and Sociobiology*, 15, 1–7.

Taylor PW & Elwood RW (2003) The mismeasure of animal contests. *Animal Behaviour*, 65, 1195–1202.

Taylor PW & Jackson RR (1999) Habitat-adapted communication in *Trite plancipes*, a New Zealand jumping spider (Araneae, Salticidae). *New Zealand Journal of Zoology*, 26, 127–154.

Taylor PW & Jackson RR (2003) Interacting effects of size and prior injury in jumping spider conflicts. *Animal Behaviour*, 65, 787–794.

Taylor PW, Hasson O & Clark DL (2001) Initiation and resolution of jumping spider contests: Roles for size, proximity, and early detection of rivals. *Behavioral Ecology and Sociobiology*, 50, 403–413.

Vollrath F (1980) Male body size and fitness in the web-building spider *Nephila clavipes*. *Zeitschrift für Tierpsychologie*, 53, 61–78.

Vollrath F (1998) Dwarf males. *Trends in Ecology and Evolution*, 13, 159–163.

Vollrath F & Parker GA (1992) Sexual dimorphism and distorted sex ratios in spiders. *Nature*, 360, 156–159.

Wells MS (1988) Effects of body size and resource value on fighting behaviour in a jumping spider. *Animal Behaviour*, 36, 321–326.

Whitehouse MEA (1997) Experience influences male–male contests in the spider *Argyrodes antipodiana* (Theriidae: Araneae). *Animal Behaviour*, 53, 913–923.

Wilder SM & Rypstra AL (2008) Sexual size dimorphism mediates the occurrence of state-dependent sexual cannibalism in a wolf spider. *Animal Behaviour*, 76, 447–454.

Chapter 7

Contest behaviour in butterflies: fighting without weapons

Darrell J. Kemp

7.1 Summary

Since the seminal work of Davies in the late 1970s, territorial male butterfly contests have offered an excellent system for the empirical scrutiny of contest theories, particularly residency-related game-theoretic principles. Because butterflies lack the obvious morphological traits usually associated with animal aggression, their extended and often spectacular aerial duels both defy simple explanation and provide unique empirical opportunities. Residency win rates often approach 100% in this group, which, coupled with the apparently 'weaponless' nature of butterflies and the non-contact nature of their disputes, provided the early impetus for tests of the 'bourgeois' resident-wins model of contest resolution. Subsequent work has emphasised how potential residency-related RHP asymmetries, including those relating to temporally variable biophysical parameters, such as body temperature, may instead contribute to high rates of residents winning. The balance of empirical work in this group, however, suggests that morphological and/or biophysical factors bear little relevance to content settlement. Contest participation is not obviously mediated by energetics, which contrasts markedly and interestingly with the aerial wars of attrition of other insects, such as odonates. More recent approaches to understanding butterfly contest resolution have led to an appreciation of how life history-level factors, such as ageing and changes in residual reproductive value, may influence aggressive motivation and subsequent levels of contest participation. These principles apply generally, thereby placing butterfly contests as a potentially important system for the empirical investigation of the broader life-historical context of animal aggression.

7.2 Introduction

In perhaps few areas of biology are theory and empiricism so closely melded as in animal contest research. As pointed out in Chapter 2, one of the very first applications of game-theoretical modelling in evolution occurred in the context of animal contests. Although a range of animal groups fuelled early tests of theory, arguably the most famous fledgling test of evolutionary game theory relied upon a rather unexpected subject: the small, forest-dwelling speckled wood butterfly (*Pararge aegeria*) (see Foreword). The study hinged around Maynard Smith and Parker's (1976) intriguing suggestion that in the absence of clear cues by which contests may be resolved, contestants could use 'arbitrary' settlement rules. Known as the 'bourgeois' strategy, this behaviour is not unlike the 'toss of a coin' as used by many human societies to quickly and easily reach a decision. Could such a thing really evolve in nature? British biologist Nicholas Davies set out to answer this question using the speckled wood butterfly, a nymphalid species in which males defend sunspot-based mating territories on the forest floor. Using a conceptually elegant experiment (section 7.3), Davies (1978) obtained results in perfect accord with the expectations of the freshly minted theory: specifically, that territory ownership appears to offer an 'uncorrelated' or arbitrary cue by which contests are decided. In fact, Davies concluded that the apparent 'contests' actually amount to little more than a polite conversation between the two butterfly rivals, with things only getting nasty when there is uncertainty about who was the resident to begin with. Davies (1978) had therefore achieved a seemingly perfect marriage of theory and empirical observation, and in doing so, established a strong precedent for the use

Animal Contests, ed. I.C.W. Hardy and M. Briffa. Published by Cambridge University Press.

Hymenopteran contests and agonistic behaviour

Ian C.W. Hardy, Marlène Goubault & Tim P. Batchelor

8.1 Summary

The insect order Hymenoptera is speciose, diverse and common. Many wasps, bees and ants are well known for their ability, and propensity, to engage in agonistic interactions via biting and stinging (chemical injection), and may also interact using chemical deposition and volatile chemical release. Such behaviours are often exhibited during acquisition and defence of resources contested either by conspecifics or by allospecific hymenopterans. Here we examine the types of contests engaged in by social and non-social hymenopterans and highlight links between these and further aspects of evolutionary and applied biology. We first consider factors influencing the outcomes of contests between pairs of females over resources for reproduction. Studies of female–female contests in bethylids and several non-aculeate species of parasitoid wasps, especially scelionids, pteromalids and eupelmids, have addressed fundamental causes of the outcomes of contest interactions and have further linked contest behaviour to strategies of patch exploitation, clutch size and parental care. Further, we review links between the study of contest behaviour in parasitoids and their use as agents of biological pest control, particularly in terms of how contest behaviour may constitute intra-guild predation and influence strategic decisions to deploy single or multiple species of natural enemies. We then consider contests between males for access to mates, especially those engaged in by fig wasps and other wasps. Male–male contests are placed in the context of the evolution of alternative male morphs, mating systems, sex ratios and social behaviours. Finally, we return to female–female contests, this time examining them in the more complex context of the social Hymenoptera, in which both between-individuals (intra-colony) and between-group (inter-colony) contests occur.

8.2 Introduction

The insect order Hymenoptera comprises two sub-orders: the ancestral Symphyta (principally the sawflies) and the wasp-waisted Apocryta (the ants, bees and wasps). We know of rather few studies of contest behaviour or aggressive behaviour among symphytans (e.g. Kudo *et al.* 1992 report aggressive brood-guarding by sawfly mothers), but there is a contrastingly enormous literature on conflicts and agonistic behaviour among apocrytans: this chapter focuses, almost unashamedly, on ants, bees and wasps. We further note that much evidence for hymenopteran contests comes from members of a monophyletic group, the Aculeata, in which many species have ovipositors modified for stinging by injecting venom. There are also species in the remainder of the Apocryta, the paraphyletic Parasitica, that engage in contest behaviour: in some of these species the adults can be heavily armoured and possess weapons, such as powerful mandibles, while in others the adult stages may be relatively weaponless.

Hymenopterans are diverse in what they do, notably showing the full range of stages of social behaviours. All ant species are eusocial (having the three defining characteristics of cooperation of brood care, reproductive division of labour and overlapping generations). Bee and wasp species span the spectrum of sociality, ranging from the reproductively solitary (lacking parental care) via sub-social (exhibiting parental care), communal species (sharing a breeding site but without cooperative brood care), quasi-social (shared breeding sites with cooperative caring)

Animal Contests, ed. I.C.W. Hardy and M. Briffa. Published by Cambridge University Press.

and semi-social (sharing breeding sites and brood care and with division of labour) to full eusociality (Choe & Crespi 1997, p. 3). There is also trophic diversity: all bees are herbivores as are the majority of symphytan species (just one symphytan family, the Orussidae, is parasitic) and ant species may be herbivorous, fungivorous, carnivorous or omnivorous. Among the wasps there are carnivorous and omnivorous eusocial species (such as *Polistes dominulus* paper wasps, section 8.5), herbivorous solitary species (such as pollinating fig wasps, section 8.4, which are mutualistic with the plant, and gall wasps which are not) and carnivorously parasitic solitary species, such as non-pollinating fig wasps (section 8.4) and other parasitoids (organisms that feed as immatures on the body of a 'host' organism, killing it during their development to a free-living adult stage, section 8.3).

Hymenopterans, particularly Apocrytans, have proven to be extraordinarily stimulating organisms for evolutionary biologists studying interactions at the behavioural, population and community levels (Godfray 1994, Hawkins 1994, Bourke & Franks 1995, Hassell 2000, Hochberg & Ives 2000, Cook 2005, Wajnberg *et al.* 2008, Hölldobler & Wilson 2009). Several of the key insights that heralded the rapid advance of behavioural ecology research that found its feet approximately half a century ago were distinctly hymenopteran-influenced in their initial formulations (especially Hamilton's models of social behaviour, 1964*a*,*b*, and sex ratios under local mate competition, 1967) and the Hymenoptera remain close to the centre of current, and intense, theoretical debate (Abbot *et al.* 2011). Meanwhile other branches of behavioural ecology were initially developed without particular reference to the Hymenoptera: these include the early game-theoretic models of dyadic contests (Chapter 2, Box 8.1, and see the Foreword) and the development of theories of clutch size and optimal foraging (Lack 1947, Stephens & Krebs 1986, Godfray *et al.* 1991). Nonetheless, hymenopterans have since been applied to testing and further developing these theories due in part to many species, such as parasitoid wasps, being well suited for use in laboratory empirical studies (Godfray 1994).

In this chapter we have twin aims: one is to provide an overview of how hymenopteran behaviour fits with expectations derived from game-theoretic models of contests, and the other is to move beyond this core to consider the ramifications of hymenopteran contests for other areas of basic and applied evolutionary ecology. We begin with dyadic interactions, treating female–female and male–male contests separately, and end with group-level interactions and further considerations of agonistic behaviour in social hymenopterans (although we do not pretend to cover all aspects of social hymenopteran aggressive behaviour comprehensively). In between, we show that exploring hymenopteran contests can aid the understanding of mating systems, foraging strategies, clutch size evolution, chemical interactions, social behaviour, population and community ecology and biological pest control.

Box 8.1 Factors predicted to influence outcomes of dyadic contests

A detailed review of theory for dyadic contests is provided in Chapter 2 and Chapter 3 expands this to group-level contest situations. Modelling predicts that two major categories of factors will influence the outcome of dyadic contests, those associated with the difference in the abilities of the contestants to acquire and retain resources and those associated with the value of the contested resource. The ability to acquire and retain resources is termed 'resource-holding potential', RHP (Parker 1974, Maynard Smith & Parker 1976, Hammerstein 1981, Chapter 2). An individual's RHP can be subdivided into a 'resource-uncorrelated' component derived from its intrinsic contest ability (e.g. fighting ability) and a 'resource-correlated' component related to whether it is the owner of the resource or a non-owner ('floater' or 'intruder'). Note that there is some variance in the literature in the usage of the terms 'correlated' and 'uncorrelated': for instance, Chapter 2 defines 'uncorrelated' as a lack of relationship with the costs of fighting, C, or resource value, V. In this chapter we use these terms in relation to intrinsic and extrinsic components of an individual's competitive ability. In the absence of further differences, the individual with the largest total RHP would be expected to win a dyadic contest. In the absence of RHP asymmetries, contests are predicted to be won by the individual that most values the resource, V (Parker 1974, Maynard Smith & Parker 1976, Enquist & Leimar 1987, Chapter 2). Asymmetries in V and RHP may also influence contest outcome concurrently.

8.3 Dyadic female–female contests

Adult female hymenopterans may enter into agonistic interactions with other females, often those belonging to the same species and competing

Box 8.2 Factors favouring the evolution of guarding and contest behaviours in female parasitoids

Although host (or patch) defence and brood-guarding are highly costly in terms of energy, time and, in some cases, risk of injury or death, many factors may have favoured the evolution of aggressive and defensive behaviours in parasitoids. The most obvious candidate explanation is a scarcity of suitable hosts, meaning that hosts will have high *V* [Chapter 2, and see Rosenheim, 2011, for an entry into literature on the relative limitations of host availably (time limitation) and reproductive ability (egg limitation)]. Data on host distribution and availability under natural conditions are usually lacking (there are a few species where partial information is available, e.g. de Jong *et al.* 2011). A rather different explanation applies to brood guarding in *Melittobia*: availability of mates. When virgin, *Melittobia* females feed on a host and then produce and guard small broods of male offspring. The mother mates with her sons on their maturity and subsequently produces larger broods, containing male and female offspring, on the same host (Matthews *et al.* 2009).

One condition for the evolution of guarding is that the resource should be defendable (van Alphen & Visser 1990; see also Moore & Greeff 2003). This is generally the case when females exploit just one host at a time, as in *G. nephantidis* (Hardy & Blackburn 1991), but when they attack clusters of hosts, patch defence behaviours are likely to be limited to species where the host masses are of a defendable size (Waage 1982, de Jong *et al.* 2011). In both cases the benefits of patch defence are strongly related to the differences in survival of offspring with and without the mother present. In the egg parasitoid *Trissolcus basalis*, the offspring of the second (superparasitising) female to exploit a patch of hosts are more likely to develop successfully than those of the initial female (Field *et al.* 1997). The clutch of the first female can even be completely destroyed by a second conspecific or allospecific female in species where females are able to commit infanticide (ovicide or larvicide; Hardy & Blackburn 1991, Pérez-Lachaud *et al.* 2002, Goubault *et al.* 2007*a*,*b*, Takai *et al.* 2008), although not all species that commit infanticide exhibit post-oviposition brood-guarding (Mayhew 1997). Females may also brood-guard to defend against potential hyperparasitism or predation, although we know of no conclusive evidence for this in parasitoids. Once broods are no longer vulnerable to the actions of intruders, guarding females may leave the host or cease to defend it as effectively as during the more vulnerable stages of brood development (Bentley *et al.* 2009).

The cost of producing eggs has also been suggested as a potential driver for the evolution of brood-guarding behaviours (Takai *et al.* 2008, Nakamatsu *et al.* 2009). Indeed, offspring defence behaviours have mainly been observed in species where females produce small numbers of large, yolky anhydropic eggs which require an exogenous source of proteins generally obtained through host-feeding (Jervis & Kidd 1986). Moreover, these types of eggs are usually produced slowly and their deposition on or in the host can take a considerable time (Harvey 2008), leading to hosts and broods having high *V*.

for the same resources. The resources females compete for are usually those connected with offspring production, such as nesting sites or food sources for offspring. As possession of such resources is usually highly influential on a female's lifetime fitness, it is expected that contest behaviours will often be selected for, even though they are likely to be costly, possibly injurious and even fatal (Chapter 2, Box 8.2). Here we consider dyadic interactions between conspecific females belonging to solitary and 'primitively' social species. These interactions in general match the situations considered by classical models of animal contests (Chapter 2, Box 8.1). Such dyadic interactions are found in a range of hymenopteran species. For instance, females of the solitary burrowing bee *Amegilla dawsoni* compete for nest burrows via grasping with mandibles and tugs-of-war: in this species the prior owner of the burrow usually wins rather than the larger individual, although body size may play a role in the earlier acquisition of ownership status (Alcock *et al.* 2006). Similar residency or positional effects are observed in other solitary nesting hymenopterans (Tepedino & Torchio 1994, Villalobos & Shelly 1996), while in others size differences between contestants determine outcomes, with larger females tending to win (Mueller *et al.* 1992), or advantages to both larger size and prior residency operate (Ghazoul 2001). Size-dependent fighting is also observed in pollinating fig wasps, the winner female lifting the loser off the substrate and thus preventing the loser from laying eggs (Moore & Greeff 2003). For the remainder of this section we focus on evidence deriving from studies on parasitoid wasps, as these provide some of the most explicit empirical assessments of contest theory (dyadic interactions in social Hymenoptera are discussed in section 8.5).

While there is an enormous literature on the behaviour, ecology and evolution of the parasitoid hymenoptera, covering virtually every aspect of their life histories (e.g. development, mating, foraging, clutch size, sex ratio, sex determination, host–parasitoid co-evolution, population dynamics and community ecology; Godfray 1994, Hochberg & Ives 2000, Jervis 2005, Wajnberg *et al.* 2008), with the majority of these studies focussing on decisions made by females rather than males, there have been comparatively few studies of dyadic contest behaviour in female parasitoids. Although a host is a reproductively crucial resource for a female parasitoid, in most species females depart from hosts once they have completed egg-laying (oviposition) and search for further oviposition opportunities elsewhere. Adult female parasitoids may thus rarely encounter each other at the resource itself and even if they do, they often do not engage in mutually agonistic behaviour (or it may instead be that relatively few scientists have thought to examine these aspects of behaviour). Inter-female competition can occur indirectly, for instance, via infanticide (killing offspring present when finding already parasitised hosts), superparasitism (conspecifics adding eggs to a parasitised host) or multiparasitism (allospecific-superparasitism). Following parasitism, superparasitism or multiparasitism, competition can occur between progeny developing from the resources of the same host, either via larvae fighting, physiological suppression, or a scramble for resources. While there is a rich theoretical and empirical literature on within-host competition between immature parasitoids (e.g. Godfray 1994, Ode & Rosenheim 1998, Harvey *et al.* 2000, Pexton *et al.* 2003, Gardner *et al.* 2007, Tena *et al.* 2008, 2009, Kapranas *et al.* 2011), we restrict our present discussion to competitive interactions between adults.

There are some parasitoid taxa in which direct, and agonistic, encounters between competing adult females appear to occur regularly (Wilson 1961, Dix & Franklin 1974, Waage 1982, J. Field 1992, Hughes *et al.* 1994, Field *et al.* 1997, 1998, S.A. Field 1998, Field & Calbert 1998, 1999, Ohno 1999, Quicke *et al.* 2005, Fernández-Arhex & Corley 2010, Mohamad *et al.* 2010, 2011, 2012, de Jong *et al.* 2011, Wyckhuys *et al.* 2011, Table 8.1). In some species contests are non-injurious (Batchelor *et al.* 2005, Goubault *et al.* 2005), but in others contests may lead to visible injury and death (Lawrence 1981, Humphries *et al.* 2006, Matthews & Deyrup 2007). Contests are promoted when foraging females remain for considerable periods with the individual hosts, or within the patches of hosts, that they find (Box 8.2). Perhaps the best examples are provided by species in the family Scelionidae and in the aculeate family Bethylidae. Contests in scelionids, which aggressively guard patches of hosts representing a potentially divisible resource, have largely been studied from the perspective of understanding patch defence strategies (section 8.3.2) while studies of bethylids have explored determinants of contests for individual hosts (section 8.3.1), which may be regarded as an essentially indivisible resource [although there are bethylid species, such as members of the genus *Scleroderma* (Chen & Cheng 2000), in which small groups of females cooperatively attack and reproduce on a single, and comparatively large, host; Baoping Li & ICWH pers. obs.].

8.3.1 Contests for hosts: influences of RHP and *V*

Bethylid females may spend several days with a paralysed host after stinging it and before laying eggs. Many species also remain with the host, and their developing broods, for days after oviposition, guarding them against the detrimental actions (superparasitism and infanticide) of other parasitoids (Goertzen & Doutt 1975, Hardy & Blackburn 1991, Takasu & Overholt 1998, Bentley *et al.* 2009, Venkatesan *et al.* 2009*a*), i.e. sub-social behaviour (section 8.2). Encounters between females during these host- and brood-guarding periods constitute classic owner–intruder contests, involving direct agonistic interactions. Agonistic behaviours include biting, using relatively large mandibles and stinging (Batchelor *et al.* 2005, Goubault *et al.* 2006, 2008, Lizé *et al.* 2012, Figure 8.1) as well as the release of volatile chemicals (Box 8.3, Chapter 4). While intraspecific contests are very rarely fatal (Humphries *et al.* 2006) and usually do not lead to visible injuries, they are usually decisive, with outcomes readily identifiable in terms of the identities of the winner and the loser of a behavioural bout and the ultimate possession, and utilisation, of the host resource.

A series of laboratory studies of factors influencing the outcomes of female–female contests in bethylids in the genus *Goniozus* has shown that asymmetries in both resource holding potential and resource value (respectively, RHP and *V*, Box 8.1), determine outcomes and that both theoretical categories are made up of several empirically identifiable components (Table 8.1, Figure 8.2a). For instance, there are at least

Table 8.1 Summary of evidence for influences on the outcomes of, and agonistic interactions within, dyadic contests between *Goniozus* females. Evidence is collated across laboratory studies on *G. nephantidis*[1–4,6,7,11] and *G. legneri*[5,8–10,12] but in all cases refers to intraspecific interactions (interspecific interactions are discussed by Hardy & Blackburn 1991, Venkatesan *et al.* 2009*b*). In nature, *Goniozus* contests are most likely to be owner–intruder (O-I) interactions but some studies have examined owner–owner (O-O) or intruder–intruder (I-I) interactions to explore effects in the absence of prior ownership asymmetries. Most staged contests result in a clear outcome but some remain unresolved during experimental observation: while asymmetries in contestant size, egg load or host size have not been found to affect the probability of contest resolution,[4,7,8,10] resolution is less common when neither contestant is a prior owner[10] and before prior owners' broods have developed into larvae.[8] In addition to the biological properties of theoretical interest, experiments have generated some evidence of side-effects of marking on contest behaviour. In most studies there was no effect of paint colour (red/yellow)[1,2,4–8,10] used to mark contestants, but in one red-marked contestants appeared advantaged.[9] Deuterium tagging of volatile biochemicals (Box 8.3) showed disadvantageous side-effects in *G. legneri*[5] but not in *G. nephantidis*:[7] it may be that deuteration affects metabolic performance (lower RHP).

Category of theoretical influence (Box 8.1)	Empirical aspect of category	Influence on outcome of resolved contests	Influence on agonistic behaviour during contests
Resource holding potential (RHP)			
Uncorrelated RHP			
	Body size	Advantage to larger contestants reported from O-I and O-O contests[2,4–6,8,11] but non-significant effects in O-I[3,7] and O-O contests[3,7] also reported *Interpretation*: Larger individuals are generally superior competitors (higher RHP). Effects sometimes not found because size asymmetry experimentally minimised[3]	Aggression during contests may be unaffected by O-I size asymmetries[2,8] but positive association between size difference and aggression found when owner wins[5] and in O-O contests.[2] Size difference can also influence effects of other correlates of aggression.[10] Larger contestants tend to initiate aggression and larger initiators are more successful than smaller initiators[2]
Correlated RHP			
	Host ownership	Owners advantaged in O-I contests[2–4,6–8,11] but *dis*advantage also reported.[5,9] *Interpretation*: Mechanistic advantages, prior experience and evolved conventions excluded or unlikely: Ownership effect may not relate to RHP as likely confounds with *V* (egg load[3])	More aggression in O-I vs. I-I contests[10] and in O-O vs. O-I contests[2] and in O-O contests with longer ownership period.[4] Owners usually chase intruders and initiate fights[2] but those intruders that initiate more agonistic behaviours than owners tend to win.[7] Effect of size difference on aggression and probability of volatile release by loser (Box 8.2) affected by prior ownership[5]
Resource Value (*V*)			
	Contestant age	Intruder success increases with age in O-I contests.[4] *Interpretation*: Older intruders place higher value (*V*) on opportunity to obtain host	More O-O aggression when owners are older[12]
	Egg load	Contestants with more mature eggs advantaged when other asymmetries minimised (O-O).[3] *Interpretation*: Higher egg loads enable faster exploitation of host resource (higher *V*). Egg load usually confounds with other aspects of *V* (e.g. host ownership) and RHP (body size)[3,4,6–8]	

(*cont.*)

Table 8.1 *(cont.)*

Category of theoretical influence (Box 8.1)	Empirical aspect of category	Influence on outcome of resolved contests	Influence on agonistic behaviour during contests
	Host size	Owner of larger host advantaged in O-O[4,6,11] but not O-I[3,4,6,8] contests. *Interpretation*: Larger hosts have higher *V* and owners have more opportunity to assess host size than intruders	Owners more likely to defend larger hosts but intruder aggression unrelated to host weight.[8] More O-O aggression when hosts are larger[12]
	Host age (post-paralysis)	Owner of younger host advantaged in O-O contests for unparasitised but paralysed hosts.[11] *Interpretation*: Metabolic changes in paralysed hosts progressively decrease nutritional quality (*V*)	
	Clutch and brood size	No effect of number of offspring developing on host in O-I[6] or O-O[4,6] contests. *Interpretation*: These expected components of owners' *V* confound with influential components of *V* (e.g. host size)	
	Brood stage	In O-I contests, owners of hosts with eggs advantaged compared to other stages in *G. nephantidis*[1,6] but not *G. legneri*.[8] *Interpretation*: Developing brood has increasing reproductive value (*V*) to owner. Owner's larvae feeding decreases host value (*V*) to intruder	Owners more likely to aggressively defend pre-larval brood hosts.[8] (Intruders progressively less likely to utilise diminishing resource if acquired[1,8])

References: [1]Hardy & Blackburn (1991), [2]Petersen & Hardy (1996), [3]Stokkebo & Hardy (2000), [4]Humphries *et al.* (2006), [5]Goubault *et al.* (2006), [6]Goubault *et al.* (2007*b*), [7]Goubault *et al.* (2008), [8]Bentley *et al.* (2009), [9]Appendix to Bentley *et al.* (2009), [10]Lizé *et al.* (2012), [11]I.C.W. Hardy, S.K. Khidr, C. Hepworth-Bell & C. Daykin (unpublished), [12]B. Stockermans & I.C.W. Hardy (unpublished).

five determinants of the value, *V*, that females place on possession of the resource, comprising aspects that are both extrinsic (host size, host age, stage of host exploitation) and intrinsic (contestant age, egg load) to the contestants themselves. The primary component of uncorrelated RHP is contestant size, with effects on contest outcome found in almost every study: indeed it is difficult to remove this RHP asymmetry in contest experiments as females must usually be very closely size-matched prior to contests to avoid finding significant effects of weight differences (e.g. Lizé *et al.* 2012).

Although prior ownership of a host clearly influences the expression of agonistic behaviour and prior owners are advantaged in terms of outcome (Table 8.1), evidence for host ownership constituting a resource-correlated component of RHP is not clear-cut: correlated RHPs can potentially operate via mechanistic advantages to ownership, such as positional effects (e.g. Tepedino & Torchio 1994), familiarity with the local environment or enhanced physical performance (Stutt & Willmer 1998, Chapter 7), but these possibilities are all considered unlikely in *Goniozus* (Stokkebo & Hardy 2000). A further potential explanation for owners tending to win is the historical accumulation of individuals with high uncorrelated RHP (e.g. Mueller *et al.* 1992) as resource owners prior to observed contests: this can be discounted as all studies reported in Table 8.1 used females with no previous contest experience. Another possibility is that host ownership is used as an arbitrary cue to contest resolution (Maynard Smith & Parker 1976, Chapter 2). Such evolved conventions can potentially favour owners (termed 'commonsense') or

Figure 8.1 Contest interaction between two female bethylid wasps. Initial stage of an owner–intruder contest between two *Goniozus legneri* females. The owner (wasp on right) has already laid a clutch and the eggs have hatched into first instar larvae and started to feed on the paralysed host caterpillar. Note the intruder's flared mandibles. (Photograph: Sonia Dourlot.) For colour version, see colour plate section.

intruders (termed 'paradoxical'), although paradoxical outcomes are expected to evolve less frequently than commonsense outcomes (Mesterton-Gibbons 1992, Kokko *et al.* 2006). While prior ownership effects, where present, conform to commonsense expectation in *G. nephantidis*, there have been two reports of owners being disadvantaged in *G. legneri* (Table 8.1). As there was no conclusive empirical evidence for paradoxical contest outcomes, unconfounded with asymmetries in uncorrelated RHP or *V*, in any other taxon, *G. legneri* contests were re-examined: owners were found to be advantaged (Bentley *et al.* 2009). It is possible that the apparently paradoxical outcomes reported by Goubault *et al.* (2006) were due to intruders experiencing a harsher (drier) environment than owners (which were confined with a moisture-transpiring host) and thus, having shorter life expectancies, placed a higher value, *V*, on possession of the host (Bentley *et al.* 2009). A further possible explanation for observed (commonsense) effects of host ownership on contest outcomes is that they are due to ownership-enhanced components of *V*, such as host-owners having more mature eggs ready to lay on the host which makes it a more valuable resource to them than to intruders (Stokkebo & Hardy 2000, Table 8.1). These latter two possibilities illustrate that empirical results appearing to fit with one

Box 8.3 Chemical applications and emissions during contests

Although the chemical composition of the hymenopteran exoskeleton is highly variable and can play a major role in mediating behavioural interactions (e.g. Gamboa *et al.* 1986, Akino *et al.* 2004, Nehring *et al.* 2010), there have been few studies of short-term changes in the chemical environment during hymenopteran interactions, possibly because chemicals may not be detectible by researchers without specialist analytical equipment. Nonetheless, chemical changes have been observed to be involved in contest and agonistic interactions in both parasitoids (Quicke *et al.* 2005; see also Gomez *et al.* 2005) and social insects (Palmer *et al.* 2000, Monnin *et al.* 2002, Wenseleers *et al.* 2002, Tarpy & Fletcher 2003, Palmer 2004, Smith *et al.* 2012).

In queenless ponerine ants the alpha (reproductive) female marks challenging females with a chemical, wiped from the stinger, and the marked individual is then punished by low-ranking females which immobilise the high-ranking challenger for up to several days, resulting in a loss of high rank and, occasionally, death (Monnin *et al.* 2002). Similarly, in the ant *Apheanogaster cockerelli*, which has a queen caste, queens attack and mark competing reproductive workers which are then attacked by nest-mate workers; sometimes the queen is also attacked and killed, possibly due to contamination from the applied chemical mark (Smith *et al.* 2012).

Using real-time mass spectrometry, Goubault *et al.* (2006, 2008) showed that a volatile spiroacetal can be actively released by female parasitoids in the genus *Goniozus*, almost certainly from their mandibles, during agonistic interactions. Chemical release is more common during more behaviourally aggressive contests. Aggression levels are greatest when prior owners successfully resist takeover by similar-sized intruders. A chemical marking technique (manipulation of molecular mass achieved by rearing females on deuterium-injected hosts) showed that volatiles released during contests are always emitted by the loser. Deuterium-marked contestants had reduced contest ability in *G. legneri* but not in *G. nephantidis*. Although marked females may be biochemically impaired (reducing their RHP) there is no effect of deuteration on a range of other *G. legneri* life-history components (Goubault & Hardy 2007). Aggression in *Goniozus* contests is reduced after spiroacetal release. Chemical release may act as a weapon of rearguard action, creating an opportunity for the releaser to retreat from a contest, rather than as a signal of submission. Distinguishing between these candidate functions is, however, challenging (Chapter 4) and, while the volatile does have insecticidal properties, no negative effects of exposure have been observed in bethylids.

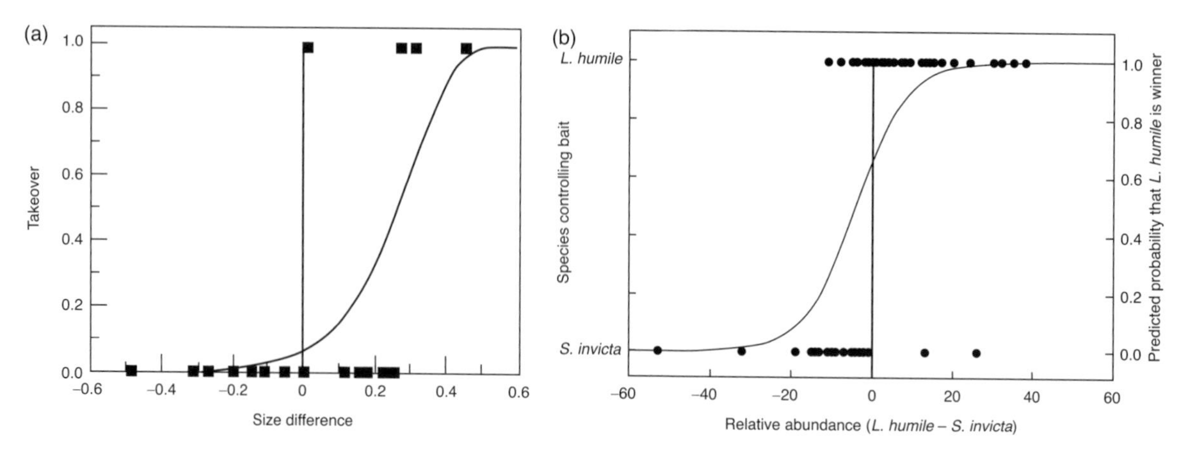

Figure 8.2 Effects of difference in individual and group size difference on dyadic and group-level contests. Panel (**a**) shows how the probability of takeover of a guarded host by an intruder female of the parasitoid wasp *Goniozus nephantidis* increases as body-size asymmetries progressively favour intruders (size difference = initial owner weight – intruder weight, mg). Heavier females have larger RHP but there is also an advantage to prior ownership as the inflection point (where takeover probability = 0.5) lies to the right of zero size difference: there are several potential explanations for this ownership effect, owner–intruder resource value asymmetries, *V*, being the most likely (section 8.3.1, Table 8.1). Panel (**b**) shows how the probability of groups of worker Argentine ants, *Linepithema humile*, controlling an experimentally provided food resource increases as numerical asymmetries progressively favour *L. humile* over the group size of its competitor, the fire ant *Solenopsis invicta*. Group size is analogous to individual RHP and in this example there is no indication that individuals of one species are competitively superior to the other as the inflection point lies close to zero. In both panels the probabilities of winning were estimated using logistic regression (Chapter 4). (*Sources*: Panel a, Petersen & Hardy 1996; Panel b, LeBrun *et al.* 2007, reproduced with permission.)

category of theoretical influence (asymmetry in RHP) may actually be due to the influence of asymmetries in another (V).

There is also evidence for the operation of asymmetries in RHP and V during intraspecific female–female contests for hosts in a range of other parasitoid species, although to date these have been explored less comprehensively than *Goniozus*. Classically, larger females are advantaged during contests (uncorrelated RHP) (Lawrence 1981, Nakamatsu *et al.* 2009, Fernández-Arhex & Corley 2010) and escalation is more common when opponents are of similar size (Nakamatsu *et al.* 2009). Small females that lose contests may, however, adopt an alternative reproductive strategy: in the hyperparasitoid *Trichomalopsis apanteloctena*, small females chew through the outer silk protection that covers the host's cluster (cocoons of *Cotesia kariyai*) and hide from larger, guarding, females in this 'refuge' (Nakamatsu *et al.* 2009). Resident females are also more likely to retain host access against intruders and the further they get in their behavioural oviposition sequence, the greater their probability of winning is (Goubault *et al.* 2007*a*, Mohamad *et al.* 2012). This can relate to an asymmetry in correlated RHP (Hughes *et al.* 1994) but also to an asymmetry in V when being the owner means having laid more eggs on a host's cluster and therefore defending more offspring (Field & Calbert 1998, Takai *et al.* 2008).

Other factors affecting the real value of the contested resource can also influence contest resolution, such as brood age (Takai *et al.* 2008), the size of a host patch (S.A. Field 1998) and more specifically the number of unparasitised hosts among a patch (Field & Calbert 1998), and see Ghazoul (2001) for how the difficulty of acquiring a nesting site affects aggression in a digger wasp. Finally, the physiological state of females and their past experience of habitat quality can modify their subjective value of the host and, as a result, their probability of winning contests: females with a larger number of mature (ready-to-lay) eggs in their ovaries have more to gain from access to hosts and are therefore more likely to win against contestants with fewer mature eggs. In *Eupelmus vuilleti*, such advantages can be generated by a single additional mature egg (Mohamad *et al.* 2010). The previous experience of a habitat poor in hosts can also increase the value that females place on the hosts: females having been deprived of hosts (and thus valuing each host more highly, V) tend to win contests (Mohamad *et al.* 2010). However, the opposite effect has also been observed in *Pachycrepoideus vindemmiae* (Goubault *et al.* 2007*a*). In this species, females are able to feed from the host (Phillips 1993). Host feeding usually provides females with considerable amounts of carbohydrate and protein (Giron *et al.* 2002) and the resulting enhanced metabolic state can possibly be used for host acquisition and defence. In host-feeding species, prior experience of a host-rich habitat can increase females' success in contests via enhanced fighting ability (RHP) (Goubault *et al.* 2007*a*).

8.3.1.1 Successive contests for hosts: influences on subsequent RHP and *V*

The majority of the above studies have examined contests between females that have not experienced contest interactions before. In other animal taxa there is increasing evidence that prior experience of winning or losing a contest affects the future contest performance of individuals: here we summarise this literature and then outline new evidence for such 'winner and loser effects' in parasitoid hymenopterans.

The experience of a prior conflict influences an animal's subsequent performance: a victory usually increases the probability of winning a subsequent conflict while a defeat decreases it (Chapters 3 and 4). Such winner and loser effects are widespread throughout the animal kingdom (Hsu *et al.* 2006, Rutte *et al.* 2006) and strongly impact hierarchy formation in socially living species (Dugatkin 1997, Dugatkin & Earley 2003, 2004, Dugatkin & Druen 2004). They are hypothesised to result from prior victory or defeat affecting contestants' estimation of their own fighting abilities, i.e. their RHPs (Hsu & Wolf 1999, Mesterton-Gibbons 1999, Fawcett & Johnstone 2010) and therefore the expected cost of engaging in a following fight (Hsu *et al.* 2006). Game-theoretic models (Mesterton-Gibbons 1999, Fawcett & Johnstone 2010) based on this assumption predict that, at least in young and naïve individuals, a loser effect can exist alone or in the presence of a winner effect, but a winner effect cannot exist alone. Moreover, when both effects coexist, the loser effect is expected to be of a greater magnitude and last longer than the winner effect. These predictions have been supported by empirical evidence from taxa other than parasitoids (Hsu *et al.* 2006, Earley & Dugatkin 2010).

There is preliminary evidence for a loser effect, but no winner effect, in the bethylid *G. nephantidis* (M.G. unpublished) and this would accord with evidence from other taxa. However, a recent exploration of possible winner and loser effects in *E. vuilleti* found a winner effect that can persist in the absence of any evident loser effects (Goubault & Decuignière 2012). This surprising result could be explained by an increased subjective host value, *V*, to winner females only. In this study system, winning the resource means parasitising the host and oviposition is known to induce an almost immediate increase in ovarian hormone (ecdysone) titres in *E. vuilleti* females (Casas *et al.* 2009). This then leads to the stimulation of oogenesis and the maturation of new eggs within 24 h (Casas *et al.* 2009). Such a rapid increase in mature eggs, only occurring in winner females, may increase their estimation of the host value, *V*, and thus their success in contests. Clearly there is a need for further empirical studies of how estimates of RHP and *V* are affected by parasitoid contest outcomes, and their subsequent influences on further contests, and there is probably also a need for formal theory of *V*-based winner–loser effects to be developed.

8.3.2 Contests for patches of hosts

The question of how an individual parasitoid should forage optimally has been well studied, both theoretically and empirically, since the development of the marginal value theorem (MVT: Charnov 1976), a static optimality approach. However, it is not uncommon that several females exploit a patch simultaneously (Godfray 1994). This may greatly affect individuals' patch exploitation strategies as, in the presence of competitors, the optimal decisions of one forager are likely to depend on those taken by the others, a game-theoretic situation (Maynard Smith 1974). When competitors do not interfere with each other, MVT can still be applied and foragers are predicted to leave simultaneously when the gain rate within the patch falls below the mean gain rate expected in the habitat (Sjerps & Haccou 1994). Because the patch depletion rate increases with the number of foragers, the more females there are foraging in a patch, the sooner the patch-leaving gain rate should be reached and the sooner the females should disperse.

Interference between females within a patch is, however, likely to occur via several mechanisms. First, superparasitism becomes advantageous when intraspecific competition increases (van Alphen & Visser 1990, Visser *et al.* 1990, 1992*a*): females should accept previously parasitised hosts, which reduces the gain rate of the first female to oviposit in the host. Second, females can interact directly with their competitors via agonistic behaviours (e.g. Field *et al.* 1997, Goubault *et al.* 2005, Fernandez-Arhex & Corley 2010). These behavioural interactions are time-consuming and tend to increase patch residence time

(Hassell & Varley 1969). Additionally, the departure of one female from the patch leads to an abrupt increase of the gain rate of the remaining foragers. Theory therefore predicts that foragers should engage in a war of attrition (Chapter 2), with some individuals leaving the patch early (when their gain rate reaches a given threshold) and other females staying longer compared to when foraging alone on a patch (Sjerps & Haccou 1994). Such a war of attrition has been observed in several parasitoid species (Visser *et al.* 1990, Pijls *et al.* 1996). In *Trissolcus basalis* females enter into a 'waiting game' for several hours where competing females stay motionless (Field *et al.* 1998, Wajnberg *et al.* 2004). This behaviour may bias the opponent's perception of the presence of a competitor and incite her to leave first (Field *et al.* 1998). Manipulating the arrival times of two females exploiting the same patch leads to an asymmetric war of attrition: as a result of an asymmetry in *V*, the first female to arrive tends to stay longer in the patch than the second female (Haccou *et al.* 2003, Le Lann *et al.* 2011). Nevertheless, there are counterexamples where the presence of competitors within patches has not been found to increase patch residence times (Goubault *et al.* 2005, Fernández-Arhex & Corley 2010). In *Pachycrepoideus vindemmiae*, a considerable proportion of females retreat before patch depletion has started, and this proportion increases with the number of conspecifics present within the patch (Goubault *et al.* 2005). Similarly, females of *Asobara citri* leave host patches shortly after engaging in a fight with a resident female, leading to the rapid establishment of a regular distribution of foragers in multipatch experiments (de Jong *et al.* 2011). The availability of other patches in nature or the potential cost of fighting might have influenced the evolution of such behaviours (Box 8.2).

Factors that classically affect optimal patch use in solitary foragers, such as habitat experience and physiological state, are also likely to influence foraging strategies under competition. For instance, experience of host scarcity or long inter-patch distance should increase the duration of the war of attrition as it increases the value of the patch, *V*. Previous experience of competition with conspecifics can also affect patch use of females even when exploiting a patch alone: these females tend to behave as if they were still in a competitive situation (Visser *et al.* 1992*b*, Goubault *et al.* 2005). Note that prior experience of contest situations can influence other aspects of oviposition decisions, and that game-theoretic considerations can apply to females even when physically separate from other decision-making competitors (Box 8.4). Female parasitoids also seem to modify their patch use strategies according to their physiological state (e.g. age, fecundity): as they become older, female *P. vindemmiae* become more aggressive and stay longer on patches in which conspecifics are present (Goubault *et al.* 2005). Finally, foragers' susceptibility to interference can be under significant genetic variation (Wajnberg *et al.* 2004). Contest and agonistic interactions can have important consequences in terms of parasitoid distribution and population dynamics.

8.3.3 Effects of female–female contests on population and community ecology

Competitive encounters between adult female parasitoids are more likely at higher parasitoid population densities. Females may respond to such encounters with conspecifics in a number of, often interrelated, ways including superparasitism (section 8.3.2), laying fewer eggs on hosts in competitive than in non-competitive situations (Box 8.4), adjusting offspring sex ratios (Waage 1982, Ode & Hardy 2008), spending time guarding hosts against possible intrusion (Box 8.2), engaging in a war of attrition (section 8.3.2) and allocating time and energy to direct agonistic interactions rather than to oviposition (e.g. Irvin & Hoddle 2005, Quicke *et al.* 2005, Goubault *et al.* 2007*a*, Nakamatsu *et al.* 2009, Mohamad *et al.* 2010). The density-dependence of such 'interference' is expected to contribute to host–parasitoid population dynamic stability, but extreme effects of interference can also lead to predictions that host populations will not be regulated by parasitism (Hassell 2000).

Despite these possible negative impacts on host parasitism efficiency, recent work has shown that female–female interactions can reduce patch residence times, at least for loser wasps (Wajnberg *et al.* 2004, Goubault *et al.* 2005, Fernández-Arhex & Corley 2010, de Jong *et al.* 2011). The direct consequence of such a dispersal effect would be a rapid distribution of parasitoids over the different patches available in the local environment. For instance, de Jong *et al.* (2009, 2011) compared female distribution of two congeneric *Drosophila* parasitoid species, *Asobara citri* and *A. tabida*, in relation to the type of female–female interactions exhibited. In *A. citri*, where females fight

Box 8.4 Effects of female–female contests on parasitoid clutch size decisions

An advantage associated with being the larger of two competing parasitoids (asymmetry in uncorrelated RHP) constitutes a component of the relationship between body size and fitness. The size–fitness relationship lies at the heart of much theory relating to parasitoid behavioural ecology, including sex ratio theory and optimal clutch size theory (Godfray 1994).

In the bethylid *G. nephantidis*, laying more eggs on a given sized host leads to a greater number of adult offspring but individual offspring are smaller. Hardy *et al.* (1992) used this species to test clutch size theory and found a mismatch between observed (about 9 eggs) and theoretically calculated (about 18 eggs) clutch sizes. These optimality calculations, however, used static estimates of fitness components (adult longevity and fecundity) associated with particular body sizes. The subsequent finding that contest outcomes were also size-dependent (Table 8.1) indicated that a game-theoretic approach is required to predict clutch size optima because success in contests (fitness) depends not just on the size of a female per se but on the size difference between a female and its opponent. Thus mothers laying eggs should take into account the clutch size decisions made by other females in the population, even though they are not laying eggs on the same hosts, because these influence the sizes of competitors that their own offspring will compete against when adult. Optimal clutch sizes were argued to be lower than when calculated assuming a static size–fitness relationship because mothers would be selected to produce smaller numbers of larger offspring (Petersen & Hardy 1996). Formal game-theoretic modelling (Mesterton-Gibbons & Hardy 2004) showed that clutch size optima should be more greatly decreased when clutch size has a stronger effect on offspring size, when contests occur more commonly, and when prior ownership (correlated RHP) confers a relatively small advantage compared to body-size asymmetry (uncorrelated RHP).

The clutch size response of *G. nephantidis* to contests was tested empirically by exposing females, just prior to egg-laying, to sequential encounters with varying numbers of conspecific intruders (which were removed after 30 min, irrespective of which female won any contests that occurred) (Goubault *et al.* 2007*c*). When hosts were relatively large there was no clutch size response to intrusion but also, within the narrow range of host sizes utilised, the assumed effect of clutch size on offspring size was absent. When hosts were smaller, females that had experienced intrusion matured fewer eggs and laid a smaller proportion of their mature eggs on their hosts, compared to control females that had been undisturbed, resulting in a reduction of clutch size of approximately two eggs. Because the weight of individual offspring was negatively correlated with clutch size, adults emerging from broods laid by mothers experiencing intrusion were larger. These results constitute evidence for an effect of contest interactions on clutch size decisions, operating via the effects of clutch size on offspring size and of body size on fitness in terms of uncorrelated RHP.

for the patch they exploit, females quickly disperse from the release point, rapidly reaching a regular distribution across experimentally provided patches. In contrast, *A. tabida* females, which do not defend patches, initially showed a clumped distribution and, after gradual dispersion, a more random distribution. By favouring dispersion and the formation of regular spatial distribution of individuals, female fighting might in such cases promote parasitoid action in suppressing host populations.

If intraspecific contest behaviour affects the host-exploitation efficiency of a parasitoid species, it may also influence the potential for ecological coexistence of several species of parasitoids attacking the same host species, even if there is no direct agonistic behaviour between females belonging to different species: for instance, a more-efficient species is generally likely to outcompete a less-efficient species (Hassell 2000). Furthermore, direct agonistic interspecific interactions do occur, with clear potential for mediating parasitoid community composition (Hokyo & Kiritani 1966, Beaver 1967, Dix & Franklin 1974, Mills 1991, Sujii *et al.* 2002, Irvin & Hoddle 2005, Venkatesan *et al.* 2009*b*, Cusumano *et al.* 2011).

The relative prevalence of intra- and interspecific agonistic interactions is not well known, but interspecific contests can be more aggressive or more injurious than intraspecific contests. For instance, females of some species of the Eulophid genus *Melittobia* are mutually non-aggressive while in other species injurious interactions are common; in intraspecific interactions the females of the less-aggressive species are killed by females of the more-aggressive species (Matthews & Deyrup 2007). Similarly, intraspecific dyadic contests between females of the three species of bethylid (*Cephalonomia stephanoderis*, *C. hyalinipennis* and *Prorops nasuta*) that attack the immature stages of the coffee better borer are agonistic but non-fatal, while interspecific interactions are more often fatal than not (Pérez-Lachaud *et al.* 2002, Batchelor *et al.* 2005). Laboratory studies found that *P. nasuta* females never killed their opponents but were killed by *C. stephanoderis* and *C. hyalinipennis*, possibly due to *P. nasuta* having a less-flexible abdominal morphology making it unable to sting and paralyse other wasps (low RHP) (Batchelor *et al.* 2005).

Although species identity probably has a major influence on factors expected to influence the outcome of intraspecific contests due to differences in strength and weaponry (RHP) and resource availability (V), outcomes may be modified by factors that are expected to correlate with RHP and V within each species. For instance, in the usually fatal interspecific interactions between *C. stephanoderis* and *C. hyalinipennis*, *C. stephanoderis* nearly always wins against the physically smaller *C. hyalinipennis* (RHP). This is the case when *C. stephanoderis* is the prior owner, whether or not *C. hyalinipennis* has been able to host-feed and mature eggs prior to the contest, and also when *C. hyalinipennis* is the prior owner and *C. stephanoderis* has host-fed, but when *C. stephanoderis* intruders have not been provided with hosts prior to the contest, *C. hyalinipennis* usually wins (Batchelor *et al.* 2005), suggesting that egg loads correlate with V (section 8.3.1), and/or that host-feeding enhances RHP, strongly enough to override intrinsic interspecific asymmetries in RHP.

As congeners, *C. stephanoderis* and *C. hyalinipennis* are somewhat similar and this could explain why intraspecific variation is able to override interspecific differences. However, equivalent results have been obtained from studying contests between much less closely related species: *Dinarmus basalis* (in the family Pteromalidae) and *Eupelmus vuilleti* (family Eupelmidae) attack the same host species and females interact agonistically. While *D. basalis*, the smaller but more aggressive species, tends to win contests, outcomes are also dependent on intra-dyadic differences in body size, egg load, foraging experience and prior owner status (Mohamad *et al.* 2011). For instance, *E. vuilleti* intruders never took over from *D. basalis* prior owners and also, when in the role of prior owner, often lost to intruding *D. basalis* females; however, losing to an intruder was less likely when the *E. vuilleti* had a higher egg load (a likely correlate of V) or had had better opportunities to feed and enhance their energetic fighting abilities (RHP) (Figure 8 3).

While contest studies might provide indicators of the potential for one species to ecologically exclude others, agonistic encounters are only one of many forms of interspecific interaction. For instance, although *P. nasuta* females are not contestants they may be able to avoid agonistic interactions by using objects to seal themselves and their hosts off from intruders (Batchelor *et al.* 2005, and see section 8.5.3 for analogous behaviour in ants). In relatively 'holistic' laboratory studies, which focussed on offspring production rather than agonistic interactions and contest outcomes, *P. nasuta* performed better than *C. stephanoderis* and *C. hyalinipennis* (Batchelor *et al.* 2006). Similarly, *C. hyalinipennis* tends to lose dyadic contests with *C. stephanoderis* but may be able to persist ecologically by acting as a facultative hyperparasitoid of both *C. stephanoderis* and *P. nasuta* (Pérez-Lachaud *et al.* 2004). In a different study system, the Scelionid *T. basalis* is aggressive towards conspecifics and the Encyrtid *Ooencyrtus telenomicida* (Cusumano *et al.* 2011) but avoids direct interactions with its more aggressive congener, *T. urichi*, and competes by exploiting patches of hosts after *T. urichi* females have left (Sujii *et al.* 2002). There are further examples in which the offspring of one parasitoid species tend to outcompete offspring of another when both develop using the resources of the same individual host (multiparasitism), but the occurrence of offspring competition is reduced via agonistic host defence by adult females of the inferior juvenile competitor species (Hardy & Blackburn 1991, Mahmoud & Lim 2008, Cusumano *et al.* 2011, Mohamad *et al.* 2011).

Interspecific agonistic encounters are only one of many means by which species interact and a wide suite of behavioural and developmental strategies will usually require investigation before population- and community-level interactions can be predicted. In the case of biological pest control applications, informed strategic decisions with respect to which and how many species of parasitoids to deploy will thus generally require multidimensional considerations. Nonetheless, consideration of direct contest interactions can play an important role in the understanding of population and community ecology and in improving biocontrol applications (Mills 1991, Batchelor *et al.* 2005, 2006, Boivin & Brodeur 2006,

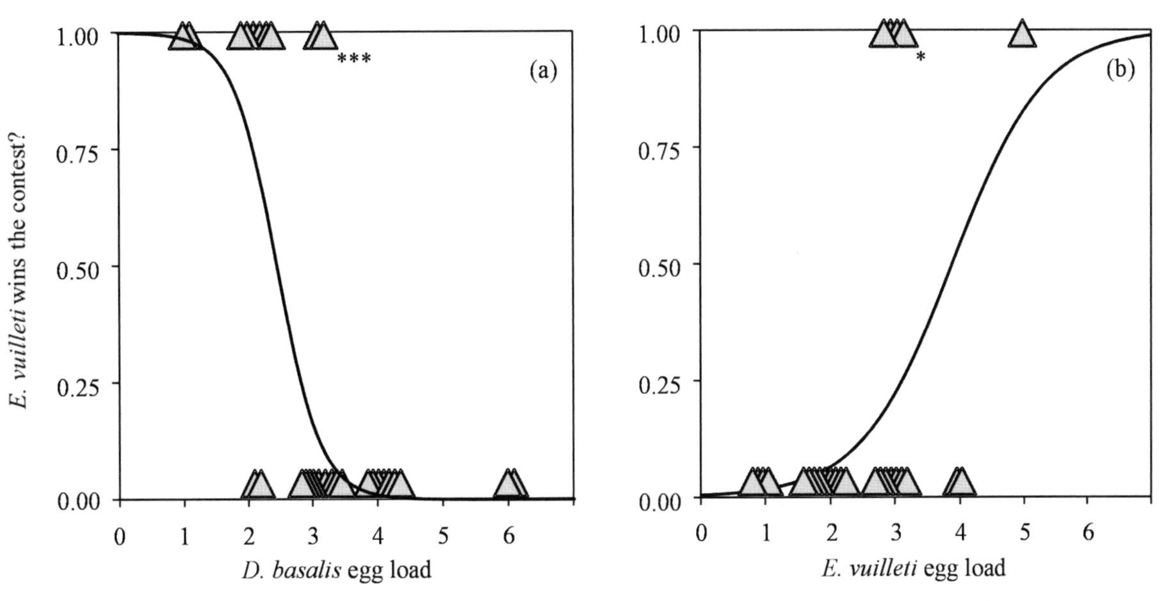

Figure 8.3 Outcomes of interspecific parasitoid contests are affected by contestant egg loads. Effects are found even though *Dinarmus basalis* (**a**) and *Eupelmus vuilleti* (**b**) belong to different hymenopteran families. Fitted probabilities of *E. vuilleti* winning derive from logistic regressions; $^{*}p < 0.005$, $^{***}p < 0.001$. (*Source:* Mohamad et al. 2011.)

Irvin *et al.* 2006, Hardy & Goubault 2007, Wyckhuys *et al.* 2011, Mohamad *et al.* 2011, Cusumano *et al.* 2012). We return to similar considerations in section 8.5.3.

8.4 Dyadic male–male contests

Males engage in agonistic interactions with other males in many species of Hymenoptera, including egg parasitoids, pollinating and non-pollinating fig wasps, solitary nesting wasps and solitary bees: we know of no case in which the contested resource is not the opportunity (direct or indirect) to mate with females. Studies of male–male contests have shown that the factors predicted to influence the outcomes of dyadic contests, RHP and *V* (Box 8.1), operate in similar vein to results for studies on female–female contests (section 8.3.1). As with females defending patches of hosts (e.g. Moore & Greeff 2003) interactions between males are not confined to dyadic situations; for instance, in parasitoid wasps belonging to the Trichogrammatid genus *Ufens*, males compete aggressively for access to newly emerged virgin females and fighting groups often consist of more than two males at a time (Al-Wahaibi *et al.* 2005) and in the solitary bee, *Amegilla dawsoni*, four or more males may struggle with each other to be at the location where a virgin female is about to emerge (Alcock 1996). Male–male contests also vary between being non-injurious (e.g. Kemp *et al.* 2006) and frequently injurious and fatal (e.g. Murray 1987, Bean & Cook 2001, Matthews *et al.* 2009, Innocent *et al.* 2011). An idea of the types of damage inflicted in some male–male contests can be obtained from Murray's (1987) injury scoring system (Table 8.2). In this

Table 8.2 Injuries incurred during male–male fighting in fig wasps. Murray (1987) developed a scoring system of the severity of contest encounters: the overall extent of injury is the sum of the points accumulated. Intact males score zero and injuries totalling 8 or more are considered severe, and will usually be fatal. This scoring system has been adopted by a number of subsequent studies (Murray 1989, West *et al.* 2001, Pereira & Prado 2005, Moore *et al.* 2009). (*Source*: Murray 1987, reproduced with permission.)

Injury	Score
Loss of part or whole of antenna	0.5
Loss of part or whole tarsus	1.0
Small bruise	1.0
Loss of part or whole tibia (plus tarsus)	2.0
Bruise with cut	2.0
Loss of part or whole femur (plus tibia and tarsus)	3.0
Large bruise, bruise plus crushed area	3.0
Loss of part or whole coax (plus femur, tibia and tarsus)	4.0
≤ Half severed abdomen or head, no evisceration	4.0
Half severed abdomen or head with evisceration	8.0

section we first summarise evidence for 'classic' influences on contest behaviour and then briefly explore links between male contest behaviour and other areas of evolutionary ecology, as the extent of male–male aggression has proved a useful tool towards understanding of the evolution of sex ratio strategies and, in particular, social behaviour.

8.4.1 Contests for mates: influences of RHP and *V*

Larger individuals are generally expected to have greater RHPs and there is evidence of large male advantage in a range of hymenopterans (e.g. solitary bees: Alcock 1996; solitary wasps: Alcock & Kemp 2006; parasitoid wasps: Hartley & Matthews 2003, Al-Wahaibi *et al.* 2005, Innocent *et al.* 2007, Reece *et al.* 2007; fig wasps: Murray 1987, Pereira & Prado 2005, Moore *et al.* 2008, 2009). Large size advantages may correlate with overall body size and/or particular morphological or physiological traits such as mandible size, head size or wing-loading (Pereira & Prado 2005, Kemp & Alcock 2003, 2008, Moore *et al.* 2008, 2009). Size advantages can operate directly via strength or persistence, without assessment prior to the fight (as in some non-pollinating fig wasps in the genus *Sycoscapter*: Moore *et al.* 2008), or pairs of males may first assess each other's size, only entering into escalated fights if neither retreats (some non-pollinating fig wasps in the genera *Idarnes*: Pereira & Prado 2005, and *Philotrypesis*: Moore *et al.* 2009). Assessment prior to escalation is generally expected only when the competitors have alternative opportunities for obtaining resources (see below). Males may also make reciprocal assessments of body size during the escalated phases of contests (Kemp *et al.* 2006). Increasing the (presumed) energetic expense of non-injurious aerial contests via wing-clipping (which may also have reduced manoeuvrability) did not influence duration in a solitary wasp (Kemp *et al.* 2006), while in some injuriously fighting fig wasps, it appears that size (and RHP) correlated energetic limitations, rather than the level of injuries sustained, determine when a loser might retreat (Moore *et al.* 2008, 2009).

Although males can obtain mating opportunities via winning contests, time spent engaging in contest behaviour can also lead to losses of reproductive opportunities: virgin females may go unnoticed by males that are engaged in fighting (Al-Wahaibi *et al.* 2005), females may be injured or even killed as a consequence of male–male struggles (Alcock 1996) or attacked and killed by males that may mistake them for other males (Matthews *et al.* 2009) and, more generally, if mating opportunities are sufficiently concentrated in time males may be selected to avoid contests and engage in scramble competition. It is also possible that 'satellite' or 'sneaker' males may obtain some matings without engaging in contests (Alcock & Kemp 2006, Kemp & Alcock 2008). Differences in mating tactics may reflect that males belong to different size classes, as in *Amegilla dawsoni*, which has 'minor' males that lose direct contests with 'major' males and adopt satellite behaviour with relatively low reproductive returns (Simmons *et al.* 2000, Tomkins *et al.* 2001, Beveridge *et al.* 2006). In *Melittobia* male size may be correlated with 'fighter' and 'non-fighter' behavioural morphs (Hartley & Matthews 2003) and, similarly, in some fig wasps there are smaller and larger males that use different competitive tactics to obtain matings (Bean & Cook 2001, Cook 2005, Cook & Bean 2006, see also Moore *et al.* 2004).

Different morphs have different RHPs: for instance, males of the 'atypical' fighter morph of *Philotrypesis* tend to incur a high level of injuries during escalated contests against 'typical' fighter morph males, even though 'atypical' males have relatively large mandibles; these apparently function as a dishonest signal of contest ability during the pre-escalation phase of the contest during which males with smaller mandibles may retreat (Moore *et al.* 2009, Chapter 4). Further, there appears to be a trade-off between fighting ability and dispersal ability across two male morphs in *Walkerella* sp. (Wang *et al.* 2010) in which all males are wingless and capable of fighting. In other fig wasp species there are clearly distinct fighting and non-fighting morphs: fighting morphs usually mate within the fig and have combat-oriented features such as winglessness, large heads and mandibles; non-fighting morphs may be winged dispersers, that seek mating opportunities outside of the natal fig or wingless forms that seek matings and avoid contests within their natal fig (e.g. Hamilton 1979, Cook *et al.* 1997, 1999, Greeff *et al.* 2003, Moore *et al.* 2004, Cook 2005).

Factors beyond body size or particular morphological characters may contribute to the RHP of male hymenopterans in particular contests. For instance, for a given size difference between fighting fig wasps, the male that attacked first is most likely to win (Moore *et al.* 2008, 2009) and males that mature earlier usually kill later-emerging males in some parasitoids: *Melittobia australica* (Abe *et al.* 2003) and *M. acasta* (Innocent *et al.* 2007), although in the latter species there is also an advantage to larger body size. Males may also utilise features of their environment that reduce the consequences of losing (fig wasps sheltering galls: Murray 1987, Pereira & Prado 2005) or correlate with the probability of winning (solitary wasps defending hilltop landmarks: Alcock & Bailey 1997, Kemp *et al.* 2006, Alcock 2008). Although the latter is a candidate example of a 'resource-correlated' RHP component (Box 8.1), it should be interpreted with caution as contest advantages to owning territories can also be due to resource-uncorrelated components of RHP, due to a historical accumulation of more powerful individuals as owners, and/or intra-dyadic asymmetries in *V*, due to information about the territory acquired during residency (Alcock & Bailey 1997, Kemp *et al.* 2006).

The value, *V*, that males place on winning a contest influences dyadic interactions in a range of hymenopterans (e.g. fig wasps: Murray 1987, Bean & Cook 2001; parasitoid wasps: Matthews *et al.* 2009, Reece *et al.* 2007; solitary wasps: Kemp *et al.* 2006). When engaging in contests incurs costs and winning a particular contest is not crucial to a male's lifetime mating success, males are expected to assess their opponent's RHP and retreat if they are unlikely to win. Moore *et al.* (2008) reported the 'assess and retreat' strategy in 5 of 6 fig wasp species examined and the species showing no assessment (*Sycoscapter* sp. A) was the one with the lowest number of females per fig, making each female an especially valuable resource. The sex ratio of adult males and females present in the fig at any one time (the operational sex ratio, OSR) is a likely correlate of *V*: if a given number of females emerge simultaneously and are available for mating (low OSR) then males will not be selected to fight intensively over each female because others will likely be available to them (scramble competition), but if the same number of females emerges over a more protracted time period (high OSR), then the waiting males must compete directly to gain access to each emerging female. Differences in the timing of female maturation provide a strong candidate explanation for the lack of male–male fighting in pollinating fig wasps with simultaneous emergence (high OSR) and the presence of fighting in non-pollinating wasps (Greeff & Fergusson 1999, see aslo Cook *et al.* 1999) and non-pollinator species with low OSR (Greeff *et al.* 2003, Nelson & Greeff 2009). Fig wasp male injury levels have been found to be higher when females were less common, both across species (West *et al.* 2001) and within some species (Bean & Cook 2001) but not others (Murray 1987).

While aspects of sex ratio can affect contest behaviour, male–male contest behaviour may also influence sex ratios (Abe *et al.* 2003, 2005, 2007, 2009, Innocent *et al.* 2007; see also Nelson & Greeff 2009): if the first males to mature kill subsequently emerging males, this can lead to an absence of the adjustment of sex ratios according to the number of mothers contributing offspring to a mating group, in contrast to the pattern that is observed in many other hymenopteran species with group-structured mating (Hamilton 1967, Ode & Hunter 2002, Ode & Hardy 2008, West 2009).

8.4.2 Kinship among contestants and the evolution of social behaviour

A further factor that may mediate agonistic interaction between males is the relatedness of the contestants: males may compete for exclusive access to mating opportunities (direct fitness benefits), but if contestant males are also relatives (e.g. brothers) the loser will gain inclusive fitness from the winner's mating success. According to Hamilton's rule, which predicts that altruistic (e.g. less aggressive) behaviour will evolve when relatedness, *r*, is closer, fitness costs, *c*, to the altruist are lower and benefits, *b*, to the recipient are greater, such that $rb - c > 0$ (Hamilton 1964*a*,*b*, 1979, Gardner *et al.* 2007), close kin might be expected to compete less aggressively than more distant relatives. Such ideas are at the core of current theory for the evolution of social behaviour (e.g. Waibel *et al.* 2011, Abbott *et al.* 2011) and thus have a far wider significance than the understanding of contests and agonistic behaviours alone. Theory also suggests that the evolution of kin-based altruism will be suppressed unless dispersal occurs, allowing the beneficiaries of altruism to compete for resources against non-relatives (Taylor 1992, Queller 1994, West *et al.* 2002) or unless mistakes in kin recognition are evolutionarily costly (Segoli *et al.* 2009).

Across Malaysian non-pollinating fig wasp species, and controlling for phylogenetic non-independence, the level of injury sustained by fighting males shows no correlation with their relatedness (estimated using the proportion of offspring that are male as an index of the number of mothers laying offspring into each fig: when more mothers contribute offspring, average relatedness between competing males will be lower) (Murray 1989, West *et al.* 2001, Cook 2005) but is negatively correlated with the degree of opportunity for future mating, estimated by counting the number of females in a mating patch (section 8.4.1). Relatedness also does not affect the likelihood of fighting in African pollinating fig wasps in which relatedness can be estimated more directly by counting the number of mothers that laid offspring into each fig because mothers remain within the fig (Nelson & Greeff 2009), although these fights may be less injurious than those observed in non-pollinating species (Greeff *et al.* 2003). In the parasitoid wasp genus *Melittobia*, variation in neither V nor relatedness influences levels of male–male aggression, which are high (Abe *et al.* 2005, Innocent *et al.* 2011).

Although we know of no studies of male–male contests that have found mediating effects of competitor relatedness, studies of female–female hymenopteran interactions have shown that relatedness can affect contest intensity, both among juveniles (polyembryonic wasps, Giron *et al.* 2004) and adults (*Goniozus legneri*, Lizé *et al.* 2012); however, both these species exhibit degrees of social behaviour (reproductive division of labour and brood care, respectively) and it may be that the ability to assess kinship is absent in non-social species such as *Melittobia* (Innocent *et al.* 2011; although virgin *Melittobia* females do exhibit brood care: Box 8.2). More empirical explorations of the importance of relatedness to male–male contests are needed, particularly comparing across situations in which males do and do not have mating opportunities beyond those in their current competitive arena and establishing whether males of non-social species are able to discriminate kin from non-kin.

8.5 Dyadic and group-level contests between social hymenopterans

Contests and agonistic interactions have been observed in a large number of social hymenopterans. These encounters can take place between colonies over resources such as food, territory in which to forage or nesting sites (e.g. Driessen *et al.* 1984). Inter-colony interactions can be similar to dyadic contests in non-social species (sections 8.3 and 8.4), in which agonistic behaviour is used to determine possession of a disputed resource, but a major difference is that resource disputes in social species can involve large numbers of individuals.

Conflicts of interest can also arise within colonies and these are likely to be over reproductive opportunities or aspects of reproductive resource-allocation (e.g. Bourke & Franks 1995); consequently, overtly aggressive interactions may often occur in female–female dyads. In many social Hymenoptera there is often clear reproductive caste differentiation of females into queens and sterile workers, although in many species workers may be capable of reproduction: workers will be selected to lay when they gain more inclusive fitness from producing their own sons than from rearing the queen's offspring. For example, if queens are polyandrous or the colony acquires a new queen, this may lead to queen–worker and worker–worker conflicts of interest (e.g. Ratnieks 1988, Ratnieks & Visscher 1989, Gobin *et al.* 1999, Toth *et al.* 2002*a*,*b*, Smith *et al.* 2012, Woyciechowski & Kuszewska 2012). In some 'casteless' (or 'queenless') species, such as *Polistes* paper wasps and ponerine ants, all nest-mates may be capable of reproduction, but egg-laying is skewed towards the alpha female in the hierarchy through agonistic interactions that suppress the egg-laying abilities or opportunities of low-ranking females, which perform roles more closely resembling those of worker castes in honeybees, bumblebees and most ants (e.g. Monnin *et al.* 2002, Cant *et al.* 2006*a*, Bridge & Field 2007, Zanette & Field 2009). Intra-colony interactions in social Hymenoptera can thus comprise queen–queen, queen–worker or worker–worker dyads.

Most research on contests and aggression in social Hymenoptera has been carried out primarily to elucidate various relationships between nest-mate recognition (e.g. Boulay *et al.* 2000, Suarez *et al.* 2002, Roulston *et al.* 2003, Vasquez & Silverman 2008), taxonomy (Mori & Lemoli 1993, Heuts *et al.* 2003), life-history decisions (Passera *et al.* 1996, Swartz 1998, Powell & Clark 2004), intra- and interspecific tolerance (Jutsum 1979, Gamboa *et al.* 1991, Boulay *et al.* 2000, Langen *et al.* 2000, Ugelvig *et al.* 2008), invasiveness (Bhatkar *et al.* 1972, Holway 1999, McGlynn 1999, Morrison 2000, Holway & Suarez 2004, Buczkowski & Bennett 2008, Ugelvig *et al.* 2008), species coexistence within ecological communities (Palmer *et al.* 2000, 2002, Stanton *et al.* 2002, 2005, Palmer 2004, Yu *et al.* 2004) and further aspects of social behaviour (Abbott *et al.* 2011). These studies have not focussed on aggression and contests per se, but have used agonistic behaviours as an indicator to test a range of evolutionary and ecological theories. However, a small number of studies have set out to test contest theory, in particular by manipulating *V* and/or correlated or uncorrelated RHP components. Here, we first discuss how contest theory has been applied to dyadic and inter-colony group contests in social hymenopterans (sections 8.5.1 and 8.5.2). We then briefly discuss the ecological consequences of aggressive interactions and contests (section 8.5.3).

8.5.1 Intra-colony interactions

The majority of aggressive interactions within colonies are between females, though there are a few reports of male–male contests. For example, in a few *Cardiocondyla* ant species males are ergatoid (non-flying) and engage in fatal contests over exclusive reproductive access to sister queens in the colony (Stuart *et al.* 1987, Yamauchi & Kawase 1992, Anderson *et al.* 2003). Males can kill juvenile male siblings that have not completely emerged (Yamauchi & Kawasa 1992, Anderson *et al.* 2003) but, when both competitors are adults, one male uses his mandibles to seize the opponent before female workers within the colony inflict the fatal attack (Yamauchi & Kawase 1992). No influences of resource-uncorrelated RHP (male body size or weaponry) or value, *V*, on outcome of adult male contests have been found (Kinomura & Yamauchi 1987, Stuart *et al.* 1987, Yamauchi & Kawase 1992).

Similar to *Cardiocondyla* males, intra-colony female–female interactions are primarily over reproduction. In bees, wasps and ants, queen–queen contests over reproductive opportunities are generally settled by differing asymmetries in factors such as body size (*Lasioglossum malachurum* and *L. pauxillum*: Smith & Weller 1989; *Solenopsis invicta*: Bernasconi & Keller 1998; *Lasius niger*: Aron *et al.* 2009), body mass (*S. invicta*: Balas & Adams 1996, Bernasconi & Keller 1998; *Apis mellifera*: Tarpy & Mayer 2009), badges of status (*Polistes dominulus*: Tibbetts & Dale 2004, Tibbetts & Shorter 2009, Tibbetts 2008), fecundity (*L. niger*: Sommer & Hölldobler 1995), age (Hughes & Strassmann 1988, Bridge & Field 2007) or order of arrival/emergence (Jerome *et al.* 1998, Seppä *et al.* 2002, Cronin & Monnin 2009). Body size, body mass and badges of status fall into the category of resource-uncorrelated components of RHP, fecundity and age may be resource-correlated RHP or *V*, and order of arrival effects appear to be due to asymmetries in resource-correlated RHP (Box 8.1).

Fatal contests (as well as executions, Box 8.3) can be common between queens in honeybees and in ants (Bernasconi & Keller 1998, Tarpy & Mayer 2009), whereas in some social wasps dominance hierarchies are established (*Polistes carloina*: Seppä *et al.* 2002; *P. dominulus*: Cant *et al.* 2006*a*,*b*; *Liostenogaster flavolineata*: Bridge & Field 2007). These generally persist with an alpha queen monopolising reproduction until she dies or is usurped by an external intruder. Dominance can be determined by resource-correlated asymmetries, such as the order of arrival at a site (Reeve 1991, Seppä *et al.* 2002) or uncorrelated asymmetries such as higher endocrine activity (Röseler *et al.* 1984) and body size; dominants are often larger than their subordinates (Dropkin & Gamboa 1981, Strassmann & Meyer 1983, Sullivan & Strassmann 1984, Dapporto *et al.* 2006, Cervo *et al.* 2008); however, dominant individuals that are smaller than their immediate subordinates generally win contests for social rank in *P. dominulus* (Cant *et al.* 2006*a*). The payoff from winning a contest, *V*, may be greater for the prior dominant than for subordinate individuals because the colony principally contains developing offspring of the dominant and a usurping subordinate would need a greater time to bring the colony to a similar value (Cant *et al.* 2006*a*), analagous to the influences of parastoid egg load on contest outcomes (section 8.3). When the dominant female dies, the beta female may assume the alpha position with little change in the order in the hierarchy and older females may be more likely to become the dominant individual (Strassmann & Meyer 1983, Tsuji & Tsuji 2005, Bridge & Field 2007, Cronin & Field 2007). Note that while contests in social insects occur over resources, these resources may not be strictly indivisible units, as often assumed in contest studies (Chapter 1): rather, dominants may retain the largest portion of the resource via sharing the remainder with their social inferiors (reproductive skew; e.g. Cant *et al.* 2006*a*). Analogous considerations apply to other hymenopterans in which the contested resource consists of the opportinity to access a group of females rather than an individual mate (e.g. fig wasps, section 8.4.1).

8.5.2 Inter-colony interactions

8.5.2.1 Contests between reproductives

Inter-colony male–male encounters in social Hymenoptera will generally involve two individuals, such as in *Polistes fuscatus* where larger male territory owners tend to win contests (Polak 1994). Inter-colony interactions between two females occur in species, such as the sweat bee *Lasioglossum malachurum*, in which nests contain a single reproductive female (gyne). Resident gynes tend to be significantly larger than floaters and win almost all encounters and thus usually retain the nest (Zobel & Paxton 2007). Larger gynes are also more aggressive and extended length contests are more common when relative size difference between combatants is small (Smith & Weller 1989). In comparison, social wasp colonies may contain a single queen or several queens in a dominance hierarchy. *Polistes dominulus*, for example, can have a varying number of queens and has been investigated to determine whether female intruders make self- and mutual assessments of RHP using clypeal spots as badges of status (Tibbetts & Dale 2004, Tibbetts 2008, Tibbetts & Lindsay 2008, Tibbetts & Shorter 2009, Tibbetts *et al.* 2010, Green & Field 2011). Relative mass was not a significant influence on outcome and contests over larger nests were more aggressive than over smaller nests (Tibbetts & Shorter 2009). Furthermore, when intruders encountered owners with high RHP badges of status, many intruders did not challenge the owner if the resource was of low V (small nest size) but as nest size increased, intruders became more likely to challenge owners (Tibbetts 2008).

8.5.2.2 Contests between workers

In honeybees and many ants, reproductive queens do not take part in inter-colony encounters: it is the workers that interact in this way (although in some ants conspecific foundress queens may engage in lethal fights for suitable locations to rear a colony: Stanton *et al.* 2002, 2005). Some worker encounters may involve dyads, but many may not immediately escalate to contact fighting, instead leading to avoidance or threat behaviours. Individuals may also seek to recruit nestmates to the contact site, at which point inter-colony interactions and fighting behaviour may take place between groups of individuals.

Similar to dyadic contests within and between colonies, the dynamics and outcomes of group contests are likely to be influenced by V, as well as resource correlated and resource uncorrelated components of RHP (Box 8.1). When groups encounter each other, however, the number of individuals in each group is likely to have an influence beyond aspects considered in classic theory for dyadic encounters. Few studies have investigated how fighting abilities (RHPs of group members) and number of individuals in groups (which could be seen as a component of 'group RHP') influence contests but Lanchester theory, which was developed to model human combat, has recently been applied to group battles in other species, in particular in ants. Lanchester theory predicts how the individual fighting abilities of group members and the number of individuals in a group may interact to influence contest dynamics, attrition rates of competing groups and eventually the outcome of fatal fights (Lanchester 1916, 1956, Adams & Mesterton-Gibbons 2003, Chapter 3). Two models have been applied: Lanchester's square attrition law and Lanchester's linear attrition law (Lanchester 1916, 1956, Chapter 3). The primary difference is that the square law assumes that groups containing more individuals are able to concentrate attacks on their less numerous opponents, hence group size is predicted to have a stronger influence on contest outcome than are the fighting abilities of individuals. The linear law, in contrast, assumes that concentration of attacks is not possible, so only one-on-one interactions may take place (Lanchester 1916, 1956, Franks & Partridge 1993, Adams & Mesterton-Gibbons 2003) and equal influences of group size and individual fighting ability on contest outcome are predicted.

Lanchester's laws have been invoked to support conclusions that different species may conform more to the square attrition law by appearing to use concentrated attack tactics or more to the linear attrition law where one-on-one duels may take place (e.g. Franks & Partridge 1993, McGlynn 2000, Pfeiffer & Linsenmair 2001, Powell & Clark 2004, Buczkowski & Bennett 2008). However, few studies have explicitly attempted to test Lanchester theory by manipulating numbers of individuals and individual body sizes (Adams & Mesterton-Gibbons 2003, Plowes & Adams 2005, Batchelor & Briffa 2010). Tests of the theory using ants found that, although attrition rates for groups are highest when they encounter opposing groups that are both numerically large and composed of large individuals (Batchelor *el al.* 2012), individual fighting behaviour has an equal, or stronger, influence on contest outcome than the number of individuals in a group (Plowes & Adams 2005, Batchelor & Briffa 2010). Furthermore, large groups do not appear to concentrate attacks effectively, thus supporting the linear law rather than the square law (Batchelor & Briffa 2010, 2011). Correlates of individual fighting ability have demonstrated that groups composed of larger (Retana & Cerda 1995, Plowes & Adams 2005) and heavier (Batchelor & Briffa 2010, Batchelor *et al.* 2012) individuals inflict higher attrition rates on opponents and that losers have lower body-sugar concentrations than non-fighting controls (Batchelor & Briffa 2010). Weapon size has not, however, been found to influence outcome (Batchelor & Briffa 2010). Further work on wood ants has found that the number of individuals in a group has a positive effect on overall group aggression levels but individuals in smaller groups appear to have higher per capita aggression levels (Batchelor *et al.* 2012). Overall group aggression relative to an opposing group has also been found to inversely correlate with attrition rates, as has relative mean mass of individuals in groups (Batchelor *et al.* 2012). These results have suggested that fight tactics and success in wood ants may vary with both individual attributes of group members and the number of individuals in the competing groups. Other studies in ants using much larger groups, but which have not measured fighting abilities and therefore not tested Lanchester theory, have found that where there are large asymmetries in the number of individuals in each group, larger groups generally win (Bhatkar *et al.* 1972, Rockwood 1973, Hölldobler & Wilson 1978, Lumsden & Hölldobler 1983, Adams 1990, Kabashima *et al.* 2007, LeBrun *et al.* 2007, Rowles & O'Dowd 2007): such results are analogous to RHP differences affecting the outcomes of dyadic contests (Figure 8.2a,b).

Research has also focussed on agonistic behaviour during encounters and found that residents are more aggressive on resident-marked territory (Hölldobler & Wilson 1977, 1978, Salzemann & Jaffe 1990, Wenseleers *et al.* 2002) and when closer to their own nest (Hölldobler 1976, Gamboa *et al.* 1991, Nowbahari *et al.* 1999, Knaden & Wehner 2004, Buczkowski & Silverman 2005, Tanner & Adler 2009). Residents are also more aggressive than intruders (Wenseleers *et al.* 2002). Elevated aggression levels in each of these situations could be influenced by correlated RHP and/or *V*, although distinguishing between these influences is difficult. Resident aggression may result in a fitness payoff to the colony of active defence of the nest (nest indicator hypothesis: Starks *et al.* 1998), which contains the queen and developing brood. Furthermore, larger groups are often more aggressive (Adams 1990, Sakata & Katayama 2001, Tanner 2006, Batchelor & Briffa 2011) which could result from a reduced individual cost to aggression (cost minimiser hypothesis: Starks *et al.* 1998, Buczkowski & Bennett 2008), positive feedback from the number of surrounding aggressive nest-mates (social facilitation: Hölldobler & Wilson 1990, Sakata & Katayama 2001) or numerical assessment that the group is large (Tanner 2006) and/or outnumbers their opponents (Hölldobler & Lumsden 1980, Lumsden & Hölldobler 1983). Larger (Smith & Weller 1989, Nowbahari *et al.* 1999) and older individuals are also often more aggressive (van Wilgenburg *et al.* 2010). Assessment of high *V* for the resource under dispute also leads to higher aggression levels (Wenseleers *et al.* 2002, Kay & Rissing 2005) and aggression may increase with individual experience (Jutsum 1979, Thomas *et al.* 2005).

8.5.3 Eco-evolutionary consequences of social insect agonistic behaviour

As with dyadic interactions between non-social species (section 8.3.3), social hymenopteran contests may be interspecific as well as intraspecific. Interspecific contests may be dyadic, between foundresses of different species within a guild attempting to initiate nests at a suitable site (Stanton *et al.* 2005), dyadic within the setting of a single colony (e.g. Ortolani & Cervo 2010, Green & Field 2011, Grüter *et al.* 2012) or between large numbers of workers belonging to colonies of different species (Palmer 2004, LeBrun *et al.* 2007).

Interspecific colony–colony contests can be both violent and last for several days as, for example, those between three congeneric members of a guild of four species of acacia ant (*Crematogaster sjostedti*, *C. mimosae*, *C. nigriceps* and *Tetraponera penzigi*) for possession of host trees (Palmer *et al.* 2000, Palmer 2004). For these four species, acacia trees are a limiting, vital and indivisible resource providing both food and colony nest sites (high *V*). Based on contest outcomes and other observations, these species can be arranged in a linear dominance hierarchy for interspecific worker–worker interactions: *C. sjostedti* > *C. mimosae* > *C. nigriceps* > *T. penzigi* (Palmer et al. 2000, 2002). For *Crematogaster* species, this ranking is dependent on numerical differences in contesting workers (group RHP) rather than differences in the fighting abilities of individuals of each species: species with larger colonies typically win by a process of attrition (Palmer 2004). Further, at the colony-founding stage, *T. penzigi* tend to win in lethal combat, and the dominance ranking of the four species is reversed (Stanton *et al.* 2002, 2005).

Exposed individual *T. penzigi* workers usually do not engage in agonistic interactions with the *Crematogaster* workers and the latter usually then return to their own colony's prior territory (Palmer *et al.* 2000). The persistence of *T. penzigi* colonies is facilitated by workers modifying the entrance holes to their colony locations (within swollen thorns) such that the more dominant *Crematogaster* workers cannot enter without expending considerable energetic and time costs enlarging the hole and also by the destruction of most of the leaf nectaries on their host tree (Palmer *et al.* 2002): both actions are expected to decrease *V* for any intruding *Crematogaster*. Further, *T. penzigi* workers possess stings which they use to kill *C. mimosae* workers that are attempting to enlarge entrance holes to *T. penzigi* colonies (Palmer *et al.* 2002), thus *T. penzigi* take advantage of their prior residents' positional advantage (resource-correlated RHP, Box 8.1). Experimentally increasing *V* for *C. mimosae* by adding nectar to trees possessed by *T. penzigi* colonies and/or decreasing colony protection by experimentally enlarging entrance holes leads to increased takeover by *C. mimosae* (Palmer *et al.* 2002).

The result of interspecific colony interactions is that a fine balance between dominance of colonies (which may change as colonies develop) and their other ecologically important attributes may result in the persistence of multi-species communities (Palmer *et al.* 2000, 2002, Stanton *et al.* 2002, 2005, Palmer 2004, Yu *et al.* 2004). Community ecology theory suggests that ecological diversity is likely to be promoted if intraspecific colony interactions are more aggressive and damaging to contestants than are interspecific interactions (e.g. Shorrocks & Sevenster 1995). While native community diversity may be balanced by competitive interactions and other factors, such as parasite–host relationships (e.g. Yu *et al.* 2004, Stanton *et al.* 2005), this balance can be severely disrupted when species are accidentally introduced into a new range. The Argentine ant *Linepithema humile*, the fire ant *Solenopsis invicta* and the invasive garden ant *Lasius neglectus* are three such species that have caused widespread reductions in native ant community diversity on introduction (e.g. Bourke & Franks 1995, Pedersen *et al.* 2006, LeBrun *et al.* 2007, Cremer *et al.* 2008, Ugelvig *et al.* 2008). This is generally a result of their colony structure: the invasive species may not engage in intraspecific aggression and huge networks of related colonies, termed supercolonies, may only exhibit interspecific competition, or at least, the prevalence of inter-supercolony interactions is reduced compared to the native range because introduced supercolonies become much larger (Pedersen *et al.* 2006; see also LeBrun *et al.* 2007). This often results in the invasive species outcompeting native species via their greater numbers, low levels of intraspecific competition and the absence of any locally present natural enemies such as parasites (e.g. Cremer *et al.* 2008, Ugelvig *et al.* 2008). However, the short-term ecological success of social hymenopterans with supercolonial structure is not thought to lead to long-term evolutionary persistence: there are no known clades containing only supercolonial ant species, probably because supercolonality is likely to promote the spread of destabilising selfish behaviours (Bourke & Franks 1995, Pedersen *et al.* 2006, Cremer *et al.* 2008).

Finally, returning to dyadic contests, Green and Field (2011) examined interactions between congeneric *Polistes*: the paper wasp *P. dominulus* and its social parasite *P. semenowi*. Individual *P. semenowi* females attempt to take over the reproductive role within a *P. dominulus* colony via aggressive and injurious contests with the prior dominant *P. dominulus* female. Larger females (RHP) tend to win these contests, irrespective of the species to which the larger contestant belonges, and apparently without size-assessment during contests, probably because *V* is high for both contestants (Green & Field 2011; see also Ortolani & Cervo 2010). More than half of the parasites that attempted nest usurpation in Green and Field's (2011) experiments were successful; such pressure from competing parasites is thought to drive co-evolutionary body size arms races between host and parasite species (Ortolani & Cervo 2010), which will probably also contribute to the maintenance of ecological species diversity.

8.6 Conclusions and prospects

The Hymenoptera are clearly very useful organisms for studying contests. In this respect they are arguably second-to-none due to their extraordinary range of variations in life-histories, making them suitable for testing classic theory for dyadic contests plus theory for group-level contests plus utilising their agonistic behaviours to gain insights into fundamental processes in social evolution. We thus have no doubts that there will be many more studies on competitive and aggressive behaviours in these species.

We suggest the following areas as among those that are likely to be productive for further investigation.

- The relationships between contest behaviour and the physiological, metabolic and hormonal states of contestants. These areas are well developed in some invertebrate taxa (e.g. Chapter 5) but in their comparative infancy among Hymenopterans. A decade ago Kravitz and Huber (2003) reviewed the literature on invertebrate aggression and showed that nearly all studies of hymenopterans were 'ethological' while for other invertebrate taxa there was a much more balanced spread of studies across their three categories of 'ethology', 'physiology' and 'amines'. We hope that a decade hence the hymenopteran spread will be more balanced. We expect recent advances in methodologies for analysing the metabolomic states of animals will facilitate the process; for example, elucidating whether different metabolites differentially affect contestant RHP and how the biochemical composition of feeding resources influence *V*. Such investigations might, for instance, help to explain winner–loser effects in parasitoids (which is also an area that may benefit from further development of theory, section 8.3.1.1) and, among the ants, both aggression and contest outcomes can be affected by carbohydrate supply (Mabelis 1979, Palmer *et al.* 2002, Grover *et al.* 2007).
- The role of kinship in mediating aggression among parasitoid wasps. Kin recognition phenomena and mechanisms are little explored among the non-social insects compared to the social insects, and yet kinship does affect contest behaviour in some parasitoids (section 8.4.2). Further studies on the commonality of such effects and the (presumably chemical, Khidr *et al.* 2013) basis of kin discrimination are required, especially if they are placed in the context of the 'scale' of the competitive environment, that is, whether kin are each other's frequent or infrequent main competitors.
- The separate and interactive influences of number of individuals and individual fighting abilities on group contests. Many ant species exhibit high inter-colony aggression levels, both in the field and laboratory, and are amenable to manipulation of both number and size of individuals in competing colonies, which is often not possible in other social animals. Ants thus have high potential for testing and honing group contest theory.
- In applied ecology, studying contest and agonistic interactions should remain useful to the understanding of invasiveness of social hymenopterans and the biological control potential of particular species and combinations of species of parasitoid wasps. Both are relevant considerations in a climatically changing world with all the associated uncertainties in global food security.

Acknowledgements

We thank our many contest-study collaborators for their various inputs to our research, Mark Briffa, James Cook and Darrell Kemp for comments on the manuscript and two UK research councils, the NERC and the BBSRC, for past funding.

References

Abbot P, Abe J, Alcock J, *et al.* (2011) Inclusive fitness theory and eusociality. *Nature*, 247 (doi: 10.1038/nature09831).

Abe J, Kamimura Y, Kondo N, *et al.* (2003) Extremely female-biased sex ratio and lethal male–male combat in a parasitoid wasp, *Mellitobia australica* (Eulophidae). *Behavioral Ecology*, 14, 34–39.

Abe J, Kamimura Y & Shimada M (2005) Individual sex ratios and offspring emergence patterns in a parasitoid wasp, *Mellitobia australica* (Eulophidae), with superparasitism and lethal combat among sons. *Behavioral Ecology and Sociobiology*, 57, 366–373.

Abe J, Kamimura Y & Shimada M (2007) Sex ratio schedules in a dynamic game: The effect of competitive asymmetry by male emergence order. *Behavioral Ecology*, 18, 1106–11115.

Abe J, Kamimura Y, Shimada M, *et al.* (2009) Extremely female-biased primary sex ratio and precisely constant male production in a parasitoid wasp *Melittobia. Animal Behaviour*, 78, 515–523.

Adams ES (1990) Boundary disputes in the territorial ant *Azteca trigona* – Effects of asymmetries in colony size. *Animal Behaviour*, 39, 321–328.

Adams ES & Mesterton-Gibbons M (2003) Lanchester's attrition models and fights among social animals. *Behavioral Ecology*, 14, 719–723.

Akino T, Yamamura K, Wakamura S, *et al.* (2004) Direct behavioral evidence for hydrocarbons as nestmate recognition cues in *Formica japonica* (Hymenoptera: Formicidae). *Applied Entomology and Zoology*, 39, 381–387.

Alcock J (1996) The relation between male body size, fighting, and mating success in Dawson's burrowing bee, *Amegilla dawsoni* (Apidae, Apinae, Anthophorini). *Journal of Zoology, London*, 239, 663–674.

Alcock J (2008) Territorial preferences of the hilltopping wasp *Hemipepsis ustulata* (Pompilidae) remain stable from year to year. *Southwestern Naturalist*, 53, 190–195.

Alcock J & Bailey WJ (1997) Success in territorial defence by male tarantula hawk wasps *Hemipepsis ustulata*: The role of residency. *Ecological Entomology*, 22, 377–383.

Alcock J & Kemp DJ (2006) The behavioral significance of male body size in the tarantula hawk wasp *Hemipepsis ustulata* (Hymenoptera: Pompilidae). *Ethology*, 112, 691–698.

Alcock J, Simmons LW, Beveridge M (2006) Does variation in female body size affect nesting success in Dawson's burrowing bee, *Amegilla dawsoni* (Apidae: Anthophorini)? *Ecological Entomology*, 31, 352–357.

Alphen JJM van & Visser ME (1990) Superparasitism as an adaptive strategy for insect parasitoids. *Annual Review of Entomology*, 35, 59–79.

Al-Wahaibi AK, Owen AK & Morse JG (2005) Description and behavioural biology of two *Ufens* species (Hymenoptera: Trichogrammatidae), egg parasitoids of *Homalodisca* species (Hemiptera: Cicadellidae) in southern California. *Bulletin of Entomological Research*, 95, 275–288.

Anderson C, Cremer S & Heinze J (2003) Live and let die: Why fighter males of the ant *Cardiocondyla* kill each other but tolerate their winged rivals. *Behavioral Ecology*, 14, 54–62.

Aron S, Steinhauer N & Fournier D (2009) Influence of queen phenotype, investment and maternity apportionment on the outcome of fights in cooperative foundations of the ant *Lasius niger*. *Animal Behaviour*, 77, 1067–1074.

Balas MT & Adams ES (1996) The dissolution of cooperative groups: Mechanisms of queen mortality in incipient fire ant colonies. *Behavioral Ecology and Sociobiology*, 38, 391–399.

Batchelor TP & Briffa M (2010) Influences on resource-holding potential during dangerous group contests between wood ants. *Animal Behaviour*, 80, 443–449.

Batchelor TP & Briffa M (2011) Fight tactics in wood ants: Individuals in smaller groups fight harder but die faster. *Proceedings of the Royal Society of London B*, 278, 3243–3250.

Batchelor TP, Hardy ICW, Barrera JF, *et al.* (2005) Insect gladiators II: Competitive interactions within and between bethylid parasitoid species of the coffee berry borer, *Hypothenemus hampei* (Coleoptera: Scolytidae). *Biological Control*, 33, 194–202.

Batchelor TP, Hardy ICW & Barrera JF (2006) Interactions among bethylid parasitoid species attacking the coffee berry borer, *Hypothenemus hampei* (Coleoptera: Scolytidae). *Biological Control*, 36, 106–118.

Batchelor TP, Santini G & Briffa M (2012) Size distribution and battles in wood ants: Group resource-holding potential is the sum of the individual parts. *Animal Behaviour*, 83, 111–117.

Bean D & Cook JM (2001) Male mating tactics and lethal male combat in the nonpollinating fig wasp *Sycoscapter australis*. *Animal Behaviour*, 62, 535–542.

Beaver RA (1967) Hymenoptera associated with elm bark beetles in Wytham Wood, Berks. *Transactions of the Society for British Entomology*, 17, 141–150.

Bentley T, Hull TT, Hardy ICW & Goubault M (2009) The elusive paradox: Owner–intruder roles, strategies and outcomes in parasitoid contests. *Behavioral Ecology*, 20, 296–304.

Bernasconi G & Keller L (1998) Phenotype and individual investment in cooperative foundress associations of the fire ant, *Solenopsis invicta*. *Behavioral Ecology*, 9, 478–485.

Beveridge M, Simmons LW & Alcock J (2006) Genetic breeding system and investment patterns within nests of Dawson's burrowing bee (*Amegilla dawsoni*) (Hymenoptera: Anthophorini). *Molecular Ecology*, 15, 3459–3467.

Bhatkar A, Whitcomb WH, Buren WF, *et al.* (1972) Confrontation behavior between *Lasius neoniger* (Hymenoptera: Formicidae) and the imported fire ant. *Environmental Entomology*, 1, 274–279.

Boivin G & Brodeur J (2006) Intra- and interspecific interactions among parasitoids: Mechanisms, outcomes and biological control. In: J Brodeur & G Boivin (eds.) *Trophic and Guild Interactions in Biological Control. Series: Progress in Biological Control*, 3, 123–144.

Boulay R, Hefetz A, Soroker V, *et al.* (2000) *Camponotus fellah* colony integration: Worker individuality necessitates frequent hydrocarbon exchanges. *Animal Behaviour*, 59, 1127–1133.

Bourke AFG & Franks NR (1995) *Social Evolution in Ants*. Princeton, NJ: Princeton University Press.

Bridge C & Field J (2007) Queuing for dominance: Gerontocracy and queue-jumping in the hover wasp *Liostenogaster flavolineata*. *Behavioral Ecology and Sociobiology*, 61, 1253–1259.

Buczkowski G & Bennett GW (2008) Aggressive interactions between the introduced Argentine ant,

Linepithema humile and the native odorous house ant, *Tapinoma sessile*. *Biological Invasions*, 10, 1001–1011.

Buczkowski G & Silverman J (2005) Context-dependent nestmate discrimination and the effect of action thresholds on exogenous cue recognition in the Argentine ant. *Animal Behaviour*, 69, 741–749.

Cant MA, English S, Reeve HK, *et al.* (2006*a*) Escalated conflict in a social hierarchy. *Proceedings of the Royal Society of London B*, 273, 2977–2984.

Cant MA, Llop JB & Field J (2006*b*) Individual variation in social aggression and the probability of inheritance: Theory and a field test. *American Naturalist*, 167, 837–852.

Casas J, Vannier F, Mandon N, *et al.* (2009) Mitigation of egg limitation in parasitoids: Immediate hormonal response and enhanced oogenesis after host use. *Ecology*, 90, 537–545.

Cervo R, Dapporto L, Beani L, *et al.* (2008) On status badges and quality signals in the paper wasp *Polistes dominulus*: Body size, facial colour patterns and hierarchical rank. *Proceedings of the Royal Society of London B*, 275, 1189–1196.

Charnov EL (1976) Optimal foraging, the marginal value theorem. *Theoretical Population Biology*, 9, 129–136.

Chen J & Cheng H (2000) Advances in applied research on *Scleroderma* spp. *Chinese Journal of Biological Control*, 16, 166–170.

Choe JC & Crespi BJ (1997) *The Evolution of Social Behavior in Insects and Arachnids*. Cambridge: Cambridge University Press.

Cook JM (2005) Alternative mating tactics and fatal fighting in male fig wasps. In: MDE Fellowes, GJ Holoway & J Rolff (eds.) *Insect Evolutionary Ecology*, pp. 83–109. Wallingford: CABI Publishing.

Cook JM & Bean D (2006) Cryptic male dimorphism and fighting in a fig. *Animal Behaviour*, 71, 1095–1101.

Cook JM, Compton SG, Herre EA, *et al.* (1997) Alternative mating tactics and extreme male dimporphism in fig wasps. *Proceedings of the Royal Society of London B*, 264, 747–754.

Cook JM, Bean D & Power S (1999) Fatal fighting in fig wasps – GBH in time and space. *Trends in Ecology and Evolution*, 14, 257–259.

Cremer S, Ugelvig LV, Drijfhout FP, *et al.* (2008) The evolution of invasiveness in garden ants. *PLoS ONE*, 3, e3838.

Cronin AL & Field J (2007) Social aggression in an age-dependent dominance hierarchy. *Behaviour*, 144, 753–765.

Cronin AL & Monnin T (2009) Bourgeois queens and high stakes games in the ant *Aphaenogaster senilis*. *Frontiers in Zoology*, 6, 24.

Cusumano A, Peri E, Vinson SB, *et al.* (2011) Intraguild interactions between two egg parasitoids exploring host patches. *Biocontrol*, 56, 173–184.

Cusumano A, Peri E, Vinson SB, *et al.* (2012) Interspecific extrinsic and intrinsic competitive interactions in egg parasitoids. *Biocontrol*, 57, 719–734.

Dapporto L, Palagi E, Cini A, *et al.* (2006) Prehibernating aggregations of *Polistes dominulus*: An occasion to study early dominance assessment in social insects. *Naturwissenschaften*, 93, 321–324.

Dix ME & Franklin RT (1974) Interspecific and intraspecific encounters of southern pine beetle parasites under field conditions. *Environmental Entomology*, 3, 131–134.

Driessen GJJ, Raalte AT van & Bryn GJ de (1984) Cannibalism in the red wood ant, *Formica polyctena*. *Oecologia*, 63, 13–22.

Dropkin JA & Gamboa GJ (1981) Physical comparisons of foundresses of the paper wasp, *Polistes metricus* (Hymenoptera, Vespidae). *Canadian Entomologist*, 113, 457–461.

Dugatkin LA (1997) Winner effects, loser effects, assessment strategies and the structure of dominance hierarchies. *Behavioral Ecology*, 8, 583–587.

Dugatkin LA & Druen M (2004) The social implications of winner and loser effects. *Proceedings of the Royal Society of London B* (Suppl), 271, S488–S489.

Dugatkin LA & Earley RL (2003) Group fusion: The impact of winner, loser, and bystander effects on hierarchy formation in large groups. *Behavioral Ecology*, 14, 367–373.

Dugatkin LA & Earley RL (2004) Individual recognition, dominance hierarchies, and winner and loser effects. *Proceedings of the Royal Society of London B*, 271, 1537–1540.

Earley RL & Dugatkin LA (2010) Behavior in groups. In: DF Westneat & VW Fox (eds.) *Evolutionary Behavioral Ecology*, pp. 285–307. Oxford: Oxford University Press.

Enquist M & Leimar O (1987) Evolution of fighting behaviour: The effect of variation in resource value. *Journal of Theoretical Biology*, 127, 187–206.

Fawcett TW & Johnstone RA (2010) Learning your own strength: Winner and loser effects should change with age and experience. *Proceedings of the Royal Society of London B*, 277, 1427–1434.

Fernández-Arhex V & Corley JC (2010) The effects of patch richness on conspecific interference in the parasitoid *Ibalia leucospoides* (Hymenoptera: Ibaliidae). *Insect Science*, 17, 379–385.

Field J (1992) Intraspecific parasitism and nest defence in the solitary pompilid wasp *Anoplius viaticus* (Hymenoptera: Pompilidae). *Journal of Zoology, London*, 228, 341–350.

Field SA (1998) Patch exploitation, patch-leaving and pre-emptive patch defence in the parasitoid wasp *Trissolcus basalis* (Insecta: Scelionidae). *Ethology*, 104, 323–338.

Field SA & Calbert G (1998) Patch defence in the parasitoid wasp *Trissolcus basalis*: When to begin fighting? *Behaviour*, 135, 629–642.

Field SA & Calbert G (1999) Don't count your eggs before they're parasitized: Contest resolution and the trade-offs during patch defense in a parasitoid wasp. *Behavioral Ecology*, 10, 122–127.

Field SA, Keller MA & Calbert G (1997) The pay-off from superparasitism in the egg parasitoid *Trissolcus basalis*, in relation to patch defence. *Ecological Entomology*, 22, 142–149.

Field SA, Calbert G & Keller MA (1998) Patch defence in the parasitoid wasp *Trissolcus basalis* (Insecta: Scelionidae): The time structure of pairwise contests, and the 'waiting game'. *Ethology*, 104, 821–840.

Franks NR & Partridge LW (1993) Lanchester battles and the evolution of combat in ants. *Animal Behaviour*, 45, 197–199.

Gamboa GJ, Reeve HK & Pfennig DW (1986) The evolution and ontogeny or nestmate recognition in social wasps. *Annual Review of Entomology*, 31, 431–454.

Gamboa GJ, Foster RL, Scope JA, *et al.* (1991) Effects of stage of colony cycle, context, and intercolony distance on conspecific tolerance by paper wasps (*Polistes fuscatus*). *Behavioral Ecology and Sociobiology*, 29, 87–94.

Gardner A, Hardy ICW, Taylor PD, *et al.* (2007) Spiteful soldiers and sex ratio conflict in polyembryonic parasitoid wasps. *American Naturalist*, 169, 519–533.

Ghazoul J (2001) Effect of soil hardness on aggression in the solitary wasp *Mellinus arvensis*. *Ecological Entomology*, 26, 457–466.

Giron D, Rivero A, Mandon N, *et al.* (2002) The physiology of host feeding in parasitic wasps: Implications for survival. *Functional Ecology*, 16, 750–757.

Giron D, Dunn D, Hardy ICW, *et al.* (2004) Aggression by polyembryonic wasp soldiers correlates with kinship but not resource competition. *Nature*, 430, 676–679.

Gobin B, Billen J & Peeters C (1999) Policing behaviour towards virgin egg layers in a polygynous ponerine ant. *Animal Behaviour*, 58, 1117–1122.

Godfray HCJ (1994) *Parasitoids: Behavioral and Evolutionary Ecology*. Princeton, NJ: Princeton University Press.

Godfray HCJ, Partridge L & Harvey PH (1991) Clutch size. *Annual Review of Ecology*, 22, 409–429.

Goertzen R & Doutt RL (1975) Ovicidal propensity of *Goniozus*. *Annals of the Entomological Society of America*, 68, 869–870.

Gómez J, Barrera JF, Rojas JC, *et al.* (2005) Volatile compounds released by disturbed females of *Cephalonomia stephanoderis*, a bethylid parasitoid of the coffee berry borer *Hypothenemus hampei*. *Florida Entomologist*, 88, 180–187.

Goubault M & Decuignière M (2012) Prior experience and contest outcome: winner effects persist in absence of evident loser effects in a parasitoid wasp. *American Naturalist*, 180, 364–371.

Goubault M & Hardy ICW (2007) Deuterium marking of chemical emissions: Detectability and fitness consequences of a novel technique for insect behavioural studies. *Entomologia Experimentalis et Applicata*, 125, 285–296.

Goubault M, Outreman Y, Poinsot D, *et al.* (2005) Patch exploitation strategies of parasitic wasps under intraspecific competition. *Behavioral Ecology*, 16, 693–701.

Goubault M, Batchelor TP, Linforth RST, *et al.* (2006) Volatile emission by contest losers revealed by real-time chemical analysis. *Proceedings of the Royal Society of London B*, 273, 2853–2859.

Goubault M, Cortesero AM, Poinsot D, *et al.* (2007*a*) Does host value influence female aggressiveness, contest outcome and fitness gain in parasitoids? *Ethology*, 113, 334–343.

Goubault M, Scott D & Hardy ICW (2007*b*) The importance of offspring value: Maternal defence in parasitoid contests. *Animal Behaviour*, 74, 437–446.

Goubault M, Mack AFS & Hardy ICW (2007*c*) Encountering competitors reduces clutch size and increases offspring size in a parasitoid with female–female fighting. *Proceedings of the Royal Society of London* B, 274, 2571–2577.

Goubault M, Batchelor TP, Romani R, *et al.* (2008) Volatile chemical release by bethylid wasps: Identity, phylogeny, anatomy and behaviour. *Biological Journal of the Linnean Society*, 94, 837–852.

Greeff JM & Fergusson JWH (1999) Mating ecology of the nonpollinating fig wasps of *Ficus ingens*. *Animal Behaviour*, 57, 215–222.

Greeff JM, Noort S van, Rasplus J-Y, *et al.* (2003) Dispersal and fighting in male pollinating fig wasps. *Comptes Rendus de Biologie*, 326, 121–130.

Green JP & Field J (2011) Interpopulation variation in status signalling in the paper wasp *Polistes dominulus*. *Animal Behaviour*, 81, 205–209.

Grover CD, Kay AD, Monson JA, *et al.* (2007) Linking nutrition and behavioural dominance: Carbohydrate scarcity limites aggression and activity in Argentine ants. *Proceedings of the Royal Society of London B*, 274, 2951–2957.

Grüter C, Menezes C, Imperatriz-Fonesca VL, *et al.* (2012) A morphologically specialized soldier caste improves colony defence in a neotropical eusocial bee. *Proceedings of the National Academy of Sciences USA*, 109, 1182–1186.

Haccou P, Glaizot O & Cannings C (2003) Patch leaving strategies and superparasitism: An asymmetric generalized war of attrition. *Journal of Theoretical Biology*, 225, 77–89.

Hamilton WD (1964*a*) The genetic evolution of social behaviour. II. *Journal of Theoretical Biology*, 7, 17–52.

Hamilton WD (1964*b*) The genetic evolution of social behaviour. I. *Journal of Theoretical Biology*, 7, 1–16.

Hamilton WD (1967) Extraordinary sex ratios. *Science*, 156, 477–488.

Hamilton WD (1979) Wingless and fighting males in fig wasps and other insects. In: MS Blum & NA Blum (eds.) *Reproductive Competition, Mate Choice and Sexual Selection in Insects*, pp 167–220. London: Academic Press.

Hammerstein P (1981) The role of asymmetries in animal contests. *Animal Behaviour*, 29, 193–205.

Hardy ICW & Blackburn TM (1991) Brood guarding in a bethylid wasp. *Ecological Entomology*, 16, 55–62.

Hardy ICW & Goubault M (2007) Wasp fights: Understanding and utilizing agonistic bethylid behaviour. *Biocontrol News and Information*, 28, 11–15.

Hardy ICW, Griffiths NT & Godfray HCJ (1992) Clutch size in a parasitoid wasp: A manipulation experiment. *Journal of Animal Ecology*, 61, 121–129.

Hartley CS & Matthews RW (2003) The effect of body size on male–male combat in the parasitoid wasp *Melittobia digitata* Dahms (Hymenoptera: Eulophidae). *Journal of Hymenopteran Research*, 12, 272–277.

Harvey JA (2008) Comparing and contrasting development and reproductive strategies in the pupal hyperparasitoids *Lysibia nana* and *Gelis agilis* (Hymenoptera: Ichneumonidae). *Evolutionary Ecology*, 22, 153–166.

Harvey JA, Corley LS & Strand MR (2000) Competition induces adaptive shifts in caste ratios of a polyembryonic wasp. *Nature*, 406, 183–186.

Hassell MP (2000) *The Spatial and Temporal Dynamics of Host–Parasitoid Interactions*. Oxford: Oxford University Press.

Hassell MP & Varley GC (1969) New inductive population model for insect parasites and its bearing on biological control. *Nature*, 223, 1133–1137.

Hawkins BA (1994) *Pattern and Process in Host–Parasitoid Interactions*. Cambridge: Cambridge University Press.

Heuts BA, Cornelissen P & Lambrechts DYM (2003) Different attack modes of *Formica* species in interspecific one-on-one combats with other ants (Hymenoptera: Formicidae). *Annales Zoologici*, 53, 205–216.

Hochberg ME & Ives AR (eds.) (2000) *Parasitoid Population Biology*. Princeton, NJ: Princeton University Press.

Hokyo N & Kiritani K (1966) Oviposition behaviour of two egg parasites, *Asolcus mitsukurii* Ashmead and *Telenomus nakagawai* Watanabe (Hymenoptera, Proctotrupoidea, Scelionidae). *Entomophaga*, 2, 191–201.

Hölldobler B (1976) Recruitment behavior, home range orientation and territoriality in harvester ants, *Pogonomyrmex*. *Behavioral Ecology and Sociobiology*, 1, 3–44.

Hölldobler B & Lumsden CJ (1980) Territorial strategies in ants. *Science*, 210, 732–739.

Hölldobler B & Wilson EO (1977) Colony-specific territorial pheromone in African weaver ant *Oecophylla longinoda* (Latreille). *Proceedings of the National Academy of Sciences USA*, 74, 2072–2075.

Hölldobler B & Wilson EO (1978) The multiple recruitment systems of African weaver ant *Oecophylla longinoda* (Latreille) (Hymenoptera: Formicidae). *Behavioral Ecology and Sociobiology*, 3, 19–60.

Hölldobler B & Wilson EO (1990) *The Ants*. Cambridge, MA: Harvard University Press.

Hölldobler B & Wilson EO (2009) *The Superorganism*. New York, NY: Norton & Co.

Holway DA (1999) Competitive mechanisms underlying the displacement of native ants by the invasive Argentine ant. *Ecology*, 80, 238–251.

Holway DA & Suarez AV (2004) Colony-structure variation and interspecific competitive ability in the invasive Argentine ant. *Oecologia*, 138, 216–222.

Hsu Y & Wolf LL (1999) The winner and loser effect: Integrating multiple experiences. *Animal Behaviour*, 57, 903–910.

Hsu Y, Earley RL & Wolf LL (2006) Modulation of aggressive behaviour by fighting experience: Mechanisms and contest outcomes. *Biological Reviews*, 81, 33–74.

Hughes CR & Strassmann JE (1988) Age is more important than size in determining dominance among workers in the primitively eusocial wasp, *Polistes instabilis*. *Behaviour*, 107, 1–14.

Hughes JP, Harvey IF & Hubbard SF (1994) Host-searching of *Venturia canescens* (Grav.) (Hymenoptera: Ichneumonidae): Interference – The effect of mature egg load and prior behaviour. *Journal of Insect Behavior*, 7, 433–454.

Humphries EL, Hebblethwaite AJ, Batchelor TP, *et al.* (2006) The importance of valuing resources: Host weight and contender age as determinants of parasitoid wasp contest outcomes. *Animal Behaviour*, 72, 891–898.

Innocent TM, Savage J, West SA, *et al.* (2007) Lethal combat and sex ratio evolution in a parasitoid wasp. *Behavioral Ecology*, 18, 709–715.

Innocent TM, West SA, Sanderson JL, *et al.* (2011) Lethal combat over limited resources: Testing the importance of competitors and kin. *Behavioral Ecology*, 22, 923–931.

Irvin NA & Hoddle MS (2005) The competitive ability of three mymarid egg parasitoids (*Gonatocerus* spp.) for glassy-winged sharpshooter (*Homalodisca coagulata*) eggs. *Biological Control*, 34, 204–214.

Irvin NA, Hoddle MS & Morgan DJW (2006) Competition between *Gonatocerus ashmeadi* and *G. triguttatus* for glassy-winged sharpshooter (*Homalodisca coagulata*) egg masses. *Biocontrol Science and Technology*, 16, 359–375.

Jerome CA, Mcinnes DA & Adams ES (1998) Group defense by colony-founding queens in the fire ant *Solenopsis invicta*. *Behavioral Ecology*, 9, 301–308.

Jervis MA (2005) *Insects as Natural Enemies: A Practical Perspective*. Dordrecht: Springer.

Jervis MA & Kidd NAC (1986) Host feeding strategies in hymenopteran parasitoids. *Biological Reviews*, 61, 395–434.

Jong PW de, Hemerik L, Gort G, *et al.* (2009) Different forms of mutual interference result in different spatial distributions of foraging *Drosophila* parasitoids *Asobara citri* and *Asobara tabida*. *Proceedings of the Netherlands Entomological Society Meetings*, 20, 17–29.

Jong PW de, Hemerik L, Gort G, *et al.* (2011) Rapid establishment of a regular distribution of adult tropical *Drosophila* parasitoids in a multi-patch environment by patch defence behaviour. *PLoS ONE*, 6, e20870.

Jutsum AR (1979) Interspecific aggression in leaf-cutting ants. *Animal Behaviour*, 27, 833–838.

Kabashima JN, Greenberg L, Rust MK, *et al.* (2007) Aggressive interactions between *Solenopsis invicta* and *Linepithema humile* (Hymenoptera: Formicidae) under laboratory conditions. *Journal of Economic Entomology*, 100, 148–154.

Kapranas A, Hardy ICW, Morse JG, *et al.* (2011) Parasitoid developmental mortality in the field: Patterns, causes and consequences for sex ratio and virginity. *Journal of Animal Ecology*, 80, 192–203.

Kay A & Rissing SW (2005) Division of foraging labor in ants can mediate demands for food and safety. *Behavioral Ecology and Sociobiology*, 58, 165–174.

Kemp DJ & Alcock J (2003) Lifetime resource utilization, flight physiology, and the evolution of contest competition in territorial insects. *American Naturalist*, 162, 290–301.

Kemp DJ & Alcock J (2008) Aerial contests, sexual selection and flight morphology in solitary pompilid wasps. *Ethology*, 114, 195–202.

Kemp DJ, Alcock J & Allen GR (2006) Sequential size assessment and multicomponent decision rules mediate aerial wasp contests. *Animal Behaviour*, 71, 279–287.

Khidr SK, Linforth RST & Hardy ICW (2013) Genetic and environmental influences on the cuticular hydrocarbon profiles of *Goniozus* wasps. *Entomologia Experimentalis et Applicata*, in press. doi:10.1111/eea.12058

Kinomura K & Yamauchi K (1987) Fighting and mating behaviors of dimorphic males in the ant *Cardiocondyla wroughtoni*. *Journal of Ethology*, 5, 75–81.

Knaden M & Wehner R (2004) Path integration in desert ants controls aggressiveness. *Science*, 305, 60.

Kokko H, López-Sepulcre A & Morrell LJ (2006) From hawks and doves to self-consistent games of territorial behaviour. *American Naturalist*, 167, 901–912.

Kravitz E & Huber R (2003) Aggression in invertebrates. *Current Opinion in Neurobiology*, 13, 736–743.

Kudo SI, Maeto K & Ozaki K (1992) Maternal care in the red headed spruce web spinning sawfly, *Cephalcia isshikii* (Hymenoptera, Pamphiliidae). *Journal of Insect Behavior*, 5, 783–795.

Lack D (1947) The significance of clutch size. *Ibis*, 89, 309–352.

Lanchester FW (1916) *Aircraft in Warfare*. New York, NY: Appleton.

Lanchester FW (1956) *Mathematics in warfare*. In: JR Newman (ed.) *The World of Mathematics*, pp. 2138–2157. New York, NY: Simon and Schuster.

Langen TA, Tripet F & Nonacs P (2000) The red and the black: Habituation and the dear-enemy phenomenon in two desert Pheidole ants. *Behavioral Ecology and Sociobiology*, 48, 285–292.

Lann C Le, Outreman Y, Alphen JJM van, *et al.* (2011) First in, last out: Asymmetric competition influences patch exploitation of a parasitoid. *Behavioral Ecology*, 22, 101–107.

Lawrence PO (1981) Interference competition and optimal host selection in the parasitic wasp, *Biosteres longicaudatus*. *Annals of the Entomological Society of America*, 74, 540–544.

LeBrun EG, Tillberg CV, Suarez AV, *et al.* (2007) An experimental study of competition between fire ants and argentine ants in their native range. *Ecology*, 88, 63–75.

Lizé A, Khidr SK & Hardy ICW (2012) Two components of kin recognition influence parasitoid aggression in resource competition. *Animal Behaviour*, 83, 793–799.

Lumsden CJ & Hölldobler B (1983) Ritualized combat and intercolony communication in ants. *Journal of Theoretical Biology*, 100, 81–98.

Mabelis AA (1979) Wood ant wars: The relationship between aggression and predation in the red wood ant (*Formica polyctena* Forst). *Netherlands Journal of Zoology*, 29, 451–620.

Mahmoud AMA & Lim UT (2008) Host discrimination and interspecific competition of *Trissolcus nigripedus* and *Telenomous gifuensis* (Hymenoptera: Scelionidae), sympatric parasitoids of *Dolycoris baccarum* (Heteroptera: Pentatomidae). *Biological Control*, 45, 337–343.

Matthews RW & Deyrup LD (2007) Female fighting and host competition among four sympatric species of *Melittobia* (Hymenoptera: Eulophidae). *Great Lakes Entomologist*, 40, 52–62.

Matthews RW, González JM, Matthews JR, *et al.* (2009) Biology of the parasitoid *Melittobia* (Hymenoptera: Eulophidae). *Annual Review of Entomology*, 54, 251–266.

Mayhew PJ (1997) Fitness consequences of ovicide in a parasitoid wasp. *Entomologia Experimentalis et Applicata*, 84, 115–126.

Maynard Smith J (1974) The theory of games and the evolution of animal conflicts. *Journal of Theoretical Biology*, 47, 209–221.

Maynard Smith J & Parker GA (1976) The logic of asymmetric contests. *Animal Behaviour*, 24, 159–175.

McGlynn TP (1999) Non-native ants are smaller than related native ants. *American Naturalist*, 154, 690–699.

McGlynn TP (2000) Do Lanchester's laws of combat describe competition in ants? *Behavioral Ecology*, 11, 686–690.

Mesterton-Gibbons M (1992) Ecotypic variation in the asymmetric Hawk–Dove game: When is bourgeois an evolutionarily stable strategy? *Evolutionary Ecology*, 6, 198–222.

Mesterton-Gibbons M (1999) On the evolution of pure winner and loser effects: A game-theoretic model. *Bulletin of Mathematical Biology*, 61, 1151–1186.

Mesterton-Gibbons M & Hardy ICW (2004) The influence of contests on optimal clutch size: A game-theoretic model. *Proceedings of the Royal Society of London B*, 271, 971–978.

Mills NJ (1991) Searching strategies and attack rates of parasitoids of the ask bark beetle (*Leperisinus varius*) and its relevance to biological control. *Ecological Entomology*, 16, 461–470.

Mohamad R, Monge JP & Goubault M (2010) Can subjective resource value affect aggressiveness and contest outcome in parasitoid wasps? *Animal Behaviour*, 80, 629–636.

Mohamad R, Monge JP & Goubault M (2011) Agonistic interactions and their implications for parasitoid species coexistence. *Behavioral Ecology*, 22, 1114–1122.

Mohamad R, Monge JP & Goubault M (2012) Wait or fight? Ownership asymmetry affects contest behaviours in a parasitoid wasp. *Behavioral Ecology*, 23, 1330–1337.

Monnin T, Ratnieks FLW, Jones GR, *et al.* (2002) Pretender punishment induced by chemical signalling in a queenless ant. *Nature*, 419, 61–65.

Moore JC & Greeff JM (2003) Resource defence in female pollinating fig wasps: Two's a contest, three's a crowd. *Animal Behaviour*, 66, 1101–1107.

Moore JC, Pienaar J & Greeff JM (2004) Male morphological variation and the determinants of body size in two *Otiteselline* fig wasps. *Behavioral Ecology*, 15, 735–741.

Moore JC, Obbard DJ, Reuter C, *et al.* (2008) Fighting strategies in two species of fig wasp. *Animal Behaviour*, 76, 315–322.

Moore JC, Obbard DJ, Reuter C, *et al.* (2009) Male morphology and dishonest signalling in a fig wasp. *Animal Behaviour*, 78, 147–153.

Mori A & Lemoli F (1993) The aggression test as a taxonomic tool – Evaluation in sympatric and allopatric populations of wood-ant species. *Aggressive Behavior*, 19, 151–156.

Morrison LW (2000) Mechanisms of interspecific competition among an invasive and two native fire ants. *Oikos*, 90, 238–252.

Mueller UG, Warneke AF, Grafe TU, *et al.* (1992) Female size and nest defense in the digger wasp *Cerceris fumipennis* (Hymenoptera, Sphecidae, Philanthinae). *Journal of the Kansas Entomological Society*, 65, 44–52.

Murray MG (1987) The closed environment of the fig receptacle and its influence on male conflicts in the Old World fig wasp, *Philotrypesis pilosa*. *Animal Behaviour*, 35, 488–506.

Murray MG (1989) Environmental constraints on fighting in flightless male fig wasps. *Animal Behaviour*, 38, 186–193.

Nakamatsu Y, Harvey JA & Tanaka T (2009) Intraspecific competition between adult females of the hyperparasitoid *Trichomalopsis apanteloctena* (Hymenoptera: Chelonidae), for domination of *Cotesia kariyai* (Hymenoptera: Braconidae) cocoons. *Annals of the Entomological Society of America*, 102, 172–180.

Nehring V, Evison SEF, Santorelli LA, *et al.* (2010) Kin-informative recognition cues in ants. *Proceedings of the Royal Society of London B*, 278, 1942–1948.

Nelson RM & Greeff JM (2009) Evolution of the scale and manner of brother competition in pollinating fig wasps. *Animal Behaviour*, 77, 693–700.

Nowbahari E, Feneron R & Malherbe MC (1999) Effect of body size on aggression in the ant, *Cataglyphis niger* (Hymenoptera; formicidae). *Aggressive Behavior*, 25, 369–379.

Ode PJ & Hardy ICW (2008) Parasitoid sex ratios and biological control. In: E Wajnberg, C Bernstein & JJM van Alphen (eds.) *Behavioral Ecology of Insect Parasitoids: From Theoretical Approaches to Field Applications*, pp. 253–291. Oxford: Blackwell Publishing.

Ode PJ & Hunter MS (2002) Sex ratios of parasitic Hymenoptera with unusual life-histories. In: ICW Hardy (ed.) *Sex Ratios: Concepts and Research Methods*, pp. 218–234. Cambridge: Cambridge University Press.

Ode PJ & Rosenheim JA (1998) Sex allocation and the evolutionary transition between solitary and gregarious parasitoid development. *American Naturalist*, 152, 757–761.

Ohno K (1999) Brood guarding in *Trissolcus plautiae* (Watanabe) (Hymenoptera: Scelionidae), an egg parasitoid of the brown-winged green bug, *Plautia crossota stali* Scott (Heteroptera: Pentatomidae). *Entomological Science*, 2, 41–47.

Ortolani I & Cervo R (2010) Intra-specific body size variation in *Polistes* paper wasps as a response to social parasite pressure. *Ecological Entomology*, 35, 352–359.

Palmer TM (2004) Wars of attrition: Colony size determines competitive outcomes in a guild of African acacia ants. *Animal Behaviour*, 68, 993–1004.

Palmer TM, Young TP, Stanton ML, *et al.* (2000) Short-term dynamics of an acacia ant community in Laikipia, Kenya. *Oecologia*, 123, 425–435.

Palmer TM, Young TP & Stanton ML (2002) Burning bridges: Priority effects and the persistence of a competitively subordinate acacia-ant in Laikipia, Kenya. *Oecologia*, 133, 372–379.

Parker GA (1974) Assessment strategy and the evolution of fighting behaviour. *Journal of Theoretical Biology*, 47, 223–243.

Passera L, Roncin E, Kaufmann B, *et al.* (1996) Increased soldier production in ant colonies exposed to intraspecific competition. *Nature*, 379, 630–631.

Pedersen JS, Krieger MJB, Vogel V, *et al.* (2006) Native supercolonies of unrelated individuals in the invasive Argentine ant. *Evolution*, 60, 782–791.

Pereira RAS & Prado APD (2005) Recognition of competitive asymmetries reduces the severity of fighting in male *Idarnes* fig wasps. *Animal Behaviour*, 70, 249–256.

Pérez-Lachaud G, Hardy ICW & Lachaud J-P (2002) Insect gladiators: Competitive interactions between three species of bethylid wasps attacking the coffee berry borer, *Hypothenemus hampei* (Coleoptera: Scolytidae). *Biological Control*, 25, 231–238.

Pérez-Lachaud G, Batchelor TP & Hardy ICW (2004) Wasp eat wasp: Facultative hyperparasitism and intra-guild predation by bethylid wasps. *Biological Control*, 30, 149–155.

Petersen G & Hardy ICW (1996) The importance of being larger: Parasitoid intruder–owner contests and their implications for clutch size. *Animal Behaviour*, 51, 1363–1373.

Pexton JJ, Rankin DJ, Dytham C, *et al.* (2003) Asymmetric larval mobility and the evolutionary transition from siblicide to non-siblicidal behavior in parasitoid wasps. *Behavioral Ecology*, 14, 182–193.

Pfeiffer M & Linsenmair KE (2001) Territoriality in the Malaysian giant ant *Camponotus gigas* (Hymenoptera: Formicidae). *Journal of Ethology*, 19, 75–85.

Phillips DS (1993) Host-feeding and egg maturation by *Pachycrepoideus vindemmiae. Entomologia Experimentalis et Applicata*, 69, 75–82.

Pijls JWAM, Poleij LM, Alphen JJM van, *et al.* (1996) Interspecific interference between *Apoanagyrus lopezi* and *A. diversicornis*, parasitoids of the cassava mealybug *Phenacoccus manihoti. Entomologia Experimentalis et Applicata*, 78, 221–230.

Plowes NJR & Adams ES (2005) An empirical test of Lanchester's square law: Mortality during battles of the fire ant *Solenopsis invicta. Proceedings of the Royal Society of London B*, 272, 1809–1814.

Polak M (1994) Large-size advantage and assessment of resource holding potential in male *Polistes Fuscatus* (F) (Hymenoptera, Vespidae). *Animal Behaviour*, 48, 1231–1234.

Powell S & Clark E (2004) Combat between large derived societies: A subterranean army ant established as a predator of mature leaf-cutting ant colonies. *Insectes Sociaux*, 51, 342–351.

Queller DC (1994) Genetic relatedness in viscous populations. *Evolutionary Ecology*, 8, 70–73.

Quicke DLJ, Laurenne NM & Barclay M (2005) Host, host location and aggressive behaviour in a tropical wood-borer parasitoid genus *Monilobracon* Quicke (Hymenoptera: Braconidae), parasitoids of Lymexylidae (Coleoptera) in Kibale Forest National Park, West Uganda. *African Entomology*, 13, 213–220.

developmental genetic tools and artificial selection. In this chapter, we first review intraspecific fighting in beetles and two of the most significant emergent properties: alternative mating tactics and the evolution of weaponry. Second, we expand on the developmental underpinnings that may have facilitated the evolution of weaponry, especially in the Scarabinae, and explore how understanding the developmental basis of weapons and weapon diversity can illuminate our understanding of the evolution and ecology of contest behaviour.

9.3 Review of contest behaviour

Fighting has been observed in at least 15 beetle families, including blister beetles (Meloidae: McLain 1982), solider beetles (Cantharidae: Mason 1980), fireflies (Lampyridae: Lloyd 1979), glowworm beetles (Phengodidae: Lloyd 1979), rove beetles (Staphylinidae: Hanley 2001), darkling beetles (Tenebrionidae: Eberhard 1980, Conner 1988, Rasa 1999, Okada *et al.* 2006), leaf beetles (Chrysomelidae: Eberhard 1980, 1981, Windsor 1987), burying beetles (Silphidae: Otronen 1988*a*, Trumbo 2007), bark beetles (Scolytidae: Rudinsky & Ryker 1976), round fungus beetles (Leiodidae: Miller & Wheeler 2005), weevils (Curculionidae: Eberhard 1980, 1983, Johnson 1982, Eberhard & Garcia-C 2000), sap beetles (Nitulidae: Okada & Mityatake 2007), 'bess bugs' (Passalidae: Schuster & Schuster 1985, King & Fashing 2007), and, of course, stag beetles (Lucanidae: e.g. Eberhard 1980) and scarabs (Scarabaeidae: e.g. Eberhard 1980). Here we provide a broad review of the functions of contests in beetles and the determinants of contest outcomes.

9.3.1 Functions of fighting in Coleoptera

Contest behaviour is energetically expensive and risky. Fights in beetles sometimes lead to injury (Howard 1955, Eberhard 1980, 1987, Siva-Jothy 1987) and even death (Hamilton 1979). Thus, fighting is predicted to evolve only when the object of the fight is of extremely high value and/or resources are clumped in a manner where defence is economical, a prediction of the basic Hawk–Dove game in game theory (Maynard Smith 1974, 1979, Hamilton 1979, Thornhill & Alcock 1983, Chapter 2). Fighting is therefore often found in taxa with increased parental care (e.g. Rasa 1999, Trumbo 2007), or when resources are rare or patchily distributed (Emlen & Oring 1977). In terms of the resources that are contested, beetle fights generally fall into two categories: the securing of food or the acquisition of mating opportunities. Fighting and weaponry have evolved many times in association with such defensible and valuable resources (reviewed in Emlen 2008).

9.3.1.1 Access to females

The vast majority of contests in Coleoptera are between males over direct or indirect access to females. Indirect fights over females occur as contests for access to locations where females are feeding or ovipositing. Males of the tropical scarab *Podischnus agenor* will defend tunnels in cane stalks, where both the males and females feed on plant juices (Eberhard 1980). Male *Dendrobias* (Cerambycidae) defend sap food sites on bushes that females visit (Goldsmith 1987). Indirect fights also occur as contests over locations where females oviposit, such as tunnels where a mated female is constructing a brood ball. For instance, males of the dung beetle *Phanaeus difformis* will defend such tunnels against intruding males by either flipping (and then sometimes pinching) rivals when outside the tunnel, or through pushing contests when inside the tunnel (Rasmussen 1994). Males will also defend 'meeting areas': for example, in the tenebrionid fungus beetle *Bolitotherus cornutus*, males will fight other males with their thoracic horn on the tops of shelf fungi, which both adults and larvae use as food sources (Conner 1988, 1989).

Males also fight directly over females. The bearded weevil *Thinostomus barbirostris* will mate with an ovipositing female (drilling a hole for eggs on a tree trunk) and then guard her by knocking rival males off the tree using his flattened, elongated rostrum (Eberhard 1980). Another weevil species, *Parisoschoenus expositus*, displays a similar behaviour where males guard ovipositing females after mating with them, by shoving opponents and, in the most intense fights, interlocking thoracic horns with thoracic grooves and pushing (Eberhard & Garcia-C 2000). Male Japanese beetles (*Popillia japonica*) will attempt physically to displace males that are riding on the backs of females (mating or mate-guarding: Kruse & Switzer 2007). Males of the brentid weevil *Brentus anchorago* use their elongated rostrum to try to dislodge males that are mating or mate-guarding by swatting or prying (Johnson 1982).

9.3.1.2 Food resources

Some beetles will fight over food resources for their own consumption. Males of the scarab *Golofa porter* use their long head horns to pry opponents off of their bamboo-like food plant, while grasping the plant with their elongated front legs for support (Eberhard 1977, 1980). In a Colombian chrysomelid closely related to *Doryphora punctatissima*, males use a horn projecting from the underside of the thorax to push and pry opponents off the vines on which they feed; females may also use their horns to defend vines from other ovipositing females (Eberhard 1980, 1981).

In other instances, beetles fight over resources for their offspring (and sometimes later actively defend their offspring). Both male and female burying beetles (*Nicrophorus* sp., Silphidae), will fight over carcasses that they later bury and feed to their offspring (Otronen 1988*a*). After securing the carcass, beetles will defend their offspring against intruders that may kill their offspring and takeover the carcass (Trumbo 2007). Similarly, monogamous pairs of passalid beetles defend tunnels (where offspring will be raised) against intruding conspecifics (e.g. Schuster & Schuster 1985) that often commit infanticide if they are able to take over the existing burrow (King & Fashing 2007). Pairs of the desert tenebrionid *Parastizopus armaticeps* divide defence of their tunnels (containing offspring and their food): the male residents attack male intruders and female residents attack female intruders (Rasa 1999).

Most examples of female–female fighting are contests over food (for themselves or offspring), similar to examples of female fighting in vertebrates (e.g. Espmark 1964). Female Japanese horned beetles *Allomyrina dichotomus* fight over sap flows but at a much lower frequency than do males (Hongo 2003). Females of the brentid weevil *Brentus anchorago* will try to take over drilled oviposition sites of other females by using their snout, head or body to uproot and dislodge resident females (Johnson 1982). Similarly, females of the cerambycid *Monochamus* sp. attempt to dislodge each other from oviposition sites in logs that take 20–60 min to drill (Hughes & Hughes 1987).

Females of several species use horns to fight over resources. In the scarab beetle *Coprophanaeus ensifer*, both males and females possess similarly sized horns, which females probably use to defend carrion (secured for feeding their very large offspring) from other females and unsuitable males (Otronen 1988*b*). In the dung beetle *Phanaeus difformis*, females possess much smaller horns than males, which they use to defend their burrows against other females attempting to steal the burrow or dung; despite involving much shorter horns, female–female contests are similar in movements to those between males, but shorter in duration (Rasmussen 1994). Such competition over dung and female defence of brood balls against parasitism of other females is likely to have driven the evolution of horns and reversed sexual dimorphism in the scarab *Onthophagus sagittarius* (Watson & Simmons 2010*a*,*b*).

9.3.2 Determinants of contest outcome

Determinants of contest outcome in beetles are similar to those in other systems, and those predicted by game theoretic models of fighting behaviour (Chapter 2). In particular, traits linked to resource holding potential, RHP, (specifically body size and relative weapon investment), are major determinants of contest outcome (Maynard Smith 1979, Hammerstein 1981). As in other study systems (e.g. crustaceans, Chapter 5; wasps, Chapter 8), resource value and ownership status are also likely to influence the outcome of fights.

9.3.2.1 Body size

Body size is a common determinant of resource holding potential in many systems (Abbott *et al.* 1985, Wells 1988, Schuett 1997) and is also an important determinant of contest outcome in beetles (Figure 9.2). If a pair of beetles is mismatched in size, the larger male almost always wins and the contest is of short duration (e.g. Eberhard 1987). The importance of body size is especially pronounced in species that do not possess weapons. For instance, body size is an advantage in both inter- and intraspecific contests in carrion beetles (*Nicrophorus* spp., Otronen 1988*a*) and body size is an important determinant of contest outcome in a weaponless grain beetle (*Tenebrio molitor*: Howard 1955) and blister beetle (*Epicauta pennsylvanica*: McLain 1982).

9.3.2.2 Weaponry

Beetles possess a wide variety of weapons (reviewed in Emlen 2008), such as enlarged mandibles, elongated forelegs, or horns on the head and thorax (section 9.4, Figure 9.1). Weapon size is often tightly correlated with body size (e.g. Emlen 1997, Hongo 2003),

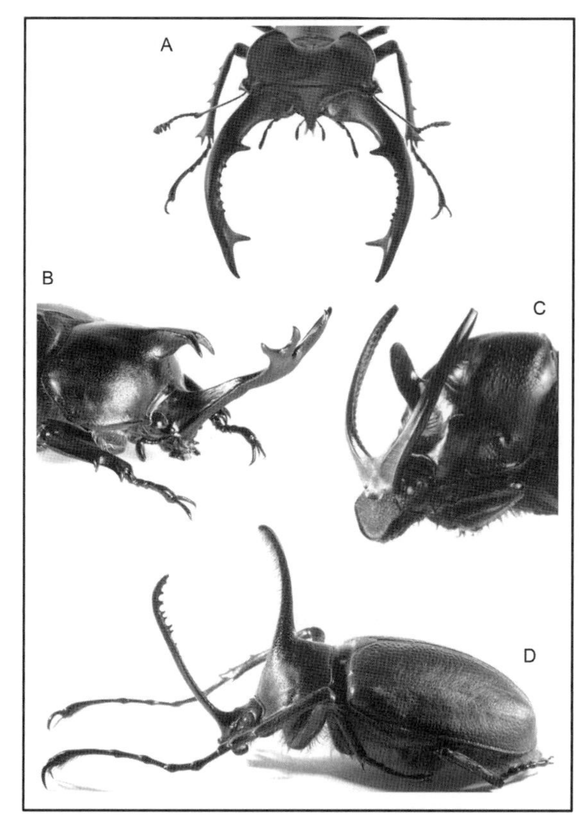

Figure 9.1 Diversity of weaponry in Coleoptera. Beetles possess a diversity of exaggerated structure used as weapons in contests, including enlarged mandibles (**A**), horns (**B**,**C**,**D**), and elongated forelegs (**D**). Weaponry has many independent evolutionary origins in beetles. Shown are (**A**) *Lucanus elaphus* (Lucanidae), (**B**) *Allomyrina dichotomus* (Dynastinae), (**C**) *Onthophagus watanabei* (Scaranaeinae), (**D**) *Golofa eacus* (Dynastinae). Photographs by Franck Simonnet (*L. elaphus*) and Armin Moczek (all others).

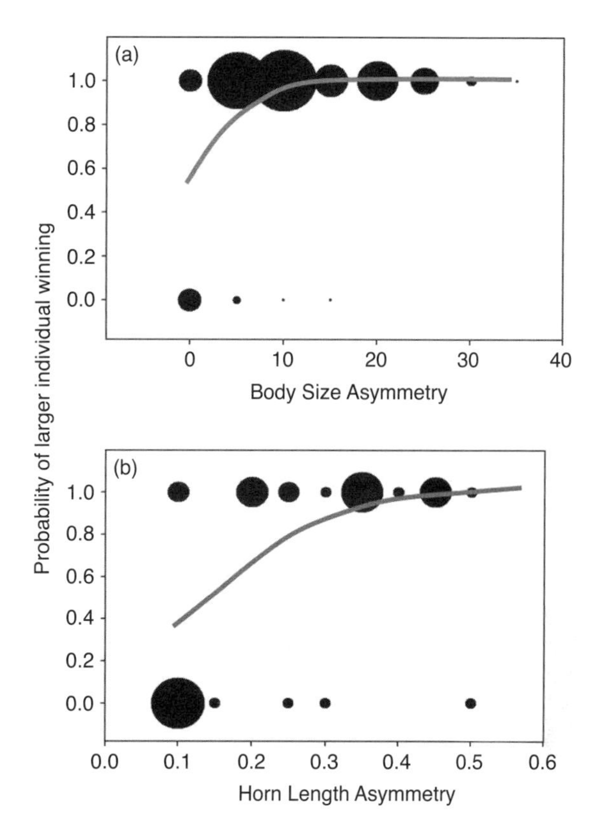

Figure 9.2 Determinants of contest outcome. Two sets of data relating asymmetries in body size and horn length in determining context outcome (1 = larger individual wins; 0 = smaller individual wins; circle size proportional to number of replicate contests). Shown are published data re-plotted as a logistic regression (Chapter 4): (**a**) data from 116 contests between *Brentus anchorago* (Johnson 1982, figure 3A); (**b**) data from 26 size-matched contests between *Onthophagus acuminatus* (Emlen 1997, figure 4).

making it difficult to determine whether body size or weapon size per se is a determinant of contest outcome. However, experiments matching beetles in size show that differences in weapon size between opponents have a major effect on contest outcome and access to females (e.g. Brown & Bartalon 1986, Conner 1989, Rasmussen 1994, Emlen 1997, Moczek & Emlen 2000, Hongo 2003, Figure 9.2).

In some systems, relative horn length appears to be even more important than relative body size in determining contest outcome (Karino *et al.* 2005). The importance of horn length in determining contest outcome may depend on body size: for instance, in *Euoniticellus intermedius*, horn length asymmetries are more important than body length asymmetries in large males but not in small males (Pomfret & Knell 2006). The importance of relative weapon size may also explain why, in some species such as the forked fungus beetle (*Bolitotherus cornutus*), selection on horn length is stronger than direct selection on body size (Conner 1988). We address the evolution of weaponry throughout the rest of the chapter, including weaponry as an evolutionary outcome of contest behaviour (section 9.4) and developmental insights into the origins of weapons (section 9.6).

9.3.2.3 Sex and pairs

In many species of beetles, contests occur between members of the same sex. However, in some contests over food or tunnels, there may be contests between males and females: in most cases the male has an advantage. For instance, unless males and females are grossly mismatched in size, male *Tenebrio* tend to win contests against females (Howard 1955). Similarly, intruding male burying beetle *Nicrophorus pustulatus* are almost twice as likely to defeat defending (single) females, relative to intruding female beetles (Trumbo 2007).

Monogamous beetles may sometimes fight as pairs and this often confers an advantage over unpaired individuals. For example, resident pairs of burying beetles (e.g. *N. pustulatus*) have an increased probability of defeating an intruder (and thus preventing infanticide) if they fight as pairs (Trumbo 2007). There are also examples of male and female cerambycid pairs fighting together to take over oviposition holes laboriously drilled by other females (Hughes & Hughes 1987). Contests involving pairs of beetles fighting as teams do not fit readily into the theoretical framework developed to study dyadic animal contests (Chapter 2) but a multi-player approach could be applied (Chapter 3).

9.3.2.4 Residency

Residence in an area has been shown to have a large effect on contest outcome in a wide range of both vertebrate and invertebrate species (e.g. Chapters 2, 7 and 8) in some cases due to the effect of residency on resource value (Enquist & Leimar 1987, Chapter 7). While asymmetries in body size and weaponry seem to outweigh the importance of residency in many beetles (e.g. Hongo 2003, Kruse & Switzer 2007), residency still exerts an effect in some systems (e.g. Hughes & Hughes 1987). In *Nicrophorus vespilloides*, individuals that arrive at a carcass first have an advantage in subsequent contests over the carcass (Otronen 1988*a*). In this particular example, learning to recognise each other may reinforce the residence advantage while fighting together (Otronen 1988*a*).

Residency can interact with other determinants to influence contest outcome. In Japanese beetles (*Popillia japonica*), males that are 'resident' on a female (mating or having just mated) have a distinct advantage over intruders that try to take over access to the female; however, greater body size of the intruder can overcome this residency advantage (Kruse & Switzer 2007). In the sap beetle *Librodor japonicus*, minor males (i.e. with relatively small mandibles, section 9.5.2) that happen to be resident at a sap site will sometimes fight major, horned males; their residency advantage can help to overcome their body-size and weapon-size disadvantage (at least in 19% of fights: Okada & Miyatake 2007).

9.3.2.5 Past experiences

The behavioural and physiological state of an individual plays a role in contest outcome in a range of systems (e.g. Chapters 5 and 10) including beetles. For example, past losing experiences can affect contest outcome in the beetle *Gnatocerus cornutus* (Okada & Miyatake 2010), similar to the 'loser effect' seen in other systems (Chase *et al.* 1994, Hsu & Wolf 2001, e.g. Chapters 3, 8 and 10). Past experience with females may also play a role: in the scarab *Canthon cyanellus*, mating status of males (virgin or mated) plays a role in contest outcome (Chamorro-Florescano & Favila 2008).

9.3.3 Assessment and escalation in contests

Given the costs of fighting and the risks of injury, contests often involve ritualised assessment behaviour and escalation to more risky behaviour (Maynard Smith 1974). As we might expect, assessment of opponents prior to escalation is an important part of beetle contests (Hongo 2003).

9.3.3.1 Escalation in contests

As expected from theory (Parker 1974, Enquist & Leimar 1983, Chapter 2) and observations in other systems (but see Chapter 4 for a discussion of how these patterns should be interpreted), fights in beetles are often settled quickly if opponents are grossly mismatched in size or weaponry (e.g. Hongo 2003). As a contest escalates, closely matched opponents will increase the use of risky, costlier fighting behaviours. For instance, prying, lifting and throwing behaviours generally precede pushing and shoving behaviours (Hongo 2003). In some species, different traits are used as weapons as a fight escalates in intensity. In the horned weevil, *Parisoschoenus expositus*, early phases of fighting consist of pushing and shoving between beetles that generally involve the elongated rostrum and sometimes 'slapping' with the forelegs. If this does not resolve the contest, beetles will interlock horns by inserting their thoracic horn into a complementary sheath in their opponent's thorax; in this way beetles are 'locked' together and push each other, often until one is raised off the substrate (Eberhard & Garcia-C 2000, Eberhard *et al.* 2000). Males of the weevil *Staminodeus vectoris* start contests for oviposition substrates by poking with the rostrum and raising and

is not sigmoidal and instead slighty changes slope at a midpoint; congruently, there are few differences between small and large males in fighting behaviour (Hongo 2003). Second, the costs of horns at the lineage level, and thus the probability of horn loss over time, may be reduced because individuals that find themselves in poor environmental conditions do not suffer the cost of growing horns. Finally, a switch-like mechanism may allow horns to be easily lost and later regained among species in a clade (West-Eberhard 2003, Whiting *et al.* 2003).

Is it possible that sigmoidal allometries may be highly likely to evolve in beetles, thus explaining the high frequency of weapons in beetles? It is possible that holometabolous development (involving complete metamorphosis and a pupal stage) favours such allometries; developing beetles have a longer time to assess what their adult body size will be prior to committing to horn growth as pupae, while hemimetabolous insects (no pupal stage), such as earwigs, grasshoppers and true bugs, start to grow many adult traits (such as compound eyes) earlier in development. This hypothesis does not, however, explain why such complex allometries are common only in some (Coleoptera, Hymenoptera) but rare or absent in other holometabolous insect orders (e.g. Diptera, Lepidoptera), and how some hemimetabolous orders have managed to overcome their developmental constraints (e.g. castes in termites and some aphids). Thus, the answer to the question of weaponry in Coleoptera may be a combination of the hypotheses discussed: beetles may be predisposed to evolve weapons due to a propensity to evolve threshold mechanisms of weapon development, combined with the frequent presence of a hardened cuticle and elytra.

For the remainder of this chapter we will shift our focus to horned scarab beetles and explore how an understanding of the developmental basis of weapons and of weapon diversity can provide important insights into the evolution and ecology of contest behaviour. This does not imply that weaponry is less common in other groups; indeed, independent evolution of exaggerated weapons and fighting behaviour may be just as common and diverse in weevils (e.g. Franz 2003), but the knowledge of developmental and genetic mechanisms underlying horn expression, and their relationship to the diversification of weaponry and contest behaviour, is by far most detailed for the scarabs.

9.7 Developmental insights into the evolution of weapons and contests

Beetle horns, particularly of species in the genus *Onthophagus*, have recently emerged as a promising model system in evolutionary developmental biology, addressing questions that centre on the origin of novel features and the connection between micro- and macro-evolution of development. In the sections that follow (sections 9.7.2 to 9.7.5) we highlight how an understanding of development can also address important questions in contest biology, such as the ancestry of alternative phenotypes, the costs of weaponry and constraints and biases in the diversification of beetle weapons. To do so we first briefly summarise the most critical aspects of beetle horn development.

9.7.1 The development of beetle horns: conservation with divergence

Generally speaking, beetle horns develop in a manner similar to more 'traditional' appendages such as legs, mouthparts, wings or antennae of most insect orders (Svacha 1992). Even though horns do not have muscles, joints or nervous tissue, they are three-dimensional outgrowths of epidermal tissue regions that undergo massive cell proliferation during late larval development as well as remodelling and sculpting during the pupal stage (Moczek 2006*a*). Horn development may thus rely at least in part on some of the same genetic and developmental mechanisms that regulate the same processes in more conventional appendages. Recent data provide strong support for this hypothesis. For example, the same transcription factors that establish the proximo-distal axis of developing legs, mouthparts, antennae and genitalia across insect orders and arthropod classes are also expressed during the development of beetle horns (Moczek & Nagy 2005, Moczek & Rose 2009), suggesting that the functions of these genes have been co-opted into a novel context: the making of weapons. Recent gene-function studies further support this conclusion and also provide some unexpected insights into the potential developmental and genetic mechanisms underlying the diversification of beetle weaponry. Specifically, gene knockdown studies (Moczek & Rose 2009) suggest that even closely related species can diverge rather significantly in exactly how these genes may regulate horn expression in a body region-, body size-, or sex-specific manner. Conservation of gene function over large phylogenetic distances apparently does not preclude diversification of genetic modifiers over shorter distances.

Similar conclusions emerge on examination of the regulation of the second major developmental period crucial to the expression of beetle horns, the often extensive sex- and species-specific remodelling during the pupal stage. Recent work now strongly implicates programmed cell death (PCD) in this process (Moczek 2006*b*, Kijimoto *et al.* 2010), an ancient physiological process employed by all metazoan organisms to dispose of cells during development (Potten & Wilson 2004). But here, too, co-option of an ancestral and highly conserved developmental mechanism appears to have been followed by the recruitment of modifier mechanisms. For example, data suggest that in some cases the activation of PCD may have come under the regulation of *Hox* genes, which organise body region development along the anterior–posterior axis (Wasik *et al.* 2010). Mechanisms such as these now permit closely related species to remodel pupal horns by utilising the same developmental machinery, PCD, in a highly species- sex-, and body region-specific manner.

More generally, these results are in line with a major theme in the evolution of novel traits: new morphologies do not require new genes or developmental pathways and instead may arise by recruiting existing developmental mechanisms into new contexts. However, these results also revealed an unexpected degree of evolutionary lability over short phylogenetic distances. In the sections that follow we use these insights into the evolutionary developmental genetics of horn formation to explore what they can teach us about four important aspects of the contest biology of beetles: the origins of alternative male morphs and sexual dimorphisms, the origins of the first weapons and the role of trade-offs in shaping weapon evolution.

9.7.2 The origins of male dimorphism in horned beetles

The horns of male beetles are not only famous for their impressive sizes and shapes but also for the remarkable *variation* in horn expression found among males of the same species (e.g. Arrow 1951). As introduced earlier, a peculiar form of variation is horn polyphenism, the existence of nutrition-induced, alternative horned (major) and hornless (minor) male morphs separated by a critical body size. Two hypotheses have been proposed to explain the evolutionary origin of polyphenic development in horned beetles (Emlen & Nijhout 2000, Moczek 2005). One hypothesis is that polyphenic development in horned beetles evolved from a *hornless* ancestor, in which genotypes acquired the ability to induce horn growth in males that exceed a certain body size. In this scenario, horn-inductive processes such as horn primodial growth or the expression of patterning genes should be visible in large males only. The other hypothesis is that, in contrast, polyphenic expression of horns evolved from obligately horned ancestors in which all males expressed horns proportional to their body size and large males were simply enlarged versions of small males. Genotypes then acquired the ability to *suppress* horn development below a certain body size threshold. In this scenario, horn-inductive processes may be observed in all males regardless of size, in addition to incomplete activation of inhibitory processes in small males only. Studies of both comparative morphology and candidate genes support the second hypothesis. Small males grow obvious, although small, horn primordia, which express many of the same patterning genes as their larger, and fully horned, counterparts (Moczek & Nagy 2005, Moczek *et al.* 2006*a*). Clearly, the absence of large horns in small males cannot be explained simply by a failure to activate horn formation. However, recent genomic approaches further complicate the picture as they show that even though the horn primordia of large and small males co-express many of the same genes there are also many important differences. For instance, genes involved in insulin signalling, sex determination and cuticle development differ in their expression patterns between horned and hornless morphs (Snell-Rood *et al.* 2011). More generally, these results highlight that transcription profiles of the same body structure in different individuals exposed to different environmental conditions can exhibit both obvious similarities and remarkable differences. An additional complication arises from the fact that the 'same' horn polyphenism in different closely related species can be generated via different developmental means. For example, male horn polyphenism in *O. taurus* and *O. nigriventris* is generated largely via dimorphic prepupal growth. In contrast, in the congener *O. binodis* prepupal growth generates monomorphic male pupae and it is the action of differential, body-size dependent, pupal remodelling that causes smaller males to resorb horn primordia nearly entirely whereas larger males convert almost all of their pupal horn into a corresponding adult structure. Clearly, there are multiple, and possibly independent, developmental routes to achieving the same pattern of morphological variation. Importantly, morphological variation among adults provides few if any insights into the developmental mechanisms responsible for this pattern. As can be seen in the next section, very similar conclusions emerge when the origins of sexual dimorphisms in horned beetles are examined.

9.7.3 The origins of sexual dimorphism in horned beetles

The same basic two hypotheses regarding the origin of male dimorphism (section 9.7.2) can be applied to the origin of sexual dimorphism. Sexual dimorphism in horn expression could have evolved from an ancestor in which both sexes lacked horns and genotypes acquired the ability to express horns in one sex only. If correct, horn-inductive processes such as growth of primordia or expression of patterning genes should be evident only in the horn-possessing sex. Alternatively, sexual dimorphism may have evolved from an ancestor in which both sexes possessed the ability to grow horns and one sex secondarily acquired the ability to inhibit horn formation. As before, under this scenario some horn-inductive processes should be evident in both sexes. In the genus *Onthophagus* the evidence is contradictory, and the only pattern that seems to emerge is one that segregates findings by horn type: in all species studied so far, a sexual dimorphism in *head* horns has been the product of sex-specific prepupal growth (Moczek 2005, 2006*b*). The sex that lacks the head horn as an adult also lacks any signs of horn primordial growth or the expression of possible horn patterning genes. Things are more complicated for sexual dimorphism in *thoracic* horn expression. Here, both sex-specific prepupal growth *and* remodelling contribute to the final sexual dimorphism evident among adults. Furthermore, the relative contributions of both processes, and the sex in which they operate, differ widely between species (Moczek *et al.* 2006*b*).

The last complication arises if we consider species, such as *O. sagittarius*, that express sexual dimorphism in both a thoracic horn *and* a head horn. Female head and thoracic horns are far larger than those of their male counterparts, but the differences emerge through different developmental means: during the prepupal stage head horns grow in a sexually dimorphic manner while thoracic horns grow similarly in both sexes. During the subsequent pupal stage it is now the thoracic horns that are remodelled in a sexually dimorphic manner while head horns are not. This is particularly intriguing because in the resulting adult females, head and thoracic horns appear to function together and form an integrated clasping device. To be functional, however, each horn must have the proper length relative to the other. Extrapolating from the adult pattern, we would thus predict that both parts should develop together in a coordinated fashion: the developmental reality is, however, different.

Combined, these results highlight how understanding the developmental underpinnings of trait diversity in horned beetles can shed light on the evolutionary history of this variation (Moczek *et al.* 2006*b*). For example, mounting evidence now suggests that head and thoracic horns evolved independently of each other and that there are multiple developmental routes to the evolution of sexual and male dimorphism. Understanding the developmental origins of adult diversity patterns puts us in a better position to understand past evolutionary trajectories and to better predict why certain evolutionary responses to changes in ecological conditions, including contests, may be manifest in some species but not others. The next two sections (9.7.4 and 9.7.5) highlight two specific examples in which developmental insights have forced a substantial revision of our understanding of the origins of horns as weapons and horn diversity and of the selective pressures shaping this diversity in past and current populations.

9.7.4 The dual use of beetle horns: implications for studying weapon evolution

In a phylogenetic analysis of 48 *Onthophagus* species, Emlen *et al.* (2005*a*,*b*) examined the phylogenetic distribution of thoracic horns and concluded that, in order to explain present-day diversity patterns, they must have originated independently 9 times in males and 7 times in females. Closer examination of horn development in a subset of species subsequently revealed that many, and possibly all, of the species included in the analysis grow thoracic horns in both sexes but resorb horn primordia in one or both sexes prior to adulthood (Moczek *et al.* 2006*a*,*b*). Phylogenetic reanalysis now suggests that pupal thoracic horns may have evolved only once, and what truly mediated the diversification of thoracic horns was a given lineage's ability to retain, reshape or resorb horns during the pupal stage. While these results illustrate the general value of developmental studies in evolutionary biology, they also raise an important question: why do one or both sexes in so many species grow what can be very substantial thoracic horn primordia during the pupal stage only to resorb them prior to adulthood? In other words, what is the adaptive significance, if any, of transient horn expression? A combination of comparative histological and surgical approaches have now revealed that pupal horns, irrespective of whether they give rise to a corresponding adult structure, play an important role during the larval-to-pupal moult and the shedding of the larval head capsule (Moczek *et al.* 2006*b*). Replicating this approach within and outside the genus *Onthophagus* further showed that the moulting function of thoracic horn primordia appears unique to onthophagine beetles. Phylogenetic analyses suggest that the moulting function of pupal horns preceded, in evolutionary time, the weapon function in adults (Moczek *et al.* 2006*b*). If correct, this would explain the maintenance of pupal thoracic horns in lineages that lack corresponding structures in the adult stage. These findings are important for research focused on the evolution of beetle contests and weapons. For example, in lineages that express both pupal horns (for moulting) and adult horns (for contests), these dual, and very diverse, functions of horns, may constrain future adaptation to one context alone, yet any constraint imposed by the pupal moulting function of horns may not be obvious from studies of adult behavior alone.

9.7.5 The role of developmental trade-offs in shaping weapon and species diversity

Section 9.6 highlighted how polyphenic expression of beetle weapons provides additional targets such as threshold values for selection to act on, thus contributing to the diversification of populations and species. We end this chapter by highlighting a third, and particularly intriguing, avenue by which polyphenic development facilitates diversification: resource allocation trade-offs during development. Resource allocation trade-offs arise during development when two or more structures compete for a shared and limited resource to sustain their growth. As such, resource allocation trade-offs not only have the potential to alter ontogenetic outcomes, as development enlargements of one structure may only be feasible at the expense of another, but also evolutionary trajectories, as development may only be able to accommodate evolutionary enlargements of one structure through compensatory reduction of another. A series of studies on horned beetles has begun to implicate resource allocation trade-offs in the diversification of horns with intriguing implications for the diversification of horned beetle species (Emlen 2001, Kawano 2002, Moczek & Nijhout 2004, Parzer & Moczek 2008).

Consider the giant rhinoceros beetle genus *Chalcosoma*, famous for the extreme size of males horns used in elaborate contest behaviour. Kawano (2002) showed that two *Chalcosoma* species had diverged in both relative horn sizes and copulatory organ sizes and that this divergence was more pronounced between sympatric (overlapping) than allopatric (geographically separated) populations. His findings were consistent with reproductive character displacement reinforced in sympatry but not allopatry. What was intriguing, however, was the direction of divergence: the species which had evolved relatively longer horns had also evolved relatively shorter copulatory organs, and vice versa. A subsequent study on *O. taurus* (Moczek & Nijhout 2004) suggested that this antagonistic coevolution may not be coincidental. In this study, surgical ablation of the genital precursor tissue during development resulted in males with disproportionally longer horns. This suggests that the development of horns and of copulatory organs is indeed connected. Enlargement of one structure may have to come at the expense of the other. This implication is intriguing for studies into the evolution of contests in general and weapons in particular, because it suggests that evolutionary changes in the expression of weaponry might directly affect the size of reproductive structures, which in turn are thought to play a major role in the evolution of reproductive isolation, and thus speciation (Eberhard 1985). To do so, however, these developmental interactions would also have to leave an evolutionary signature.

The strongest evidence to date documenting exactly that kind of signature comes from a recent study that examined both within- and between-species co-variation in the relative investment into horns versus copulatory organs (Parzer & Moczek 2008). Morphometric quantification of this investment across four geographically isolated populations revealed a near perfect negative correlation: populations in which males invest more into horns invest less into copulatory organs and vice versa. When the same approach was applied to 10 different *Onthophagus* species, the same highly significant negative correlation between relative investment into horns versus copulatory organ size emerged. Combined, these data suggest that copulatory organ size trades-off with horn size, that the former may diverge as a by-product of evolutionary changes in the latter, and that the resulting signatures of antagonistic co-evolution are detectable both over micro- as well as macro-evolutionary time scales. Given the significance of copulatory organ morphology for reproductive isolation these findings also begin to raise the possibility that diversification of contest weaponry may secondarily promote incipient reproductive isolation.

9.8 Conclusions: beetles as a system for studying contest behaviour

The study of contests in beetles has provided complementary and novel insights relative to fighting behaviour in other systems. Many observations of beetle fighting support theory and observations from other taxa. For instance, body size and relative weapon investment are major determinants of contest outcome. In addition, the evolution of fighting and weaponry is associated with valuable and defendable resources, such as tunnels or food patches.

Beetles as a system come with several advantages that allow several unique opportunities for studies of the evolution of contest behaviour. First, there have been multiple independent evolutionary origins of weaponry in beetles, allowing for tests of associations between weaponry and ecological variables (e.g. Emlen & Philips 2006, Emlen 2008). Second, beetles (in particular the genera *Tribolium* and *Onthophagus*) are emerging model systems in evolutionary developmental biology (Angelini *et al.* 2009, Moczek & Rose 2009), which permits enquiry into the evolution of fighting and weaponry from perspectives not available in other taxa. For instance, as outlined in the above section, an understanding of the development of beetle horns has revealed that multiple origins of adult horns (Emlen *et al.* 2005*a*,*b*) may have been facilitated by the exaptation of pupal horns, which are important for cuticle shedding during metamorphosis (Moczek 2006*b*). Third, beetles, like many invertebrate systems, are often amenable to large laboratory studies of genetics and artificial selection. Such experiments have yielded unique insights into the evolution of fighting. For instance, artificial selection on relative horn length has revealed that fighting behaviour and morphological traits associated with horn length are genetically correlated and thus likely co-evolve (Okada & Miyatake 2009).

In conclusion, the study of contest behaviour in beetles provides many opportunities to test hypotheses concerning the evolution of fighting behaviour. Taking advantage of the diversity and power of beetle systems provides many exciting opportunities for future research, such as:

1. investigating developmental biases that may promote and shape 'independent' evolutionary origins of weaponry;
2. testing the effect of sexual selection intensity on evolutionary changes in fighting behaviour (similar to Simmons & Garcia-Gonzalez 2008);
3. determining the mechanisms that facilitate diversification in weaponry (as discussed in Emlen 2008);
4. testing links between sexual selection and speciation (e.g. West-Eberhard 1983) for systems dominated by male–male contests relative to female choice.

Acknowledgements

We are grateful to the editors and two anonymous reviewers for comments that helped to improve this manuscript. This chapter was written while ESR was supported by NIH NRSA F32GM083830; the content does not necessarily represent the official views of the National Institutes of Health. Additional support was provided by National Science Foundation grants IOS 0445661 and IOS 0718522 to APM.

References

Abbott JC, Dunbrack RL & Orr CD (1985) The interaction of size and experience in dominance relationships of juvenile steelhead trout *(Salmo gairdneri)*. *Behaviour*, 92, 241–253.

Andersson M (1994) *Sexual Selection*. Princeton, NJ: Princeton University Press.

Angelini DR, Kikuchi M & Jockusch EL (2009) Genetic patterning in the adult capitate antenna of the beetle *Tribolium castaneum*. *Developmental Biology*, 327, 240–251.

Arrow GJ (1951) *Horned Beetles*. The Hague: Junk Publishers.

Brown L & Bartalon J (1986) Behavioral correlates of male morphology in a horned beetle. *American Naturalist*, 127, 565–570.

Chamorro-Florescano IA & Favila ME (2008) Male reproductive status affects contest outcome during nidification in *Canthon cyanellus cyanellus* LeConte (Coleoptera: Scarabaeidae). *Behaviour*, 145, 1811–1821.

Chase ID, Bartolomeo C & Dugatkin LA (1994) Aggressive interactions and inter-contest interval – How long do winners keep winning? *Animal Behaviour*, 48, 393–400.

Conner J (1988) Field measurements of natural and sexual selection in the fungus beetle *Bolitotherus cornutus*. *Evolution*, 42, 736–749.

Conner J (1989) Density-dependent sexual selection in the fungus beetle, *Bolitotherus cornutus*. *Evolution*, 43, 1378–1386.

Convey P (1989) Influences on the choice between territorial and satellite behaviour in male *Libellula quadrimaculata* Linn. (Odonata: Libellulidae). *Behaviour*, 109, 125–141.

Cook DF (1990) Differences in courtship, mating and postcopulatory behaviour between male morphs of the dung beetles *Onthophagus binodis* Thunberg (Coleoptera: Scarabaeidae). *Animal Behaviour*, 40, 428–436.

Crespi BJ (1988) Adaptation, compromise, and constraint: the development, morphometrics, and behavioral basis of a fighter–flier polymorphism in male *Hoplothrips karnyi* (Insecta: Thysanoptera). *Behavioural Ecology and Sociobiology*, 23, 93–104.

Darwin C (1874) *The Descent of Man, and Selection in Relation to Sex*. New York: A. L. Burt.

Dominey WJ (1984) Alternative mating tactics and evolutionarily stable strategies. *American Zoologist*, 24, 385–396.

Eberhard WG (1977) Fighting behavior of *Golofa porter* beetles (Scarabeidae: Dynastinae). *Psyche*, 83, 292–298.

Eberhard WG (1979) The function of horns in *Podischnus agenor* (Dynastinae) and other beetles. In: MS Blum & NA Blum (eds.) *Sexual Selection and Reproductive Competition in Insects*, pp. 231–258. New York, NY: Academic Press.

Eberhard WG (1980) Horned beetles. *Scientific American*, 242, 166–182.

Eberhard WG (1981) The natural history of *Doryphora* sp. (Coleoptera, Chrysomelidae) and the function of its sterna horn. *Annals of the Entomological Society of America*, 74, 445–448.

Eberhard WG (1982) Beetle horn dimorphism: Making the best of a bad lot. *American Naturalist*, 119, 420–426.

Eberhard WG (1983) Behavior of adult bottle brush weevils (*Rhinostomus barbirostris*) (Coleoptera: Curculionidae). *Revista de Biologia Tropical*, 31, 233–244.

Eberhard WG (1985) *Sexual Selection and Animal Genitalia*. Cambridge, MA: Harvard University Press.

Eberhard WG (1987) Use of horns in fights by the dimorphic males of *Ageopsis nigricollis* (Coleoptera, Scarabeidae, Dynastinae). *Journal of the Kansas Entomological Society*, 60, 504–509.

Eberhard WG & Garcia-C JM (2000) Ritual jousting by horned *Parisoschoenus expositus* weevils (Coleoptera: Curculionidae, Baridinae). *Psyche*, 103, 55–84.

Eberhard WG & Gutiérrez EE (1991) Male dimorphism in beetles and earwigs and the question of developmental constraints. *Evolution*, 45, 18–28.

Eberhard WG, Garcia-C JM & Lobo J (2000) Size-specific defensive structures in a horned weevil confirm a classic battle plan: avoid fights with larger opponents. *Proceedings of the Royal Society of London B*, 267, 1129–1134.

Emlen DJ (1994) Environmental control of horn length dimorphism in the beetle *Onthophagus acuminatus* (Coleptera: Scarabaeidae). *Proceedings of the Royal Society of London B*, 256, 131–136.

Emlen DJ (1996) Artificial selection on horn length–body size allometry in the horned beetle *Onthophagus acuminatus* (Coleoptera: Scarabaeidae). *Evolution*, 50, 1219–1230.

Emlen DJ (1997) Alternative reproductive tactics and male-dimorphism in the horned beetle *Onthophagus acuminatus* (Coleoptera: Scarabaeidae). *Behavioral Ecology and Sociobiology*, 41, 335–341.

Emlen DJ (2001) Costs and the diversification of exaggerated animal structures. *Science*, 291, 1534–1536.

Emlen DJ (2008) The evolution of animal weapons. *Annual Review of Ecology Evolution and Systematics*, 39, 387–413.

Emlen DJ & Nijhout HF (1999) Hormonal control of male horn length dimorphism in the dung beetle *Onthophagus taurus* (Coleoptera: Scarabaeidae). *Journal of Insect Physiology*, 45, 45–53.

Emlen DJ & Nijhout HF (2000) The development and evolution of exaggerated morphologies in insects. *Annual Review of Entomology*, 45, 661–708.

Emlen DJ & Nijhout HF (2001) Hormonal control of male horn length dimorphism in *Onthophagus taurus* (Coleoptera: Scarabaeidae): A second critical period of sensitivity to juvenile hormone. *Journal of Insect Physiology*, 47, 1045–1054.

Emlen DJ & Philips TK (2006) Phylogenetic evidence for an association between tunneling behavior and the evolution of horns in dung beetles (Coleoptera:

Scarabaeidae: Scarabaeinae). *Coleopterists Society Monograph*, 5, 47–56.

Emlen DJ, Hunt J & Simmons LW (2005*a*) Evolution of sexual dimorphism and male dimorphism in the expression of beetle horns: Phylogenetic evidence for modularity, evolutionary lability, and constraint. *American Naturalist*, 166, S42–S68.

Emlen DJ, Marangelo J, Ball B & Cunningham CW (2005*b*) Diversity in the weapons of sexual selection: Horn evolution in the beetle genus *Onthophagus* (Coleoptera: Scarabaeidae). *Evolution*, 59, 1060–1084.

Emlen ST & Oring LW (1977) Ecology, sexual selection, and the evolution of mating systems. *Science*, 197, 215–223.

Enquist M & Leimar O (1983) Evolution of fighting behaviour, decision rules and assessment of relative strength. *Journal of Theoretical Biology*, 102, 387–410.

Enquist M & Leimar O (1987) Evolution of fighting behaviour – The effect of variation in resource value. *Journal of Theoretical Biology*, 127, 187–205.

Enquist M & Leimar O (1990) The evolution of fatal fighting. *Animal Behaviour*, 39, 1–9.

Espmark Y (1964) Studies in dominance–subordinate relationships in a group of semi-domestic reindeer (*Rangifer tarandus* L.). *Animal Behaviour*, 12, 420–426.

Franz NM (2003) Mating behaviour of *Staminodeus vectoris* (Coleoptera: Curculionidae), and the value of systematics in behavioral studies. *Journal of Natural History*, 37, 1727–1750.

Gadgil M (1972) Male dimorphism as a consequence of sexual selection. *American Naturalist*, 106, 574–580.

Goldsmith SK (1987) The mating system and alternative reproductive behaviours of *Dendrobias mandibularis* (Coleoptera: Cerambycidae). *Behavioural Ecology and Sociobiology*, 20, 111–115.

Gross MR (1996) Alternative reproductive strategies and tactics: Diversity within sexes. *Trends in Ecology and Evolution*, 11, 92–98.

Hamilton WD (1979) Wingless and fighting males in fig wasps and other insects. In: MS Blum & NA Blum (eds.) *Sexual Selection and Reproductive Competition in Insects*, pp. 167–220. New York, NY: Academic Press.

Hammerstein P (1981) The role of asymmetries in animal contests. *Animal Behaviour*, 29, 193–205.

Hanley RS (2001) Mandibular allometry and male dimorphism in a group of obligately mycophagous beetles (Insecta: Coleoptera: Staphylinidae: Oxyporinae). *Biological Journal of the Linnean Society*, 72, 451–459.

Hongo Y (2003) Appraising behavior during male–male interaction in the Japanese horned beetle *Trypoxylus dichotomus septentrionalis* (Kono). *Behaviour*, 140, 501–517.

Howard RS (1955) The occurrence of fighting behavior in the grain beetle *Tenebrio molitor* with the possible formation of a dominance hierarchy. *Ecology*, 36, 281–285.

Hsu YY & Wolf LL (2001) The winner and loser effect: What fighting behaviours are influenced? *Animal Behaviour*, 61, 777–786.

Hughes AL & Hughes MK (1987) Asymmetric contests among sawyer beetles (Cerambycidae: *Monochamus notatus* and *Monochamus scutellatus*). *Canadian Journal of Zoology*, 65, 823–827.

Hunt J & Simmons LW (2000) Maternal and paternal effects on offspring phenotype in the dung beetle *Onthophagus taurus*. *Evolution*, 54, 936–941.

Hunt J & Simmons LW (2001) Status-dependent selection in the dimorphic beetle *Onthophagus taurus*. *Proceedings of the Royal Society of London B*, 268, 2409–2414.

Johnson LK (1982) Sexual selection in a brentid weevil. *Evolution*, 36, 251–262.

Judge KA & Bonanno VL (2008) Male weaponry in a fighting cricket. *PLoS ONE*, 3, 10.

Karino K, Niiyama H & Chiba M (2005) Horn length is the determining factor in the outcomes of escalated fights among male Japanese horned beetles, *Allomyrina dichotoma* L. (Coleoptera : Scarabaeidae). *Journal of Insect Behavior*, 18, 805–815.

Kawano K (2002) Character displacement in giant rhinoceros beetles. *American Naturalist*, 159, 255–271.

Kawano K (2006) Sexual dimorphism and the making of oversized male characters in beetles (Coleoptera). *Annals of the Entomological Society of America*, 99, 327–341.

Kijimoto T, Andrews J & Moczek AP (2010) Programmed cell death shapes the expression of horns within and between species of horned beetles. *Evolution and Development*, 12, 449–458.

King A & Fashing N (2007) Infanticidal behavior in the subsocial beetle *Odontotaenius disjunctus* (Illiger) (Coleoptera: Passalidae). *Journal of Insect Behavior*, 20, 527–536.

Kruse KC & Switzer PV (2007) Physical contests for females in the Japanese beetle, *Popillia japonica*. *Journal of Insect Science*, 7, 34.

Longair RW (2004) Tusked males, male dimorphism and nesting behavior in a subsocial Afrotropical wasp, *Synagris cornuta*, and weapons and dimorphism in the genus (Hymenoptera: Vespidae: Eumeninae). *Journal of the Kansas Entomological Society*, 77, 528–557.

Lloyd JE (1979) Sexual selection in luminescent beetles. In: MS Blum & NA Blum (eds.) *Sexual Selection and Reproductive Competition in Insects*, pp. 293–342. New York, NY: Academic Press.

Madewell R & Moczek AP (2006) Horn possession reduces maneuverability in the horn-polyphenic beetle, *Onthophagus nigriventris. Journal of Insect Science*, 6, article 21.

Mason LG (1980) Sexual selection and the evolution of pair bonding in solider beetles. *Evolution*, 34, 174–180.

Maynard Smith JM (1974) Theory of games and evolution of animal conflicts. *Journal of Theoretical Biology*, 47, 209–221.

Maynard Smith JM (1979) Game theory and the evolution of behaviour. *Proceedings of the Royal Society of London B*, 205, 475–488.

McLachlan AJ & Allen DF (1987) Male mating success in Diptera: Advantages of small size. *Oikos*, 48, 11–14.

McLain DK (1982) Behavioral and morphological correlates of male dominance and courtship persistence in the blister beetle *Epicauta pennsylvanica* (Coleoptera: Meoloidae). *American Midland Naturalist*, 107, 396–403.

Miller KB & Wheeler QD (2005) Asymmetrical male mandibular horns and mating behavior in *Agathidium* Panzer (Coleoptera: Leiodidae). *Journal of Natural History*, 39, 779–792.

Moczek AP (2003) The behavioral ecology of threshold evolution in a polyphenic beetle. *Behavioral Ecology*, 14, 841–854.

Moczek AP (2005) The evolution and development of novel traits, or how beetles got their horns. *BioScience*, 11, 935–951.

Moczek AP (2006*a*) Integrating micro- and macroevolution of development through the study of horned beetles. *Heredity*, 97, 168–178.

Moczek AP (2006*b*) Pupal remodeling and the development and evolution of sexual dimorphism in horned beetles. *American Naturalist*, 168, 711–729.

Moczek AP (2009) Developmental plasticity and the origins of diversity: a case study on horned beetles. In: TN Ananthakrishnan & D Whitman (eds.) *Phenotypic Plasticity in Insects: Mechanisms and Consequences*, pp. 27–80. Plymouth, UK: Science Publishers Inc.

Moczek AP & Emlen DJ (1999) Proximate determination of male horn dimorphism in the beetle *Onthophagus taurus* (Coleoptera: Scarabaeidae). *Journal of Evolutionary Biology*, 12, 27–37.

Moczek AP & Emlen DJ (2000) Male horn dimorphism in the scarab beetle, *Onthophagus taurus*: Do alternative reproductive tactics favour alternative phenotypes? *Animal Behaviour*, 59, 459–466.

Moczek AP & Nagy LM (2005) Diverse developmental mechanisms contribute to different levels of diversity in horned beetles. *Evolution and Development*, 7, 175–185.

Moczek AP & Nijhout HF (2002) Developmental mechanisms of threshold evolution in a polyphenic beetle. *Evolution and Development*, 4, 252–264.

Moczek AP & Nijhout HF (2003) Rapid evolution of a polyphenic threshold. *Evolution and Development*, 5, 259–268.

Moczek AP & Nijhout HF (2004) Trade-offs during the development of primary and secondary sexual traits in a horned beetle. *American Naturalist*, 163, 184–191.

Moczek AP & Rose DJ (2009) Differential recruitment of limb patterning genes during development and diversification of beetle horns. *Proceedings of the National Academy of Sciences USA*, 106, 8992–8997.

Moczek AP, Brühl CA & Krell F-T (2004) Linear and threshold-dependent expression of secondary sexual traits in the same individual: Insights from a horned beetle (Coleoptera: Scarabaeidae). *Biological Journal of the Linnean Society*, 83, 473–480.

Moczek AP, Rose D, Sewell W, & Kesselring BR (2006*a*) Conservation, innovation, and the evolution of horned beetle diversity. *Development Genes and Evolution*, 216, 655–665.

Moczek AP, Cruickshank, TE, & Shelby JA (2006*b*) When ontogeny reveals what phylogeny hides: Gain and loss of horns during development and evolution of horned beetles. *Evolution*, 60, 2329–2341.

Nijhout HF & Emlen DJ (1998) Competition among body parts in the development and evolution of insect morphology. *Proceedings of the National Academy of Sciences USA*, 95, 3685–3689.

Okada K & Miyatake T (2007) Ownership-dependent mating tactics of minor males of the beetle *Librodor japonicus* (Nitidulidae) with intra-sexual dimorphism of mandibles. *Journal of Ethology*, 25, 255–261.

Okada K & Miyatake T (2009) Genetic correlations between weapons, body shape and fighting behaviour in the horned beetle *Gnatocerus cornutus. Animal Behaviour*, 77, 1057–1065.

Okada K & Miyatake T (2010) Effect of losing on male fights of broad-horned flour beetle, *Gnatocerus cornutus. Behavioral Ecology and Sociobiology*, 64, 361–369.

Okada K, Miyanoshita A & Miyatake T (2006) Intra-sexual dimorphism in male mandibles and male aggressive behavior in the broad-horned flour beetle *Gnatocerus cornutus* (Coleoptera: Tenebrionidae). *Journal of Insect Behavior*, 19, 457–467.

Okada K, Miyatake T, Nomura Y & Kuroda K (2008) Fighting, dispersing, and sneaking: Body-size dependent mating tactics by male *Librodor japonicus* beetles. *Ecological Entomology*, 33, 269–275.

O'Neill KM, Evans HE & O'Neill RP (1989) Phenotypic correlates of mating success in the sand wasp

Bembecinus quinquespinosus (Hymenoptera: Sphecidae). *Canadian Journal of Zoology*, 67, 2557–2568.

Otronen M (1988*a*) The effect of body size on the outcome of fights in burying beetles (*Nicrophorus*). *Annales Zoologici Fennici*, 25, 191–201.

Otronen M (1988*b*) Intra- and intersexual interactions at breeding burrows in the horned beetle, *Coprophanaeus ensifer*. *Animal Behaviour*, 36, 741–748.

Palmer TJ (1978) A horned beetle which fights. *Nature*, 274, 583–584.

Panhuis TM, Butlin R, Zuk M, *et al.* (2001) Sexual selection and speciation. *Trends in Ecology and Evolution*, 16, 364–371.

Parker GA (1974) Assessment strategy and evolution of fighting behavior. *Journal of Theoretical Biology*, 47, 223–243.

Parzer HF & Moczek AP (2008) Rapid antagonistic coevolution between primary and secondary sexual characters in horned beetles. *Evolution*, 62, 2423–2428.

Pizzo A, Roggero A, Palestrini C, *et al.* (2008) Rapid shape divergences between natural and introduced populations of a horned beetle partly mirror divergences between species. *Evolution and Development*, 10, 166–175.

Pomfret JC & Knell RJ (2006) Sexual selection and horn allometry in the dung beetle *Euoniticellus intermedius*. *Animal Behaviour*, 71, 567–576.

Potten C & Wilson J (2004) *Apoptosis: The Life and Death of Cells*. Cambridge: Cambridge University Press.

Rasa OAE (1999) Division of labour and extended parenting in a desert tenebrionid beetle. *Ethology*, 105, 37–56.

Rasmussen JL (1994) The influence of horn and body size on the reproductive behavior of the horned rainbow Scarab beetle *Phanaeus difformis* (Coleoptera: Scarabaeidae). *Journal of Insect Behavior*, 7, 67–82.

Rowland JM & Emlen DJ (2009) Two thresholds, three male forms result in facultative male trimorphism in beetles. *Science*, 323, 773–776.

Rubenstein DI (1984) Resource acquisition and alternative mating strategies in water striders. *American Zoologist*, 24, 345–353.

Rudinsky JA & Ryker LC (1976) Sound production in Scolytidae: rivalry and pre-mating stridulation of male Douglas-fir beetles. *Journal of Insect Physiology*, 22, 997–1003.

Schuett GW (1997) Body size and agonistic experience affect dominance and mating success in male copperheads. *Animal Behaviour*, 54, 213–224.

Schuster JC (1983) Acoustical signals of Passalid beetles: Complex repertoires. *Florida Entomologist*, 66, 486–496.

Schuster JC & Schuster LB (1985) Social behavior in Passalid beetles (Coleoptera: Passalidae): Cooperative brood care. *Florida Entomologist*, 68, 266–272.

Setsuda K, Tsuchida K, Watanabe H, *et al.* (1999) Size dependent predatory pressure in the Japanese horned beetle, *Allomyrina dichotoma* L. (Coleoptera; Scarabaeidae). *Journal of Ethology*, 17, 73–77.

Shelby JA, Madewell R & Moczek AP (2007) Juvenile hormone mediates sexual dimorphism in horned beetles. *Journal of Experimental Zoology*, 308B, 417–427.

Simmons LW & Emlen DJ (2006) Evolutionary trade-off between weapons and testes. *Proceedings of the Royal Society of London B*, 103, 16346–16351.

Simmons LW & Garcia-Gonzalez F (2008) Evolutionary reduction in testes size and competitive fertilization success in response to the experimental removal of sexual selection in dung beetles. *Evolution*, 62, 2580–2591.

Simmons LW, Tomkins JL & Hunt J (1999) Sperm competition games played by dimorphic male beetles. *Proceedings of the Royal Society of London B*, 266, 145–150.

Siva-Jothy MT (1987) Mate securing tactics and the cost of fighting in the Japanese Horned Beetle, *Allomyrina dichotoma* L. (Scarabaeidae). *Journal of Ethology*, 5, 165–172.

Snell-Rood EC, Cash A, Han MV, *et al.* (2011) Developmental decoupling of alternative phenotypes: Insights from the transcriptomes of horn-polyphenic beetles. *Evolution*, 65, 231–245.

Svacha P (1992) What are and what are not imaginal discs: Reevaluation of some basic concepts (Insecta, Holometabola). *Developmental Biology*, 154, 101–117.

Tomkins JL, Kotiaho JS & LeBas NR (2005) Matters of scale: Positive allometry and the evolution of male dimorphisms. *American Naturalist*, 165, 389–402.

Thornhill R & Alcock J (1983) *The Evolution of Insect Mating Systems*. New York, NY: iUniverse.com, Inc.

Trumbo ST (2007) Defending young biparentally: Female risk-taking with and without a male in the buyring beetle, *Nicrophorus pustulatus*. *Behavioral Ecology and Sociobiology*, 61, 1717–1723.

Wasik BR, Rose DJ & Moczek AP (2010) Beetle horns are regulated by the *Hox* gene, sex combs reduced, in a species- and sex-specific manner. *Evolution and Development*, 12, 353–362.

Watson NL & Simmons LW (2010*a*) Mate choice in the dung beetle *Onthophagus sagittarius*: Are female horns ornaments? *Behavioral Ecology*, 21, 424–430.

Watson NL & Simmons LW (2010*b*) Reproductive competition promotes the evolution of female weaponry.

Proceedings of the Royal Society of London B, 277, 2035–2040.

Wells MS (1988) Effects of body size and resource value on fighting behaviour in a jumping spider. *Animal Behaviour*, 36, 321–326.

West-Eberhard MJ (1983) Sexual selection, social competition, and speciation. *Quarterly Review of Biology*, 58, 155–183.

West-Eberhard MJ (2003) *Developmental Plasticity and Evolution*. Oxford: Oxford University Press.

Whiting MF, Bradler S & Maxwell T (2003) Loss and recovery of wings in stick insects. *Nature*, 421, 264–267.

Windsor DM (1987) Natural history of a subsocial tortoise beetle *Acromis sparsa* Boheman (Chrysomelidae: Cassidinae) in Panama. *Psyche*, 94, 127–150.

Yamane T, Okada K, Nakayama S, *et al.* (2010) Dispersal and ejaculatory strategies associated with exaggeration of weapon in an armed beetle. *Proceedings of the Royal Society of London B*, 277, 1705–1710.

Chapter

10 Contest behaviour in fishes

Ryan L. Earley & Yuying Hsu

10.1 Summary

Fishes have been central to our understanding of many of the major aspects of contest behaviour, extending from Tinbergen's early work on social releasers to some of the initial tests of assessment models and now to the neuroendocrine and genomic regulation of aggression and dominance. In this chapter, we focus on some exciting areas of research in fish contest behaviour that promise to shed light on the multidimensionality of resource holding potential (RHP), sex- and size-related differences in decision-making during contests, whole-organism performance and fight outcomes, selection and potential constraints on contest behaviour; and the role of developmental plasticity in driving RHP-related phenotypic variation. We have developed this chapter more as a prospectus than a review, using the concrete foundation laid down by numerous researchers to highlight areas that could be of great import in the years to come. This approach, of course, leaves us with many unanswered questions that we hope will serve as a springboard for rigorous hypothesis testing using an integrative framework for fish contest behaviour.

10.2 Introduction

The formal study of fish aggression has a long and prolific history dating back at least 70 years to a curious observation of three-spined sticklebacks, *Gasterosteus aculeatus*, responding intensely to a red postal truck that would occasionally pass the window of Niko Tinbergen's laboratory (Kruuk 2003, p. 87). What triggered aggression in the sticklebacks, of course, was not the truck but rather the colour red, a trait that males boast on their throat and ventral surface during the breeding season (ter Pelwijk & Tinbergen 1937), and that might indicate an imminent threat to a resident male's territory (Bolyard & Rowland 1996). Tinbergen and his contemporaries subsequently made significant efforts to identify behavioural, morphological and chromatic 'releasers' of aggression (e.g. Seitz 1940, Tinbergen 1948 and references in Earley *et al.* 2000). These classic studies would serve as a springboard for fishes to be recognised as excellent models in which to investigate aggression and dominance. The field has since progressed beyond viewing fish as automatons that respond instinctually to 'social releasers'. Indeed, the behaviour that fish exhibit, and the decisions that they make during contests, are shaped in complex ways by social context, information gleaned through a battery of sensory modalities (e.g. olfactory, visual, electric, acoustic), prior fighting experience, motivational state (e.g. hunger, resource value), predation risk and physiological condition (Hsu *et al.* 2006). Contest performance, 'winning', 'losing', 'escalating' and so forth, then ultimately feeds back on individual fitness (e.g. LaManna & Eason 2011). Fitness consequences come in the form of differential access to sought-after resources such as mates, food and refuge, either through direct interference or through persistent experience-induced changes in behaviour. In some fish, the stakes are higher as dominance can dictate reproductive strategy, sexual maturity and even sex itself (Borowsky 1973, Rodgers *et al.* 2007, Desjardins & Fernald 2008).

Aggression and contest performance in fishes is not, however, entirely at the mercy of moment-by-moment fluctuations in internal, physical, social or community environments (Figure 10.1). Variation in behavioural and physiological responses to contests is linked to variation in breeding seasonality, parental care and social system, suggesting that the

Animal Contests, ed. I.C.W. Hardy and M. Briffa. Published by Cambridge University Press.

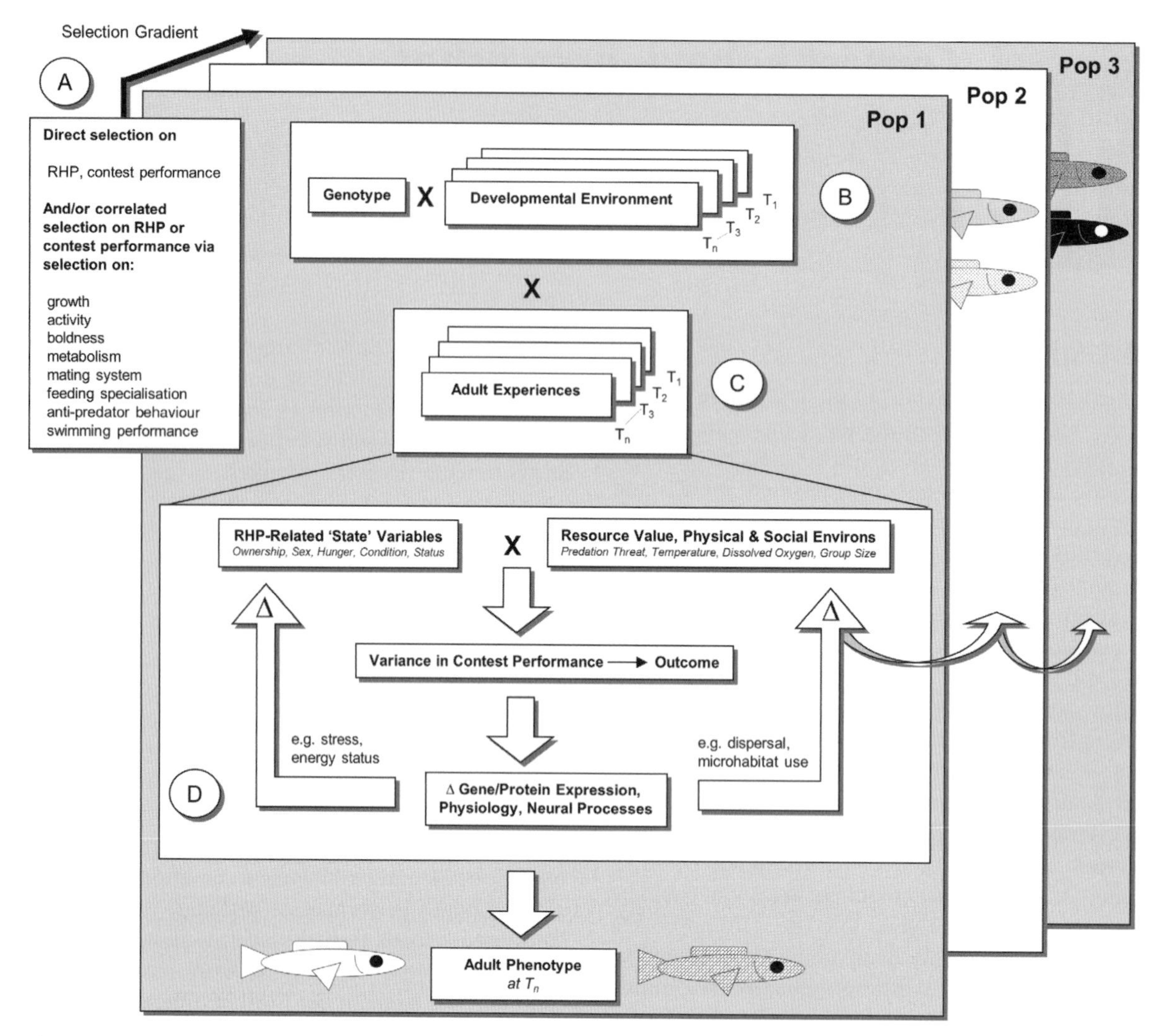

Figure 10.1 Summarising the complexities of contest behaviour in fishes. (**A**) divergent selection regimes result in population-level differences in RHP-related traits, (**B**) gene × environment interactions (developmental plasticity) shape the morphological, physiological, and behavioural phenotype in tandem with (**C**) adult experiences to give a 'snapshot' adult phenotype at any given moment. (**D**) Among-individual variance in contest performance will result from A to C plus the current state of the fish and the environmental and social context in which the fight occurs. The dynamics and outcome of the contest result in manifold changes in gene/protein expression, physiology and neural processes, which then feed back on the fish's 'state'. The fish might also disperse or alter microhabitat use following a contest, which will affect the environmental and social context in which future contests occur, or expose the fish and its progeny (if any) to different selection pressures. Δ refers to 'change', and T refers to a particular time point.

mechanisms of response are evolutionarily rooted in lifestyle (Hirschenhauser *et al.* 2004). Stable inter-individual and population-level differences in behaviour are well documented and result from genotypic variation among individuals, local adaptation of populations to historically prevailing environmental conditions and from plasticity during both development and adulthood (Brelin *et al.* 2008, Archard *et al.* 2012, reviewed by Foster & Endler 1999, Sih *et al.* 2004, Ghalambor *et al.* 2007, Sih & Bell 2008, Dingemanse *et al.* 2009, Stamps & Groothuis 2010*a*,*b*). Inter-individual differences in aggression and contest performance also co-vary predictably with numerous physiological (e.g. stress responsiveness: Øverli *et al.* 2007) and behavioural (e.g. boldness: Bell 2005) traits. Such co-variance can emerge from traits being under common genetic, cellular or neuroendocrine regulatory control. The result is that individual fish possess definable suites of phenotypic traits (commonly referred to as 'coping styles', 'personalities' or 'syndromes'), which can translate into individual variation in contest performance (Øverli *et al.* 2005) and underscore plasticity in aggressive responses across contexts (Briffa *et al.* 2008, Natarajan *et al.* 2009).

A diversity of assessment strategies, the potent effects of prior experience on future contest outcomes, the importance of social and community context on decision-making during contests and the links between physiology and contest behaviour have been illuminated through decades of research on fishes. We will use much of this work as a foundation for exploring questions that have received very little attention to date but that have the potential to advance the field in new and exciting ways. Some of these questions delve deeper into our understanding of assessment and experience effects to elucidate some complexities that deserve further attention. For instance, do individuals employ singular assessment strategies or can they switch between strategies depending on the phase and intensity of the contest? Have we underestimated the persistence of prior social experience effects on future contest performance? Other questions have a strong conceptual and theoretical foundation in other areas of behaviour research but have yet to be addressed in any significant way with respect to contest performance in fishes. Does developmental plasticity in response to either the physical or social environment influence the phenotype in ways that affect contest performance? Is contest success contingent on the match between the environment in which an individual is raised and the environment in which a contest occurs? Still other questions have gained some traction in other study systems but have been neglected in fishes. What can the biomechanics of performance traits, such as bite force and display behaviours, tell us about inter-individual variation in fighting success, constraints on behavioural performance and the evolution of contest behaviour?

We have structured this chapter as a prospectus rather than as a review. There is an enormous literature on contest behaviour in fishes, and a number of influential papers have appraised specific topics related to contest theory, physiological and behavioural correlates of winning and losing, and correlations between aggression and other phenotypic traits in much greater detail than we can here (Hsu *et al.* 2006, Rutte *et al.* 2006, Briffa & Sneddon 2007, Arnott & Elwood 2009*a*,*b*, Oliveira 2009, Conrad *et al.* 2011). We thus both direct the reader to those reviews and here move towards generating a platform for future work that integrates developmental context, functional morphology, physiology and learning with our established understanding of fish contest dynamics and outcome.

10.3 Fighting ability and assessment

10.3.1 Perspectives on resource holding potential

Resource holding potential (RHP, Chapter 2) is a multidimensional 'trait' that reflects an animal's overall ability to secure access to quality mates, prime locations (e.g. shelter, territories), and food: i.e. it can be thought of as an animal's fighting ability (Figure 10.1D). For the sake of simplicity and experimental tractability, researchers often use measures of body size as proxies for RHP, which probably grossly oversimplifies what it takes for an animal to secure limited resources. Large animals win a numerical majority of contests especially when pitted against much smaller opponents, but factors such as ownership and resource value can override body size effects in fish contests (Dugatkin & Ohlsen 1990, Dugatkin & Biederman 1991). Surprisingly few recent studies, however, have assessed body size alongside other factors, such as aggression, metabolic rate, hormone concentrations, condition, sex, age and experience, to determine which traits or combinations of traits most accurately predict contest success (e.g. Beacham 1988, Wazlavek & Figler 1989, Beaugrand *et al.* 1996, Cutts *et al.* 1999, Draud & Lynch 2002); resolving such 'traditional' issues remains essential for the study of contest behaviour in fishes and other taxonomic groups.

Fishes possess diverse signalling and tactical repertoires, including a series of increasingly intense displays, attack-bite (often circling) sequences and mouth-locking (e.g. Simpson 1968, Dow *et al.* 1976, Jakobsson *et al.* 1979). Escalated fights can be fast and furious; thus, basic video recording is probably inadequate for identifying and quantifying the fine nuances of contest behaviour. One of these nuances, skill, can be defined as the ability of an animal to perform precise and/or challenging feats of motor coordination. Skill is beginning to be studied intensively in the context of male courtship and female mate choice (e.g. Clark 2009, Byers *et al.* 2010, Pérez-Barbería &

Yearsley 2010, Barske *et al.* 2011) but, to our knowledge, has yet to be studied in the context of aggressive contests, or as a component of RHP. To investigate the relationship between skill and contest success we need to illuminate individual variation in the performance of both ritualised and escalated fighting tactics. By capturing, on high-speed video, fight sequences and/or individual responses to standard aggression-eliciting stimuli we can resolve the kinematics of rapid or subtle movements with high precision, even those that escape the most careful observer. We may even identify motor patterns that are used by only some individuals or in only some contexts. Skill might also be repeatable within individuals, either with respect to individual proficiency or the diversity of tactics employed during a fight, allowing for contests to be staged between opponents that are matched as evenly as possible for everything except skill. These types of studies will help to reveal the functional significance of certain displays, their production costs, and how behavioural displays combine with other attributes (e.g. colour, fin size) to convey information to opponents during contests (section 10.6).

Contest success can also be influenced by residency status (e.g. Kokko *et al.* 2006): residents typically beat intruders. This could be due to convention, where intruders defer to owners even when fighting ability asymmetries are weighted in favour of the intruder (Maynard Smith & Parker 1976, but see anti-bourgeois ESS in Maynard Smith 1982, Chapter 2). Alternatively, residents might win *because* they have greater fighting ability, possess more information about the contested resource, or have greater motivation to maintain the resource (Kokko *et al.* 2006). Residency effects could also result from animals selecting or modifying habitats/territories in ways that provide some tactical advantage or that favourably alter the cost structure of contests.

This is in many ways similar to the concepts of niche picking and niche construction, which recently have been forwarded as important in governing the ontogeny of individual differences in behaviour (Stamps & Groothuis 2010*a*,*b*). Might animals select territories that provide the best vantage point for detecting intruders, or modify the habitat in ways that reduce their own fighting costs or increase costs for an intruder? We can draw an analogy to human military operations here: it might be best to erect a fort with a panoramic view of the surrounding area rather than in a closed canopy forest where intruders could capitalise on the element of surprise (niche-picking), and it might also be best to dig a moat around the fort to increase the effort required to breach the territory walls (niche construction). It is well established that animals select habitats/territories by integrating numerous information sources (e.g. profitability, defensibility: Stamps 1994, Piper 2011) and that site familiarity may, in and of itself, provide some tactical advantage (e.g. Piper 2011). However, studies on residency effects in fishes rarely: (1) provide options (e.g. territories with different characteristics), (2) examine individual differences in the types of habitat selected, (3) evaluate the relative defensibility of these territories, (4) determine which qualities of the territory might skew the economics of fighting in favour of the owner and (5) relate all of the aforementioned issues to actual contest success. Furthermore, fish will use such things as landmarks to reduce defence costs (e.g. LaManna & Eason 2003, Smith 2011), construct elaborate structures uniquely suited to prevailing environmental conditions (e.g. nest-building strategies vary with flow regime in sticklebacks: Rushbrook *et al.* 2010), and strategically use structures within their environment (e.g. concealing courtship in the presence of rivals: Dzieweczynski & Rowland 2004). Whether fishes are capable of utilising and/or modifying the habitat in ways that reduce fighting costs remains to be explicitly tested.

10.3.2 Assessment during contests

Selection should favour strategies that minimise the net costs of fighting (Enquist & Leimar 1983, Arnott & Elwood 2009*a*; Chapters 1 and 2). In fishes, these costs come in a variety of guises, including injury (e.g. Barlow *et al.* 1986, Neat *et al.* 1998*b*, Maan *et al.* 2001), energy depletion or the accumulation of metabolic by-products such as lactate that limit performance (e.g. Haller 1991*a*,*b*, Campbell *et al.* 2005, Abrahams *et al.* 2005, Ros *et al.* 2006, Briffa & Sneddon 2007, Guderley 2009, Copeland *et al.* 2011) and decreased vigilance (e.g. Jakobsson *et al.* 1995, Brick 1999). Assessment during contests can provide information about relative fighting ability, time and energy costs accrued, or injury sustained and can minimise costs if the information is used to make prudent decisions about when to continue to fight and when to flee. Fishes utilise a variety of signalling modalities during contests and therefore likely combine streams of sensory information to reduce errors in assessment (e.g. acoustic: Ladich 1997, 1998, Amorim & Almada 2005, Remage-Healey & Bass 2005, Amorim & Neves 2008, Johnston *et al.* 2008, Vasconcelos *et al.* 2010; chemical: Oliveira *et al.* 1996, Giaquinto & Volpato 1997, Barata *et al.* 2007, 2008, Hirschenhauser *et al.* 2008, Maruska & Fernald 2012; electrical: Westby 1975, Hagedorn & Zelick 1989, Dunlap 2002, Dunlap *et al.* 2002, Terleph & Moller 2003, Triefenbach & Zakon 2008, Salazar & Stoddard 2009, Fugère *et al.* 2011, Batista *et al.* 2012; 'fixed' visual: Evans & Norris 1996, Moretz 2005, Dijkstra *et al.* 2010; 'dynamic' visual: O'Connor *et al.* 1999, 2000, Miyai *et al.* 2011).

Three major assessment models have been frequently evaluated (Chapters 1 and 2). The diversity of signalling modalities used by fishes has understandably led researchers to focus on mutual assessment (Table 10.1), which occurs when animals gauge relative fighting ability through a process analogous to statistical sampling. Opponents begin with relatively uninformative displays and escalate to more costly forms of combat if additional information is needed to settle the dispute (sequential assessment model, SAM: Enquist & Leimar 1983, 1987, Chapters 1 and 2). Self-assessment occurs when animals pay no attention to the fighting ability of the opponent but, rather, gauge their own time or energy costs. When these costs cross some threshold, the animal should give up (wars of attrition without assessment: Mesterton-Gibbons *et al.* 1996; energetic wars of attrition, EWOA: Payne & Pagel 1996, Chapters 1 and 2). Cumulative assessment, CAM, is somewhat of a hybrid strategy because animals keep tabs on how much injury they have sustained; they do not directly assess their opponent and pay closer attention to their own state but injury results from the actions of their opponent (Payne 1998, Chapters 1 and 2).

There has been intense interest in devising methods to discriminate these strategies (e.g. Gammell & Hardy 2003, Taylor & Elwood 2003, Arnott & Elwood 2009*a*, Briffa & Elwood 2009, Chapter 4). For instance, because the relationship between RHP asymmetry and contest duration (Enquist & Leimar 1983) does not allow one to distinguish between the various forms of assessment, Taylor and Elwood (2003) encouraged the community to plot the RHP of both the smaller and larger contestant against contest duration. In doing so, mutual and self-assessment can be distinguished but not mutual and cumulative assessment (Briffa & Elwood 2009). Therefore, recent studies have emphasised that we should 'go back to the basics' and begin looking more carefully at behavioural patterns employed during contests and the dynamics of each phase of the contest (e.g. Egge *et al.* 2011, Rudin & Briffa 2011). In fishes, it might be worthwhile to revisit some of the early studies on contest behaviour, many of which provided remarkably detailed ethograms (e.g. see Table 10.1 legend) or to use these foundational studies as templates for obtaining the data needed to discriminate the aforementioned assessment strategies in other fish species.

We have tabulated those studies that explicitly test one or more of the assessment strategies in fishes (Table 10.1). Only seven studies (see Table 10.1, Method of Evaluation 2) have used the methods described in Taylor and Elwood (2003) and no study in fishes has coupled these statistical methods with detailed behavioural analysis during each stage of conflict (Briffa & Elwood 2009). In addition, the table reveals virtually no convincing demonstrations of mutual assessment, and only sparse support for self-assessment using the appropriate methodologies in fishes. Hsu *et al.* (2008) documented perhaps the best evidence of mutual assessment using Taylor and Elwood's statistical approach for the display phase of contests between opponents of a cyprinodont fish. In that study, the relationships between small/large opponents' body size and contest resolution were in the predicted direction (Chapter 4), but trends were not significant for the smaller opponent. Other studies have revealed relationships between size asymmetries and contest duration but not for the absolute sizes of the smaller/larger opponent (e.g. Prenter *et al.* 2008, Batista *et al.* 2012). In some ways, we are back to square one when it comes to assessment strategies in fishes. The vast majority of studies have relied upon the original method (e.g. plotting size asymmetry against contest duration or intensity: Enquist & Leimar 1983), which cannot discriminate mutual assessment, self-assessment without inflicted costs (sometimes referred to as 'pure' self-assessment) and self-assessment with inflicted costs (i.e. as envisaged under the CAM, see Chapters 1, 2 and 4). Therefore, it might be worth compiling the raw data from those earlier studies for re-examination. Certainly, for us to move forward in our understanding of assessment in fishes there is great need for additional studies that integrate the approaches described in the previous paragraph.

Table 10.1 A summary of studies investigating assessment tactics in fishes, arranged alphabetically by species. Type of assessment was based on the authors' determination, although some studies provide some evidence for other assessment strategies (e.g. Maan *et al.* 2001 measured energy and injury costs). Method of evaluation was assigned based upon the following criteria: (1) RHP asymmetries were plotted against fight duration (Enquist & Leimar 1983), (2) RHP asymmetries and absolute sizes plotted against fight duration to distinguish mutual and self-assessment (Taylor & Elwood 2003), (3) RHP asymmetries and absolute sizes plotted against fight duration *and* contest phases evaluated to distinguish mutual and cumulative assessment (Briffa & Elwood 2009), (4) other methods that support a given assessment model are employed (e.g. reduction in contest duration after previewing opponents); here, the method is noted. √ indicates evidence for a particular type of assessment and the use of a given method of evaluation √? indicates partial evidence for a particular type of assessment (e.g. negative relationship between winner size and contest duration but no relationship between loser size and contest duration). X indicates that a particular type of assessment was reported not to occur. X? indicates that evidence for a particular assessment strategy was lacking but that small asymmetries were employed, thus reducing effect size. MUTUAL = mutual/sequential assessment (Enquist & Leimar 1983, 1987), SELF = self-assessment (war of attrition without assessment: Mesterton-Gibbons *et al.* 1996, energetic war of attrition: Payne & Pagel 1996), CAM = cumulative assessment (Payne 1998), Δ = difference. This table was compiled based on *Web of Science* searches for 'contest and ('mutual assessment' or 'self-assessment' or 'cumulative assessment' or 'sequential assessment')' and by scanning the literature that cites Enquist and Leimar (1983), Taylor and Elwood (2003), Gammell and Hardy (2003), Briffa and Elwood (2009) and Arnott and Elwood (2009*a*) for studies that employ fish models. Foundational studies, such as Baerends and Baerends-von Roon (1950), Braddock and Braddock (1955), Simpson (1968), Ewing (1975), Dow *et al.* (1976), Jakobsson *et al.* (1979), discuss patterns of behavioural change during fights but were excluded because they predate the formalisation of assessment models; one exception is Frey and Miller (1972) because they plotted contest duration versus size asymmetries.

	Characteristics		Type of assessment			Method of evaluation				
Species	**Size/age/sex**	**Contest type**	**MUTUAL**	**SELF**	**CAM**	**1**	**2**	**3**	**4**	**Reference and notes**
Aequidens rivulatus	80–170 mm (adult, male)	Dyadic, symmetric (< 8% Δ in total length)	X?			√				Maan *et al.* (2001) Winners/losers of symmetric contests accrue similar injury costs, losers have greater respiration frequencies, highly structured contests
Amatitlania nigrofasciata	48.3–69.9 mm 4–14 g (adult, male)	Dyadic across partition, asymmetric (random pairings)		√ 1st pairing					√	Arnott & Elwood (2010) Startle duration protocol for evaluation of assessment strategies, evidence for opponent-only assessment in third pairing
Amatitlania nigrofasciata	43.3–71.7 mm SL, 2.95–16.1 g (adult, male)	Dyadic, matched on average, < 5% Δ in mass		√		√	√			Copeland *et al.* (2011) Winners/losers accrue similar metabolic costs
Apteronotus leptorhynchus	17.6–23.0 cm body length 19.7–28.6 g (adult, male)	Dyadic, asymmetric: from 0 to 3.5 cm Δ	√			√				Triefenbach & Zakon (2008) Electric modality

*Archocentrus nigrofasciatum**	30.7–75.7 mm SL (adult, male–female pairs)	Between male–female pairs, asymmetric: males 15–20% Δ, females 12–18% Δ	√			√			Draud & Lynch (2002)
*Archocentrus nigrofasciatus**	5–12 g: small pairings 15–30 g: large pairings (adult, male)	Dyadic, matched < 10% Δ in mass and total length	√					√	Leiser *et al.* (2004) Contests between large males involve more tailbeats, bites and chases, behaviour patterns consistent with sequential assessment
Cichlasoma citrinellum	Relative mass reported (adult, male)	Dyadic, size matched with Δ up to 9% but differed in aggression scores	×?		√?	√	√		Barlow *et al.* (1986)[1]
Cichlasoma meeki	16.44 ± 5.38–18.39 ± 4.59 g (adult, male)	Dyadic, asymmetric in both size and carotenoid treatment: opponent up to 2× greater mass	√ White light × Green light			√			Evans & Norris (1996) As cumulative mass + carotenoid asymmetry increase, no. contests/trial decreases
*Cichlasoma nigrofasciatum**	% asymmetry reported (adult male, adult female)	Same-sex dyadic, size-matched with Δ between 0.04 and 5.59%	√					√	Keeley & Grant (1993) Shorter contest durations when opponents have visual access prior to contest
*Cichlasoma nigrofasciatum**	2.1–19.1 g (adult male, adult female)	Same-sex dyadic, gradient of Δ from 1.02 to 4.86 larger/smaller ratio	√			√			Koops & Grant (1993) Discrete contest phases lacking, no sex differences in duration
Gymnotus omarorum	7–78.4 g 14.5–30 cm body length (adult male, adult female)	Same-sex and opposite-sex dyadic, asymmetric with gradient of Δ from 0.2 to 50% mass and 0–23.3% body length	√ mass Δ ×? Absolute masses	×?		√	√		Batista *et al.* (2012) Electric modality, phasic contests. Mass Δ negatively related to contest duration but neither dominant nor subordinate absolute masses related to duration, no sex differences in outcome/behaviour
Herichthys cyanoguttatum	Volume calculations 65–230 cm^3 (males) 68–154 cm^3 (females)	Same-sex dyadic (round-robin), asymmetric: from 0 to 75% volume Δ	√ male √ female			√			Draud *et al.* (2004) No sex differences in behaviour but sex differences in size Δ versus duration/escalation

(*cont.*)

Table 10.1 *(cont.)*

Species	Characteristics: Size/age/sex	Contest type	Type of assessment: MUTUAL	SELF	CAM	Method of evaluation: 1	2	3	4	Reference and notes
Kryptolebias marmoratus	Random-size pairings 18.87–29.64 mm: larger 17.20–25.33 mm: smaller Matched pairings 16.29–27.72 mm (adult, hermaphrodite)	Dyadic asymmetric or symmetric (< 1 mm Δ)	√ in display phases	√ in escalated phases			√			Hsu *et al.* (2008) Evidence for switching assessment strategies
Nannacara anomala	3.2–4.75 g (adult, male)	Dyadic, asymmetric: opponent 39–73% smaller	√			√				Enquist & Jakobsson (1986)
Nannacara anomala	1–6 g (adult, male)	Dyadic, across partition asymmetric: opponents 20–70% smaller	√			√				Enquist *et al.* (1987)
Nannacara anomala	1–9 g (adult, male)	Dyadic, asymmetric: mass ratio from 1.0 to 4.95	√			√				Enquist *et al.* (1990)
Nannacara anomala	1–4.8 g (adult, male)	Dyadic, size-matched with < 5% mass Δ	√						√	Brick (1998) Δ in behaviour in response to increased cost of sampling (predation)
Nannacara anomala	1–2.5 g (adult, male)	Dyadic, size-matched with 1–9% Δ	√						√	Brick (1999) Δ in behaviour in response to increased cost of sampling (predation)
Neolamprologus pulcher	42.7–71.5 mm SL (adult male, adult female, all subordinate helpers)	Same-sex dyadic, size-matched by eye, Δ from 0 to > 15%	√?	×			√			Reddon *et al.* (2011) No sex differences

Onchorhynchus mykiss	85.4 ± 1.9–9.58 ± 1.8 (juvenile)	Triadic/dyadic, symmetric (< 5% Δ)	√					√	Johnsson & Åkerman (1998) Less aggression between familiar versus unfamiliar pairs
Oreochromis mossambicus	45–90 mm SL (adult, male)	Dyadic, asymmetric: SL ratios from 1 to 1.4	√			√			Turner & Huntingford (1986)
Tilapia zillii	20–140 g (1 year old, adult male)	Dyadic, asymmetric: mass Δ range from 0 to 0.24 [$\ln(mass_a/mass_b)$]	√ Δ Mass √ Δ GSI[2]			√			Neat *et al.* (1998*a*) Relationship between Δ in mass/GSI[2] and duration existed only for carouselling phase
Trichogaster trichopterus	49–66 mm (adult, male)	Dyadic, size Δ 1–5 mm	√?			√			Frey & Miller (1972) Bout durations associated with relative but not absolute size
Trichopsis vittata	42.7–63.3 mm total BL 0.67–1.8 g (adult, male)	Dyadic, size Δ up to 0.3 mm, SPL[3] Δ up to 1.5 decibels	√ Δ BL √ Δ SPL[3]			√			Ladich (1998) Acoustic modality
Xiphophorus helleri	% asymmetry reported (adult, male)	Dyadic, size-matched with Δ up to 13.8%	×?			√			Franck & Ribowski (1989)
Xiphophorus helleri	% asymmetry reported (adult, male)	Dyadic, size-matched with Δ up to 15%, mismatched with Δ up to 46%	√			√			Ribowski & Franck (1993)
Xiphophorus helleri	Lateral surface area 4.35–9.22 cm^2 (adult, male)	Dyadic, size-matched (< 0.2 cm^2)	√					√	Earley *et al.* (2003) Shorter contest durations when opponents have visual access prior to contest
Xiphophorus helleri	Various size measures reported in the authors' table 1 (adult, male)	Dyadic, matched for body length (BL, mean Δ 5.83%)	×? Δ BL √? Δ sword	×		√	√		Prenter *et al.* (2008)
Xiphophorus helleri	Various size measures, ranges not reported (adult, male)	Dyadic, range of asymmetries [$\ln(mass_a/mass_b)$]	√		√?	√	√		Wilson *et al.* (2011)

(*cont.*)

Table 10.1 *(cont.)*

	Characteristics		Type of assessment			Method of evaluation				
Species	Size/age/sex	Contest type	MUTUAL	SELF	CAM	1	2	3	4	Reference and notes
Xiphophorus multilineatus	Absolute size differences reported: 0–16.9 mm (adult, male)	Dyadic, asymmetric: Δ from 0 to 1.9 mm	√			√				Morris *et al.* (1995) Only bites/min (intensity) correlates negatively with size asymmetry, not duration, absolute number of aggressive acts
Xiphophorus nigrensis	Absolute size differences reported: 0–18.9 mm (adult, male)	Dyadic, asymmetric: Δ from 0 to 1.9 mm	√			√				Morris *et al.* (1995) All aggression measures but not duration correlate negatively with size asymmetry

* Species name since changed to *Amatilania nigrofasciata.*

[1] Partial tests of assessment models. For instance, examined relationship between scores of most aggressive animal and display phase duration (e.g. similar to what Taylor and Elwood (2003) encourage with regressions of absolute body size or larger/smaller against contest duration). Relationship between size asymmetry (albeit small) and display phase duration was different for long-term versus short-term experiments, no relationship in former, positive relationship in the latter. Authors suggest no evidence for injury-based assessment but this is equivocal.

[2] GSI refers to gonadosomatic index (testis mass as a percentage of body mass).

[3] SPL refers to sound pressure level (decibels).

Furthermore, a recent study in the mangrove rivulus, *Kryptolebias marmoratus*, emphasised that fishes can *switch* assessment strategies at different phases of the contest; in non-escalated phases, mutual assessment appears to be used while in escalated phases, the animals resort to self-assessment (Hsu *et al.* 2008). Such reports challenge the assumption of fixed assessment strategies within a species, and encourage future work that evaluates behaviour in different phases or different types of contests. In addition, Arnott and Elwood (2009*d*) describe a method for evaluating assessment strategies that, to some degree, divorces the animals' motivation to fight from fight dynamics. Here, the experimenters allow opponents to interact across a clear partition, drop a marble into the focal animals' tank (a perturbation that cannot be perceived by the non-focal animal), and then ask how long it takes the focal animal to recover from being startled. If shorter startle durations reflect the fish's motivation to fight, and if motivation is based on the information gained during the phases of the fight that precede startle, then one can evaluate how information about the opponent and self influences behaviour and fight dynamics (e.g. Arnott & Elwood 2009*a*; a test of the method is provided in Arnott & Elwood 2010). Advantages to this technique include that the experimenter can probe the eventual winner and eventual loser (or larger versus smaller) independently and at any point during the contest. The latter advantage might help to reveal whether fishes switch assessment strategies by appraising the relationship between startle duration and measures of RHP as contests proceed from ritualised to more intense phases. The only disadvantage to probing motivation is that the fish must be separated by a partition, which limits to visual cues/signals the type of information available for assessment.

To a large extent, we still lack a comprehensive understanding of assessment in fishes. This stems in part from the emphasis on overt motor patterns such as fin erection, opercular flaring, mouth-locking, gaping and carouselling. All of these behaviours are likely to be important for contest settlement and have different associated costs (e.g. Maan *et al.* 2001), but they represent only a fraction of a fish's display repertoire. Fishes probably use many, perhaps independently informative, signals and cues during contests (see above). It will be important to understand how different signalling modalities reflect RHP, whether variance in contest performance and outcome is better explained by an interaction between signalling modalities, whether there is a hierarchy of signals employed during different phases of the contest, and how the diverse signalling repertoires can be used to evaluate assessment (e.g. by measuring the costs of vocalisations or the opponent response to variation in signal form).

Furthermore, there has been an obvious bias towards male–male contests: only 5 of the 31 studies on assessment investigated contests between females (Table 10.1). Sex differences in aggression and assessment are negligible in some species but pronounced in others (Table 10.1). In species with female harems, reversed sexual size dimorphism or sex roles, cooperatively breeding groups, or pairbonding and cooperative nest defence, females can be quite aggressive (e.g. Beeching *et al.* 1998, Brandtmann *et al.* 1999, Desjardins *et al.* 2006, Arnott & Elwood 2009*c*, Taves *et al.* 2009, Archard & Braithwaite 2011, Schumer *et al.* 2011). It is therefore likely that the economics of female aggression and assessment depends critically on social and/or mating systems, and that selection might act differently on males and females in some cases but not others (e.g. Holder *et al.* 1991). However, our understanding of the evolution of assessment strategies in the fishes is severely limited by sparse phylogenetic coverage and limited attempts to connect fighting strategies with other aspects of the animals' life history. Of the 31 studies shown in Table 10.1, 25 were conducted on 12 species in just 2 taxonomic families (3 poeciliids, 9 cichlids). This is a situation where the 'model system' approach, although powerful because it allows us to delve deeper into the mechanisms of contest behaviour, might impair our ability to place aggression, dominance, assessment and the neural and physiological mechanisms associated with decision-making in a comparative evolutionary context (e.g. Blumstein *et al.* 2010). At present, we know little about the social, ecological and evolutionary contexts in which assessment takes place. This marks an exciting direction for future research, and one that holds much promise for revealing, for instance, how flexible (or fixed) the assessment strategies might be among individuals, populations, or across species, how the economics of assessment shapes neural architecture and function, or perhaps how existing neural architecture, which arises through selection on other aspects of natural and life histories, drives and/or constrains assessment strategies, signal diversity, or signal perception (e.g. Pollen *et al.* 2007, Sylvester *et al.* 2010).

10.4 Winner and loser effects

Recent contest experience can modify a fish's fighting decisions (Hsu *et al.* 2006, Figure 10.1C). After a recent victory, individuals often become more aggressive, persist longer before retreating and as a result have a higher chance of winning a subsequent contest ('winner effect': Chapters 3, 4 and 8). On the other hand, after a recent defeat, individuals become more submissive, being likely to retreat immediately when challenged and as a result have a lower chance of winning the next contest ('loser effect': Chapters 3, 4 and 8). Recent winning and losing experiences probably modify an individual's contest decisions by adjusting its assessment of its own fighting ability and thus its expected cost of engaging in a subsequent contest (Hsu *et al.* 2009).

The changes in the probability of winning a new contest after a recent win or loss could also, however, partly result from opponents detecting status-related cues released by individuals with different contest experiences and altering their behaviour toward these individuals (social-cue mechanism: Rutte *et al.* 2006). Dominant and subordinate fish have been reported to release different chemicals or different quantities of the same chemicals into water (Oliveira *et al.* 1996, Barata *et al.* 2007, 2008). In Mozambique tilapia, *Oreochromis mossambicus*, for instance, dominant males produce more urine than subordinates and their urine elicits strong electro-olfactogram responses from conspecifics (Barata *et al.* 2007, 2008; see Chapter 5 for the role of urine-borne cues in contests between crustaceans). The olfactory system of female tilapia was more sensitive to dominants' urine, thus the authors concluded that dominant tilapia males use the urinary odorant to signal status to the females (Barata *et al.* 2008). It is conceivable that fish use chemical cues to communicate their dominance status in contests. In a mangrove killifish, *Kryptolebias marmoratus*, contests between individuals with a recent victory, a recent defeat, or no recent contest experience (a total of 6 combinations) showed that contestants changed their contest behaviour after a recent winning or losing experience, but they did not respond differently toward opponents with different recent contest experiences (Hsu *et al.* 2009). In Siamese fighting fish, *Betta splendens*, although individuals with a prior winning experience win more contests against individuals with a prior losing experience (Wallen & Wojciechowski-Metzlar 1985), bystanders do not respond to prior winners and losers that they did not observe fighting (Oliveira *et al.* 1998). These results suggest that the experience effects in these fish mainly result from individuals modifying their contest strategy after winning or losing a fight, but do not offer supporting evidence for fish assessing their opponents' recent contest experiences through chemical cues. Given fish's ability to release status-related chemicals into the water and to detect these chemicals, more research on the relationship between chemical cues and contest behaviour is warranted.

Fish are capable of integrating information from multiple contest experiences. The contest behaviour of individuals of *K. marmoratus* is affected by the combined effect of the most recent and the penultimate contest experiences, although the influence of a penultimate experience is less significant than that of a recent experience (Hsu & Wolf 1999, 2001). It is not clear, however, whether this is caused by the penultimate experience decaying with time, decaying with the acquisition of a new experience, or both. Each of these types of decay could make the information from a past experience less valuable. The effect of a winning or losing experience decaying with time has been demonstrated in some fish species (paradise fish, *Macropodus opercularis*: Francis 1983; stickleback fish, *Gasterosteus aculeatus*: Bakker *et al.* 1989; pumpkinseed sunfish, *Lepomis gibbosus*: Chase *et al.* 1994). The possibility of the effect of an older experience decaying with the acquisition of a new experience is yet to be explored.

The importance of experience effects on contest behaviour appears to vary between species, with some species showing both winner and loser effects (e.g. sticklebacks, *Gasterosteus aculeatus*: Bakker *et al.* 1989; killifish, *Kryptolebias marmoratus*: Hsu & Wolf 1999, 2001), some the loser effect only (e.g. green sunfish, *Lepomis cyanellus*: McDonald *et al.* 1968; paradise fish, *Macropodus opercularis*: Francis 1983), and others neither effect (e.g. green swordtail, *Xiphophorus helleri*: Earley & Dugatkin 2002). Overall, the loser effect is more frequently detected, is stronger and lasts longer than the winner effect. The asymmetry of winner and loser effects within species and the differences in their importance between species could partially result from the experimental procedures employed for detecting these effects and the differences in the methodology used in different studies (Hsu *et al.* 2006). Asymmetries in these experience effects also could be due to differences in the behaviour measured in contests that follow prior wins and losses. For instance, Huang *et al.* (2011) showed that prior winning/losing affected future escalation rates for 2–4 days/1–2 days but winning probabilities for < 1 day/2–4 days, respectively. These data suggest that focusing on winning probabilities might underestimate the effects of previous winning experiences on future contest behaviour. Winner–loser effect asymmetries could also be consequences of the information from winning and losing experiences having different values to the individuals of the same species and to those of different species. For instance, if engaging in contests but losing incurs much higher costs (time, energy, injuries) than retreating without confrontation, individuals may adopt a more conservative strategy and be more responsive to a losing experience than to a winning experience. The interspecific difference in the importance of experience effects may reveal a difference in the usefulness and the reliability of the information for future contests. Species characteristics that influence the usefulness and the reliability of the information such as the frequency of aggressive interaction and growth rate may contribute to differences among species in the importance or persistence of experience effects (Hsu *et al.* 2006). For instance, during rapid growth, past experiences might not accurately reflect current RHP. Thus, information gained from previous wins and losses might be less likely to be used by species that grow quickly relative to those that grow slowly or during stages of ontogeny characterised by rapid versus slow growth. From a mechanistic perspective, asymmetries in the persistence of winner and loser effects might reflect status differences in metabolic recovery rates or short-term pulses of steroid hormone that are released in a status-dependent manner and that activate or depress behavioural responsiveness to varying degrees (Hsu *et al.* 2011).

An individual could potentially obtain information about its fighting ability relative to that of its opponent from various sources, and the acquisition and maintenance of information from those sources may impose different costs on it. For instance, an individual could engage in a long and potentially costly interaction with its opponent, gradually accumulating more accurate information about the opponents' relative fighting ability as the contest advances (Enquist *et al.* 1990). Another individual could avoid repeatedly going through that costly process and remember the outcome of its previous contest and/or its opponent's identity and past performance, which would require it to possess and maintain the physiological mechanisms of memory, which could involve some other costs. Difference in the importance of experience between different species may reflect the interspecific differences in acquiring and maintaining information from different sources. Understanding how information from different sources is combined to influence contest behaviour should allow us to have a better overview of decision-making in animal contests.

This may require more explicit infusion of learning theory into our studies of experience effects and contest behaviour. A number of studies have employed classical conditioning paradigms to determine whether fishes are capable of learning the association between certain features of the environment (e.g. light, colour, water flow) and territorial intrusion or conspecific aggression (e.g. Demarest 1992, Hollis *et al.* 1995, Jenkins & Rowland 1996, Hollis 1999, Carpenter & Summers 2009, Antunes *et al.* 2009). In many of these studies, fishes showed dramatic behavioural and/or hormonal responses to the conditioned stimulus alone following some period of conditioned stimulus–unconditioned stimulus pairing. This could reflect past selection on the capacity to learn, which might be favoured because it allows the fish to better anticipate rival presence, mount an immediate hormonal/behavioural response, and thus reduce contest costs by more quickly deterring an opponent. However, we are likely to have to entertain more complex learning mechanisms (e.g. extinction, forgetting functions, spontaneous recovery) to gain a comprehensive understanding of how fighting experience affects future behaviour. For instance, despite the vast literature suggesting that winner and loser effects persist for relatively short time frames, Lan and Hsu (2011) showed that individuals obtaining a losing experience one month prior to a focal contest are more sensitive to recent wins and losses than individuals obtaining a winning experience at the earlier time point. What predisposes prior losers to this increased sensitivity? Which neural mechanisms drive these types of 'carry-over' experience? Are these neural mechanisms associated with learning? And, might the magnitude and persistence of winner and loser effects as well as the response of underlying neural and hormonal mechanisms depend on the type of experience gained (e.g. intense versus benign contests: Beaugrand & Goulet 2000)? These are the types of questions that could be very important to pursue over the next several years.

10.5 Mechanisms and performance characteristics

10.5.1 Hormones, metabolism and gene expression profiling

Aggressive contests can trigger dramatic changes in steroid hormone and neurotransmitter concentrations, which in turn govern behavioural plasticity in response to winning, losing or simply being involved in a fight (e.g. Summers & Winberg 2006, Earley & Hsu 2008, Oliveira 2009, Oliveira *et al.* 2009, Dijkstra *et al.* 2011). For instance, 11-ketotestosterone (the primary fish androgen: Borg 1994) levels rise within 30 min of a subordinate male cichlid, *Astatotilapia burtoni*, ascending to dominant status (Maruska & Fernald 2010). Dominant Arctic charr, *Salvelinus alpinus*, have significantly higher testosterone and 11-ketotestosterone and lower cortisol levels than subordinates (Elofsson *et al.* 2000). Further, brain serotonergic activity tends to be higher in subordinates, particularly after 24 h of persistent social interaction (e.g. Winberg & Nilsson 1993, Øverli *et al.* 1999). These neural and endocrine responses are hypothesised to modulate future behaviour in ways that match the demands of being a winner or loser. Higher androgen levels might prime the fish for future combat, while higher cortisol levels and serotonergic activity might force the body into a state of metabolic recovery and behavioural depression. Studies that manipulate steroid hormone levels or serotonergic activity seem to support these hypotheses. For example, supplementing Atlantic cod, *Gadus morhua*, with L-tryptophan, the precursor to serotonin, increases serotonergic activity and suppresses aggressive behaviour (Höglund *et al.* 2005), cortisol administration to rainbow trout, *Oncorhynchus mykiss*, predisposes to subordinate status (DiBattista *et al.* 2005, but see Øverli *et al.* 2002 for increased aggression after acute cortisol administration), and treatment with 11-ketotestosterone increases metabolism in cichlids (*Oreochromis mossambicus*: Ros *et al.* 2004) and aggressive vocalisations in toadfish (*Opsanus beta*: Remage-Healey & Bass 2006).

Steroids can exact their behavioural effects in two major ways: (1) by binding receptors located in the nucleus or cytosol and initiating a slow (e.g. hours) 'genomic' process that involves hormone–receptor complexes acting as transcription factors, which alter gene expression patterns and ultimately some aspect of the phenotype, or (2) by binding membrane G-protein coupled receptors and initiating a fast (e.g. seconds to minutes) 'non-genomic' cellular response that culminates in such things as neurotransmitter release. Studies on fish aggression point to rapid, non-genomic behavioural responses to androgen administration (e.g. Remage-Healey & Bass 2006, Oliveira 2009), but the precise mechanisms involved remain largely unknown because identification and functional characterisation of membrane-bound receptors in fishes is still under way (Thomas 2012). Nevertheless, there are ways to begin addressing hypotheses regarding the relative roles of membrane versus nuclear/cytosolic receptors in the short- and long-term behavioural responses to contests. For instance, hormones linked to bovine serum albumin (BSA) are unable to cross cell membranes and thus, any behavioural response to hormone–BSA administration (relative to hormone that is not conjugated to BSA) can localise the effects to a membrane receptor. These are the types of studies that will benefit from our more detailed understanding of receptor distributions in the fish brain (e.g. Munchrath & Hofmann 2010) and areas of the brain that constitute the social behaviour network (e.g. Goodson 2005, Desjardins & Fernald 2010, O'Connell & Hofmann 2011).

These studies and the techniques they employ have the potential to launch us towards a greater understanding of the mechanisms that underlie aggression, experience effects, and the maintenance of dominance status in fishes. However, the aforementioned mechanisms are but a fraction of possible neuroendocrine factors involved in the regulation of fighting behaviour. For instance, the neuropeptide arginine vasotocin (AVT), which is known to modulate the stress response, influences aggression and dominance in rainbow trout (e.g. Backström & Winberg 2009) and damselfish (Santangelo & Bass 2006), but its precise effects seem to depend on the mode of administration (e.g. intramuscular versus intracerebroventricular). Advanced methods in gene expression profiling, such as real-time quantitative polymerase chain reaction (PCR), suppression subtractive hybridisation, microarray and RNAseq have ushered in a new era for understanding reciprocal relationships between mechanisms and aggression, contest decisions, and dominance (e.g. Sneddon *et al.* 2005, Renn *et al.* 2008, Filby *et al.* 2010*a*, Aubin-Horth *et al.* 2012). For instance, using microarrays, Renn *et al.* (2008) identified 'modules' of socially regulated genes that were differentially expressed between dominant and subordinate cichlids. Sneddon *et al.* (2005) identified 1165 genes that were differentially expressed in dominant and subordinate rainbow trout using suppressive subtractive hybridisation. Both Filby *et al.* (2010*a*) and Aubin-Horth *et al.* (2012) used targeted gene expression assays to identify, for example, aspects of the neuroendocrine stress and gonadal axes, histamine pathways, serotonin, dopamine and AVT as important in the regulation of aggressive behaviour. Establishing a causal relationship between contest behaviour and these identified genes will, however, rely on the application of knockdown and transgenic technologies to both model (e.g. zebrafish) and non-model systems. Fortunately, a number of research groups have characterised such things as winner–loser effects, behavioural syndromes, and the neuroendocrine correlates of aggression and dominance in zebrafish, providing the foundation for a powerful integration of behavioural ecology and genomics (e.g. Norton *et al.* 2011, Oliveira *et al.* 2011, Dahlbom *et al.* 2012).

As with assessment strategies, we know relatively little about female hormonal responses to aggressive contests, but some studies suggest sex-specific patterns of endocrine change (e.g. Taves *et al.* 2009). Although males and females might exhibit similar behaviour patterns, the sexes might invoke different 'circuitry' (both endocrine and neural) to either regulate contest behaviour or respond to contest dynamics/outcome. By resolving such issues, we might position ourselves to understand the diversity of mechanisms that can be co-opted for similar behavioural functions, or identify new regulatory pathways that emerge during sexual differentiation. Another important finding is that contest dynamics often predict individual variation in hormone titres better than status (Earley *et al.* 2006, Earley & Hsu 2008), which suggests that animal physiology might be more sensitive to contest proceedings than contest outcomes. If hormonal responses direct energy expenditure or metabolic recovery (e.g. Ros *et al.* 2004, Milligan 2003), and if metabolic parameters change as a function of contest intensity (e.g. Copeland *et al.* 2011), we might predict that hormones also would correlate with contest dynamics. However, the interrelationships between contest behaviour, hormones and metabolism are not well understood and deserve further attention.

We know that contests can be metabolically costly in terms of increases in respiration rates, lactate accumulation, and energy depletion (Chellappa & Huntingford 1989, Neat *et al.* 1998*b*, Guderley & Couture 2005, Ros *et al.* 2006, Copeland *et al.* 2011). These metabolic costs limit performance during contests and dominance ability: under hypoxic conditions, aggression can decline precipitously (Abrahams *et al.* 2005, Marks *et al.* 2005) and dominance hierarchies destabilise (Sneddon & Yerbury 2004). However, apart from hypoxia exposure, no study has manipulated metabolic parameters (e.g. increasing lactate load) to estimate 'giving-up' thresholds, which would provide unequivocal support for the predictions of self-assessment (section 10.3). Furthermore, tolerance for anaerobic activity scales positively with body size (e.g. Almeida-Val *et al.* 2000, Nilsson & Östlund-Nilsson 2008), leading to the prediction that different metabolic 'rules' might govern contest decisions in small and large animals of the same species (e.g. Leiser *et al.* 2004), a prediction that remains untested.

10.5.2 Morphology, kinematics and biomechanics

Recent studies have highlighted performance-based traits as key to our understanding of dominance outcomes, constraints on signal form and signal divergence in lizards, birds and invertebrates (e.g. Lailvaux & Irschick 2006, Vanhooydonck *et al.* 2007, Herrel *et al.* 2009, Hall *et al.* 2010). In various lizard species, bite force and gaping displays predict fighting success and convey information about weapon performance, respectively (e.g. Huyghe *et al.* 2005, Husak *et al.* 2006, Lappin *et al.* 2006, Henningsen & Irschick 2012, Chapter 12). Links among performance traits also change as a function of the animal's ecology and behavioural habits. For instance, dewlap size is an honest indicator of bite force, and bite force predicts contest success in highly territorial, sexually size-dimorphic *Anolis* lizards: in anoles with less territoriality and sexual size dimorphism, dewlap size alone predicts contest success and is unrelated to bite force (Lailvaux & Irschick 2007). In fishes, the tools for examining the kinematics of movement, fluid mechanics and morphology–force relationships have been developed and applied to such things as feeding, locomotion (in both aquatic and terrestrial habitats) and escape responses (e.g. Westneat 2003, Holzman *et al.* 2007, Peng *et al.* 2007, Mehta 2009, Cooper *et al.* 2011, Gibb *et al.* 2011, Parsons *et al.* 2012). We see the field advancing in at least two specific ways by applying these principles to contest behaviour in fishes (e.g. McDonald *et al.* 2007).

First, describing variance in the kinematics of display behaviour (e.g. opercular flaring, lateral display) among individuals, populations, or species will provide the impetus for investigating underlying anatomical differences (e.g. skeletal arrangements, lever systems). This, in turn, will promote comparative evolutionary analysis of structure–function relationships; as has been done effectively with, for instance, jaw morphology, swimming performance and manoeuvrability, and feeding habits in cichlids and eels, sticklebacks and gobies, respectively (Hulsey *et al.* 2008, Mehta & Wainwright 2008, Maie *et al.* 2009, Hendry *et al.* 2011). In addition, understanding contest behaviour kinematics will undoubtedly shed light on the fine motor patterns executed during a fight, and allow us to evaluate among-individual differences in 'skill'.

Second, by quantifying the forces applied to an opponent (e.g. with a bite) or to the water (e.g. undulating displays), and resultant velocities, we can better understand the signalling value of these behaviours. We are still at the point where terms like 'strength' are rather ill-defined and we assume that information about strength is being conveyed by behaviours such as displays and mouth-locking. For instance, the undulating lateral display, termed tailbeating, where individuals appear to push water at the lateral line of the opponent, has been hypothesised to function as a signal of strength: stronger animals push a larger column of water with greater velocity. This hypothesis and others that extend from assessment models (e.g. that individuals vary in their ability to inflict costs: Payne 1998) can be tested using the principles of biomechanics and some technical ingenuity. For instance, Figure 10.2 shows a hypothetical contest arena in which two opponents can interact across glass partitions; these animals might vary in age, size, sex, shape, experience, and so forth. If the partitions separating the two contestants doubled as force transducers, we could estimate the force associated with bites or any other contact behaviour. In addition, there have been remarkable advances in our ability to visualise flow fields in two or three dimensions using particle image velocimetry and volumetric three-component velocimetry (e.g. Flammang *et al.* 2011, Borazjani *et al.* 2012). These techniques seed the fluid in which the fishes behave using particles whose motion (e.g. speed and direction of movement) can be digitised, which could portray the hydrodynamics of a given movement

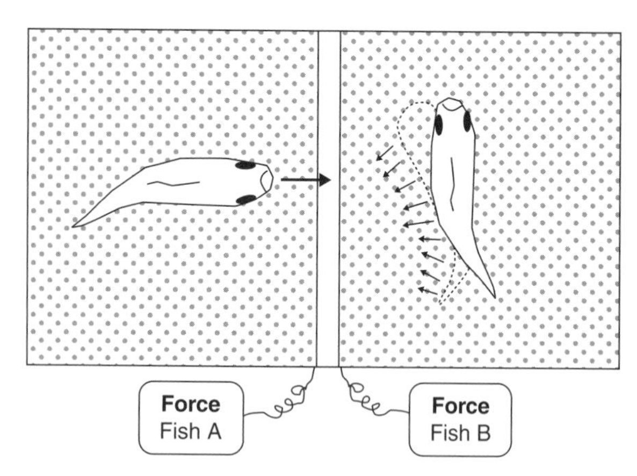

Figure 10.2 A hypothetical set-up for examining the biomechanics of contest behaviour in fishes. The partition separating the opponents would measure the force of contact aggression while movement of the particles (patterned dots) would be digitised to estimate the hydrodynamics of display behaviours using, for instance, particle image velocimetry.

during fish contests, ultimately informing hypotheses about signal function. The repeatability of, and individual differences in, the ability to exert force or generate water flow could be estimated using standard aggression-eliciting stimuli (e.g. mirror or stationary model) linked to force transducers and/or velocimetry systems. By evaluating the dynamics and outcome of contests staged between opponents with a gradient of force/velocity asymmetries, we could then estimate in controlled fashion the relationships between performance traits and fighting ability (e.g. Henningsen & Irschick 2012).

10.6 Selection, plasticity and dominance

10.6.1 Selection regimes, integrated phenotypes and contest behaviour

Blows (2007) emphasises that 'adaptation is an inherently multivariate process' and that 'natural selection often acts upon sets of functionally related traits, rather than one-dimensional phenotypes' (see also Berg 1960, Lande & Arnold 1983). We should thus expect to witness changes in suites of correlated traits, not singular traits, in response to artificial and natural selection. Our goal in this section is to highlight correlational selection as a potentially potent force driving variance in fighting ability among individuals derived from populations experiencing divergent selection regimes (Figure 10.1A). We also refer the interested reader to a rich literature addressing the multivariate approaches used to resolve behavioural correlations, genetic/phenotypic variance–co-variance matrix evolution, and the evolution of individual traits in a multivariate context (e.g. Uher 2008, Hine *et al.* 2009, Budaev 2010, McGuigan & Blows 2010), as well as mechanisms such as genetic pleiotropy and hormonal pleiotropy, which have been proposed as powerful agents of phenotypic integration (e.g. Falconer & Mackay 1996, McGlothlin & Ketterson 2008, Ketterson *et al.* 2009).

Some of the classic examples of how selection can shape suites of phenotypic traits come from investigations in fish species. Guppies, *Poecilia reticulata*, derived from high- and low-predation communities show remarkable divergence in colour, an arsenal of life-history traits, behaviour and burst swimming performance (e.g. Ghalambor *et al.* 2003, Magurran 2005). Similarly, threespine sticklebacks inhabiting shallow lakes, deep lakes, streams and marine environments differ in size, body shape, armour, colour, metabolic physiology and behaviour (e.g. Sneddon & Yerbury 2004, Foster *et al.* 2008, Schluter & Conte 2009). Investigation of correlated traits has intensified over the past decade, particularly with respect to the modularity of morphological form (e.g. Parsons *et al.* 2012), stress coping styles (e.g. Øverli *et al.* 2005, 2007) and behavioural syndromes (e.g. Sih *et al.* 2004, Sih & Bell 2008, see also Dochtermann 2011). Many of these correlations involve traits known to influence contest performance, including body size, aggression and hormone profiles. For instance, rainbow trout, *Oncorhynchus mykiss*, artificially selected for divergent stress responses also show marked variation in a number of other phenotypic traits, including fighting ability (e.g. Øverli *et al.* 2005, 2007, but see Ruiz-Gomez *et al.* 2008). Artificial selection for large body size in domesticated fishes, and various aspects of the aquaculture environment, such as stocking density and feeding regime, may also result in correlated selection for increased aggression (Huntingford 2004). Rapid divergence between hatchery and wild populations in RHP-related traits has major implications for the implementation of effective restocking strategies and for understanding the potential impact of unintentional releases of hatchery stock into wild populations (Huntingford 2004). For instance, in some cases, fish from hatcheries dominate their wild counterparts but, in other cases, hatchery fish compete poorly (Berejikian *et al.* 1996, Pearsons *et al.* 2007, Saikkonen *et al.* 2011).

These examples of artificial and/or inadvertent selection for aggression and RHP-related traits suggest that we should be paying more attention to the geographic origins of and selection regimes experienced by wild populations. Indeed, we stand to gain much information by conducting population-level comparisons of aggression, assessment strategies, RHP-related traits including fine-scale anatomy and physiology and dominance outcomes. For instance, it seems possible that selection for particular trophic morphologies (e.g. suction-feeding versus biting: Roberts *et al.* 2011) might affect bite force or the ability to establish or maintain mouth locks during a contest, as suggested by Turner (1994). Similarly, predators might impose selection on body shape, fin shape or muscle performance (e.g. Langerhans 2009) that either facilitates or constrains some of the locomotor patterns exhibited during contests. As an example of such trade-offs, Blob *et al.* (2010) demonstrated that selection for predator escape responses in Hawaiian stream gobies, *Sicyopterus stimpsoni*, resulted in fish with deeper bodies that were less able to climb waterfalls, which they must do to lay eggs. Understanding how selection has, literally, shaped fishes should allow us to generate exciting hypotheses about population- and species-level differences in contest behaviour repertoires, the function of these behaviours, the energy costs associated with muscle performance, divergent energy/metabolic thresholds, and the capacity to deliver or withstand costly blows during a contest.

10.6.2 Developmental plasticity and dominance

The environment that an animal is exposed to during early development can dictate, at least in part, its phenotypic trajectory, which may or may not be reversible. Developmental plasticity can therefore generate a diverse spectrum of phenotypes whereby genetically identical individuals exposed to a gradient of early developmental environments will vary in their phenotypic characteristics: the pattern of phenotypic diversity across environments is the reaction norm (West-Eberhard 2003, Figure 10.1B). Several recent reviews have highlighted the importance of examining behavioural plasticity, and adopting ‘reaction norm’ approaches to the study of both individual behaviours and suites of correlated behaviours: animal personalities and behavioural syndromes (e.g. Dingemanse *et al.* 2009, Stamps & Groothuius 2010*a*,*b*). A study by Chapman *et al.* (2008), for instance, showed that predator escape response latency, body depth and body length changed as a function of early rearing environment (with juveniles only, with adults partitioned off or with the opportunity to receive aggression from adults). Juveniles that received aggression from adults were quicker to respond to a predator and had deeper but shorter bodies, suggesting that rearing with adults can induce anti-predator defences. With respect to contest behaviour in fishes, relatively few studies have investigated the effects of early developmental environment on aggression and fighting ability, although studies of this sort in species of economic importance (e.g. salmonids) have vastly outpaced studies in non-aquaculture species (e.g. Berejikian *et al.* 1996, Sundström *et al.* 2003, Pearsons *et al.* 2007, Brockmark & Johnsson 2010, Saikkonen *et al.* 2011). Here, we highlight studies that have unveiled early social, nutritional and endocrine 'experiences' as potentially potent mediators of adult RHP and aggression in fishes.

Cooperatively breeding cichlids, *Neolamprologus pulcher*, form complex groups with adults, helpers and young wherein aggressive interactions establish within-group dominance hierarchies and 'social competence' (e.g. behaving submissively when appropriate) is critical for achieving helper status (Taborsky 1985). Young *N. pulcher* raised in a social group with older family members show more appropriate behavioural responses as owners/intruders of a territory than young cichlids raised only with siblings (Arnold & Taborsky 2010). These effects persisted for at least 1 month after the early social environment treatments had concluded, suggesting that some aspect of early rearing conditions caused the young to be more socially competent. Interestingly, because adults rarely engage directly with fry, behavioural plasticity in the young was not precipitated by physical interactions with older members of the group. Arnold and Taborsky (2010) forward a number of testable alternative hypotheses, including that the behavioural phenotype of the young might be shaped by early life social learning or by sensory stimulation in the acoustic or chemical channels.

Food availability during development also can affect fighting ability later in life. For instance, Royle *et al.* (2005) demonstrated in swordtails, *Xiphophorus helleri*, that early food restriction between 2 and 6 months of age resulted in a reduction in growth compared to fish fed a normal diet over the same span. The food-restricted animals exhibited compensatory growth such that, at 10 months, their size was indistinguishable from the unrestricted animals. However, despite the similarity in size, animals that were food-restricted showed less aggression, more submission and less capacity to achieve dominance when pitted against opponents that received the normal diet. The decrease in RHP evident in food-restricted fish could result from irreversible changes in metabolic capacity that affect contest success. Alternatively, the results could reflect the persistent costs associated with compensatory growth, which the animals might be able to recoup later in life: to test this hypothesis one could establish restricted–unrestricted dyads at some later point and evaluate whether RHP asymmetries remain. Other studies have shown negligible effects of early life food ration on male–male contest outcomes but strong effects on other aspects of behaviour, such as courtship and female-directed aggression (e.g. bluefin killifish: McGhee & Travis 2011). Kolluru and Grether (2005) also showed that 'stream of origin' characteristics (e.g. low or high food availability) interact with experimental manipulation of food availability (from 0 to 15 weeks of age) to drive differences in behaviour in guppies. For example, males from high-resource streams showed greater plasticity in their aggressive behaviour when raised in low versus high food availability treatments. These studies suggest that early life nutritional status can drive plasticity in aggression and contest success in various ways, and that there could be some exciting differences among fish species with different life-history strategies (e.g. livebearers versus external fertilisers).

Growing evidence in birds and mammals points to maternal hormones as a potent mediator of offspring phenotype, including RHP and RHP-related traits such as standard metabolic rate (e.g. Dloniak *et al.* 2006, Tobler *et al.* 2007). Far less is known about how hormone allocation to eggs affects offspring phenotype in fishes. Two recent studies applied cortisol to the eggs of brown trout (*Salmo trutta*) and revealed contrasting trends. Sloman (2010) demonstrated that fish derived from cortisol-treated eggs were significantly more aggressive towards a mirror image at 194 days post-fertilisation than untreated eggs. Burton *et al.* (2011), however, found that fish derived from cortisol- and testosterone (T)-treated eggs were less aggressive (marginally so for T-treated eggs) and achieved lower social status than fish from untreated eggs; in this study, behaviour was quantified in social groups consisting of one fry (76–118 days post-hatch) from each treatment. Given the experimental differences – age at testing, the method for evaluating aggression and the cortisol dosage (e.g. 500 versus 200 $\mu g\ l^{-1}$) – the polarity between these two studies is not entirely unexpected. Both provide convincing evidence that egg hormones affect the behavioural phenotype of the offspring. Although direct maternal transfer of hormones is possible (e.g. see Andersson *et al.* 2011 for correlation between maternal and egg cortisol), fishes 'leak' hormones across their gills (Scott *et al.* 2008). Thus, parental (male or female) substrate or cavity spawners might create a hormonal microenvironment wherein the eggs are essentially bathed in testosterone, 11-ketotestosterone, oestradiol, or cortisol. If the hormone profile of the parents somehow reflects predation regime (e.g. Archard *et al.* 2012) or competition (e.g. Oliveira 2009) and if these parameters are relatively stable, we might hypothesise that 'egg bathing' would prime the offspring to exhibit a phenotype with high fitness in a similar environment ('predictive adaptive responses': Gluckman *et al.* 2005). To test this hypothesis would require that investigators manipulate predation and competition intensities, assess parental hormones and interactions with eggs/offspring (e.g. parental care, proximity), and determine both offspring phenotype and fitness along a predation/competition gradient.

All of the studies mentioned thus far have concentrated on how the developmental environment affects offspring phenotypes when the behaviour tests are conducted under standardised conditions. In the wild, animals might find themselves in environments that remain relatively stable or predictable for long periods, even many generations, but it is also quite possible that adult animals encounter environments that deviate to some degree from the environment their parents experienced or that they experienced during early life. In these situations, it is essential to understand the ramifications of developmental plasticity on adult fitness along an environmental gradient (e.g. Bateson *et al.* 2004, Gluckman *et al.* 2005, Monaghan 2008). The ability to secure resources through aggression and dominance ability constitutes one critical aspect of adult fitness, yet only one study in fishes has examined adult aggression in conditions that either match or oppose the developmental environment

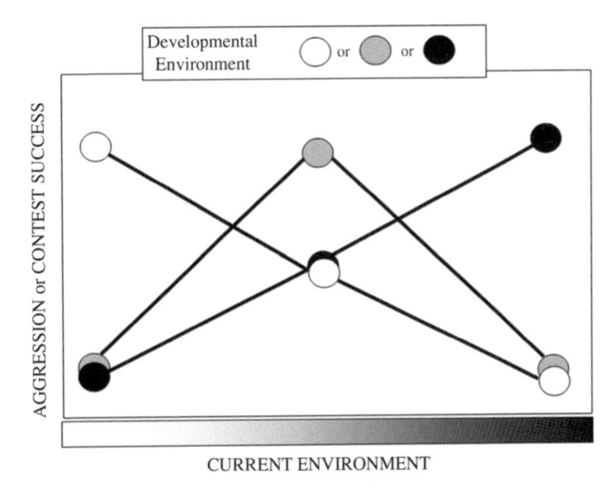

Figure 10.3 A hypothetical set of three genetically identical fish exposed to different environments during development. As adults, the fish are more aggressive or perform best in contests that occur in environments (white, grey or black circles) that match their developmental environment. The graph depicts early environment × current environment interactions for a single genotype (e.g. Marks *et al.* 2005). Different genotypes are expected to show different reaction norms, which would then reflect gene × early environment × current environment interactions.

(Figure 10.3). Marks *et al.* (2005) raised zebrafish, *Danio rerio*, in either normoxic or hypoxic conditions for ~75 days prior to conducting mirror aggression tests. In these tests, first from both developmental environments were either acclimated or not to hypoxic and normoxic conditions prior to quantifying behaviour. Without acclimation, rearing regime had no effect on aggression under hypoxia, but normoxia-reared fish were more aggressive than hypoxia-reared fish under normoxic testing conditions. With 16 h of acclimation, individuals were more aggressive when the testing environment matched their developmental environment, i.e. hypoxia-reared fish were more aggressive under hypoxic testing conditions, and normoxia-reared fish were more aggressive under normoxic conditions. Studies like these reveal complex gene × developmental environment × current environment interactions that govern contest behaviour, and should encourage researchers to begin testing hypotheses that evaluate how development influences contest performance.

10.7 Conclusions

Our primary goal was to provide fodder for future research in the area of fish contest behaviour. While preparing this chapter, it became clear that recent infusions of developmental biology, biomechanics and genomics into our experimental toolbox promise to reveal some of the intricacies of resource holding potential and individual variation in contest behaviour. We also identified some obvious deficiencies in our understanding of fish contest behaviour; these include the sparse accounts of female decision-making during (and hormonal responses to) aggressive interactions, the lack of taxonomic breadth (which limits comparative evolutionary analysis) and the application of learning theory to understand how prior fighting experiences are integrated at the neurobiological level and subsequently translate into future behavioural change. We focused very little on key mediators of contest behaviour that have been reviewed elsewhere or that are in the early stages of being integrated into contest theory. For instance, opponent and kin recognition, eavesdropping and audience effects have, and will continue to be, fruitfully integrated with the topics discussed here (e.g. Ward & Hart 2003, Höjesjö *et al.* 2007, McGregor 2005, Earley 2010, Grether 2011). In addition, recent evidence points to parasites as important mediators of fighting ability (e.g. Hamilton & Poulin 1995, Mikheev *et al.* 2010), perhaps through modulation of steroid hormones or brain monoamines (e.g. Shaw *et al.* 2009, Kittilsen *et al.* 2012). Further, the relationship between immune function and RHP in fishes remains something of a mystery (e.g. Filby *et al.* 2010*b*, Dijkstra *et al.* 2011). What we hope the reader will take away from this chapter is that the field is teeming with unanswered questions and that future discoveries will emerge from integrative experiments that test new hypotheses or address some historically rooted assumptions that, to date, have gone unchallenged.

Acknowledgements

We thank Gareth Arnott and Marlène Goubault for their reviewers' comments on this chapter.

References

Abrahams MV, Robb TL & Hare JF (2005) Effects of hypoxia on opercular displays: evidence for an honest signal? *Animal Behaviour*, 70, 427–432.

Almeida-Val VMF, Val AL, Duncan WP, *et al.* (2000) Scaling effects on hypoxia tolerance in the Amazon fish *Astronotus ocellatus* (Perciformes: Cichlidae): contribution of tissue enzyme levels. *Comparative Biochemistry and Physiology B*, 125, 219–226.

Amorim MCP & Almada VC (2005) The outcome of male–male encounters affects subsequent sound production during courtship in the cichlid fish *Oreochromis mossambicus*. *Animal Behaviour*, 69, 595–601.

Amorim MCP & Neves ASM (2008) Male painted gobies (*Pomatoschistus pictus*) vocalize to defend territories. *Behaviour*, 145, 1065–1083.

Andersson MA, Silva PIM, Steffensen JF, *et al.* (2011) Effects of maternal stress coping style on offspring characteristics in rainbow trout (*Oncorhynchus mykiss*). *Hormones and Behaviour*, 60, 699–705.

Antunes RA, Oliveira RF & Kravitz EA (2009) Hormonal anticipation of territorial challenges in cichlid fish. *Proceedings of the National Academy of Sciences USA*, 106, 15985–15989.

Archard GA & Braithwaite VA (2011) Variation in aggressive behaviour in the poeciliid fish *Brachyrhaphis episcopi*: population and sex differences. *Behavioural Processes*, 86, 52–57.

Archard GA, Earley RL, Hanninen AF, *et al.* (2012) Correlated behaviour and stress physiology in fish exposed to different levels of predation pressure. *Functional Ecology*, 26, 637–645.

Arnold C & Taborsky B (2010) Social experience in early ontogeny has lasting effects on social skills in cooperatively breeding cichlids. *Animal Behaviour*, 79, 621–630.

Arnott G & Elwood RW (2009*a*) Assessment of fighting ability in animal contests. *Animal Behaviour*, 77, 991–1004.

Arnott G & Elwood RW (2009*b*) Information gathering and decision making about resource value in animal contests. *Animal Behaviour*, 76, 529–542.

Arnott G & Elwood RW (2009*c*) Gender differences in aggressive behaviour in convict cichlids. *Animal Behaviour*, 78, 1221–1227.

Arnott G & Elwood RW (2009*d*) Probing aggressive motivation in a cichlid fish. *Biology Letters*, 5, 762–764.

Arnott G & Elwood RW (2010) Startle durations reveal visual assessment abilities during contests between convict cichlids. *Behavioural Processes*, 84, 750–756.

Aubin-Horth N, Deschênes M & Cloutier S (2012) Natural variation in the molecular stress network correlates with a behavioural syndrome. *Hormones and Behavior*, 61, 140–146.

Backström T & Winberg S (2009) Arginine-vasotocin influence on aggressive behavior and dominance in rainbow trout. *Physiology and Behavior*, 96, 470–475.

Baerends GP & Baerends-Roon JM van (1950) An introduction to the study of the ethology of cichlid fishes. *Behaviour*, (Suppl.) 1, 1–243.

Bakker ThCM, Bruijn E & Sevenster P (1989) Asymmetrical effects of prior winning and losing on dominance in sticklebacks (*Gasterosteus aculeatus*). *Ethology*, 82, 224–229.

Barata EN, Hubbard PC, Almeida OG, *et al.* (2007) Male urine signals social rank in the Mozambique tilapia (*Oreochromis mossambicus*). *BMC Biology*, 5, 54.

Barata EN, Fine JM, Hubbard PC, *et al.* (2008) A sterol-like odorant in the urine of Mozambique tilapia males likely signals social dominance to females. *Journal of Chemical Ecology*, 34, 438–449.

Barlow GW, Rogers W & Fraley N (1986) Do Midas cichlids win through prowess or daring? It depends. *Behavioural Ecology and Sociobiology*, 19, 1–8.

Barske J, Schlinger BA, Wikelski M, *et al.* (2011) Female mate choice for male motor skills. *Proceedings of the Royal Society of London B*, 278, 3523–3528.

Bateson P, Barker D, Clutton-Brock T, *et al.* (2004). Developmental plasticity and human health. *Nature*, 430, 419–421.

Batista G, Zubizarreta L, Perrone R, *et al.* (2012) Non-sex-biased dominance in a sexually monomorphic electric fish: fight structure and submissive electric signalling. *Ethology*, 118, 398–410.

Beacham JL (1988) The relative importance of body size and aggressive experience as determinants of dominance in pumpkinseed sunfish, *Lepomis gibbosus*. *Animal Behaviour*, 36, 621–623.

Beaugrand JP & Goulet C (2000) Distinguishing kinds of prior dominance and subordination experiences in males of green swordtail fish (*Xiphophorus helleri*). *Behavioural Processes*, 50, 131–142.

Beaugrand JP, Payette D & Goulet C (1996) Conflict outcome in male green swordtail fish dyads (*Xiphophorus helleri*): interaction of body size, prior dominance/subordination experience, and prior residency. *Behaviour*, 133, 303–319.

Beeching SC, Gross SH, Bretz HS, *et al.* (1998) Sexual dichromatism in convict cichlids: the ethological significance of female ventral coloration. *Animal Behaviour*, 56, 1021–1026.

Bell AM (2005) Behavioral differences between individuals and two populations of stickleback (*Gasterosteus aculeatus*). *Journal of Evolutionary Biology*, 18, 464–473.

Berejikian BA, Mathews SB & Quinn TP (1996) Effects of hatchery and wild ancestry and rearing environments on the development of agonistic behavior in steelhead trout (*Oncorhynchus mykiss*) fry. *Canadian Journal of Fisheries and Aquatic Sciences*, 53, 2004–2014.

Berg RL (1960) The ecological significance of correlational Pleiades. *Evolution*, 14, 171–180.

Blob RW, Kawano SM, Moody KN, *et al.* (2010) Morphological selection and the evaluation of potential trade-offs between escape from predators and the climbing of waterfalls in the Hawaiian stream goby *Sicyopterus stimpsoni*. *Integrative and Comparative Biology*, 50, 1185–1199.

Blows MW (2007) A tale of two matrices: multivariate approaches in evolutionary biology. *Journal of Evolutionary Biology*, 20, 1–8.

Blumstein DT, Ebensperger LA, Hayes LD, *et al.* (2010) Toward an integrative understanding of social behavior: new models and new opportunities. *Frontiers in Behavioral Neuroscience*, 4, 34.

Bolyard K & Rowland WJ (1996) Context-dependent response to red coloration in stickleback. *Animal Behaviour*, 52, 923–927.

Borazjani I, Sotiropoulos F, Tytell ED, *et al.* (2012) Hydrodynamics of the bluegill sunfish C-start escape response: three-dimensional simulations and comparison with experimental data. *Journal of Experimental Biology*, 215, 671–684.

Borg B (1994) Androgens in teleost fish. *Comparative Biochemistry and Physiology C*, 109, 219–245.

Borowsky RL (1973) Social control of adult size in males of *Xiphophorus variatus*. *Nature*, 245, 332–335.

Braddock JC & Braddock ZJ (1955) Aggressive behaviour among females of the Siamese fighting fish (*Betta splendens*). *Physiological Zoology*, 28, 152–172.

Brandtmann G, Scandura M & Trillmich F (1999) Female–female conflict in the harem of a snail cichlid (*Lamprologus ocellatus*): behavioural interactions and fitness consequences. *Behaviour*, 136, 1123–1144.

Brelin D, Petersson E, Dannewitz J, *et al.* (2008) Frequency distribution of coping strategies in four populations of brown trout (*Salmo trutta*). *Hormones and Behavior*, 53, 546–556.

Brick O (1998) Fighting behaviour, vigilance and predation risk in the cichlid fish *Nannacara anomala*. *Animal Behaviour*, 56, 309–317.

Brick O (1999) A test of the sequential assessment game: the effect of increased cost of sampling. *Behavioral Ecology*, 10, 726–732.

Briffa M & Elwood RW (2009) Difficulties remain in distinguishing between mutual and self-assessment in animal contests. *Animal Behaviour*, 77, 759–762.

Briffa M & Sneddon LU (2007) Physiological constraints on contest behaviour. *Functional Ecology*, 21, 627–637.

Briffa M, Rundle SD & Fryer A (2008) Comparing the strength of behavioural plasticity and consistency across situations: animal personalities in the hermit crab *Pagurus bernhardus*. *Proceedings of the Royal Society of London B*, 275, 1305–1311.

Brockmark S & Johnsson JI (2010) Reduced hatchery rearing density increases social dominance, postrelease growth, and survival in brown trout. *Canadian Journal of Fisheries and Aquatic Sciences*, 67, 288–295.

Budaev SV (2010) Using principal components and factor analysis in animal behaviour research: caveats and guidelines. *Ethology*, 116, 472–480.

Burton T, Hoogenboom MO, Armstrong JD, *et al.* (2011) Egg hormones in a highly fecund vertebrate: do they influence offspring social structure in competitive conditions? *Functional Ecology*, 25, 1379–1388.

Byers J, Hebets E & Podos J (2010) Female mate choice based upon male motor performance. *Animal Behaviour*, 79, 771–778.

Campbell HA, Handy RD & Sims DW (2005) Shifts in a fish's resource holding power during a contact paired interaction: the influence of a copper-contaminated diet in rainbow trout. *Physiology and Biochemical Zoology*, 78, 706–714.

Carpenter RE & Summers CH (2009) Learning strategies during fear conditioning. *Neurobiology of Learning and Memory*, 91, 415–423.

Chapman BB, Morrell LJ, Benton TG, *et al.* (2008) Early interactions with adults mediate the development of predator defenses in guppies. *Behavioral Ecology*, 19, 87–93.

Chase ID, Bartolomeo C & Dugatkin LA (1994) Aggressive interactions and inter-contest interval: how long do winners keep winning? *Animal Behaviour*, 48, 393–400.

Chellapa S & Huntingford SA (1989) Depletion of energetic reserves during reproductive aggression in male three-spined stickleback, *Gasterosteus aculeatus*. *Journal of Fish Biology*, 35, 275–286.

Clark CJ (2009) Courtship dives of Anna's hummingbirds offer insights into flight performance limits. *Proceedings of the Royal Society of London B*, 276, 3047–3052.

Conrad JL, Weinersmith KL, Brodin T, *et al.* (2011) Behavioural syndromes in fishes: a review with implications for ecology and fisheries management. *Journal of Fish Biology*, 78, 395–435.

Cooper WJ, Wernle J, Mann K, *et al.* (2011) Functional and genetic integration in the skulls of Lake Malawi cichlids. *Evolutionary Biology*, 38, 316–334.

Copeland DL, Levay B, Sivaraman B, *et al.* (2011) Metabolic costs of fighting are driven by contest performance in male convict cichlid fish. *Animal Behaviour*, 82, 271–280.

Cutts CJ, Metcalfe NB & Taylor AC (1999) Competitive asymmetries in territorial juvenile Atlantic salmon, *Salmo salar*. *Oikos*, 86, 479–486.

Dahlbom SJ, Backstrom T, Lundstedt-Enkel K, *et al.* (2012) Aggression and monoamines: effects of sex and social rank in zebrafish (*Danio rerio*). *Behavioural Brain Research*, 228, 333–338.

Demarest J (1992) Reassessment of socially mediated learning in Siamese fighting fish (*Betta splendens*). *Journal of Comparative Psychology*, 106, 150–162.

Desjardins JK & Fernald RD (2008) How do social dominance and social information influence reproduction and the brain? *Integrative and Comparative Biology*, 48, 596–603.

Desjardins JK & Fernald R (2010) What do fish make of mirror images? *Biology Letters*, 6, 744–747.

Desjardins JK, Hazelden MR, Kraak GJ van der, *et al.* (2006) Male and female cooperatively breeding fish provide support for the 'challenge hypothesis'. *Behavioral Ecology*, 17, 149–154.

DiBattista JD, Anisman H, Whitehead M, *et al.* (2005) The effects of cortisol administration on social status and brain monoaminergic activity in rainbow trout *Oncorhynchus mykiss*. *Journal of Experimental Biology*, 208, 2707–2718.

Dijkstra PD, Linström J, Metcalfe NB, *et al.* (2010) Frequency-dependent social dominance in a color polymorphic cichlid fish. *Evolution*, 64, 2797–2807.

Dijkstra PD, Wiegertjes GF, Forlenza M, *et al.* (2011) The role of physiology in the divergence of two incipient cichlid species. *Journal of Evolutionary Biology*, 24, 2639–2652.

Dingemanse NJ, Kazem AJN, Réale D, *et al.* (2009) Behavioural reaction norms: animal personality meets individual plasticity. *Trends in Ecology and Evolution*, 25, 81–89.

Dloniak SM, French JA & Holekamp KE (2006) Rank-related maternal effects of androgens on behaviour in wild spotted hyaenas. *Nature*, 440, 1190–1193.

Dochtermann NA (2011) Testing Cheverud's conjecture for behavioral correlations and behavioral syndromes. *Evolution*, 65, 1814–1820.

Dow M, Ewing AW & Sutherland J (1976) Studies on the behaviour of Cyprinodont fish. III. The temporal patterning of aggression in *Aphyosemion striatum* (Boulenger). *Behaviour*, 59, 252–268.

Draud M & Lynch PAE (2002) Asymmetric contests for breeding sites between monogamous pairs of convict cichlids (*Archocentrus nigrofasciatum*, Cichlidae): pair experience pays. *Behaviour*, 139, 861–873.

Draud M, Macías-Ordóñez R, Verga J, *et al.* (2004) Female and male Texas cichlids (*Herichthys cyanoguttatum*) do not fight by the same rules. *Behavioral Ecology*, 15, 102–108.

Dugatkin LA & Biederman L (1991) Balancing asymmetries in resource holding power and resource value in the pumpkinseed sunfish. *Animal Behaviour*, 42, 691–692.

Dugatkin LA & Ohlsen SR (1990) Contrasting asymmetries in value expectation and resource holding power: effects on attack behaviour and dominance in the pumpkinseed sunfish, *Lepomis gibbosus*. *Animal Behaviour*, 39, 802–804.

Dunlap KD (2002) Hormonal and body size correlates of electrocommunication behavior during dyadic interactions in a weakly electric fish, *Apternonotus leptorhynchus*. *Hormones and Behaviour*, 41, 187–194.

Dunlap KD, Pelczar PL & Knapp R (2002) Social interactions and cortisol treatment increase the production of aggressive electrocommunication signals in male electric fish, *Apteronotus leptorhynchus*. *Hormones and Behavior*, 42, 97–108.

Dzieweczynski TL & Rowland WJ (2004) Behind closed doors: use of visual cover by courting male three-spined stickleback, *Gasterosteus aculeatus*. *Animal Behaviour*, 68, 465–471.

Earley RL (2010) Social eavesdropping and the evolution of conditional cooperation and cheating strategies. *Philosophical Transactions of the Royal Society B*, 365, 2675–2686.

Earley RL & Dugatkin LA (2002) Eavesdropping on visual cues in green swordtail (*Xiphophorus helleri*) fights: a case for networking. *Proceedings of the Royal Society of London B*, 269, 943–952.

Earley RL & Hsu Y (2008) Reciprocity between endocrine state and contest behavior in the killifish, *Kryptolebias marmoratus*. *Hormones and Behavior*, 53, 442–451.

Earley RL, Hsu Y & Wolf LL (2000) The use of standard aggression testing methods to predict combat behaviour and contest outcome in *Rivulus marmoratus* dyads (Teleostei: Cyprinodontidae). *Ethology*, 106, 743–761.

Earley RL, Tinsley M & Dugatkin LA (2003) To see or not to see: does previewing a future opponent affect the contest behavior of green swordtail males (*Xiphophorus helleri*)? *Naturwissenschaften*, 90, 226–230.

Earley RL, Edwards JT, Aseem O, *et al.* (2006) Social interactions tune aggression and stress responsiveness in a territorial cichlid fish (*Archocentrus nigrofasciatus*). *Physiology and Behavior*, 88, 353–363.

Egge AR, Brandt Y & Swallow JG (2011) Sequential analysis of aggressive interactions in the stalk-eyed fly *Teleopsis dalmanni*. *Behavioral Ecology and Sociobiology*, 65, 369–379.

Elofsson UOE, Mayer I, Damsgård B, *et al.* (2000) Intermale competition in sexually mature arctic charr: effects on brain monoamines, endocrine stress responses, sex hormone levels, and behavior. *General and Comparative Endocrinology*, 118, 450–460.

Enquist M & Jakobsson S (1986) Decision making and assessment in the fighting behaviour of *Nannacara anomala* (Cichlidae, Pisces). *Ethology*, 72, 143–153.

Enquist M & Leimar O (1983) Evolution of fighting behavior – decision rules and assessment of relative strength. *Journal of Theoretical Biology*, 102, 387–410.

Enquist M & Leimar O (1987) Evolution of fighting behavior – the effect of variation in resource value. *Journal of Theoretical Biology*, 127, 187–205.

Enquist M, Ljungberg T & Zandor A (1987) Visual assessment of fighting ability in the cichlid fish *Nannacara anomala*. *Animal Behaviour*, 35, 1262–1264.

Enquist M, Leimar O, Ljungberg T, *et al.* (1990) A test of the sequential assessment game: fighting in the cichlid fish *Nannacara anomala*. *Animal Behaviour*, 40, 1–14.

Evans MR & Norris K (1996) The importance of carotenoids in signalling during aggressive interactions between male firemouth cichlids (*Cichlasoma meeki*). *Behavioral Ecology*, 7, 1–6.

Ewing AW (1975) Studies on the behaviour of Cyprinodont fish. II. The evolution of aggressive behaviour in Old World rivulins. *Behaviour*, 52, 172–195.

Falconer DS & Mackay TFC (1996) *Introduction to quantitative genetics*. Harlow: Longman.

Filby AL, Paull GC, Hickmore TFA, *et al.* (2010*a*) Unravelling the neurophysiological basis of aggression in a fish model. *BMC Genomics*, 11, 498.

Filby AL, Paull GC, Bartlett EJ, *et al.* (2010*b*) Physiological and health consequences of social status in zebrafish (*Danio rerio*). *Physiology and Behavior*, 101, 576–587.

Flammang BE, Lauder GV, Troolin DR, *et al.* (2011) Volumetric imaging of fish locomotion. *Biology Letters*, 7, 695–698.

Foster SA & Endler JA (1999) *Geographic Variation in Behavior: Perspectives on Evolutionary Mechanisms*. Oxford: Oxford University Press.

Foster SA, Shaw KA, Robert KL & Baker JA (2008) Benthic, limnetic and oceanic threespine stickleback: profiles of reproductive behaviour. *Behaviour*, 145, 485–508.

Francis RC (1983) Experiential effects on agonistic behavior in the paradise fish, *Macropodus opercularis*. *Behaviour*, 85, 292–313.

Franck D & Ribowski A (1989) Escalating fights for rank-order position between male swordtails (*Xiphophorus helleri*): effects of prior rank-order experience and information transfer. *Behavioral Ecology and Sociobiology*, 24, 133–143.

Frey DF & Miller RJ (1972) The establishment of dominance relationships in the blue gourami, *Trichogaster trichopterus* (Pallas). *Behaviour*, 42, 8–62.

Fugère V, Ortega H & Krahe R (2011) Electrical signalling of dominance in a wild population of electric fish. *Biology Letters*, 7, 197–200.

Gammell MP & Hardy ICW (2003) Contest duration: sizing up the opposition? *Trends in Ecology and Evolution*, 18, 491–493.

Ghalambor GK, Walker JA & Reznick DN (2003) Multi-trait selection, adaptation, and constraints on the evolution of burst swimming performance. *Integrative and Comparative Biology*, 43, 431–438.

Ghalambor GK, McKay JK, Carroll SP, *et al.* (2007) Adaptive versus non-adaptive phenotypic plasticity and the potential for contemporary adaptation in new environments. *Functional Ecology*, 21, 394–407.

Giaquinto PC & Volpato GL (1997) Chemical communication, aggression, and conspecific recognition in the fish Nile tilapia. *Physiology and Behavior*, 62, 1333–1338.

Gibb AC, Ashley-Ross MA, Pace CM, *et al.* (2011) Fish out of water: terrestrial jumping by fully aquatic fishes. *Journal of Experimental Zoology*, 315, 649–653.

Gluckman, PD, Hanson MA & Spencer HG (2005) Predictive adaptive responses and human evolution. *Trends in Ecology and Evolution*, 20, 528–533.

Goodson JL (2005) The vertebrate social behavior network: evolutionary themes and variations. *Hormones and Behavior*, 48, 11–22.

Grether G (2011) The neuroecology of competitor recognition. *Integrative and Comparative Biology*, 51, 807–818.

Guderley H (2009) Muscle metabolic capacities and plasma cortisol levels of the male three-spined stickleback *Gasterosteus aculeatus*: are there 'femme fatale' or 'macho male' effects? *Physiological and Biochemical Zoology*, 82, 653–661.

Guderley H & Couture P (2005) Stickleback fights: why do winners win? Influence of metabolic and morphometric parameters. *Physiological and Biochemical Zoology*, 78, 173–181.

Hagedorn M & Zelick R (1989) Relative dominance among males is expressed in the electric organ discharge characteristics of a weakly electric fish. *Animal Behaviour*, 38, 520–525.

Hall MD, McLaren L, Brooks RC, *et al.* (2010) Interactions among performance capacities predict male combat outcomes in the field cricket. *Functional Ecology*, 24, 159–164.

Haller J (1991*a*) Muscle metabolic changes during the first six hours of cohabitation in pairs of male *Betta splendens*. *Physiology and Behavior*, 49, 1301–1303.

Haller J (1991*b*) Biochemical cost of a fight in fed and fasted *Betta splendens*. *Physiology and Behavior*, 49, 79–82.

Hamilton WJ & Poulin R (1995) Parasites, aggression and dominance in male upland bullies. *Journal of Fish Biology*, 47, 302–307.

Hendry AP, Hudson K, Walker JA, *et al.* (2011) Genetic divergence in morphology–performance mapping between Misty Lake and inlet stickleback. *Journal of Evolutionary Biology*, 24, 23–35.

Henningsen JP & Irschick DJ (2012) An experimental test of the effect of signal size and performance capacity on dominance in the green anole lizard. *Functional Ecology*, 26, 3–10.

Herrel A, Podos J, Vanhooydonck B, *et al.* (2009) Force–velocity trade-off in Darwin's finch jaw function: a biomechanical basis for ecological speciation? *Functional Ecology*, 23, 119–125.

Hine E, Chenoweth SF, Rundle HD, *et al.* (2009) Characterizing the evolution of genetic variance using genetic covariance tensors. *Philosophical Transactions of the Royal Society B*, 364, 1567–1578.

Hirschenhauser K, Taborsky M, Oliveira T, *et al.* (2004) A test of the 'challenge hypothesis' in cichlid fish: simulated partner and territory intruder experiments. *Animal Behaviour*, 68, 741–750.

Hirschenhauser K, Canàrio AVM, Ros AFH, *et al.* (2008) Social context may affect urinary excretion of 11-ketotestosterone in African cichlids. *Behaviour*, 145, 1367–1388.

Höglund E, Bakke MJ, Øverli Ø, *et al.* (2005) Suppression of aggressive behaviour in juvenile Atlantic cod (*Gadus morhua*) by L-tryptophan supplementation. *Aquaculture*, 249, 525–531.

Höjesjö J, Andersson P, Engman A, *et al.* (2007) Rapid bystander assessment of intrinsic fighting ability: behavioural and heart rate responses in rainbow trout. *Animal Behaviour*, 74, 1743–1751.

Holder JL, Barlow GW & Francis RC (1991) Differences in aggressiveness in the Midas cichlid fish (*Cichlasoma citrinellum*) in relation to sex, reproductive state and the individual. *Ethology*, 88, 297–306.

Hollis KL (1999) The role of learning in the aggressive and reproductive behavior of blue gouramis, *Trichogaster trichopterus*. *Environmental Biology of Fishes*, 54, 355–369.

Hollis KL, Dumas MJ, Singh P, *et al.* (1995) Pavlovian conditioning of aggressive behavior in blue gourami fish (*Trichogaster trichopterus*): winners become winners and losers stay losers. *Journal of Comparative Psychology*, 109, 123–133.

Holzman R, Day SW & Wainwright PC (2007) Timing is everything: coordination of strike kinematics affects the force exerted by suction feeding fish on attached prey. *Journal of Experimental Biology*, 210, 3328–3336.

natural conditions (e.g. Schwartz 1986, Gerhardt *et al.* 1987, Wells & Bard 1987, Arak 1988, Passmore *et al.* 1992, Sullivan 1992, Jennions *et al.* 1995, Schwartz *et al.* 1995, Smith & Roberts 2003), so the high calling rates of many species in dense choruses probably reflect competition among males for access to females. However, increases in rate may reflect the tendency for some species to alternate calls with those of other males, which leads to the matching of calling rates by interacting individuals (Wells 2007). In addition to increasing rate, there is a tendency for some frogs to increase call duration in choruses. In frogs that produce calls consisting of a series of repeated notes, duration is increased by adding notes to the call. A well-studied example of this behaviour is seen in the eastern grey treefrog (*Hyla versicolor*): males often double the duration of their calls in response to calls of other males. In this and in other species an increase in call duration is usually accompanied by a decrease in call rate (Wells & Taigen 1986, Schwartz *et al.* 2002). This trade-off between rate and duration probably reflects the high energetic cost of producing advertisement calls. Wells and Taigen (1986) found that increasing call duration and decreasing rate does not appear to increase the aerobic costs of calling in *H. versicolor*. However, Grafe (1997) showed that call duration is weakly correlated with the rate of lipid depletion and that lipids are the main source of energy for calling in this species. In addition, Sullivan and Hinshaw (1992) found that the nightly calling period and chorus tenure of males producing long calls is significantly shorter than those that produce shorter calls at a faster rate. These findings, together with the finding that females prefer long calls delivered at a slow rate over short calls delivered at a fast rate (Klump & Gerhardt 1987), suggest that males alter their calling behaviour in dense choruses in ways that enhance their attractiveness to females.

In species that produce single-note calls when calling in isolation, males often add notes to the call when calling in choruses. The added notes can simply be a repeat of the single note (as above) (e.g. *Rana clamitans*, Bee & Perrill 1996) or individual notes may vary in structure (Bevier *et al.* 2004). Other species produce calls with distinct primary and secondary notes and call duration is increased by adding secondary notes in response to calls of other males (e.g. *Dendropsophus* (*Hyla*) *ebraccatus*: Wells & Schwartz 1984*a*; *D. microcephalus*: Schwartz & Wells 1985; *D. phlebodes*: Schwartz & Wells 1984; *Engystomops* (*Physalaemus*) *pustulosus*: Rand & Ryan 1981). Under experimental conditions, females of several species express preferences for complex, multi-note calls over simple calls (e.g. Rand & Ryan 1981, Wells & Schwartz 1984*a*, Schwartz 1986, Pallett & Passmore 1988). The reason for these preferences is unclear but could indicate female choice of males that sustain high levels of signal production because this indicates superior genetic quality (Wells 2007), or it could simply indicate a preference for calls that are conspicuous because of their high-energy output. Whatever the reason, males would probably benefit from producing long and/or complex calls all the time. However, the high energetic cost of producing such calls would make this impossible so males only do so when competition intensifies and background noise levels increase.

11.3.2.1.2 Call matching

Call matching, where the number and complexity of calls that are produced by a male are approximately matched by his neighbour, has only been described in a small number of anuran species (Arak 1983, Schwartz 1986, Pallett & Passmore 1988, Jehle & Arak 1998, Gerhardt *et al.* 2001), but may be more widespread. Arak (1983) first described call matching in the treefrog *Pseudophilautus* (*Philautus*) *leuchorhinus*: males that call in isolation produce a single-note advertisement call but switch to multi-note calls when interacting with a neighbour. They show a tendency to match the number of multi-note calls produced by the rival. Arak (1983) considered three hypotheses for call matching in *P. leucorhinus*. First, asymmetric matching by pairs of males could indicate dominance: this was rejected because males matched their rivals. Asymmetric matching was also dismissed as a possible adaptive reason for males of the Australian frog *Crinia georgiana* matching the number of notes in neighbours' advertisement calls as size-dependent matching is absent (Gerhardt *et al.* 2001). Second, call matching may be a mechanism to direct aggression toward particular males, grading the response by the precision of matching. Arak (1983) considered this to be the mostly likely explanation for call matching in *P. leucorhinus*. The third hypothesis relates to female choice: if females respond preferentially to calls with multiple notes, males who call-match ensure their call is at least as attractive to females as those of their neighbours without wasting energy. In other words, call matching is an attempt to preserve energy levels but remain effective competitors for females (Gerhardt *et al.* 2001), a theory that makes sense given the high cost of call production in frogs.

11.3.2.1.3 Changes in call timing

Fine-scale adjustments in call timing during social interactions have been found in many species of anurans (Gerhardt & Huber 2002, Wells 2007). In the large aggregations that often occur in these animals, masking interference of signals used to attract mates is likely to occur and fine-scale timing relationships between signals may be a means of reducing this interference (e.g. Schwartz & Wells 1985, Grafe 2003). However, adjustments in timing may also represent competition between males if changes in timing result in one male's call being more conspicuous to females than the calls of his near neighbours. One case where this may occur is when changes in timing by one male result in his calls overlapping those of a neighbour so that his calls either lead or lag the neighbouring male's call.

A preference by females for leading calls has been found in several species of anurans (Wells & Schwartz 1984*a*, Grafe 1996*a*, Whitney & Krebs 1975, Klump & Gerhardt 1992, Howard & Palmer 1995, Bosch & Marquez 2002) and in some cases these preferences override preferences based on other call properties (Dyson & Passmore 1988*a*,*b*). Goin (1949) reported that *Pseudacris crucifer* choruses are organised into duets and trios and that certain individuals are persistent leaders: similar phenomena have been reported in other anuran species (e.g. Bratstrom & Yarnell 1968, Whitney & Krebs 1975). There is, however, little evidence from natural choruses of frogs that males consistently serve as leaders or that call leadership is determined through competitive interactions with nearby rivals.

There is some evidence that males time their calls to be following or lagging calls (e.g. *Smilisca sila*: Tuttle & Ryan 1982; *Dendropsophus ebraccatus*: Wells & Schwartz 1984*a*). Grafe (1999) found that, under natural calling conditions and in response to playbacks, male *Kassina fusca* overlapped calls by between 20% and 21.5%. As these lagging calls do not appear to be generated by a calling rhythm to which other males entrain, Grafe suggested that this sort of timing results from attempts by males to mask the calls of neighbours by overlapping their calls in a competitive fashion. He also showed that females preferred following calls when calls overlapped by 10–25% (as in natural choruses), but this preference switched to leading calls when the overlap increased to 75% and 90%. This suggests that males of this species can outcompete rival individuals in the chorus with respect to attracting females if they time their calls in response to neighbours' calls. Grafe (1999) also notes that this kind of interaction may result in a war of attrition with no males calling but, as changes in factors such as calling rate affect female mate preferences for following calls, this is unlikely to happen. In *Dendropsophus ebraccatus*, rapid male vocal responses result in masking of shorter secondary notes of a leading conspecific male with the longer primary note of a following male. Tests with females demonstrated that following males are favoured under such circumstances (Wells & Schwartz 1984*a*).

11.3.2.2 Defence of territories and calling sites

In addition to attracting females, the advertisement call serves to advertise a male's position in the chorus to other competing males. Males of many species of anurans defend territories or calling sites from which other males are excluded. The maintenance of territories or calling sites is mediated by the intensity of a neighbour's advertisement calls. Males tolerate the intensity of a neighbour's call at or below a certain threshold (e.g. Wilczynski & Brenowitz 1988, Brenowitz 1989, Gerhardt *et al.* 1989, Marshall *et al.* 2003), and calls perceived by a resident male above this threshold elicit an aggressive response from the resident towards an intruder (Fellers 1979, Telford 1985, Wells 1988, Wells & Schwartz 1984*b*, Schwartz 1989, Bee *et al.* 1999).

In species with an aggressive call that is distinct from the advertisement call, the switch from advertisement to aggressive calling is an all-or-nothing response (Robertson 1984, Telford 1985). For example, in *Hyperolius marmoratus*, males switch from producing a tone-like advertisement call to producing a trilled aggressive call in response to intrusion by a neighbour (Telford 1985). In contrast, in a number of hylid species from Central and South America the switch from advertisement to aggressive calling is more graded. Males gradually increase the proportion of aggressive calls as they approach one another so that it is difficult to define a threshold at which males switch from advertisement to aggressive calling (e.g. Wells & Schwartz 1984*b*, Schwartz 1986, 1989). Male *Dendropsophus ebraccatus* calling in isolation produce advertisement calls with an introductory note with a species-specific pulse rate followed by a few secondary click notes. The aggressive call also consists of a pulsed primary note, given with or without clicks but with a much higher pulse repetition rate (Wells & Greer 1981). In response to playbacks simulating an approaching intruder, males switch to aggressive calling and as the rival approaches, increase the duration and rise time of the introductory note while gradually dropping the secondary click notes until at very close distances a single-note long aggressive call is given (Wells & Schwartz 1984*b*). Graded aggressive calls have also been found in *Dendropsophus microcephalus* (Schwartz & Wells 1985), *D. phlebodes* (Schwartz & Wells 1984) and *Pseudacris crucifer* (Schwartz 1989) and have been described in several myobatrachid, hyperoliid, ranid and rhacophorid frogs (Wells 2007).

Unlike species where the advertisement and aggressive calls are structurally distinct, cricket frogs (*Acris crepitans*) have a simple advertisement call composed of a variable number of pulses and pulsed groups (Wagner 1989*a*). Rather than produce a discrete aggressive call, males respond aggressively by increasing the length and complexity of the advertisement call by increasing the number of pulses per call, the number of pulse groups per call and the number of calls per group. Wagner (1989*a*) showed that this represents a graded system, with males increasing the length of the call as they approach one another.

In section 11.3.2.2.1 we explore the kinds of information that may be available to males during aggressive encounters. Vocal responses of males to playbacks of conspecific advertisement and aggressive calls and the use of staged aggressive encounters between males provide some insight into whether males communicate their RHP to rival males or information about motivation or aggressive intent.

11.3.2.2.1 Assessment of fighting ability and/or aggressive intent

Studies of aggressive behaviour in anurans provided some of the earliest support for game theoretical models of aggressive signalling (e.g. Davies & Halliday 1978, Arak 1983). The exchanges of energetically costly vocalisations that take place during anuran contests seem to be prototypical aggressive signals that neatly fit into models in which signaller quality is encoded in signals that are exchanged with, and assessed by, the contestants (Chapter 2). Indeed, as described below, several studies have examined the characteristics that may determine RHP in anurans, and in many cases the components of RHP can be related to characteristics of the vocalisations that are exchanged by competitors during contests. Nonetheless, there are additional complications in anuran aggressive behaviour that may challenge the predictions of current models of aggressive signalling, and remarkably few studies of anurans have taken the steps necessary to link directly aggressive signalling behaviour with RHP and assessment during contests. Furthermore, mutual assessment of aggressive call characteristics is often assumed, yet few studies have performed the manipulations necessary to rule out alternative assessment strategies or even to confirm that assessment of aggressive calls is taking place at all.

Particularly challenging to explain are cases in which assessment apparently proceeds through the exchange of distinctive aggressive vocalisations. At first glance, it appears relatively straightforward to assign to anuran aggressive calls the role typically ascribed to aggressive signals in most other animals; i.e. that these signals communicate information about the signaller's relative fighting abilities, motivation, or aggressiveness (e.g. Maynard Smith & Price 1973, Maynard Smith & Parker 1976, Enquist 1985, Hurd 2006). However, these predictions have seldom been rigorously tested for anuran aggressive calling behaviour and there are several reasons why the occurrence of aggressive calling is challenging to explain. Essentially, as described in detail in section 11.3.2.2.2, there are distinct costs associated with the production of aggressive calls. In order for the communication system to remain evolutionarily stable, these costs must be offset somehow by benefits associated with the production of aggressive calls during contests. The benefits of signalling during contests are generally described as a reduction in the costs associated with intense physical fighting by giving less costly aggressive signals (Maynard Smith & Price 1973). However, with the exception of species with weapons (section 11.3.1.), physical fights in anurans are often characterised by a lack of potential for serious injury (Shine 1979) and their relative lack of intensity suggests low energetic investment relative to the high energetic cost of producing vocalisations. Presumably aggressive signals often provide information on a contestant's RHP. However, as described below, even in species with distinctive aggressive calls, the characteristics of advertisement calls often have equally strong relationships with potential determinants of RHP as the characteristics of aggressive calls, and males are known to be sensitive to variation in these advertisement call characteristics (e.g. Wells & Schwartz 1984*a*, Wells & Taigen 1986, Wagner 1989*b*). If likely determinants of RHP can be assessed by the advertisement calls that males are already giving to attract females, why should an additional call type be used? Indeed, most of the studies that clearly demonstrate a role of calls in contest assessment or outcome in frogs involve advertisement calling, as described below.

In anurans, the shape and mass of the laryngeal apparatus is related to body size and determines the fundamental frequency of signals (Duellman & Trueb 1986). Fundamental frequency of the advertisement call is negatively related to body size in anurans and is a better predictor of body size than any other call property (Ramer *et al.* 1983, Robertson 1986, Wagner 1989*b*, Bee *at al.* 1999, Bee & Gerhardt 2001*a*). Also, in many species large body size confers an advantage in physical fights between males (Davis & Halliday 1978, Howard 1978, Wells 1978, Sullivan 1982, Arak 1983, Given 1988, Wagner 1989*b*, Dyson & Passmore 1992). Therefore, accurate assessment of an opponent's relative size, and hence fighting ability, based on call fundamental frequency would allow male frogs to gauge the likelihood of winning an escalated contest. It has been demonstrated in several species that frogs can use the frequency of calls to assess an opponent's size (Davies & Halliday 1978, Arak 1983, Ramer *et al.* 1983, Robertson 1986, Wagner 1989*b*). In *Bufo bufo* the call by which size is assessed is a release call produced during fighting (Davies & Halliday 1978). In response to the playback of advertisement calls with low dominant frequency indicative of a large male, male natterjack toads (*Epidalea* (*Bufo*) *calamita*) are more likely to abandon calling or retreat, whereas they are more likely to approach and attack the speaker when calls of high dominant frequency indicative of a small male are broadcast (Arak 1983). Similar results were found in cricket frogs (Wagner 1992). As in most other frogs, in cricket frogs, call dominant frequency is negatively correlated with size with larger males having lower frequency calls than smaller males (Wagner 1989*a*). Larger males win more fights over calling sites and males tend to retreat from broadcast of low-frequency calls but are more likely to attack loudspeakers playing high-frequency calls (Wagner 1989*b*). These and other studies suggest, therefore, that decisions about whether to physically challenge an intruder may, in the first instance, depend on the size of the intruding male and hence his RHP. However, they say little about how contests are resolved once a resident male elects to escalate an encounter rather than retreat. Several studies have looked at the vocal responses of males to playbacks of conspecific calls, in an attempt to determine whether males communicate fighting ability or RHP to rivals during encounters. These have shown that a common response is for the resident male to lower the dominant frequency of his own call (Lopez *et al.* 1988, Wagner 1989*b*, 1992, Bee & Perrill 1996, Bee *et al.* 1999, Given 1999, Bee 2002, Bee & Bowling 2002). Wagner (1992) found that male cricket frogs respond in this way to playbacks of low-frequency calls, and suggested that in doing so they gain an advantage in aggressive interactions. This suggestion was supported by playback experiments showing that calls that decreased in frequency were more effective in repelling opponents than calls that either increased, or did not change, in dominant frequency (Wagner 1992).

Wagner (1992) proposed three hypotheses for why males may benefit from lowering call dominant frequency: first, by lowering call dominant frequency males provide information about size to an opponent. Second, by lowering dominant frequency males signal size-independent fighting ability to opponents; and third, males decrease dominant frequency to deceive opponents about their size. This latter hypothesis was ruled out because the extent to which males lowered frequency was positively related to probability of attack. Wagner (1992) reasoned that because animals that exaggerate their fighting ability should not attack opponents, the benefit of a decrease in call dominant frequency does not appear to be derived from the production of deceptive signals. The hypothesis that changes in call dominant frequency represent honest signals of size was also rejected because size becomes less predictable when males change their dominant frequency and because the extent of the change in dominant frequency is not related to size. Wagner (1992) therefore suggested that lowering call frequency in response to a simulated intrusion may function primarily to advertise size-independent fighting ability and hence RHP is most likely based on physiological condition or motivation. Because calling is energetically expensive for male frogs (section 11.3.2.2.2), and as deviations from the fundamental frequency may be especially energetically costly, the extent to which a male lowers his dominant frequency, and not the initial call frequency of the caller, may contain information about his physiological condition or the probability that he is sufficiently motivated to attack an opponent.

Further playback experiments (Burmeister *et al.* 2002) provided support for Wagner's (1992) suggestion that in cricket frogs, changes in call dominant frequency are not used to assess the size of a competitor. They showed that although absolute size or relative size did not appear to influence a resident male's decision as to whether to attack, retreat from or tolerate a simulated intruder, the vocal response of resident males to a simulated intruder differed. During stimulus presentation, males that abandoned or tolerated an intruder produced higher frequency calls than males that attacked even though pre-stimulus call frequencies did not differ between the three groups of males. Also, males that tolerated the opponent did not, on average, change their dominant frequency while males that attacked and those that abandoned lowered their dominant frequency during stimulus presentation. Finally, males that attacked lowered their dominant frequency to a greater extent than did males that abandoned. Because in all groups the magnitude of the frequency change was not related to size, Burmeister *et al.* (2002) concluded that although spectral changes in calls are important in cricket frog contests, these changes are probably not used to assess opponent size. However, as changes in dominant frequency did provide information about which males would attack, tolerate or abandon a call site in response to an intruder, a decrease in call dominant frequency may provide information about intention. Burmeister *et al.* (2002) also suggested that variation in intention may be a reflection of resource value, *V* (section 11.3.2.2.2, Chapter 2).

The possibility that changes in call dominant frequency represent a dishonest signal of size (and hence RHP) and that small males bluff by lowering call frequency has been investigated in male green frogs (*Rana clamitans*: Bee & Perrill 1996, Bee *et al.* 1999, 2000). Males of this species defend territories that contain oviposition sites required by females for reproduction. Body size and the dominant frequency of calls are strongly negatively correlated and larger male green frogs win fights more often than small males (Wells 1978). In a series of playback experiments, Bee and Perrill (1996) demonstrated that male green frogs lower the frequency of their calls during simulated territorial intrusions. In further studies, Bee *et al.* (1999, 2000) showed that males modify their vocal responses in different ways or to different degrees depending on the dominant frequency of the opponent's call. In response to a large-male stimulus, resident males lowered the dominant frequency of their response calls to a greater extent than to small-male and medium-male stimuli. Therefore, information about an opponent's size is transmitted to receivers during aggressive interactions and the possibility exists that males use this information to assess the fighting ability of opponents. However, these studies also showed that small males produced relatively lower-frequency calls in response to a large-male stimulus but large males did not. Therefore, they suggested that the frequency alteration of resident male's calls may represent a dishonest signal of size where small males 'bluff' about their size when confronted with large male opponents.

Although call dominant frequency has been implicated in many studies as a means of assessing male size and hence RHP, few studies have investigated the role of temporal features of calls in information transfer during contests. Playbacks of aggressive calls to calling males have shown that the threshold intensity required to elicit an aggressive response from a focal male is lower than that for advertisement calls and males usually give more aggressive responses to aggressive call playbacks (e.g. Brzoska *et al.* 1982, Robertson 1984, Telford 1985, Lopez *et al.* 1988, Rose & Brenowitz 1991). Frogs may also change the temporal features of their calls in response to playbacks of aggressive calls or during aggressive interactions and this often involves a dramatic increase in call duration (e.g. Wells & Schwartz 1984*b*, Schwartz 1989, Wells 1989). For example, Schwartz (1989) found that male *Pseudacris crucifer* can detect differences in the duration of aggressive calls given by opponents and, in response, increase the duration and rate of their own calls. He suggested that by increasing call duration, males signal a higher level of aggressive motivation and probability of physical attack, or that calls contain information about the physical strength of the male and so facilitate assessment of opponents. Schwartz (1989) reasoned that if the production of longer calls leads to the depletion of glycogen reserves, longer calls may communicate potential endurance in a physical encounter by indicating a male with high levels of glycogen reserves. The suggestion that males may evaluate fighting ability of opponents was further supported by Schwartz (1993) using a computerised system to broadcast aggressive calls in response to spontaneous aggressive calls given by male *D. microcephalus* in a chorus. The computer was programmed to produce calls that were shorter, of equal length or longer than the calls of the test male. When the test male was presented with a call the same length as his own, he responded by increasing his own call duration, but this response was not made if playback calls were longer or shorter than those of the test male. Wells (2007) suggests that this may indicate that males escalate their aggressive responses to elicit information about whether an opponent is capable of producing even longer calls.

Burmeister *et al.* (2002) investigated the function of temporal changes in cricket frog (*A. crepitans*) calls during aggressive interactions. In this species, changes in temporal call characters are believed to represent graded levels of aggression (section 11.3.2.2.2; Wagner 1989*a*, Burmeister *et al.* 1999*a*) and may represent intention to respond to an opponent. In a series of playback experiments, they found that temporal changes in the calls of responding males indicate whether a male will tolerate (ignore) or respond to an opponent (either by attacking or abandoning his calling site). Males that tolerated intruders produced shorter calls with fewer pulse groups and higher pulse rates than males that abandoned or attacked. However, these changes did not predict the type of response a male would make; males that attacked the simulated intruder or abandoned their calling site in response to the intruder changed temporal features of their calls in similar ways. Burmeister *et al.* (2002) also found, however, that subjects were more likely to attack if the stimulus call had more pulses than the subjects' call and more likely to tolerate if the stimulus had more pulse groups than the subject. They therefore suggested that pulse groups and pulse rate influence the decision to tolerate, while the number of pulses influences the decision to attack.

As an alternative to playbacks, Reichert and Gerhardt (2011, 2012) used a series of staged aggressive encounters between calling male *Hyla versicolor* in an attempt to determine if, and how, assessment of RHP occurs during aggressive encounters between males, and to examine the role of calling behaviour on contest outcome. In the first study, Reichert and Gerhardt (2011) investigated three potential components of RHP: mass, length and body condition. They found that when all levels of aggressive escalation were combined (from aggressive calling by one male to physical combat), larger (heavier, longer and better condition) males were significantly more likely to win. However, when the analysis was restricted to contests that involved physical fighting, larger males did not have an advantage over small males. Reichert and Gerhardt (2011) concluded that if RHP is defined as a measure of fighting ability (as is typically the case), body size is a minor component of RHP in *H. versicolor*. However, they did not rule out the possibility that assessment of RHP during aggressive encounters occurs in *H. versicolor* and suggested that other measures, such as energetic state, might be more related to contest outcome. Two further studies appear to support this suggestion. In the first, Reichert and Gerhardt (unpublished) examined the role of specific aggressive call characteristics on the outcome of contests. They analysed the vocalisations of winners and losers in staged aggressive interactions to determine whether characteristics of males' aggressive calls influenced the outcome of aggressive interactions and whether the characteristics of winners' calls could relay information about RHP. They concluded that aggressive calling interactions in *H. versicolor* are based on mutual assessment of aggressive call characteristics related to the energetic cost of calling rather than on body size; RHP is thus likely to be based on an individual's underlying physiological state. In the second study, Reichert and Gerhardt (2012) provide further evidence that this is indeed the case. Previous studies using playbacks and observation in natural choruses (Wells & Taigen 1986, Schwartz *et al.* 2002) revealed a trade-off in temporal parameters of the advertisement call in response to changes in male chorus density. With increasing male density and hence competition, males increase the duration of their calls while decreasing call rate (section 11.3.2.1). In this species, call rate and call duration are positively correlated with energy expenditure (Taigen & Wells 1985), so by increasing duration and decreasing rate, the total amount of calling energy (duty cycle) remained constant. Reichert and Gerhardt (2012) tested whether the trade-off between rate and duration persisted under high levels of acoustic competition where males were calling at very close range during contests. They found that at extreme levels of competition, some males increased both calling rate and duration thereby increasing overall calling effort. If calling performance is linked to some underlying quality of the signaller, this study raises the possibility that this information is relayed to the receiver during aggressive interactions.

11.3.2.2.2 Costs and benefits of aggressive signalling and contests

Choruses represent a classic example of a communication network in which large numbers of both signallers and receivers gather and communicate within close range of one another. The challenge to male frogs communicating in large choruses is to balance the costs and benefits of attracting females, repelling rival males and avoiding predators and/or parasites (Grafe 2005). Aggressive signalling has important costs and benefits. On the one hand, production of both advertisement and aggressive calls enables males to maintain some minimum distance between themselves and rival males (e.g. Whitney & Krebs 1975, Robertson 1984, Telford 1985, Wilczynski & Brenowitz 1988, Brenowitz 1989, Murphy & Floyd 2005) or to maintain a territory, both of which improve a male's chance of attracting a female. The ability of females to recognise and locate potential mates and exercise preferences for particular call properties has been shown to be facilitated by spacing between males (Telford 1985, Dyson & Passmore 1992, Schwartz & Gerhardt 1995). On the other hand, aggression incurs costs including reduced mate attraction during episodes of aggression since females prefer advertisement over aggressive calls (Oldham & Gerhardt 1975, Schwartz 1986, Wells & Bard 1987, Brenowitz & Rose 1999, Marshall *et al.* 2003), increased risk of injury, increased energy expenditure and decreased vigilance against predators.

Given the trade-off between attracting females and repelling rivals, one would expect natural selection to favour the evolution of flexible calling strategies that optimise the ratio of benefits to costs of aggressive calling within a chorus. Male anurans indeed exhibit such flexibility and they do so in several ways. In species where the aggressive call is distinct from the advertisement call, both in structure and in function, the cost of aggressive calling relates principally to female attraction. Because females generally prefer advertisement calls over aggressive calls (e.g. Oldham & Gerhardt 1975, Schwartz 1986, 1987, Wells & Bard 1987, Grafe 1995, Brenowitz & Rose 1999, Marshall *et al.* 2003), any time spent in aggressive interactions reduces a male's chance of attracting a female. Several studies have shown that local caller density affects the threshold at which males respond aggressively to a neighbouring male. As male density increases, the intensity of a neighbour's call required to elicit an aggressive response from a resident male also increases (Rose & Brenowitz 1991, 1997, Marshall *et al.* 2003) and nearest-neighbour distances decrease (Brenowitz & Rose 1994, Gerhardt *et al.* 1989, Dyson & Passmore 1992). The increase in tolerance to neighbours at closer distances in high-density choruses may represent a trade-off between vocalising to attract females and maintaining an optimum call-site area. Persistent aggressive responses at a given fixed threshold or distance regardless of male density would be maladaptive because males would be engaged in an increasing number of aggressive interactions with neighbours at the expense of advertisement calling at higher chorus densities. This would be especially costly if aggression peaked at the time when females were actively searching for mates: evidence from a few species suggests that flexibility in aggressive thresholds reflects the increase in the cost of aggressive calling as the evening progresses and more females are searching for mates. Studies on *Hyperolius marmoratus* (Dyson & Passmore 1992, Grafe 1995). Both studies showed that males establish spatial relationships early in the evening so that early evening choruses are characterised by high levels of aggression (see also Garton & Brandon 1975, Wells 1978, Robertson 1984, Given 1987, Wells & Bard 1987, Reichert 2010). After securing a calling site, males begin producing advertisement calls, resulting in several hours of advertisement calling when aggressive calls are heard only occasionally, despite the fact that more males join the chorus as the nightly chorusing activity proceeds. Dyson and Passmore (1992) found that the number of aggressive interactions reaches a minimum at peak chorusing time when male density is highest. Females arrive at the breeding site later than males, with a peak in mating activity at peak chorusing time (Dyson & Passmore 1992) or at the time when male aggression is lowest (Grafe 1995). These data therefore indicate that, when females are likely to be present, maintaining a large individual distance through the continual production of aggressive calls rather than advertisement calls is too costly to be worthwhile, and this is likely to be mediated by plasticity in the aggressive response threshold.

The flexibility in calling strategies and aggression in response to changes in the social environment described above suggests that a male's assessment of whether to escalate an interaction is not always based on the RHP of rivals, but rather on the perceived value, V, of a calling site or territory.

Evidence of this comes from a study on cricket frogs (Burmeister *et al.* 1999*a*,*b*). In that study, the most important factors affecting the decision to fight, flee or tolerate an intruder were local competition and time within a season. When local caller density was high, residents were more likely to ignore or tolerate intruders and continue to call whereas when local caller densities were low, and nearest-neighbour call amplitude was high, the resident was likely to abandon calling. These results show that the costs and benefits of defending a site are not constant and that decisions about whether to escalate an aggressive encounter depend to some extent on the resident's perceived value of his calling site or calling space. At high local caller densities, calling sites may be limited so the cost of losing a site may be higher than the cost of sharing acoustic space and males tend to tolerate neighbours and continue calling. At lower caller densities, the cost of fighting may be more than the cost of finding a new site from which to call. Burmeister *et al.* (1999*a*,*b*) also speculate that if cricket frog signals provide information about intention and contain little information about RHP (as suggested by previous studies, see above), then local competition and season may constrain individuals from bluffing the intention to attack. Shared knowledge of resource value, *V*, based on local competition and season would allow intentional signalling to be an evolutionarily stable strategy.

Plasticity in aggression in *Pseudacris crucifer* was investigated by Humfeld *et al.* (2009) by determining how differences in the type of signalling interactions experienced by calling males influence how they respond aggressively towards nearby neighbours in a chorus. They tested the hypothesis that in addition to local density, plasticity in aggressive signalling in *P. crucifer* is influenced by the type of signals produced by near neighbours and by the persistence with which neighbours produced those signal types. Playback experiments tested how males respond to a new neighbour that was either initially aggressive (producing both aggressive and advertisement calls) or non-aggressive (producing only advertisement calls). Presentation of either playback signal produced the same response in the test male: an initial shift from advertisement to aggressive calling to a mixture of advertisement and aggressive calling followed by a gradual return to advertisement calling. However, if the playback stimulus then simulated an escalatory shift in aggressiveness of a neighbour from that of a non-aggressive male producing only advertisement calls to an aggressive male producing both aggressive and advertisement calls, the aggressive response of the test male was renewed. Humfeld *et al.* (2009) proposed that the plasticity in aggression exhibited resembles the expectation for short-term forms of habituation and that this serves as a proximate mechanism allowing males to fine-tune their aggressive interactions to local conditions adaptively. Rapid-response decrements would allow males to avoid paying the costs associated with aggressive signalling by accommodating new and persistently calling neighbours as male density increases. However, a rapid recovery of aggression would allow males to maximise the benefits of defending a calling site in response to a new calling neighbour or changes in signalling behaviour of an established neighbour.

The cost, in terms of reduced mating success, of producing aggressive calls may be reduced by a graded signalling system. Males may reduce the costs of aggressive calling by producing calls that vary continuously between attractive and aggressive and escalate agonistic encounters when other males are nearby without rendering their calls completely unattractive to females (Grafe 1995). Wells and Schwartz (1984*b*) proposed that graded signals in *Dendropsophus ebraccatus* function to balance the competing demands of mate attraction and agonistic behaviour, a proposal supported by female preference data. As males approach one another, the introductory notes of the aggressive call become progressively longer and the secondary click notes, which make the call more attractive to females, are dropped (Wells & Schwartz 1984*b*, Wells & Bard 1987, Wells 1989). Males therefore adjust the relative attractiveness of their calls depending on how close their opponent approaches. A graded signalling system has been found in several other species and, in general, as males approach one another, the temporal features of their calls change. These changes include increasing the duration of introductory notes (*Geocrinia victoriana*; Littlejohn & Harrison 1985) and increasing duration by adding pulses (*Hyperolius marmoratus broadleyi*; Grafe 1995) or by adding notes (*Hylarana* (*Rana*) *nicobariensis*; Jehle & Arak 1998).

Wagner (1989*b*) suggested that the communication system of the northern cricket frog (*Acris crepitans*), in which males respond to a decrease in distance between neighbours by gradually increasing the aggressive content of calls, may similarly reflect a trade-off between the advertisement call's function in mate attraction and aggression. He also suggested an alternative but not necessarily exclusive hypothesis that males change the spectral and temporal parameters of their calls in ways that increase their relative attractiveness to females when other males are nearby (Wagner 1991). This hypothesis was tested by Kime *et al.* (2004), who found that females either preferred or showed no preference for temporally escalated advertisement calls associated with aggression over normal advertisement calls. Therefore, there is no evidence that there is a conflict between the different functions of the call (aggression and mate attraction). Similar results have been found in several other frogs (Rand & Ryan 1981, Pallett & Passmore 1988, Howard & Young 1998, Jehle & Arak 1998, Benedix & Narins 1999).

In a recent study on *Dendropsophus* (*Hyla*) *ebraccatus*, Reichert (2011) proposed that males may use aggressive calls to reduce their disadvantage in call timing interactions. In an earlier study, Reichert (2010) observed high levels of aggressive calling in this species and also found that, although males displayed an increase in aggressive thresholds in response to repeated presentation of advertisement calls above initial threshold levels, they lowered their aggressive thresholds in response to the presentation of aggressive calls. In other words, on hearing an aggressive call, males switch to producing aggressive calls of their own at lower threshold intensities than for advertisement calls. Females of this species prefer lagging calls when the advertisement calls of neighbouring males overlap (Wells & Schwartz 1984*a*). Aggressive calls are longer than advertisement calls and this makes it highly likely that aggressive calls will end after advertisement calls when there is overlap, even when given in the leading position. Reichert (2011) thus suggested that males in a leading position switch to aggressive calling to mitigate the disadvantage of leading and found, in support of this hypothesis, that although females prefer advertisement over aggressive calls when calls alternate, this preference is absent when these calls overlap and where the aggressive call begins and ends after the advertisement call. Together these studies suggest a novel explanation for aggressive calling: in addition to functioning in agonistic encounters between males, it serves as a strategy by which leading males reduce their disadvantage in call timing interactions. The production of aggressive calls is therefore initiated as a consequence of timing interactions and not necessarily by the intrusion of rivals into the call site territory of residents.

Plasticity in male aggression and the increased tolerance to neighbours in high-density choruses could also function as a way of conserving energy to sustain long periods of calling. Calling in male frogs is an extremely energetically expensive behaviour and can increase aerobic metabolism to more than 25 times the resting rate (Taigen & Wells 1985, Wells & Taigen 1986, 1989). In addition, energy expenditure is positively related to calling effort (Taigen & Wells 1985, Wells & Taigen 1989, Grafe *et al.* 1992, Grafe 1996*b*, Wells *et al.* 1996). Studies have also shown that both seasonal and nightly calling activity is constrained by glycogen and lipid reserves (Pough *et al.* 1992, Runkle *et al.* 1994, Schwartz *et al.* 1995, Bevier 1997) and in one species at least, energetic limitations play a role in determining a male's chorus tenure (Murphy 1994*a*,*b*). These studies together suggest that if aggressive calls are more energetically expensive to produce than advertisement calls, then by engaging in high levels of aggressive calling males could severely limit their calling activity both within nights and over the breeding season. Both these factors, and in particular chorus tenure, have been shown to affect male mating success in frogs (e.g. Forester & Czarnowsky 1985, Forester *et al.* 1989, Sullivan & Hinshaw 1992, Murphy 1994*a*, Dyson *et al.* 1998, Passmore *et al.* 1992). There has been little work done on estimating the cost of aggressive calls relative to the production of advertisement calls in anurans. However, Marshall *et al.* (2003) proposed that prolonged bouts of energetically expensive aggressive calling could potentially limit the extent to which male spring peepers (*Pseudacris crucifer*) maintain high levels of advertisement calling and time spent calling within nights and also constrain chorus attendance over the season. In playback experiments of advertisement calls at 4 dB above threshold, male calling effort during the first 15 s of playback nearly doubled relative to the pre-stimulus period (392.5 $s\,h^{-1}$ to 783.7 $s\,h^{-1}$) resulting primarily from the production of aggressive calls (0% pre-stimulus to 38.8% post-stimulus). They estimated on the basis of these results that if aggressive and advertisement calls are equivalent in energetic cost, that the metabolic costs imposed by aggressive calling increased from 27.1 $J\,h^{-1}$ to more than 50J/h, which is equivalent to a rate of oxygen consumption of 2.5ml O_2/g/h, close to the maximum aerobic capacities reported in anurans (Wells & Taigen 1984, 1989, Taigen & Wells 1985, Grafe *et al.* 1992, Grafe 1996*b*).

In addition to flexibility in maintaining a calling area, frogs can decrease the costs of aggressive calling by being selective about which individuals they initiate an interaction with. In many territorial vertebrate species, territory holders exhibit a form of social recognition where they display lower levels of aggression to familiar neighbouring territory holders than they do to unfamiliar conspecifics. This phenomenon, known as the 'dear-enemy effect' or 'neighbour recognition' (Chapters 2 and 3), is believed to be an adaptation that allows males to reduce the costs of physical combat that may be incurred during repeated aggressive interactions with individuals that pose little threat to territory ownership. In bullfrogs (*Rana catesbiana*) territorial aggression has been well characterised and consists of advertisement calling, stereotyped movements, the production of a distinct aggressive vocalisation termed 'the encounter call' (Bee *et al.* 2000) and physical fighting. During aggressive encounters, males direct their advertisement calls and encounter calls towards conspecifics that encroach onto their territory. However, studies of bullfrog aggression have shown that territorial males display lower levels of aggression towards vocalisations of familiar neighbouring males broadcast from the direction of a neighbour's territory than they do towards vocalisations of neighbouring males broadcast from novel positions or vocalisations of unfamiliar males (Davis 1987). Similar results were found by Bee and Gerhardt (2001*a,b*) who suggested that reduced aggression between territorial males could be mediated by habituation to an individually distinct property of a communication signal (in this case fundamental frequency) and the location from which the signal originates. Bee and Gerhardt (2001*b*) further showed that males habituate to neighbouring male vocalisations and that this habituation is specific to spectral properties of advertisement calls that are individually distinct. In other words, bullfrog males can learn to discriminate between two signals that differ in fundamental frequency, which is an individually distinctive call parameter, in the absence of any other sensory or spatial cues. Lesbarreres and Lode (2002) found that, in *Rana dalmatina*, playback of unfamiliar calls to resident males resulted in an increase in pulse number and duration of resident's calls. No such changes were seen in response to familiar calls. They concluded that males are able to discriminate familiar from unfamiliar calls and that males' costs of producing energetically expensive calls are reduced by tolerating familiar neighbours.

Variation in circulating hormone levels may interact with energetic state to underlie the expression of aggressive behaviours in anurans. Emerson (2001) proposed the energetics–hormone vocalisation model, which describes the expected links and feedback loops between levels of hormones and energetically costly call characteristics and predicts that increases in nightly calling activity correspond with increased levels of steroid hormones. While this model was developed to explain variation in anuran advertisement calling, it is also likely to apply to variation in the propensity to give aggressive calls, and to variation in aggressive call characteristics themselves. A related model, the challenge hypothesis (Wingfield *et al.* 1990), predicts that increased testosterone levels are related to increased expression of aggressive behaviours. While the latter model was originally proposed for birds, its predictions have been tested in many other taxa and are potentially of importance for amphibian aggressive behaviour. Nonetheless, few studies have examined the role of hormones in aggressive behaviour in amphibians. Indeed, a recent meta-analysis of studies that tested the challenge hypothesis in vertebrates includes only five studies of amphibians (Hirschenhauser & Oliveira 2006).

Although there have been few studies of the effects of hormones on aggressive behaviour in amphibians, the evidence suggests that hormones may be an important factor regulating the expression of aggressive behaviours. In the cricket frog, *Acris crepitans*, males injected with the peptide hormone arginine vasotocin (AVT) and then returned to natural choruses gave calls with characteristics of less-aggressive males (Marler *et al.* 1995). Thus, Marler *et al.* (1995) hypothesised that AVT may reduce aggression in this species. In a follow-up study, Chu *et al.* (1998) examined whether the effects of AVT were dependent upon the social context by injecting males with AVT and then exposing them to a playback simulating an agonistic encounter. While males responded as aggressively to the simulated competitor as control males, their calling was less aggressive than control males in the period immediately following the simulated encounter (Chu *et al.* 1998). In sum, these results suggest the intriguing possibility that AVT may be involved in regulating changes in call characteristics associated with the trade-off between mate attraction and male aggression discussed above.

Hormones may also regulate the outcome and intensity of contests. In the grey treefrog, *Hyla versicolor*, Semsar *et al.* (1998) staged contests between males injected with AVT and males that had not been injected but that were resident on their calling site. Compared to control males, AVT-injected males were more likely to begin calling when placed near the resident, and also were more likely to be victorious in contests with the resident male, and thus take over its calling site (Semsar *et al.* 1998). This result is particularly interesting because resident males normally have a strong advantage in *H. versicolor* (Fellers 1979). That AVT can partially override the effects of ownership suggests that proximate factors can interact with other asymmetries in RHP to determine contest outcome. Tito *et al.* (1999) examined the effects of AVT injection on the characteristics of both advertisement and aggressive calls in *H. versicolor*. While AVT affected advertisement call rate and frequency, there were no effects of AVT injection on the propensity to give aggressive calls or on specific aggressive call characteristics (Tito *et al.* 1999). However, this study examined aggressive calls given by males in natural choruses that were not necessarily engaged in an intense contest, and variation in aggressive calls related to hormone levels may still arise in the latter situation. For now, it can be concluded that variation in AVT levels appears to have an effect on the motivation to call, and this motivation in turn leads to an increased success in contests.

In general, little is known about the role of hormones in amphibian aggression. The few studies that have been performed have largely focused on treefrogs and on the effects of a single hormone (AVT). Other hormones such as testosterone may play a role in aggressive behaviour, and the role of these hormones may differ among amphibian taxa. Future study on the effects of hormone treatments on aggressive behaviour, calling characteristics and contest outcome is thus needed for a wider taxonomic range. At the same time, effects of aggressive behaviour on circulating hormone levels remain entirely unexplored. Such studies are especially important because if contest experience affects hormonal state, then hormonal state is likely to feed back onto subsequent social behaviours including mate attraction and agonistic behaviour. Comparisons between the hormone profiles of individuals that won or lost contests or that engaged in contests of high or low intensity may reveal important differences between these contexts that ultimately affect individuals' reproductive success. These comparisons would also provide tests of the predictions of general models of hormones and aggressive behaviour (Wingfield *et al.* 1990, Emerson 2001).

11.3.2.2.3 Vocal contests in anurans: conclusions

Recordings of advertisement calls in playback experiments have shown that males change not only the timing of their signals in response to calls of other males but also the structure of their own calls. Many of these changes are clearly an attempt to make calls more attractive or evident to females, but whether such changes indicate RHP and whether size and/or energetic considerations are better indicators of RHP is as yet unclear and the answer may differ across species. It does appear that in some species the frequency of a male's advertisement call may indeed indicate RHP in that in several species a calling male's decision about whether to escalate an encounter is influenced by the dominant frequency of a rival's signal. However, the type of information that is transferred once a male frog elects to escalate and respond with an aggressive call, either by lowering the dominant frequency of his own call or by changing temporal parameters, is still unclear. Some studies suggest that information about male RHP is contained in the change in frequency of males' calls once they elect to escalate an encounter while others suggest that temporal changes indicate male motivation or the probability of escalation.

The calls associated with aggression carry two important costs. The first is that females are less attracted to aggressive calls than advertisement calls so by calling aggressively a male reduces his chance of attracting a female. The second is that calls associated with aggression may often be more energetically expensive than advertisement calls and so may reduce male advertisement calling within nights or constrain male calling activity over the season. A positive relationship between chorus tenure and mating success has been found in more than 20 species (Wells 2007). Within nights, mating success is greater for males that call for longer time periods and produce calls that are longer, louder and produced at a faster rate (i.e. are more energetically costly). Thus, energy spent on aggressive calling is expected to reduce the pool of energy available for advertisement calling, which may in turn hamper an individual's ability to attract a mate. As both these costs may have severe implications for male mating success, for aggressive communication to be evolutionarily stable the information content of aggressive calls must be above and beyond that of advertisement calls and it must be sufficiently beneficial to overcome potential costs.

The use of vocalisations to mediate aggressive interactions in frogs and toads has been the subject of extensive research over the past three decades (Wells 2007). However, our current knowledge of contest behaviour in frogs and toads is still relatively poor, particularly with respect to whether aggressive calls are used in contests to assess rivals and how they affect contest outcome. Our knowledge of whether aggressive calls function as signals of RHP is not straightforward and even less is known about whether they function as signals of motivation or aggressive intent. Success during contests is not always determined by size and studies looking at factors such as prior residency, previous experience, motivation and subjective resource value, V, are still needed to understand aggressive behaviour in frogs. In addition, we know little about how asymmetry in size or other correlates of RHP between contestants affects the duration of interactions. Vocal exchanges between males during aggressive encounters often vary in duration and fights can last from a few seconds to 10 min or more (MSR, MLD, pers. obs.). What determines this variation is unknown. With respect to proximate causes of aggression, we still know little about the energetic cost of aggressive signals and of fighting in frogs. We know that hormones modulate aggression in frogs but the knowledge about how aggressive experience may alter levels of hormones is limited, as is the extent to which hormones influence the intensity or duration of interactions. Finally, little is known about how aggressive calls are processed in the anuran auditory system. Until studies demonstrate a clear role for aggressive calls above and beyond that of advertisement calls, then the appropriateness of current theoretical models of aggressive signalling to describe such behaviour in anurans is questionable. To address this issue, more studies that directly test the role of aggressive calls in mediating contest outcome and assessment, as well as tests of the relationship between RHP and aggressive call characteristics, are needed. These studies will not only inform our knowledge of amphibian behaviour, but also can potentially provide a useful system for the testing of theoretical models of aggressive signalling and contest behaviour. Anuran amphibians are tractable model systems for the study of communication and sexual selection (Gerhardt & Huber 2002), thus future research efforts examining aggressive behaviour and aggressive signalling are likely to be fruitful.

The types of studies we suggest above are necessary to provide a broader understanding of contest behaviour in anurans. In addition, however, these studies can inform more general questions on the evolutionary consequences of aggressive behaviour. Comparative studies of aggressive calling across species that vary in life-history traits related to the expected consequences of aggressive behaviour on fitness can provide important tests of game-theoretic models that predict different levels of aggressive signalling depending on the costs and benefits of such signals. For example, contest structure and the role of aggressive signals within contests is likely to be different in species that differ in breeding season length, degree of territoriality, relative energetic costs of advertisement and aggressive calls and the attractiveness of advertisement and aggressive calls to females. These differences in life-history traits can be mapped onto a phylogeny along with aggressive call characteristics such as the propensity to give aggressive calls and the relationship between RHP and call characteristics to test whether life-history traits may be driving differences in the use of aggressive calls between species. In general, knowledge of the links between aggressive behaviour and fitness in anurans are tenuous at best. While this is tacitly assumed in most studies of contest behaviour, studies that demonstrate that variation in aggressive behaviours, including aggressive calling, have demonstrable effects on individual fitness are largely lacking for amphibians. To address this difficult question, the various costs and benefits associated with contest behaviour need to be more thoroughly quantified where previously they have often been assumed. Specifically, do individuals that are successful in aggressive interactions gain fitness benefits that are sufficient to offset the various costs of aggression? To quantify these costs and benefits, additional field studies of the relationship between contest success and mating success, as well as laboratory studies of the energetic costs of calling, the nature of the interactions between hormones, calling and contest intensity, and better estimates of the actual costs of physical combat in species both with and without weapons are much needed.

11.3.3 Non-vocal contests in anurans

Because vocalisations are such a prominent feature of frog behaviour, and because so many frogs mate in darkness, it is widely assumed that other sensory modalities are not involved in their social interactions. It is becoming clear, however, that the sensory world of frogs is much richer than previously supposed. Most notably, visual signals are widespread and diverse among anurans, have evolved independently in several families (Hödl & Amézquita 2001) and are used in a variety of contexts, including courtship and competition. For example, the bright orange vocal sac of male European treefrogs (*Hyla arborea*), a species with nocturnal mating, has been shown to be an important cue in female choice (Gomez *et al.* 2009, Richardson *et al.* 2009). Indeed, for some females in this species, visual cues appear to take preference over acoustic cues in mate choice (Richardson *et al.* 2010).

The evolution of visual signalling in anurans is strongly associated with two ecological factors: a noisy environment due to fast-flowing water and diurnal activity (Hödl & Amézquita 2001). These factors occur together in the Brazilian torrent frog *Hylodes asper*, for which the splash zone near waterfalls and torrents creates both noise and constant humidity, reducing the risk of desiccation by day (Haddad & Giaretta 1999). This species has a complex visual communication system, involving foot-flagging (Figure 11.2A). The Panamanian golden frog (*Atelopus zeteki*) is active by day, brightly coloured and toxic and lives close to fast-flowing streams. It uses semaphore-like movements of the forelimbs in both male–male and male-female encounters (Lindquist & Hetherington 1996, 1998, Figure 11.2B). Forelimb movements by males are performed in response to the calls of other males. The Bornean black-spotted rock frog (*Staurois guttatus*) is diurnal, lives by rocky streams and performs waving movements with its hindlimbs that expose contrastingly coloured feet (Grafe & Wanger 2007). This display is performed in conjunction with calling and other visual displays, including arm-waving, vocal-sac pumping and an open-mouth display, in a multimodal communication system in which calling alerts male and female recipients to visual displays.

While most reports of visual signalling by frogs refer to diurnally active species, there is increasing evidence that it may be common among species that mate at night, especially in the neotropics (Toledo *et al.* 2007). To date, observations of visual signalling are essentially anecdotal and there have been no studies that have investigated how visual signals might resolve contests. A more detailed study has been made of the Amazonian frog *Dendropsophus* (*Hyla*) *parviceps*, a species that has a repertoire of several calls as well as both foot-flagging and arm-waving displays (Amézquita & Hödl 2004). Males were experimentally introduced into the territories of 13 males in their natural habitat. Foot-flagging was performed more often by resident males and at a greater distance from the intruder than arm-waving, suggesting that these two displays represent escalating levels of aggression.

The aposematic strawberry poison frog (*Oophaga* (*Dendrobates*) *pumilio*) varies in the brightness of its red dorsal colouration (Crothers *et al.* 2010). Males are highly territorial and territory owners approach and call more frequently in response to more brightly coloured intruders. More brightly coloured males are themselves more aggressive, approaching intruding frogs more quickly, directing more calls to brighter intruders and lowering the pulse rate of their advertisement calls. Brighter male strawberry frogs are also more attractive to females (Maan & Cummings 2009). In captivity, territorial residents were consistently dominant over intruders, irrespective of their relative body size (Baugh & Forester 1994). A study of the non-toxic dendrobatid frog *Allobates femoralis* sought to determine the relative importance of acoustic and visual cues in contests between males (de Luna *et al.* 2010). Using dummy frogs, it was found that the probability of attack is related to the movement and size of a dummy but not to its colour.

Perhaps the most puzzling form of visual communication in a frog is provided by the diurnally active African frog *Phrynobatrachus krefftii*. When calling to attract females, males inflate a bright yellow subgular vocal sac but, when interacting with other males, they usually (77% of observations) inflate the vocal sac without producing any sound (Hirschmann & Hödl 2006). The fact that vocal sac inflation during calling provides a visual stimulus has been studied in the dart-poison frog *Epipedobates femoralis*: physical attacks by territorial males are elicited only by a combination of acoustic and visual cues, not by either cue on its own (Narins *et al.* 2003).

Even less common than visual signalling among frogs is olfactory communication. Males of the Australian terrestrial toadlet *Pseudophryne bibroni* use a combination of calls and chemical signals to enable females to locate them (Byrne & Keogh 2007). If a

Figure 11.2 (**A**) Foot-flagging in the Brazilian torrent frog (*Hylodes asper*) (Drawing by Tim Halliday from Amézquita & Hödl 2004.) (**B**) Forelimb semaphoring in the Panamanian golden frog (*Atelopus zeteki*). (Drawing by Tim Halliday from Lindquist & Hetherington 1998.)

male detects the odour of another male, he switches from producing advertisement calls to territorial calls. The fact that a very primitive frog, *Leiopelma hamiltoni* from New Zealand, can discriminate between individuals on the basis of odour cues alone may suggest that chemical signalling could occur in other anurans (Waldman & Bishop 2004).

The use of another sensory modality has been reported in the red-eyed treefrog (*Agalychnis callidryas*). In addition to a variety of calls and hind-limb waving, competing males in this species vibrate the foliage on which they are perched with a stereotyped frequency, a behaviour called tremulation (Caldwell *et al.* 2010). This is visually conspicuous but has been shown experimentally to be a vibrational stimulus that provides information about the motivation and size of the performer. Tremulation is the most frequently used display in male–male interactions and ultimately victorious males tremulate more often and for longer than losing males; however, body size does not (directly) predict the outcome of contests in this species.

11.4 Territorial contests in salamanders

Salamanders are small, secretive animals and very little is known about their social behaviour. An exception is the red-backed salamander (*Plethodon cinereus*) which has been studied in both the laboratory and the field by Jaeger and collaborators (e.g. Jaeger 1981, 1984, Jaeger *et al.* 1982, 1986, 1995, 2002). This work has revealed that both sexes of *P. cinereus* are strongly territorial, that they sometimes engage in seriously injurious fights and that these are often averted by visual and olfactory displays. Red-backed salamanders inhabit deciduous forests throughout eastern North America, where they can reach population densities on the forest floor of several individuals per square metre (Burton & Likens 1975). As a result, competition for both food and hiding places becomes intense at those times of year when food is scarce and the weather is dry.

Mild aggression involves a rapid nip with the front of the mouth, which causes no damage to the receiver. Strong aggression consists of a full mouth hold which, because these species possess numerous sharp teeth, causes lacerations of the recipient's skin. Bites are usually aimed at the tail or the snout, where they cause serious damage. Bites on the tail cause tail autotomy, leading to a loss of the victim's fat reserves. Bites on the snout may damage the nasolabial grooves, important chemosensory organs, causing victims to suffer a reduced ability to find mates and food (Jaeger 1981). Such potentially costly fights are typically avoided by means of exchanges of visual and olfactory signals, intruders leaving territories before biting occurs (Jaeger *et al.* 1982).

Visual signalling involves an 'all-trunk raised' (ATR) posture in which the signaller raises its body clear of the ground and looks towards its opponent; conversely, submission is signalled by pressing the body against the ground and facing away (Jaeger *et al.* 1982, Jaeger 1984, Figure 11.3). In laboratory encounters, both resident and intruder spent 50% of their time in the ATR posture but residents showed more biting behaviour. Territories were successfully defended in 74% of trials, lost in 18%, with 8% ending as a draw.

Male and female red-backed salamanders set up territories under cover objects in defence of exclusive feeding areas, prey being limited during dry periods. Territory quality (resource value, *V*) varies as a function of the type of insect food available, termites being a higher quality food than ants, for example. In laboratory experiments, territorial residents fed primarily on termites were more aggressive in staged encounters with intruders than those fed on ants. Moreover, they elicited higher levels of aggression from the intruders (Gabor & Jaeger 1995). The food content of a territory is signalled through faecal pellets, which territorial residents deposit around their territories (Jaeger *et al.* 1986). Intruding males and females sample these by tapping them with their snouts and, sometimes, squashing them vigorously. Faecal pellets thus provide females with a quick way to assess the food quality of a male's territory (Jaeger *et al.* 1995). Faecal pellets also contain pheromones, enabling them to be used to identify territory holders individually (Jaeger *et al.* 1986, Simons *et al.* 1994). In a detailed study of the function of scent marks in red-backed salamanders, Simons *et al.* (1997) found that scent marks served neither to repel nor to intimidate intruders. Rather, they provide information that allows intruders to assess the competitive ability of the resident and thus adjust their aggressive or submissive behaviour.

Escalated contests between red-backed salamanders are potentially very costly and the risk of such contests is reduced among territorial individuals, of both sexes, by 'dear enemy' recognition (Jaeger 1981, Jaeger & Peterson 2002). Odour cues are used to differentiate between familiar and unfamiliar individuals and individuals are less aggressive and more submissive towards familiar territorial neighbours than towards strangers.

There is considerable variation among red-backed salamanders in terms of body size, a factor likely to be important in the context of aggression (RHP). The role of body size has been studied, in nature and in the laboratory, by Mathis (1990*a*, 1991). In the field, if resident animals are experimentally removed from under cover

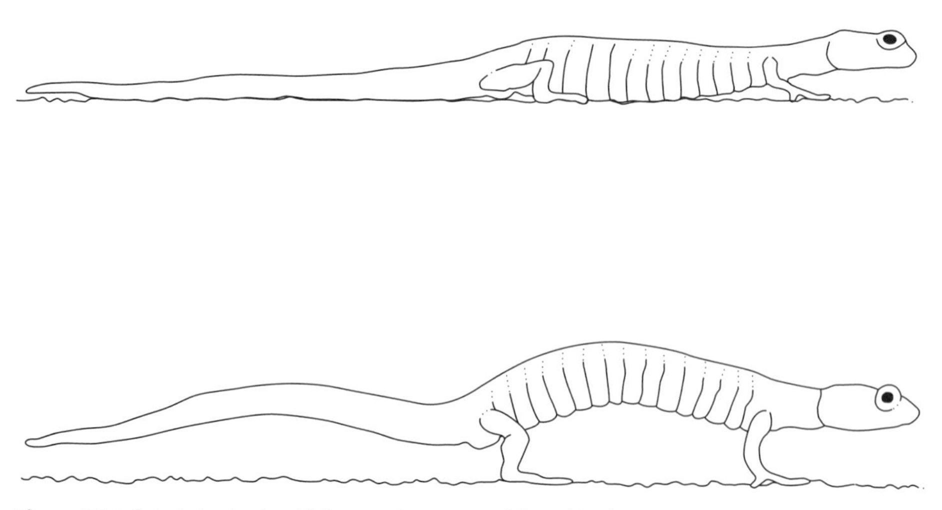

Figure 11.3 Submissive (top) and full aggressive posture of the red-backed salamander (*Plethodon cinereus*). (Drawing by Tim Halliday from Wells 2007.)

objects, such as rocks and logs, they are soon replaced by smaller individuals intruding naturally. There is also a positive correlation between the body size of a resident and the size of its cover object. In choice tests in both laboratory and field, larger cover objects were preferred to smaller ones. The adaptive basis of this preference is that temperatures are lower under larger objects. Assessment of body size by opponents is partly by means of olfactory cues (Mathis 1990*b*). In laboratory experiments, male intruders were significantly more aggressive when exposed to the odour of residents of similar body size and more submissive when exposed to the odours of individuals larger than them. Visual cues in the form of displays are also important, in both aggressive interactions and courtship (Mathis 1991).

Body size assessment during contests has been studied in detail in males of the Ozark zigzag salamander (*Plethodon angusticlavius*): an individual's behaviour is a function of his relative size, his absolute size and his prior experience (Mathis & Britzke 1999). Both larger and smaller males perform more threat displays when paired with an opponent smaller than themselves. Large males perform more threat displays when paired with a same-size opponent than small males. Small males performed fewer threat displays towards a same-sized opponent if they had had no recent experience of agonistic encounters. Further experimental studies of this species show that, for both sexes, prior residence in a territory gives an individual an advantage in contests that is independent of its body size (Mathis *et al.* 2000). A study of the plethodontid salamander *Ensatina eschscholtzii* suggests that territorial residence has a greater impact on the outcome of contests than asymmetries in size between territorial residents and intruders (Wiltenmuth 1996).

As noted above, tail autotomy, which may arise through intraspecific contests or encounters with predators, reduces fitness and is a potential source of asymmetry that could affect the outcome of contests. Wise and Jaeger (1998) found that tail loss had no effect on the behaviour of residents, but that intruders were more aggressive towards tailless residents. Another asymmetry that can affect the outcome of contests in Ozark zigzag salamanders is parasitism: males heavily infected with ectoparasitic mites are less aggressive than those with low infestations (Maksimowich & Mathis 2000).

11.4.1 Sexual interference in urodeles

In all but the most primitive urodeles, fertilisation is internal, allowing females greater control over choosing a mate and controlling the timing and location of egg-laying than is possible in anurans. Male urodeles, however, lack a penis and sperm transfer is achieved indirectly, by means of a spermatophore, a structure secreted by the male (Halliday 1990). This process is open to a particular form of disruption by rival males, called sexual interference, while behaviour by courting males that counters sexual interference is called sexual defence (Arnold 1976). During sexual

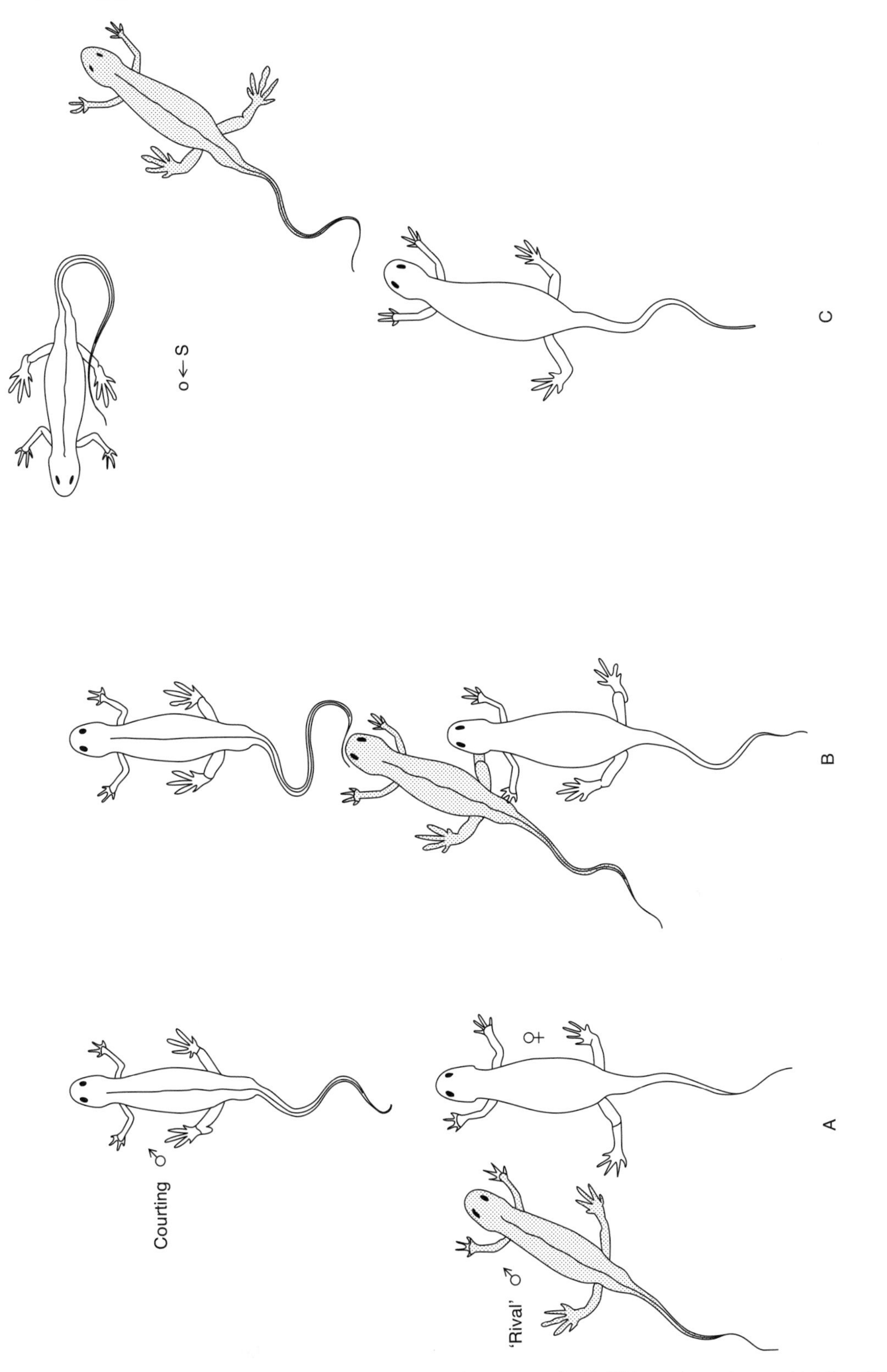

Figure 11.4 Sexual interference in the European smooth newt (*Lissotriton vulgaris*). (**A**) A courting male waves his tail, waiting for the following female to touch it. (**B**) A rival male moves in front of the female and touches the courting male's tail before she can. (**C**) The courting male responds to the tail touch by depositing a spermatophore (S) and the rival male leads the female away, initiating a spermatophore transfer of his own. (Drawing by Tim Halliday.)

interference, a rival male prevents a female picking up the spermatophore of a courting male and attempts to persuade her to pick up his (Figure 11.4). Halliday and Tejedo (1995) reviewed examples of sexual interference and defence in 13 species, belonging to 3 urodele families: its taxonomic distribution suggests that sexual interference is as ancient as sperm transfer by means of a spermatophore. In the European smooth newt (*Lissotriton* (*Triturus*) *vulgaris*), sexual interference is also performed by females, who sometimes steal spermatophores from one another (Waights 1996).

While not contests in a conventional sense, in that protagonists do not directly confront one another, sexual interference is a form of competition in the context of mating, and is an alternative mating strategy (Halliday & Tejedo 1995). It is an opportunistic behaviour pattern and there is no indication that any category of animal, such as larger individuals, is at an advantage.

11.5 Conclusions

This chapter has highlighted a number of general and specific areas where further research is needed. In general, our knowledge of the social behaviour of salamanders and caecilians is woefully deficient and we need to find ways to study these taxa more extensively. Some progress has been made in studying aggressive interactions and courtship behaviour in salamanders and newts, but much of the literature on these topics is based on laboratory studies, raising the possibility that artificial environments produce atypical behaviour. Biologists need to find ways to study urodeles and caecilians in the field. Given the kind of environment in which these organisms live, this is never likely to be an agreeable experience!

In contrast, our understanding of acoustic interactions in frogs has progressed remarkably in recent years and it is particularly striking how what we once regarded as highly stereotyped signals are actually rich in subtle variation that enables frogs to alter their behaviour according to their circumstances. There are, however, many questions about frog calls that remain to be answered. A way forward that is likely to be profitable is to pursue long-term studies of individual frogs; not an easy task but one that becomes more feasible as techniques for marking and tracking individuals improve. Only a few studies have followed the behaviour of individuals from one night to the next within a chorus, and none has followed individuals from one year to the next. Such studies are essential if we are to understand the fitness consequences of the trade-offs that frogs appear to be facing during calling contests.

Acknowledgements

We would like to thank Martin Gammell for his comments on an earlier version of this chapter.

References

Amézquita A & Hödl W (2004) How, when and where to perform visual displays: the case of the Amazonian frog *Hyla parviceps*. *Herpetologica*, 60, 420–429. AmphibiaWeb: http://amphibiaweb.org

Arak A (1983) Vocal interactions, call matching and territoriality in a Sri Lankan treefrog, *Philautus leucorhinus* (Rhacophoridae). *Animal Behaviour*, 31, 292–302.

Arak A (1988) Female mate selection in the natterjack toad: active choice or passive attraction? *Behavioural Ecology and Sociobiology*, 22, 317–327.

Arnold SJ (1976) Sexual behaviour, sexual interference and sexual defense in the salamanders *Ambystoma maculatum*, *Ambystoma tigrinum* and *Plethodon jordani*. *Zeitschrift für Tierpsychologie*, 42, 247–300.

Balinsky BI & Balinsky JB (1954) On the breeding habits of the South African bullfrog, *Pyxicephalus adspersus*. *South African Journal of Science*, 51, 55–58.

Baugh JR & Forester DC (1994) Prior residence effect in the dart-poison frog, *Dendrobates pumilio*. *Behaviour*, 131, 207–224.

Bee MA (2002) Territorial male bullfrogs (*Rana catesbeiana*) do not assess fighting ability based on size related variation in acoustic signals. *Behavioral Ecology*, 13, 109–124.

Bee MA & Bowling AC (2002) Socially mediated pitch alteration by territorial male bullfrogs, *Rana catesbeiana*. *Journal of Herpetology*, 36, 140–143.

Bee MA & Gerhardt HC (2001*a*) Neighbour–stranger discrimination by territorial male bullfrogs (*Rana catesbeiana*) I. Acoustic basis. *Animal Behaviour*, 62, 1129–1140.

Bee MA & Gerhardt HC (2001*b*) Habituation as a mechanism of reduced aggression between territorial male bullfrogs, *Rana catesbiana*. *Journal of Comparative Psychology*, 115, 68–82.

Bee MA & Perrill SA (1996) Responses to conspecific advertisement calls in the green frog (*Rana clamitans*) and their role in male–male competition. *Behaviour*, 133, 283–301.

Bee MA, Perrill SA & Owen PC (1999) Size assessment in simulated territorial encounters between male green frogs (*Rana clamitans*). *Behavioral Ecology and Sociobiology*, 45, 177–184.

Bee MA, Perrill SA & Owen PC (2000) Male green frogs lower the pitch of defensive signals in defence of territories: a possible dishonest signal of size? *Behavioral Ecology*, 11, 169–177.

Benedix JH & Narins PM (1999) Competitive calling behaviour by male treefrogs, *Eleutherodactylus coqui* (Anura: Leptodactylidae). *Copeia*, 1999, 1118–1122.

Bevier CR (1997) Utilization of energy substrates during calling activity in tropical frogs. *Behavioral Ecology and Sociobiology*, 41, 343–352.

Bevier CR, Larson K, Reilly K, *et al.* (2004) Vocal repertoire and calling activity of the mink frog, *Rana septentrionalis*. *Amphibia-Reptilia*, 25, 255–264.

Biegler R (1966) A survey of recent longevity records for reptiles and amphibians in zoos. *International Zoo Yearbook*, 6, 487–493.

Bosch J & Marquez R (2002) Acoustic competition in male midwife toads *Alytes obstetricans* and *Alytes cisternasii*: response to neighbour size and calling rate. Implications for female choice. *Ethology*, 102, 841–855.

Brattstrom BH & Yarnell RM (1968) Aggressive behaviour in two species of leptodactylid frogs. *Herpetologica*, 24, 222–228.

Brenowitz EA (1989) Neighbour call amplitude influences aggressive behaviour and intermale spacing in choruses of the Pacific treefrog (*Hyla regilla*). *Ethology*, 83, 69–70.

Brenowitz EA & Rose GJ (1994) Behavioural plasticity mediates aggression in choruses of the Pacific treefrog. *Animal Behaviour*, 47, 633–641.

Brenowitz EA & Rose GJ (1999) Female choice and plasticity of male calling behaviour in the Pacific treefrog. *Animal Behaviour*, 57, 1337–1342.

Brzoska J, Schneider H & Nevo E (1982) Territorial behavior and vocal response in male *Hyla arborea savignyi* (Amphibia: Anura). *Israel Journal of Zoology*, 31, 27–37.

Burmeister SJ, Konieczka J & Wilczynski W (1999*a*) Temporal call changes and prior experience affect graded signalling in the cricket frog. *Animal Behaviour*, 57, 611–618.

Burmeister SJ, Konieczka J & Wilczynski W (1999*b*) Agonistic encounters in a cricket frog (*Acris crepitans*) chorus: behavioural outcomes vary with local competition and within the breeding season. *Ethology*, 105, 335–347.

Burmeister SJ, Ophir AG, Ryan MJ, *et al.* (2002) Information transfer during cricket frog contests. *Animal Behaviour*, 64, 715–725.

Burton TM & Likens GE (1975) Salamander populations and biomass in the Hubbard Brook Experimental Forest, New Hampshire. *Copeia*, 1975, 541–546.

Bush SJ (1997) Vocal behaviour of males and females in the Majorcan midwife toad. *Journal of Herpetology*, 31, 251–257.

Bush SL & Bell DJ (1997) Courtship and female competition in the Majorcan midwife toad, *Alytes muletensis*. *Ethology*, 103, 292–303.

Byrne PG & Keogh JS (2007) Terrestrial toadlets use chemosignals to recognize conspecifics, locate mates and strategically adjust calling behaviour. *Animal Behaviour*, 74, 1155–1162.

Caldwell MS, Johnston GR, McDaniel JG, *et al.* (2010) Vibrational signalling in the agonistic interactions of red-eyed treefrogs. *Current Biology*, 20, 1012–1017.

Channing A, Preez L du & Passmore N (1994) Status, vocalization and breeding biology of two species of African bullfrogs (Ranidae: *Pyxicephalus*). *Journal of Zoology*, 234, 141–148.

Chu J, Marler CA & Wilczynski W (1998) The effects of arginine vasotocin on the calling behavior of male cricket frogs in changing social contexts. *Hormones and Behavior*, 34, 248–261.

Crothers L, Gering E & Cummings M (2010) Aposematic signal variation predicts male–male interactions in a polymorphic poison frog. *Evolution*, 65, 599–605.

Davies NB & Halliday TR (1978) Deep croaks and fighting assessment in toads, *Bufo bufo*. *Nature*, 274, 683–685.

Davis MS (1987) Acoustically mediated neighbour recognition in the North American bullfrog, *Rana catesbeiana*. *Behavioural Ecology and Sociobiology*, 21, 185–190.

Deullman WE & Trueb L (1986) *Biology of Amphibians*. New York, NY: McGraw-Hill.

Dyson ML & Passmore NI (1988*a*) Two choice phonotaxis in *Hyperolius marmoratus* (Anura: Hyperolidae): the effect of temporal variation in presented stimuli. *Animal Behaviour*, 36, 648–652.

Dyson ML & Passmore NI (1988*b*) The combined effect of intensity and the temporal relationships of stimuli on phonotaxis in the painted reed frog, *Hyperolius marmoratus*. *Animal Behaviour*, 36, 648–652.

Dyson ML & Passmore NI (1992) Inter-male spacing and aggression in African painted reed frogs, *Hyperolius marmoratus*. *Ethology*, 91, 237–247.

Dyson ML, Henzi SP, Halliday TR, *et al.* (1998) Success breeds success in mating male reed frogs (*Hyperolius marmoratus*). *Proceedings of the Royal Society of London B*, 265, 1417–1421.

Emerson SB (1994) Testing pattern predictions of sexual selection: a frog example. *American Naturalist*, 143, 848–869.

Emerson SB (2001) Male advertisement calls: behavioral variation and physiological processes. In: MJ Ryan (ed.) *Anuran Communication*, pp. 36–44. Washington, DC: Smithsonian Institution Press.

Emerson SB & Voris H (1992) Competing explanations for sexual dimorphism in a Bornean voiceless frog. *Functional Ecology*, 6, 654–660.

Enquist M (1985) Communication during aggressive interactions with particular reference to variation in choice of behaviour. *Animal Behaviour*, 33, 1152–1161.

Fabrezi M & Emerson SB (2003) Parallelism and convergence in anuran fangs. *Journal of Zoology*, 260, 41–51.

Fellers GM (1979) Aggression, territoriality and mating behaviour in North American treefrogs. *Animal Behaviour*, 27, 107–119.

Forester DC & Czarnowsky R (1985) Sexual selection in the spring peeper, *Hyla crucifer* (Amphibia, Anura): role of the advertisement call. *Behaviour*, 92, 112–128.

Forester DC, Lykens DV & Harrison WK (1989) The significance of persistent vocalization by the spring peeper, *Pseudacris crucifer* (Anura: Hylidae). *Behaviour*, 108, 197–208.

Gabor CR & Jaeger RG (1995) Resource quality affects the agonistic behavior of territorial salamanders. *Animal Behaviour*, 49, 71–79.

Garton JS & Brandon RA (1975) Reproductive ecology of the green treefrog, *Hyla cineria*, in southern Illinois (Anura: Hylidae). *Herpetologica*, 31, 150–161.

Gerhardt HC & Huber F (2002) *Acoustic Communication in Insects and Frogs: Common Problems and Diverse Solutions*. Chicago, IL: University of Chicago Press.

Gerhardt HC, Daniel RE, Perrill SA, *et al.* (1987) Mating behaviour and male mating success in the green treefrog. *Animal Behaviour*, 35, 1490–1503.

Gerhardt HC, Diekamp B & Ptacek M (1989) Inter-male spacing in choruses of the spring peeper, *Pseudacris (Hyla) crucifer*. *Animal Behaviour*, 38, 1012–1024.

Gerhardt HC, Roberts JD, Bee MA, *et al.* (2001) Call matching in the quacking frog (*Crinia georgiana*). *Behavioral Ecology and Sociobiology*, 48, 243–251.

Given MF (1987) Vocalizations and acoustic interactions of the carpenter frog, *Rana virigatipes*. *Herpetologica*, 43, 467–481.

Given MF (1988) Territoriality and aggressive interactions of male carpenter frogs, *Rana virgatipes*. *Copeia*, 1988, 411–421.

Given MF (1999) Frequency alteration of the advertisement call in the carpenter frog, *Rana virgatipes*. *Herpetologica*, 55, 304–317.

Goin CJ (1949) The peep order of spring peepers, a swamp serenade. *Quarterly Journal of the Florida Academy of Science*, 2, 59–61.

Gomez D, Richardson C, Lengagne T, *et al.* (2009) The role of nocturnal vision in mate choice: females prefer conspicuous males in the European tree frog (*Hyla arborea*). *Proceedings of the Royal Society of London B*, 276, 2351–2358.

Grafe TU (1995) Graded aggressive calls of the African painted reed frog, *Hyperolius marmoratus* (Hyperoliidae). *Ethology*, 101, 67–81.

Grafe TU (1996*a*) The function of call alternation in the African reed frog (*Hyperolius marmoratus*): precise call timing prevents auditory masking. *Behavioral Ecology and Sociobiology*, 38, 149–158.

Grafe TU (1996*b*) Energetics of vocalization in the African reed frog (*Hyperolius marmoratus*). *Comparative Biochemistry and Physiology A*, 114, 235–243.

Grafe TU (1997) The use of metabolic substrates in the gray treefrog, *Hyla versicolor*: implications for calling behaviour. *Copeia*, 1997, 356–362.

Grafe TU (1999) A function of synchronous chorusing and a novel female preference shift in an anuran. *Proceedings of the Royal Society of London B*, 226, 2331–2336.

Grafe TU (2003) Synchronized interdigitated calling in the Kuvangu running frog, *Kassina kuvangensis*. *Animal Behaviour*, 66, 127–136.

Grafe TU (2005) Anuran choruses as communication networks. In: PK McGregor (ed.) *Communication Networks*, pp. 227–299. Cambridge: Cambridge University Press.

Grafe TU & Wanger TC (2007) Multimodal signalling in male and female foot-flagging frogs *Staurois guttatus* (Ranidae): an alerting function of calling. *Ethology*, 113, 772–781.

Grafe TU, Schmuck R & Linsenmair KE (1992) Reproductive energetics of the African reed frogs, *Hyperolius viridiflavus* and *Hyperolius marmoratus*. *Physiological Zoology*, 65, 153–171.

Grobler JH (1972) Observations on the amphibian *Pyxicephalus adspersus* Tschudi in Rhodesia. *Arnoldia (Rhod.)*, 6, 1–4.

Haddad CFB & Giaretta AA (1999) Visual and acoustic communication in the Brazilian torrent frog, *Hylodes asper* (Anura: Leptodactylidae). *Herpetologica*, 55, 324–333.

Halliday TR (1990) The evolution of courtship behaviour in newts and salamanders. *Advances in the Study of Behaviour*, 19, 137–169.

Halliday TR & Adler K (2002) *The New Encyclopedia of Reptiles and Amphibians*. Oxford: Oxford University Press.

Halliday TR & Tejedo M (1995) Intrasexual selection and alternative mating behavior. In: H Heatwole & BK Sullivan (eds.) *Amphibian Biology 2: Social Behaviour*, pp. 419–468. Chipping Norton: Surrey Beatty & Sons.

Halliday TR & Verrell PA (1986) Sexual selection and body size in amphibians. *Herpetological Journal*, 1, 86–92.

Halliday TR & Verrell PA (1988) Body size and age in amphibians and reptiles. *Journal of Herpetology*, 22, 253–265.

Hayes T & Licht P (1992) Gonadal involvement in sexual size dimorphism in the African bullfrog (*Pyxicephalus adspersus*). *Journal of Experimental Zoology*, 264, 130–135.

Hirschenhauser K & Oliveira RF (2006) Social modulation of androgens in male vertebrates: meta-analyses of the challenge hypothesis. *Animal Behaviour*, 71, 265–277.

Hirschmann W & Hödl W (2006) Visual signalling in *Phrynobatrachus krefftii* Boulenger, 1909 (Anura: Ranidae). *Herpetologica*, 62, 18–27.

Höbel G (2000) Reproductive ecology of *Hyla rosenbergi* in Costa Rica. *Herpetologica*, 56, 446–454.

Hödl W & Amézquita A (2001) Visual signalling in anuran amphibians. In: MJ Ryan (ed.) *Anuran Communication*, pp. 121–141. Washington, DC: Smithsonian Institution Press.

Howard RD (1978) The evolution of mating strategies in bullfrogs, *Rana catesbeiana*. *Evolution*, 32, 850–871.

Howard RD & Palmer JD (1995) Female choice in *Bufo americanus*: effects of dominant frequency and call order. *Copeia*, 1995, 212–217.

Howard RD & Young JR (1998) Variation in male vocal traits and female mating preferences in *Bufo americanus*. *Animal Behaviour*, 55, 1165–1179.

Humfeld SC, Marshall VT & Bee MA (2009) Context-dependent plasticity of aggressive signalling in a dynamic social environment. *Animal Behaviour*, 78, 915–924.

Hurd PL (2006) Resource holding potential, subjective resource value, and game theoretical models of aggressiveness signalling. *Journal of Theoretical Biology*, 241, 639–648.

Jaeger RG (1981) Dear enemy recognition and the costs of aggression between salamanders. *American Naturalist*, 117, 962–974.

Jaeger RG (1984) Agonistic behavior of the red-backed salamander. *Copeia*, 1984, 309–314.

Jaeger RG & Peterson MG (2002) Familiarity affects agonistic interactions between female red-backed salamanders. *Copeia*, 2002, 865–869.

Jaeger RG, Kalvarsky D & Shimuzu N (1982) Territorial behaviour of the red-backed salamander: expulsion of intruders. *Animal Behaviour*, 30, 490–496.

Jaeger RG, Goy JM, Tarver M & Márquez CE (1986) Salamander territoriality: pheromonal markers as advertisement by males. *Animal Behaviour*, 34, 860–864.

Jaeger RG, Schwarz J & Wise SE (1995) Territorial male salamanders have foraging tactics attractive to gravid females. *Animal Behaviour*, 49, 633–639.

Jaeger RG, Prosen ED & Adams DC (2002) Character displacement and aggression in two species of terrestrial salamanders. *Copeia*, 2002, 391–401.

Jehle R & Arak A (1998) Graded call variation in the Asian cricket frog *Rana nicrobariensis*. *Bioacoustics*, 9, 35–48.

Jennions MD, Bishop PJ, Backwell PRY, *et al.* (1995) Call rate variability and female choice in the African frog, *Hyperolius marmoratus*. *Behaviour*, 132, 709–720.

Katsikaros K & Shine R (1997) Sexual dimorphism in the tusked frog, *Adelotus brevis* (Anura: Myobatrachidae): the roles of natural and sexual selection. *Biological Journal of the Linnean Society*, 60, 39–51.

Kime NM, Burmeister SS & Ryan MJ (2004) Female preferences for socially variable call characters in the cricket frog, *Acris crepitans*. *Animal Behaviour*, 68, 1391–1399.

Kluge AG (1981) The life history, social organization, and parental behavior of *Hyla rosenbergi* Boulenger, a nest-building gladiator frog. *Miscellaneous Publications of the Museum of Zoology, University of Michigan*, 160, 1–170.

Klump GM & Gerhardt HC (1987) The use of non-arbitrary acoustic criteria in mate choice by female gray treefrogs. *Nature*, 326, 286–288.

Klump GM & Gerhardt HC (1992) Mechanisms and function of call timing in male–male interactions in frogs. In: PK McGregor (ed.) *Playback and Studies of Animal Communication*, pp. 153–174. New York, NY: Plenum.

Lesbarreres D & Lode T (2002) Variations in male calls and response to unfamiliar advertisement calls in a territorial breeding anuran, *Rana dalmatina*: evidence for the 'dear enemy' effect. *Ethology, Ecology and Evolution*, 14, 287–295.

Lindquist ED & Hetherington TE (1996) Field studies on visual and acoustic signalling in the 'earless' Panamanian golden frog, *Atelopus zeteki*. *Journal of Herpetology*, 30, 347–354.

Lindquist ED & Hetherington TE (1998) Semaphoring in an earless frog: the origin of a novel visual signal. *Animal Cognition*, 1, 83–87.

Littlejohn MJ & Harrison PA (1985) The functional significance of the diphasic advertisement call of *Geocrinia victoriana* (Anura: Leptodactylidae). *Behavioural Ecology and Sociobiology*, 16, 363–373.

Lopez PT, Narins PM, Lewis ER, *et al.* (1988) Acoustically induced call modification in the white-lipped frog, *Leptodactylus albilabris*. *Animal Behaviour*, 36, 1295–1308.

Lucas JR, Howard RD & Palmer JG (1996) Callers and satellites: chorus behaviour in anaurans as a stochastic dynamic game. *Animal Behaviour* 51, 501–518.

Luna AG de, Hödl W & Amézquita A (2010) Colour, size and movement as visual subcomponents in multimodal communication by the frog *Allobates femoralis*. *Animal Behaviour*, 79, 739–745.

Maan ME & Cummings ME (2009) Sexual dimorphism and directional sexual selection on aposematic signals in a poison frog. *Proceedings of the National Academy of Sciences USA*, 106, 19072–19077.

Maksimowich DS & Mathis A (2000) Parasitized salamanders are inferior competitors for territories and food resources. *Ethology*, 106, 319–329.

Marler CA, Chu J & Wilczynski W (1995). Arginine vasotocin injection increases probability of calling in cricket frogs, but causes call changes characteristic of less aggressive males. *Hormones and Behavior*, 29, 554–570.

Marshall VT, Humfeld SC & Bee MA (2003) Plasticity of aggressive signalling and its evolution in male spring peepers, *Pseudacris crucifer*. *Animal Behaviour*, 65, 1223–1234.

Martins M, Pombal J & Haddad CFB (1998) Escalated aggressive behaviour and facultative parental care in the nest building gladiator frog, *Hyla faber*. *Amphibia–Reptilia*, 19, 65–73.

Mathis A (1990*a*) Territoriality in a terrestrial salamander: the influence of resource quality and body size. *Behaviour*, 112, 162–175.

Mathis A (1990*b*) Territorial salamanders assess sexual and competitive information using chemical signals. *Animal Behaviour*, 40, 953–962.

Mathis A (1991) Large male advantage for access to females: evidence of male–male competition and female discrimination in a territorial salamander. *Behavioral Ecology and Sociobiology*, 29, 133–138.

Mathis A & Britzke E (1999) The roles of body size and experience in agonistic displays of the Ozark zigzag salamander, *Plethodon angusticlavius*. *Herpetologica*, 55, 344–352.

Mathis A, Jaeger RG, Keen WH, *et al.* (1995) Aggression and territoriality by salamanders and a comparison with the territorial behaviour of frogs. In: H Heatwole & BK Sullivan (eds.) *Amphibian Biology 2: Social Behaviour*, pp. 633–676. Chipping Norton: Surrey Beatty & Sons.

Mathis A, Schmidt DW & Medley KA (2000) The influence of residency status on agonistic of male and female Ozark zigzag salamanders *Plethodon angusticlavius*. *American Midland Naturalist*, 143, 245–249.

Maynard Smith J & Parker GA (1976) The logic of asymmetric contests. *Animal Behaviour*, 24, 159–175.

Maynard Smith J & Price GR (1973) The logic of animal conflict. *Nature*, 246, 15–18.

Murphy CG (1994*a*) Chorus tenure of male barking treefrogs, *Hyla gratiosa*. *Animal Behaviour*, 48, 763–777.

Murphy CG (1994*b*) Determinants of chorus tenure in barking treefrogs (*Hyla gratiosa*). *Behavioral Ecology and Sociobiology*, 34, 285–294.

Murphy CG & Floyd SB (2005) The effect of call amplitude on male spacing in choruses of barking treefrogs, *Hyla gratiosa*. *Animal Behaviour*, 69, 419–426.

Narins PM, Hödl W & Grabul DS (2003) Bimodal signal requisite for agonistic behavior in a dart-poison frog, *Epipedobates femoralis*. *Proceedings of the National Academy of Sciences USA*, 100, 577–580.

Navas CA, Gomes FR & Carvalho JE (2008) Thermal relationships and exercise physiology in anuran amphibians: integration and evolutionary implications. *Comparative Biochemistry and Physiology A*, 151, 344–362.

Oldham RS & Gerhardt HC (1975) Behavioural isolating mechanisms of the treefrogs, *Hyla cineria* and *H. gratiosa*. *Copeia*, 1975, 223–231.

Pallett JR & Passmore NI (1988) The significance of multi-note advertisement calls in the reed frog, *Hyperoilus tuberilinguis*. *Bioacoustics*, 1, 13–23.

Passmore NI & Carruthers VC (1979) *South African Frogs*. Johannesburg: Witwatersrand University Press.

Passmore NI, Bishop PJ & Caithness N (1992) Calling behaviour influences mating success in ale painted reed frogs, *Hyperolius marmoratus*. *Ethology*, 92, 227–241.

Peterson CL, Wilkinson RF, Topping MS & Metter DE (1983) Age and growth of the Ozark hellbender (*Cryptobranchus alleganiensius bishopi*). *Copeia*, 1983, 225–231.

Pough FH, Magnusson WE, Ryan MJ, *et al.* (1992) Behavioral energetics. In: ME Feder & WW Burggren (eds.) *Environmental Physiology of the Amphibians*, pp. 395–436. Chicago, IL: University of Chicago Press.

Prestwich KN (1994) The energetics of acoustic signalling in anurans and insects. *American Zoologist*, 34, 625–643.

Ramer JD, Jenssen TA & Hurst CJ (1983) Size-related variation in the advertisement call of *Rana clamitans* (Anura: Ranidae), and its effect on conspecific males. *Copeia*, 1983, 141–155.

Rand AS & Ryan MJ (1981) The adaptive significance of a complex repertoire in a neotropical frog (*Physalaemus pustulosus*). *Zeitschrift für Tierpsychologie*, 57, 721–736.

Reichert MS (2010) Aggressive thresholds in *Dendropsophus ebraccatus*: habituation and sensitization to different call types. *Behavioral Ecology and Sociobiology*, 64, 529–539.

Reichert MS (2011) Aggressive calls improve leading callers' attractiveness in the treefrog *Dendropsophus ebraccatus*. *Behavioral Ecology*, 22, 951–959.

Reichert MS & Gerhardt HC (2011) The role of body size on the outcome, escalation and duration of contests in the grey treefrog, *Hyla versicolor*. *Animal Behaviour*, 82, 1357–1366.

Reichert MS & Gerhardt HC (2012) Trade-offs and upper limits to signal performance during close-range vocal competition in gray treefrogs, *Hyla versicolor*. *American Naturalist*, 180, 425–437.

Richardson C, Popovici A, Bellbert F, *et al.* (2009) Conspicuous colouration of the vocal sac of a nocturnal chorusing treefrog: carotenoid-based? *Amphibia-Reptilia*, 30, 576–580.

Richardson C, Gomez D, Burleux R, *et al.* (2010) Hearing is not necessarily believing in nocturnal anurans. *Biology Letters*, 6, 633–635.

Robertson JGM (1984) Acoustic spacing by breeding males of *Uperoleia rugosa* (Anura: Leptodactylidae). *Zeitschrift für Tierpsychologie*, 64, 283–293.

Robertson JGM (1986) Territoriality, fighting and assessment of fighting ability in the Australian frog *Uperoleia rugosa*. *Animal Behaviour*, 34, 763–772.

Rose GJ & Brenowitz EA (1991) Aggressive thresholds of male Pacific treefrogs for advertisement calls vary with amplitude of neighbour's calls. *Ethology*, 89, 244–252.

Rose GJ & Brenowitz EA (1997) Plasticity of aggressive thresholds in *Hyla regilla*: discrete accommodation to encounter calls. *Animal Behaviour*, 53, 353–361.

Runkle LS, Wells KD, Robb CC, *et al.* (1994) Individual, nightly, and seasonal variation in calling behaviour in the gray treefrog, *Hyla versicolor*: implications for energy expenditure. *Behavioral Ecology and Sociobiology*, 5, 318–325.

Savage JM (2002) *The Amphibians and Reptiles of Costa Rica*. Chicago, IL: University of Chicago Press.

Schwartz JJ (1986) Male calling behaviour and female choice in a neotropical treefrog, *Hyla microcephala*. *Ethology*, 73, 116–127.

Schwartz JJ (1987) The importance of spectral and temporal properties in species and call recognition in a neotropical treefrog with a complex vocal repertoire. *Animal Behaviour*, 35, 340–347.

Schwartz JJ (1989) Graded aggressive calls of the spring peeper, *Psuedacris crucifer*. *Herpetologica*, 45, 172–181.

Schwartz JJ (1993) Male calling behaviour, female discrimination and acoustic interference in the neotropical treefrog, *Hyla microcephala* under realistic acoustic conditions. *Behavioural Ecology and Sociobiology*, 32, 401–414.

Schwartz JJ & Gerhardt HC (1995) Directionality of the auditory system and call pattern recognition during acoustic interference in the gray tree frog, *Hyla versicolor*. *Auditory Neuroscience*, 1, 195–206.

Schwartz JJ & Wells KD (1984) The vocal behaviour of the neotropical treefrog *Hyla phlebodes*. *Herpetologica*, 40, 452–463.

Schwartz JJ & Wells KD (1985) Intraspecific and interspecific vocal behaviour of the neotropical treefrog *Hyla microcephala*. *Copeia*, 1985, 27–38.

Schwartz JJ, Ressel SJ & Bevier CR (1995) Carbohydrates and calling: depletion of muscle glycogen and the chorusing dynamics of the neotropical treefrog *Hyla microcephala*. *Behavioral Ecology and Sociobiology*, 37, 125–135.

Schwartz JJ, Buchannan B & Gerhardt HC (2002) Acoustic interactions among gray treefrogs, *Hyla versiclor*, in a chorus setting. *Behavioral Ecology and Sociobiology*, 53, 9–19.

Semsar K, Klomberg KF & Marler C (1998) Arginine vasotocin increases calling-site acquisition by nonresident male grey treefrogs. *Animal Behaviour*, 56, 983–987.

Shine R (1979) Sexual selection and sexual dimorphism in the Amphibia. *Copeia*, 1979, 297–306.

Simons RR, Felgenhauer BE & Jaeger RG (1994) Salamander scent marks: site of production and their role in territorial defence. *Animal Behaviour*, 48, 97–103.

Simons RR, Jaeger RG & Felgenhauer BE (1997) Competitor assessment and area defense by territorial salamanders. *Copeia*, 1997, 70–76.

Smith MJ & Roberts JD (2003) Call structure may affect male mating success in the quacking frog *Crinia georgiana* (Anura: myobatrachidae). *Behavioral Ecology and Sociobiology*, 53, 221–226.

Stewart MM & Bishop PJ (1994) Effects of increased sound level on advertisement calls on calling male frogs, *Eleutherodactylus coqui*. *Journal of Herpetology* 28, 46–53.

Sullivan BK (1982) Sexual selection in woodhouse's toad (*Bufo woodhousei*): I chorus organisation. *Animal Behaviour*, 30, 680–686.

Sullivan BK (1992) Sexual selection and calling behavior in the American toad (*Bufo americanus*). *Copeia*, 1992, 1–7.

Sullivan BK & Hinshaw SH (1992) Female choice and selection on male calling behaviour in the gray treefrog *Hyla versicolor*. *Animal Behaviour*, 44, 733–744.

Taigen TL & Wells KD (1985) Energetics of vocalization by an anuran amphibian (*Hyla versicolor*). *Journal of Comparative Physiology B*, 155, 163–170.

Telford SR (1985) Mechanisms and evolution of intermale spacing in the painted reed frog (*Hyperolius marmoratus*). *Animal Behaviour*, 33, 1353–1361.

Tito MB, Hoover MB, Mingo AM, *et al.* (1999). Vasotocin maintains multiple call types in the gray treefrog, *Hyla versicolor*. *Hormones and Behavior*, 36 166–175.

Toledo LF, Araújo OGS, Guimarães LD, *et al.* (2007) Visual and acoustic signalling in three species of Brazilian nocturnal tree frogs (Anura, Hylidae). *Phyllomedusa*, 6, 61–68.

Tsuji H (2004) Reproductive ecology and mating success of male *Limnonectes kuhlii*, a fanged frog from Taiwan. *Herpetologica*, 60, 155–167.

Tsuji H & Matsui M (2002) Male-male combat and head morphology in a fanged frog (*Rana kuhlii*) from Taiwan. *Journal of Herpetology*, 36, 520–526.

Tuttle MD & Ryan MJ (1982) The role of synchronized calling, ambient light, and ambient noise in anti-bat predator behaviour of a treefrog. *Behavioural Ecology and Sociobiology*, 11, 125–131.

Verrell PA & Brown LE (1993). Competition among females for mates in a species with male parental care, the midwife toad *Alytes obstetricans*. *Ethology* 93, 247–257.

Wagner WE Jr (1989*a*) Graded aggressive calls in Blanchard's cricket frog: vocal responses to opponent proximity and size. *Animal Behaviour*, 38, 1025–1038.

Wagner WE Jr (1989*b*) Fighting, assessment and frequency alternation in Blanchard's cricket frog. *Behavioural Ecology and Sociobiology*, 25, 429–436.

Wagner WE Jr (1991) Social selection on male calling behaviour in Blanchard's cricket frog. PhD thesis, University of Texas.

Wagner WE Jr (1992) Deceptive or honest signalling of fighting ability? A test of alternate hypotheses for the function of changes in call dominant frequency by male cricket frogs. *Animal Behaviour*, 44, 449–462.

Waights V (1996) Female sexual interference in the smooth newt, *Triturus vulgaris vulgaris*. *Ethology*, 102, 736–747.

Waldman B & Bishop PJ (2004) Chemical communication in an archaic anuran amphibian. *Behavioral Ecology*, 15, 88–93.

Wells KD (1977) Territoriality and male mating success in green frog (*Rana clamitans*). *Ecology*, 58, 750–762.

Wells KD (1978) Territoriality in the green frog (*Rana clamitans*): vocalizations and aggressive behaviour. *Animal Behaviour*, 26, 1051–1063.

Wells KD (1988) The effects of social interactions on anuran vocal behaviour. In: B Fritszch, MJ Ryan, W Wilczynski, TE Hetherington & W Walkowiak (eds.) *The Evolution of the Amphibian Auditory System*, pp. 433–454. New York, NY: John Wiley & Sons.

Wells KD (1989) Vocal communication in a neotropical frog, *Hyla ebraccata*: responses of males to graded aggressive calls. *Copeia*, 1989, 461–466.

Wells KD (2007) *The Ecology and Behavior of Amphibians*. Chicago, IL: University of Chicago Press.

Wells KD & Bard KM (1987) Vocal communication in a neotropical treefrog, *Hyla ebraccata*: responses of females to advertisement and aggressive calls. *Behaviour*, 101, 200–210.

Wells KD & Greer BJ (1981) Vocal responses to conspecific calls in a neotropical hylid frog, *Hyla ebraccata*. *Copeia*, 1981, 615–624.

Wells KD & Schwartz JJ (1984*a*) Vocal communication in a neotropical treefrog, *Hyla ebraccata*: advertisement calls. *Animal Behaviour*, 32, 405–420.

Wells KD & Schwartz JJ (1984*b*) Vocal communication in a neotropical treefrog *Hyla ebraccata*: aggressive calls. *Behaviour*, 91, 128–145.

Wells KD & Schwartz JJ (2007) The behavioural ecology of anuran communication. In: PM Narins, AS Feng, RR Ray & AN Popper (eds.) *Hearing and Sound Communication in Amphibians*, pp. 44–86. New York, NY: Springer.

Wells KD & Taigen TL (1984) Reproductive behavior and aerobic capacities of male American toads (*Bufo americanus*): is behavior constrained by physiology? *Herpetologica*, 40, 292–298.

Wells KD & Taigen TL (1986) The effect of social interactions on calling energetics in the gray treefrog (*Hyla versicolor*) *Behavioural Ecology and Sociobiology*, 19, 9–18.

Wells KD & Taigen TL (1989) Calling energetics of a neotropical treefrog, *Hyla Microcephala*. *Behavioural Ecology and Sociobiology*, 25, 13–22.

Wells KD, Taigen TL & O'Brien JA (1996) The effect of temperature on calling energetics of the spring peeper (*Pseudacris crucifer*). *Amphibia-Reptilia*, 17, 149–158.

Weygoldt P (1981) Beobachtungen zur Fortpflanzungsbiologie von *Phyllodytes luteolus* (Wied 1824) im Terrarium. *Salamandra*, 15, 171–181.

Whitney CL & Krebs JR (1975) Mate selection in Pacific treefrogs. *Nature*, 255, 325–326.

Wilczynski W & Brenowitz EA (1988) Acoustic cues mediate inter-male spacing in a neotropical frog. *Animal Behaviour*, 36, 1054–1063.

Wiltenmuth EB (1996) Agonistic and sensory behaviour of the salamander *Ensatina eschsholtzii* during asymmetrical contests. *Animal Behaviour*, 52, 841–850.

Wingfield JC, Hegner RE, Dufty AM, *et al.* (1990) The challenge hypothesis – theoretical implications for patterns of testosterone secretion, mating systems, and breeding strategies. *American Naturalist*, 136, 829–846.

Wise SE & Jaeger RG (1998) The influence of tail autotomy on agonistic behaviour in a territorial salamander. *Animal Behaviour*, 55, 1707–1716.

Chapter 12

Lizards and other reptiles as model systems for the study of contest behaviour

Troy A. Baird

12.1 Summary

Reptiles, especially sexually selected lizards, have proven to be good model systems for studying the evolution of contest competition. Aggression, especially by males, plays an important role in structuring reptilian social systems often characterised by territorial defence and both fixed and plastic alternative male tactics. Reptilian aggressive behaviour ranges from overt physical attacks that are likely to require significant energy expenditure and risk of injury to less costly signalling using visually conspicuous stereotypical motor patterns, striking colouration and chemical cues. Morphological traits that promote success in aggressive contests involve development of exaggerated and specialised physical armaments, colour conspicuousness, as well as whole animal performance traits and large overall size. Aggressive behaviour patterns may also be influenced by body temperature, prior social experience and other social variables that affect the context of social interactions. Lizards especially are important models for tests of both the proposed influence of hormones on aggression, as well as the possible effects of aggression on hormone levels. Lastly, lizards have also been used to test the extent to which game-theoretic models can be used to explain the evolutionary maintenance of alternative colour/behaviour morphs and the outcome of dyadic aggressive contests.

12.2 Introduction

Extant vertebrates commonly known as 'reptiles' actually include three lineages that are related only distantly (Zug *et al.* 2001, Vitt & Caldwell 2009). Although the fossil record suggests fascinating hypotheses pertaining to intraspecific aggression in some extinct reptiles (e.g. Hieronymus *et al.* 2009, Peterson *et al.* 2009), for obvious reasons, research on contest behaviour has focused on extant forms. For this chapter, 'reptiles' thus refers to the extant members of three clades: Crocodilians (alligators and crocodiles), Chelonians (turtles) and, especially, the Lepidosauria (tuataras, lizards and snakes).

Reptiles were once thought to display only rudimentary social behaviour relative to endothermic vertebrates. It has, however, become clear during the past 30 years that the diversity, complexity and subtlety of social behaviour in reptiles rivals that of any other vertebrate taxon. My overall objective is to review selectively studies of the social behaviour in extant reptiles that pertain to the broad theoretical framework of this book, i.e. factors that influence the nature and outcome of agonistic contests. A more specific focus of the book is to consider animal aggression in the context of game-theoretic models of extended contests (Parker 1974, Chapter 2). Tests designed to evaluate the predictions of these models in reptiles are too few to organise this chapter around these alone, but the literature pertaining to reptilian aggressive contests is large and diverse. Therefore, the chapter is organised to review empirical studies of the highly varied aspects of reptilian biology that are related to the resolution of competitive contests, a portion of which use game-theoretical models. The reptiles in which contest behaviour has been examined most thoroughly are those that are under strong sexual selection, and lizards are especially important study models (Fox *et al.* 2003). Therefore, my primary focus is on studies that examine contest competition among males, often lizards. However, I also review studies of female aggression, where data are available.

Animal Contests, ed. I.C.W. Hardy and M. Briffa. Published by Cambridge University Press.

Because the pertinent literature is very large, my review is admittedly selective. I begin by characterising the diverse nature of intraspecific aggressive behaviour patterns that have been described in extant reptilian clades. Individuals that enter into aggressive contests are hypothesised to incur many potential costs such that aggression is only predicted to evolve when individuals accrue compensatory benefits. Therefore, I next review studies that have used reptilian models empirically to test the extent to which aggression is costly, and studies that examine the diverse benefits acquired through aggression. Similar to other taxa, reptilian aggression is most usually linked to sexual selection and its powerful influence on social and mating systems. I thus follow with a review of the types of social systems where aggression is involved in structuring relationships among individuals. I then review the large literature that tests the role of highly variable phenotypic traits hypothesised to influence resource holding potential (RHP) through aggression, particularly among lizards, the best studied group. Lizards are again the leading taxon for experimental tests of the hypothesised interaction between hormones and aggression, as well as hypotheses predicting that the costly nature of aggression has resulted in context-/condition-dependent mechanisms, perhaps influenced by experience, that adaptively influence the occurrence and intensity of aggressive contests. Therefore, I follow with a review of pertinent studies in both of these areas. Next, I focus on the relatively few studies that employ game theoretic models to understand the evolution of lizard social systems characterised by alternative behavioural and morphological morphs, and those in which extended contests between lizard combatants have been used to test specific predictions of game-theoretic models. I conclude by discussing some directions for future studies to prompt continued research in areas where numerous questions remain.

12.3 Reptilian aggressive behaviour patterns

Reptiles show rich variety in the behaviour patterns employed during aggressive interactions because both the habitats occupied by different species and the sensory modalities used in communication with conspecifics vary substantially. That reptilian aggression has been most thoroughly studied in lizards (Sauria), especially diurnally active, visually oriented species

Figure 12.1 Territorial male collared lizard beginning a full show display characterised by lateral compression and elevation of the torso and dewlap extension (photograph by Teresa D. Baird). For colour version, see colour plate section.

(Iguania), is certainly related to their use of relatively open habitats that are conducive to observation and their reliance on visually conspicuous social displays. Visually conspicuous lizard displays are thought to promote assessment of opponents during contests (Ord *et al.* 2001). Lizard aggressive display patterns typically involve a variety of stereotypical postures/movements (Figure 12.1) including lateral compression of the torso (Baird *et al.* 2007), extension of the dewlap (Echelle *et al.* 1971), bobbing the head vertically (Decourcy & Jenssen 1994), 'pushups' achieved by dorso-ventral flexion of the legs (Martins 1993), thrashing of the tail (Peters & Evans 2003), circumduction of the front limbs (Carpenter & Ferguson 1977), and holding the mouth in an open gape (Gier 2003, Lappin *et al.* 2006). In many species, these movements/postures advertise brightly coloured areas of integument (Martins 1993, Whiting *et al.* 2003; section 12.6.2). More intense aggressive encounters involve chases of opponents which, if neither lizard retreats, may result in direct physical contact including ramming, wrestling and biting (Baird *et al.* 2003, Beck 2005, Lappin & Husak 2005).

Among well-studied saurians, contests between male Gila monsters, beaded lizards (Helodermatidae) and monitor lizards (Varanidae), appear to involve the greatest degree of physical contact that ultimately determines social dominance (Carpenter & Ferguson 1977, Beck 2005). Contests between helodermatid males, for example, involve a variety of highly ritualised but physically intense postures and holds, the ultimate objective of which is to press the opponent to the ground and/or to force it onto its back (Beck 2005; Figure 12.2). Bouts of strenuous combat usually last for 4–15 min, but successive bouts between the same two opponents may extend over 16 h (Beck 2005). Aggressive contests involving prolonged overt physical attacks (wrestling, biting) also occur between male marine iguanas (*Amblyrhynchus cristatus*) (Parteke *et al.* 2002; Figure 12.3a) and some species of agamidae (Figure 12.3b). For example, two male eastern water dragons (*Physignathus leseueri*) that were already engaged in combat when discovered fought continuously for an additional 2 h before one male won and the other retreated (Baird *et al.* 2012). Although work on most lizards has focused on visually conspicuous displays and physically overt aggressive behaviour patterns, chemical signals also appear to play an important role in social relationships, especially in active-foraging scleroglossan lizards (Cooper 2004, López & Martín 2004, section 12.6.3), and acoustic signals may play a display role in some Gekkonids (Marcellini 1977).

Similar to many lizards, territorial male tuataras (*Spenodon punctatus*) use visual displays involving inflating the lungs, elevating the dorsal crest, darkening the integument, and shaking the head laterally. When encounters escalate, male tuataras snap their mouths shut while croaking intermittently, chase, bite, and use their tails to whip opponents (Gillingham *et al.* 1995).

Documented instances of aggression in snakes consist of combat between two males, usually during the breeding season over access to females (Schuett 1996). Male–male combat has been described in Viperids, Colubrids, Boiids and Elapids (Gillingham 1987). The occurrence of male combat appears positively correlated with both male-biased sexual size dimorphism (Shine 1978) and subjugation of prey by either constriction or envenomation (Schuett *et al.* 2001). Although experimental studies blocking the vomeronasal organ indicate that chemoreception plays an important role in contests between males (Glaes 1982) and in some snakes males have more deeply forked tongues than females (perhaps to detect social chemical cues: Smith *et al.* 2008), tactile communication via physical contact is paramount in snake combat (C.C. Carpenter 1977). Social dominance is achieved by maintaining a superior position of the head over that of the opponent (topping) while orienting horizontally (Elapidae, Colubridae), or raising vertically off of the ground while intertwined (Viperidae) (C.C. Carpenter 1977, Gillingham 1987). Once one male establishes dominance, subordinate individuals abruptly retreat and do not initiate further interactions (Gillingham 1987, Schuett 1996).

In turtles, both the frequency of aggression and the types of aggressive behaviour appear to vary greatly depending upon habitat. Based on the limited number of species for which data were available, Berry and Shine (1980) concluded that the frequency of aggression, particularly among males, is positively correlated with the degree of terrestriality, perhaps because visual and olfactory cues transmit better in terrestrial than aquatic habitats. Both types of cues play an important role in interactions between males of terrestrial species. Following the exchange of visual cues communicated by head bobbing, and of chemical cues detected by sniffing, encounters between males may escalate to ramming, biting and overturning opponents (Ruby & Nublick 1994), which is facilitated by dimorphic gular projections of the plastron in males of some species (Johnson *et al.* 2009). Although less frequent, aggression by both sexes occurs in semi-aquatic species when they emerge to bask, and involves open-mouthed gesturing, biting and displacement of opponents (Bury & Wolfheim 1973, Lovich 1988). If interactions result in turtles entering the water, aggression involves following, chasing, biting the limbs and shell margins, and rapidly vibrating the foreclaws against the head of their opponent (titillation) (Rundquist 1985, Babb 2004).

Among crocodilians, aggressive behaviour patterns have been described most thoroughly in the American alligator (*Alligator mississippiensis*), and to a lesser extent in two species of crocodiles (*Crocodylus*) (Garrick & Lang 1977). Social displays in American alligators consist of loud vocalisations (bellowing) and headslaps, which are complex (up to eight component parts) assertion displays that vary greatly in intensity (Vliet 1989). Bellowing is exhibited by both sexes, primarily during the breeding season, but is most intense

multivariate analyses of several morphological and behavioural traits. This design generates larger data sets, but results are also potentially influenced by subjects gaining social experience from participation in several contests (section 12.7). A second common approach is to evaluate correlations between/among one or more phenotypic characters and estimates of male reproductive success, which is presumably linked with the success of males in contests with same-sex rivals. Male reproductive success, however, may be influenced by female preferences for one or more traits of males, and those traits may or may not be the same attributes that promote success in intrasexual contests (Baird *et al.* 1997, López *et al.* 2002).

12.6.1 Morphometric traits

Size-based attributes are often difficult to signal deceptively (Whiting *et al.* 2003, but see Jenssen *et al.* 2005). It is thus not surprising that large size, usually measured as length, is among the phenotypic attributes that are most frequently postulated to correlate with social dominance in reptiles. There is widespread evidence that intrasexual competition favours large male size, especially in lizards, which has resulted in size being used as a proxy for fighting ability, even though this assumption may not always be valid (Stuart-Fox 2006). Sexual size dimorphism has evolved in some lizards by mechanisms not related to intrasexual selection (e.g. Anderson & Vitt 1990, Watkins 1996, van Wyk & Mouton 1998, Kratochvíl & Frynta 2002). Although reptilian sexual dimorphism is usually male-biased (Shine 1978, Berry & Shine 1980) there are interesting exceptions where females are the larger sex (Shine *et al.* 1998, Zamudio 1998). Nonetheless, the frequency that contests are initiated and/or the frequency of winning contests are positively correlated with relative size of contestants in turtles (Lovich 1988, Lindeman 1999), snakes (Schuett 1997) and, especially, in several phylogenetically diverse lizards (e.g. Agamidae: Radder *et al.* 2001; Crotaphytidae: Baird *et al.* 1997; Chameleonidae: Karsten *et al.* 2009; Iguanidae: Tokarz 1985, Stamps & Krishnan 1994*b*, Jenssen *et al.* 2005; Lacertidae: López & Martín 2001, Aragón *et al.* 2004; Phrynosomatidae: Hews 1990; Scincidae: Cooper & Vitt 1987; Teidae: Anderson & Vitt 1990, Censky 1995). Indeed, among lizards there appears to be only one documented example of males that were smaller in overall size attaining social dominance in dyadic contests: OB-male tree lizards were socially dominant in contests with O-males. However, OB-males were also stockier for their length than were the more slender O-males (Hews *et al.* 1997, Moore *et al.* 1998).

Although aggression in lizards is certainly more frequent in males, it is also well documented among females of some species (Manzur & Fuentes 1979, Carothers 1981, Mahrt 1998, Woodley & Moore 1999), and large size may promote success in female contests. For example, in one population of collared lizards where elevated rock perches (that this species uses to scan for prey) were limiting, larger (also older) females maintained core areas surrounding these perches from which they repelled non-territorial females that were smaller and younger (Baird & Sloan 2003).

Success in aggressive contests may also be promoted by disproportionate enlargement of particular body dimensions/structures. Because aggressive lizards may grasp their opponents in their jaws (Anderson & Vitt 1990), enlarged head dimensions have been hypothesised to function as armaments that promote dominant social status (Carothers 1984). Exploitation of different prey by the sexes may also explain differences in head dimensions (Anderson & Vitt 1990). However, in several lizards where males had larger heads, the size of food items taken by the sexes was not limited by head dimensions (Vitt & Cooper 1985, Perry 1996, Herrel *et al.* 1999) and several lizard studies have presented indirect evidence that supports the sexual selection hypothesis. For example, Carothers (1984) examined male aggression and head size dimorphism among 11 species of herbivorous lizards in which head dimensions are not expected to be influenced by diet. Three species having little or no dimorphism in head dimensions showed only low levels of male aggression whereas the other eight species displayed both dimorphic head dimensions and high levels of male aggression (Carothers 1984). In other lizards with male-biased head size dimorphism, males were more likely to enter into agonistic contests (Perry 1996, Baird *et al.* 2003), showed a higher incidence of scarring (presumably as a consequence of intrasexual contests: Cooper & Vitt 1987, Mouton *et al.* 1999, Lappin & Husak 2005) and social status was positively related to both head dimensions and body size (Gier 2003). In dyadic contest trials using males with different head sizes, males with larger head dimensions usually won (Vitt & Cooper 1985, Molina-Borja *et al.* 1998, Hews 1990, López *et al.* 2002, Gvoždík & van Damme 2003, Karsten *et al.* 2009), supporting the hypothesis that enlarged heads are at least correlated with fighting ability and may promote it directly. Moreover, in the lacertid, *Gallotia galloti*, rates of aggression increased when contestant males were more similar in both total size and head size, and the probability of winning was related to fighting ability (Molina-Borja *et al.* 1998). Fights between *Podarcis hispanica* males were also more intense when opponents were similar in size (López & Martín 2001).

Displays in some lizards involve stereotypical movements of the tail (Ord & Evans 2003), whereas the tail is used directly during fights in other species (Beck 2005). Hence, lizard tail length is another body dimension that has been hypothesised to function either as a social signal or during fighting. Lizard tail length can be particularly labile owing to loss of portions to predators. Studies on at least two species suggest that loss of a portion of the tail reduces social status: in *Uta stansburiana*, removal of tails from dominant juveniles decreased social status in both sexes, whereas experimental restoration of the tail restored social status in females but not in males (Fox *et al.* 1990). These results suggest that tail length is a status signal in juvenile females of this species (Fox *et al.* 1990). Similarly, in *Lacerta monticola*, males lost social status following experimental tail reduction (Martín & Salvador 1993) and tail length may be important during the combat postures of some Helodermatids (Beck 2005). In contrast, there was no influence of tail loss on social status or aggressiveness in *Anolis sagrei* (Kaiser & Mushinsky 1994), suggesting that the potential social role of tails may be highly variable, perhaps as a function of the types of displays that are used by different species during social interactions.

In Chameleonid lizards a variety of other male morphological characters show an interesting mosaic of positive and negative correlations with success in intrasexual contests. In Cape dwarf chameleon males, both the height of head casques and the size of lateral colour patches were positively correlated with fighting ability (Stuart-Fox & Johnstone 2005). In contrast, in one chameleonid from Madagascar, male fighting ability was negatively correlated with the length of rostral appendages, and there was strong intrasexual selection for fewer dorsal cones in a second species, the latter of which may correlate with age and experience (Karsten *et al.* 2009), factors that influence the intensity of intrasexual aggression in other lizards (section 12.7).

12.6.2 Colouration

Among reptiles, a role of colour variation in social interactions has been considered primarily in lizards and turtles. The best-known examples of intraspecific variation in turtle colouration involve the degree of melanism in freshwater pond species. In temperate populations, male red-eared pond sliders (*Trachemys scripta*) deposit melanin in both the carapace and the integument as they age, rendering older males darker than both females and younger males (Lovich *et al.* 1990, Babb 2004). Based on anecdotal observations, numerous authors have suggested that melanism in male pond sliders promotes social dominance over non-melanistic males (Lovich *et al.* 1990, Tucker *et al.* 1995). Empirical evidence to address this hypothesis is inconclusive. Individual space use is an important variable in social systems structured by asymmetries in social status. Most studies of spatial and social behaviour in relation to melanism in freshwater turtles are plagued by small sample sizes and other methodological problems that render the results equivocal. In the most thorough study to date, melanistic and non-melanistic males did not differ in either home-range area or inter- and intrasexual home range overlap (Babb 2004). Moreover, in controlled experiments pitting size-matched melanistic and non-melanistic males against one another, males of neither group were aggressive. Instead, males frequently approached and followed one another non-aggressively, non-melanistic males more frequently so than melanistic males (Babb 2004). These results do not support the hypothesis that melanism in freshwater turtles functions as a sexually selected social signal in agonistic contests.

There are many examples of marked between- and within-sex colour variation in lizards (Cooper & Greenberg 1992), including species displaying both fixed (Moore *et al.* 1998, Calsbeek & Sinervo 2002*a*, Zamudio & Sinervo 2003) and plastic (Zucker 1989, Baird *et al.* 2003, Aragón *et al.* 2004, Baird 2008) alternative male tactics. Experimental evidence supports multiple signalling functions of colouration in lizards including sex recognition (Cooper & Burns 1987, Cooper & Vitt 1988, Cooper & Greenberg 1992) and female choice of male mates (Baird *et al.* 1997, Kwiatkowski & Sullivan 2002). Conspicuous colouration has also been implicated as a signal of RHP in several male lizards. For example, in laboratory studies of collared lizards from three

Oklahoma populations, more brightly coloured males from one population had higher agonistic scores and prevailed in staged contests over less brightly coloured, size-matched opponents, whereas colouration did not influence contest outcome in males from the other two populations (Baird *et al.* 1997). Colouration in male tree lizards is highly variable within and among populations (Figure 12.5), and this variation may be complexly related to intrasexual male interactions (Hews *et al.* 1997). In a well-studied Arizona population, characterised by two fixed male morphs that differed in throat coloration, OB-throated males were more aggressive and dominant over males having solid orange (O) throat colouration (Moore *et al.* 1998), and among OB-males the size of the blue spot was positively correlated with the probability of winning contests (Thompson & Moore 1991). Throat colouration was the best predictor of social dominance among males from one New Mexico tree lizard population in which colouration was polymorphic, whereas social experience (section 12.7) was a better predictor among males from a nearby monomorphic population (G.C. Carpenter 1995). In addition to throat colouration, especially in dense populations, male tree lizards appear to signal dominance by darkening their dorsal surface (Zucker 1989, 1994).

Perhaps the best evidence supporting the hypothesis that colouration plays a role in social status derives from studies that have manipulated colouration patches (hypothesised badges of social status) and then examined the social consequences for individuals. For example, Olsson (1994) experimentally manipulated the green lateral patches that male sand lizards (*Lacerta agilis*) develop during the reproductive season. Males with more colour-saturated patches were more likely to initiate and win contests with consexuals. When the size of manipulated badges was not equal, contests were settled faster than when male contestants had equally sized badges, suggesting that green lateral colouration functions to advertise RHP (Olsson 1994).

Similarly, in male flat lizards (*Platysaurus broadleyi*), which display remarkably conspicuous and complex colouration, both badge size and reflectance in ultraviolet wavelengths appear to influence fighting ability (Whiting *et al.* 2003, 2006). Males of this species develop orange–yellow abdominal patches that are bordered anteriorly by black on the chest. There are also stripes running along both sides of the lateral torso that match the colour of the abdominal patch. These badges are displayed when males elevate the side of the abdomen that faces approaching rivals (ventral displays). Using males matched for snout-to-vent length (SVL) and head dimensions, Whiting *et al.* (2003) painted males to manipulate badge size. Success in contests was higher in males having enlarged badges than in those with their lateral stripes removed, but the latter group won more contests than males in which badges were masked. These results indicate that abdominal patches and lateral stripes function as honest signals of RHP (Whiting *et al.* 2003). Male *P. broadleyi* with enlarged badges did not initiate more agonistic contests, whereas enhancement of badges increased the initiation of contests in male sand lizards (Olsson 1994). If males are aware of the size of their colour badge, then theory predicts that those having enhanced badges should behave congruently by initiating more aggression (Whiting *et al.* 2003). The observation that the behaviour of badge-enhanced males was consistent with this hypothesis in sand lizards, but not in flat lizards, may be related to the respective locations (lateral versus ventral) of the badges in these two species (Whiting *et al.* 2003). In addition to the signalling function of visible spectrum colouration of flat lizard abdominal patches and lateral stripes, the total ultraviolet reflectance of throat colouration in this species was also positively related with success in aggressive contests (Whiting *et al.* 2006). Moreover, ultraviolet throat colouration was related to social tactics, with territorial males having higher UV reflectance from the throat patch than conspecific floater males (Whiting *et al.* 2006).

12.6.3 Chemical signals

Many reptiles have sensory systems that are well adapted to detect chemicals in the environment, use chemical cues to detect prey, and/or have glands that secrete odiferous substances (Zug *et al.* 2001). Nonetheless, the potential role of chemical signals in aggressive interactions has been investigated only in a few lizard species. Studies focusing primarily on active-foraging lizards that locate their prey chemically have provided much support for the hypothesis that chemical signals may influence a variety of social interactions including distinguishing same from opposite-sex conspecifics, female preferences for male mates, recognition of familiar versus non-familiar same-sex rivals, recognition of individuals, and the mediation of aggressive interactions between males (e.g. Cooper 2004, López & Martín 2004, Martín *et al.* 2007, Carazo *et al.* 2008). Chemical cues that allow discrimination of the sexes and/or individual same-sex rivals may allow for adaptive variation in behavioural responses to conspecifics that reduce the overall costs of aggression (section 12.7).

Most of the experimental studies on chemical signalling in lizards have focused on the Lacertidae and the Scincidae and the cues often involve compounds that are present in secretions emitted by femoral pores/glands, faecal pellets, or both (Cooper 2004, López & Martín 2004). The sexes are dichromatic in the model species used for these studies, hence both colouration and chemical signals are possible and both appear to play a role in sex recognition. For example, in Iberian rock lizards, males attacked females that were painted to mimic conspecific males but courted males that were painted to resemble females, indicating that colouration plays an important role in sexual identification. However, male attacks on females painted to mimic males ceased after females were tongue-flicked by these males, indicating that chemical cues override visual cues for sex recognition (López & Martín 2004). Similarly, irrespective of colouration, male *Podarcis hispanica* reacted more aggressively to same-sex intruders treated with male odours than to males treated using scent from conspecific females, whereas males reacted more aggressively towards females treated with male odours than to those treated with female scent (López & Martín 2004). Chemical cues also allowed adult male *Psammodromus algirus* to distinguish small males, even though their colouration mimicked that of females (López *et al.* 2003). Together these results indicate a role of chemical signals in sex recognition and aggression even in dichromatic species. Where both types of cues are available, colouration may function at longer distances, whereas chemical cues may be more important over short distances, or perhaps when low light levels prevent effective transmission of colour-based signals. In species where colouration does not differ between the sexes, chemical cues may play an even stronger role in sex recognition and aggression (López & Martín 2004).

The ability to use chemical cues to differentiate individual competitors may be highly adaptive because it potentially allows behavioural modifications, such as avoidance of encounters with superior competitors, or dear-enemy recognition (section 12.7), both of which may reduce the costs of aggression. In support of this hypothesis, López *et al.* (1998) used laboratory trials to show that male Iberian rock lizards discriminate between their own faecal pellets and those of competitor males, and avoid areas marked with the pellets of competitors. Field observations indicate that faecal pellets are placed in prominent, elevated positions close to refuges, suggesting that they may function as 'composite signals' that convey both visual information because they are easy to locate from a distance, and chemical compounds that function as signals at a close range (López *et al.* 1998, Duvall *et al.* 1987, Alberts 1992). Male Iberian rock lizards also discriminate chemically between familiar competitors (that have overlapping home ranges) and unfamiliar males (living at least 50 m away). Similarly, male *P. hispanica* varied the intensity of aggressive responses to intrusions based upon chemical cues (López & Martín 2004). In staged encounters between male *L. monticola*, large males were socially dominant over smaller males (Martín *et al.* 2007). When body size was controlled statistically, however, males that were socially dominant were also found to produce more hexadecanol and octadecanol in their femoral pore secretions than subordinate males. Experiments during which tongue-flicks were recorded as the response variable revealed that males may reduce aggression toward individuals that produce secretions rich in hexadecanol, suggesting that the concentration of chemicals in femoral pore secretions function as a signal that advertises RHP (Martín *et al.* 2007).

Laboratory trials with *P. hispanica* also provide strong evidence that males can use scent markings on the substrate to recognise individual rivals (Carazo *et al.* 2008). In this study, resident males were first allowed to habituate to scent marks deposited by two different rival donor males (size-matched with residents); one scent was placed at the centre of the resident's territory and the other at the periphery. The location of scent produced by the two donors was then reversed, following which contests were staged between the resident and each of the donors separately. Results showed that resident males discriminated between individual rivals based on scent alone, and during staged contests, residents were more aggressive to rivals that had marked the core of their territories than to those marking the periphery (Carazo *et al.* 2008). These data appear to be the strongest evidence to date that lizards distinguish individual rivals rather than merely differentiating between classes of rivals (section 12.7).

12.6.4 Whole-organism performance traits

One very fast-growing area of research on factors that may influence the outcome of animal competitive contests, including those of reptiles, is focused on whole animal performance traits. These are often defined as the ability to conduct ecologically relevant tasks, and studies have generally focused on the maximal performance of physically challenging aggressive or locomotor behaviour patterns (Irschick *et al.* 2008, Hare & Miller 2009). Although reptiles have provided important study models of performance traits, most studies thus far examine the influence of performance on survival and evasion of predators (Irschick *et al.* 2008). Far fewer studies have determined the extent to which performance traits are linked with success in contests or hypothesised correlates of male competitive ability such as reproductive success. One paradigm is to correlate traits such as bite force or sprint speed that are postulated to promote success in fights (Vanhooydonck *et al.* 2005, Anderson *et al.* 2008, Le Galliard & Ferrière 2008) with measures of reproductive success. A potential weakness of these studies is that they assume that male reproductive success is a consequence of winning contests with same-sex rivals even though there is evidence of female choice in some lizards (Baird *et al.* 1997, Censky 1997), and/or that male traits promoting fighting ability may not be the same traits that females prefer (López & Martín 2004).

In collared lizards, the number of hatchlings sired by territorial males was positively correlated with both bite force and sprint speed corrected for head dimensions and total body size, but neither was correlated with plasma levels of testosterone (Lappin & Husak 2005, Husak *et al.* 2006*a*). Bite force was also positively correlated with the size of UV-reflective patches of epithelium that are exposed at the corners of the mouth when collared lizards gape, suggesting that these patches function to advertise body size-independent bite force (Lappin *et al.* 2006). Because large males in this species access females though territory defence (Baird *et al.* 2003, Baird & Curtis 2010), both bite force and sprint speed may promote male reproductive success because they enhance the ability of territory owners to repel intruders (Lappin & Husak 2005; Husak *et al.* 2008).

Evidence that individual performance traits influence success in aggressive contests has also been revealed by studies that make direct measurements of one or more performance traits that are then associated with success in contests staged experimentally. Bite force was the most important predictor of male contest outcome in both *Gallotia galloti* (Huyghe *et al.* 2005, 2010) and venerable collared lizards (Husak *et al.* 2006*b*). Similarly, in contests between *Anolis carolinensis* males that were classified as 'heavyweights', those having stronger bite force won more frequently (Lailvaux *et al.* 2004). Development of the muscles that regulate lizard bite force may be modulated hormonally. Huyghe *et al.* (2010) showed that basal testosterone level in male *G. galloti* was correlated with bite force, and implantation of testosterone increased the size of jaw closing muscles. Surprisingly, bite force did not increase in hormone-implanted males, perhaps because they were housed alone during these experiments which prevented males from using their increased muscle mass (Huyghe *et al.* 2010).

Locomotor performance has also been correlated with winning staged contests between size-matched opponents. 'Lightweight' male green anoles (*Anolis carolinensis*) having greater jumping velocity and acceleration won more contests (Lailvaux *et al.* 2004). In *Sceloporus occidentalis* males, winners sprinted faster than losers but stamina did not differ (Garland *et al.* 1990). Both sprint speed and endurance of winners exceeded that of losers in male tree lizards (Robson & Miles 2000). Studies by Perry *et al.* (2004) on male *Anolis cristellus* also indicate that enhanced locomotor performance increases success in competitive contests. Frequency of assertion displays was first measured in the field, followed by laboratory measurements of body size and locomotor ability as well as staged contests between size-matched opponents. Winners of staged contests had higher frequencies of assertion displays in the field, higher endurance in laboratory performance trials and significantly deeper heads. These data suggest that locomotor performance may influence social status directly, or that both of these traits may be linked to an underlying physiological mechanism (perhaps hormone levels) that influences several aspects of male phenotype (Perry *et al.* 2004).

12.6.5 Preferred body temperature

Because reptiles are ectothermic and temperature regulation is primarily behavioural, maintenance of preferred body temperature (PBT) may have an important influence on a variety of other factors that influence fitness (Angilletta *et al.* 2002, Martin & Huey 2008), perhaps including aggression and the outcome of contests. The most prevalent general effect of temperature on aggressive behaviour appears to be a transition from fleeing simulated predator threat stimuli to becoming aggressive as temperature decreases (Crowley & Pietruska 1983, Hertz *et al.* 1982, Mautz *et al.* 1992). There are a few studies that suggest a relationship between aggression with conspecifics and maintenance of PBT. In two species of *Crocodylus*, aggressive exclusion from basking sites may prevent some individuals from maintaining body temperature in the preferred range (Seebacher & Grigg 1997, Grigg *et al.* 1998). In two lizard species, males warmed their bodies faster than females (Ibargüengoytía 2005, Vidal *et al.* 2008) and cooling rates were slower in males of a third species (Fraser 1985). Reduced time for warming and slower cooling rates may promote the ability of males to defend territories without compromising their maintenance of PBT (Vidal *et al.* 2008). The most thorough study of aggression and maintenance of PBT appears to be that by Stapley (2006) on mountain log skinks (*Pseudemoia entrecasteauxii*). Males of this species develop either orange or white venters, and orange males are more aggressive and socially dominant over white males (Stapley & Keogh 2004, Stapley 2006). Levels of both aggression and courtship were positively correlated with PBT, suggesting that thermoregulatory ability may influence both intra- and intersexual behavioural parameters that are potentially related to fitness (Stapley 2006).

How the sexes expose themselves to thermal challenges as a consequence of emergence from hibernacula may also be related to levels of intrasexual competition. For example, territorial collared lizard males emerge earlier in the spring than non-territorial females (Baird *et al.* 2003). Earlier emergence probably exposes males to cooler temperatures which may influence vulnerability to predators (Braun *et al.* 2010), but may also allow reacquisition of territories that have promoted successful reproduction during previous seasons (Schwartz *et al.* 2007) without high levels of aggression (Baird *et al.* 2001).

12.7 Influence of social context and experience on aggressive interactions in lizards

Because aggressive interactions impose one or more costs (section 12.4), it is frequently hypothesised that selection has acted on the manner in which individuals enter into aggression to reduce the cumulative costs of these activities. Many studies on lizards support this general hypothesis by showing variability in the aggressive responses of territory defenders depending upon individual recognition of rivals and the location at which rivals are encountered. The most widely recognised formulation of this concept is the dear-enemy hypothesis, first proposed for birds (Krebs 1982, Ydenberg *et al.* 1988) but now tested in several lizards. This hypothesis proposes that once individuals establish territories within a neighbourhood, the cumulative costs of defence will be reduced if established neighbours that encounter one another frequently do not continue to engage in intense aggression along their shared borders. Hence, established territorial neighbours are expected to respond less aggressively to one another than they do to unfamiliar intruders. Consistent with this hypothesis, field experiments testing the responses of resident males to tethered intruders showed more intense responses to unfamiliar strangers than to neighbours in species that utilise both visual (Fox & Baird 1992, Whiting 1999, Van Dyk & Evans 2007) and chemical cues to distinguish classes of rivals (Font & Desfilis 2002, Cooper 2004, López & Martín 2004). In one species, resident males used chemical cues to distinguish individual rivals (Carazo *et al.* 2008, section 12.6.3).

Because recognition of familiar versus unfamiliar competitors may be influenced by their relative positions within territorial neighbourhoods, responses may vary depending upon where competitors intrude. To test this hypothesis, Husak and Fox (2003) used tethering experiments in the field to examine whether responses by collared lizard male residents were influenced by both individual recognition and the location of the intrusion. Results revealed that unfamiliar intruders elicited more intense responses than did familiar neighbours, but familiar neighbours introduced on an inappropriate border elicited more intense aggression than did the same intruder on the shared appropriate border. Similarly, male *Anolis sagrei* responded differently to competitors intruding on the ground than those intruding from elevated perches which, in this species, are likely to be valuable resources (Calsbeek & Marnocha 2005). Results of these studies provide strong support for the influence of location as an important component of the dear-enemy hypothesis in lizards.

Several studies suggest that a variety of other social context variables may influence lizard aggressive behaviour. In sagebrush lizards (*Sceloporus graciosus*) the type of display given depends upon the nature of social interactions: displays used during direct challenges are highly stereotypical and reveal both gular and abdominal colour patches, whereas those given in the absence of direct challenges are much more variable (Martins 1993). In trials with *Anolis sagrei*, the type of aggressive displays by residents during establishment of territories was influenced both by distance from competitors and by residency time on the territory (McMann 2000). The amount of activity by intruders was an important predictor of variability in aggressive responses among residents in *Lacerta vivipara* (Cote *et al.* 2008). Because females are a valuable resource for males, their presence may also influence the behaviour of males and the outcome of aggressive contests. In tree lizards, males used darkened dorsal colouration to signal territorial social status only when females were present (Zucker 1994). Male leopard geckos (*Eublepharis macularius*) that had been previously housed with females showed more territorial and marking behaviour than did naïve males, even following their castration (Sakata *et al.* 2002). Similarly, in laboratory trials with *Podarcis hispanica*, males increased fighting effort when novel females were present, with the increase being more marked for males that did not have a female in their prior cage compared to those that had already established a relationship with a female (López & Martín 2002).

Prior experience with rivals may also influence the outcome of aggressive contests in lizards. For example, in one population of tree lizards that was monomorphic for colouration, social experience was the best predictor of contest outcome (G.C. Carpenter 1995), whereas prior experience plus body mass was a powerful predictor of outcome in a colour polymorphic population (Zucker & Murray 1996). In green anoles (*Anolis carolinensis*), male competitors showed a transient increase in aggressiveness in response to repetitive video stimuli from competitors but they remained aggressive when these video competitors were novel (Yang *et al.* 2001). These results suggest that both habituation to known competitors and stimulation by novel competitors influence aggressive responses, results which are entirely consistent with predictions of the dear enemy hypothesis.

Social experience may also allow a reduction in aggressive costs because prior occupancy of the same territories provides advanced knowledge of the locations and phenotypic qualities of competitors as well as specific attributes of the local terrain and other resources. This hypothesis was tested in collared lizard males by comparing the aggressive responses of adult males defending a territory for the first time (inexperienced) with those having at least one previous season of territory occupancy (experienced). Experienced males maintained larger territories and interacted with more females. Moreover, when challenged by tethered, size- and experienced-matched intruders, experienced males responded less intensely than did inexperienced males. These results support the hypothesis that experience in social interactions on the same territory allows males to increase the net benefit of territorial defence by adjusting their behaviour to reduce costs (Schwartz *et al.* 2007).

12.8 Interaction between hormones and aggression in lizards

Lizards are important empirical study systems in the investigation of the widely hypothesised association between steroid hormones and aggression. The proposed influence of steroid hormones on animal phenotypes is usually broken down into organisational versus activational effects (Adkins-Regan 2005). Organisational effects primarily involve the non-reversible actions of hormones on the development of phenotypes prior to sexual maturity, such as morphological differentiation of either primary or secondary characters of the sexes. In contrast, hormones may also exert activational effects which generally occur later in life and stimulate phenotypic changes, often in behaviour.

One major focus of endocrine–behavioural research in lizards is the role of steroid hormones in the organisation or activation of aggressive tactics. To provide a general framework for predicting patterns of hormone levels in vertebrates that display fixed versus plastic ART, Moore (1991) extended the organisational–activational model to formulate the relative plasticity hypothesis (RPH). For species having fixed ART, if alternative male types result from organisational effects that occur during early development, then once they are mature males displaying alternative behavioural tactics should not differ hormonally. In contrast, in species having plastic ART the transition between morphs is predicted to be activated by hormonal secretion. Therefore, mature males displaying different tactics should have different levels of circulating steroids (Moore 1991). Results of many, but not all empirical studies on lizards support the predictions of the RPH. In tree lizards, levels of testosterone were similar in males displaying different tactics, which is consistent with the predictions of the RPH because in this population, the alternative male tactics are fixed (Moore 1991). Higher circulating levels of testosterone in territorial marine iguanas compared with those of satellite and sneaker males are also consistent with RPH predictions because the three male tactics are plastic in this species (Wikelski *et al.* 2005). In contrast, even though male collared lizards show behavioural tactics that are clearly plastic, there were no differences in the steroid profiles of territorial and non-territorial males (Baird & Hews 2007). Moreover, in *Uta stansburiana*, elevated levels of testosterone in highly territorial orange-throated males relative to both yellow- and blue-throated males (Sinervo *et al.* 2000) that display different tactics (Sinervo & Lively 1996, Zamudio & Sinervo 2003) is inconsistent with predictions of the RPH, assuming that male morphs are indeed fixed (Sinervo & Lively 1996, Zamudio & Sinervo 2003). However, Sinervo *et al.* (2000) reported that some yellow-throated males transformed to the blue-throated morphology late in the season, and that these transforming males had elevated testosterone relative to their levels earlier in the season. These results suggest that tactics utilised by male *Uta* may not always be fixed, and that androgens could play a role in activating changes in behavioural tactics.

Experiments involving supplementation of circulating androgens clearly show the activational effects of hormones on aggression in several lizard species. Increasing androgens increased both general activity (DeNardo & Sinervo 1994, Lovern *et al.* 2001, Olsson *et al.* 2000) and aggression (Cooper *et al.* 1987, Salvador *et al.* 1997, Rhen & Crews 2000, Kabelik *et al.* 2008). In large, 'heavyweight' male green anoles, circulating testosterone level was positively correlated with both dewlap size and bite force (Husak *et al.* 2007), the latter promoting success in contests in this species (Lailvaux *et al.* 2004). Activation of aggression by steroids is not limited to lizard males: experimental increase of at least oestradiol, and perhaps testosterone, increased aggression in territorial female mountain spiny lizards (Woodley & Moore 1999). Perhaps the most dramatic demonstration of the activation of lizard aggression is the result of experiments on Galapagos marine iguanas (Wikelski *et al.* 2005): in this species (that displays three plastic male mating tactics, section 12.5) increasing or chemically blocking circulating testosterone causes males to adopt alternative aggressive tactics. Although these studies provide strong support for steroids activating aggression, the effect is not common to all lizards, as androgen levels are not positively correlated with aggression in several species (e.g. Moore 1991, Whiting *et al.* 2006, Baird & Hews 2007), suggesting that other factors may also interact with circulating steroid hormone levels to influence aggressive behaviour.

Despite an early bias toward examining the influence of hormones on behaviour, causation in the opposite direction may also be important: social interactions (including aggression) may influence hormone production, which may then influence subsequent behavioural interactions (Adkins-Regan 2005). The leading theoretical framework for predicting the influence of behavioural interactions on endocrine responses is the challenge hypothesis, originally proposed for endotherms (Wingfield *et al.* 1990). The general formulation of this hypothesis makes opposing predictions about the influence of aggression on hormones depending upon the social system and the presence or absence of male parental care. In species where males are monogamous and participate in the care of offspring, testosterone levels are expected to be elevated early in the reproductive season (when territories are being contested and settled) but then decline as males become involved in incubating eggs, and protecting and feeding offspring. During this latter period, territorial challenges are predicted to prompt rapid but temporary increases in circulating testosterone. In contrast, in polygynous endotherms, in which only females provide parental care, males are expected to maintain testosterone at physiologically maximal levels throughout the breeding season in order to garner additional matings through heightened territorial activity. In such species, challenges are not expected to increase circulating testosterone because receptors are saturated and production of additional hormone should not influence behaviour (Wingfield *et al.* 1990).

Squamates, especially lizards, are interesting comparative model systems in which to test the general applicability of the challenge hypothesis because males are often polygynous and generally do not provide parental care. Hence, males are not expected to increase steroid levels in response to intrasexual social challenges. Consistent with this prediction of the challenge hypothesis, even though males increased aggression in response to challenges, they did not increase testosterone levels in tree lizards (Moore *et al.* 1998), mountain spiny lizards (Moore 1988), northern fence lizards (Klukowski & Nelson 1998), jacky dragons (Watt *et al.* 2003), collared lizards (Curtis 2010) and the only snake investigated, the copperhead (Schuett *et al.* 1996).

Other results on polygynous lizards are less consistent with predictions of the challenge hypothesis. In at least three lizard studies that have monitored changes in male hormone levels throughout the breeding season, males showed significant temporal fluctuations (Moore 1986, Klukowski & Nelson 1998, Curtis 2010), which is not consistent with the expectation that polygynous males maintain androgens at maximal levels throughout the season. Four days of successive daily intrusions during the breeding season prompted significant increases in testosterone in male fence lizards, followed by a decrease even though intrusions were continued for 10 days in total (Smith & John-Alder 1999). Male green anoles that won contests had higher androgen levels relative to both males that lost contests and control males, suggesting that high levels of aggression required to win contests may have prompted secretion of androgens (Greenberg & Crews 1990). In marine iguanas, chemically blocking the effects of high levels of circulating testosterone in territorial males caused them to become less aggressive, and less able to repel neighbouring competitors that were not treated with the testosterone-blocking agent. As a consequence, these non-treated males became more aggressive and their circulating levels of testosterone increased (Wikelski *et al.* 2005). Once the effects of the testosterone-blocking agent administered to territorial males wore off, they increased levels of aggression to re-establish control of their territories, and their testosterone levels increased to higher than before they were treated with blocking agents (Wikelski *et al.* 2005). In female marine iguanas, aggression toward males appeared to decrease testosterone, but increased both oestradiol and progesterone (Rubenstein & Wikelski 2005). Lastly, testosterone levels decreased following aggression in male green sea turtles (Jessop *et al.* 1999). Together these studies suggest that the influence of aggression on hormones is highly variable in polygynous lizards and other reptiles, and that the challenge hypothesis may not provide an adequate general framework (Hirschenhauser & Oliveira 2006).

The relationship between vertebrate aggression and androgens has also prompted much interest in stress hormones (Adkins-Regan 2005). Corticosterone, for example, is often predicted to increase as a consequence of physiological/psychological stress, which may be a product of participating in aggressive contests especially for individuals that are defeated, and there is some evidence that corticosterone and testosterone act antagonistically to one another (Moore & Jessop 2003). Results of empirical studies pertaining to the proposed relationship(s) among aggression, corticosterone and androgens in reptiles are highly variable. Studies on lizards in particular indicate some interesting interactive relationships, but general patterns are elusive. In mountain spiny lizards, although introduction of intruders prompted high levels of aggression by male territorial defenders, Moore (1987) discovered no changes in either corticosterone or testosterone in territorial males sampled at 15-min intervals during aggressive encounters. Green sea turtle males that experienced aggression from consexual rivals had lower androgen levels but there was no change in levels of corticosterone (Jessop *et al.* 1999). Among male collared lizards, baseline corticosterone levels were negatively correlated with testosterone in males that defended territories by using high rates of advertisement display, but no such correlation was observed among non-territorial males that did not display frequently (Baird & Hews 2007). In tree lizards, non-territorial (O) males that had won an aggressive encounter on the previous day exhibited both elevated levels of corticosterone and decreased levels of testosterone, whereas levels of both hormones were unchanged in OB-males that had won encounters on the previous day (Knapp & Moore 1996). This difference in the hormonal responses to aggression by the two male morphs may be explained by the finding that in territorial males the binding capacity of steroid proteins (which bind both testosterone and corticosterone) is higher in territorial male tree lizards (Jennings *et al.* 2000). Similarly, plasma binding proteins may be involved in brown anoles, where treatment with exogenous corticosterone decreased both plasma testosterone and aggression (Tokarz 1987).

In contrast, in male fence lizards that received long-term challenges, circulating levels of corticosterone increased even though testosterone levels did not (Klukowski & Nelson 1998). In side-blotched lizards, experimental increase of circulating corticosterone appears to decrease home range size and activity level (DeNardo & Sinervo 1994), whereas in *Podarcis muralis*, juveniles given increased corticosterone experimentally showed higher rates of activity (Belliure & Clobert 2004). These mixed findings clearly indicate that the relationship among aggression, androgens and corticosterone is far from clear, and merits further investigation among lizards and other reptiles.

12.9 Tests of game-theoretic models using lizards

As described in Chapters 2 and 3, game theory provides a set of powerful mathematical models that can be used to examine the net payoff of behavioural tactics used by individuals during agonistic contests. To my knowledge, game-theoretic models have not been used to address any aspect of aggression and social system evolution in non-lizard reptiles. Probably the best-known application of game theory to questions about lizard social behaviour is that used to explain the evolution and maintenance of different colour morphs that display alternative behavioural tactics. Both side-blotched and European common lizards are characterised by three alternative male colour morphs, the maintenance of which is hypothesised to be a consequence of a negative frequency-dependent 'rock–paper–scissors game'. According to this model, males are more successful competing against rivals using different tactics than they are competing against males adopting their same tactic, and males playing the rarest tactic during any particular breeding season obtain the highest fitness (Sinervo & Lively 1996, Sinervo *et al.* 2007; section 12.5). The relative fitness of three female common lizard colour morphs has also been examined using game theory. Yellow females are more aggressive and socially dominant in the laboratory (Vercken & Clobert 2008). Field manipulation of the frequency of colour morphs indicates that yellow females behave as aggressive 'hawks' whereas orange females behave as non-aggressive 'doves'. Females having mixed colouration employ a conditional 'bully' strategy (Maynard-Smith 2005) involving the use of dove tactics against hawk opponents, but hawk tactics against dove opponents (Vercken *et al.* 2010).

There appear to be relatively few studies designed explicitly to test the extent to which dyadic contests between combatants conform to predictions of game-theoretic models and, once again, the few available studies involve lizards. Of these, results of at least five support some predictions of the sequential assessment game (Chapter 2). This hypothesis proposes that contestants assess their own competitive ability relative to that of their opponents as contests progress, and it predicts that one contestant will retreat when it assesses that the benefits of continuing the contest are exceeded by the costs. Therefore, the sequential assessment game model predicts that contest outcome will be determined by fighting ability (Parker 1974, Maynard-Smith 2005).

Consistent with predictions of the sequential assessment game model, rates of aggression between Tenerife lizard males were higher when contestant males were more similar in both total size and head size, and the probability of winning was related to asymmetries in fighting ability (Molina-Borja *et al.* 1998). Similarly, in both Iberian wall lizards and snow skinks (*Niveoscincus microlepidotus*), the intensity of fights between males was negatively correlated with the size disparity between opponents (Olsson & Shine 2000, López & Martín 2001). Because monitor lizards (Varanidae) engage one another in prolonged contests, they may be well suited for tests of game-theoretic predictions, and Earley *et al.* (2002) reviewed the available literature on 17 species: Varanid contests were organised into distinct phases during which combatants gave repetitive displays, both features that probably allowed opponents to assess one another. Consistent with the sequential assessment model, both contest duration and outcome were determined by asymmetries between males (Earley *et al.* 2002).

Jenssen *et al.* (2005) staged contests between pairs of male green anoles that were allowed to establish territories in adjacent (but separated) semi-natural enclosures matched for habitat resources and the presence of a mate. In one treatment group, contestants were matched for size, whereas males were size-mismatched in a second treatment group. Once both males had established territory occupancy, the partition was removed which allowed either male to enter his opponent's territory. Similar to contests in varanids, contests between green anoles involved distinct phases. Within phases, males matched aggressive tactics and signals and winning depended on larger size. Both of these results are also consistent with predictions of the sequential assessment game model. Because either male was able to intrude to engage their opponents, asymmetric contests were also used to test whether smaller males would assess their size deficit and adopt 'bluffing' tactics, or instead utilise aggressive 'hawk' tactics despite their size disadvantage ('Napoleon complex', Chapter 2). Even though they lost 90% of fights, smaller males initiated encounters by invading territories held by larger males and continued to signal aggressively and engage physically with larger males as contests intensified (Jenssen *et al.* 2005). These results demonstrate that, at least when both male contestants have achieved territorial status, male green anoles continue to use aggressive hawk tactics even when they are competitively inferior as a consequence of smaller size.

In contrast with the studies described above, Stuart-Fox (2006) found no support for predictions of either the 'sequential assessment' or 'war of attrition' (Chapter 2) models in explaining the outcome of contests between Cape-dwarf chameleon males. Intra-sexual male contests in this species vary substantially in duration and intensity (Stuart-Fox 2006). Using a tournament of staged interactions between a series of different male opponents, Stuart-Fox (2006) investigated the influence of both morphological and behavioural traits on winning, as well as the extent to which the outcome of previous contests supported predictions of several game theoretic models. Because males engaged in multiple contests sequentially, they acquired fighting experience as they progressed through the tournament. Multivariate analyses revealed that relative fighting ability was predicted by head-casque height, the area of colourful patches on the flanks, and experience (winning or losing) in prior contests. Results revealed that eventual winners were more aggressive and consistently outperformed eventual losers, rather than each male matching its opponent's behaviour as predicted by war of attrition models. Moreover, contest intensity and duration were determined by the threshold fighting ability of male losers rather than asymmetries between contestants as predicted by the sequential assessment model. Overall, results of contests between Cape-dwarf chameleons were most consistent with the 'cumulative assessment' model (Stuart-Fox 2006, Chapter 2). However, there were no differences in three contest parameters (initial behaviour patterns, time to escalate to physical contact, rate of display) that this model predicts should differ in winner and loser males. Therefore, additional tests, particularly those that evaluate externally derived costs of contests, are required to test more fully the applicability of the cumulative assessment model in this species (Stuart-Fox 2006).

12.10 Prospects for future studies

The gap between what we know and what we would like to know about the evolution of reptilian aggression is large, leaving many opportunities for productive research. Obviously, the saurian-biased imbalance of this field needs to be remedied by studies on other reptilian taxa. Such expansion will require overcoming serious, taxon-specific, logistical obstacles guided by careful *a priori* analyses of both the limitations and the potential that each system holds for hypothesis testing. Judging from the studies that have been most informative in the past, this approach is best begun by recording a broad and thorough database, examination of suggestive correlations in these data, followed by formulation of meaningful hypotheses and either comparative or experimental tests. Because a wide variety of attributes is likely to contribute to individual competitive ability in reptiles, future studies of aggression will be promoted by the integration of expertise and techniques employed by a variety of specialists. To the extent possible, field studies on free-ranging animals should be used to inform and guide design of empirical tests. Technical advancements that allow recording of behaviour under natural conditions, and application of contemporary molecular techniques to determine reproductive success in the field may facilitate future studies. Currently, application of an integrative approach in reptilian studies may be exemplified best by the relatively new focus on performance traits. Because individual performance of behaviours that influence competitive success is likely to be constrained or promoted by multiple attributes that are inter-correlated, multivariate analyses of fighting ability appear to be a necessity. A promising approach appears to be application of game theoretic model predictions to experimentally induced agonistic contests that employ a tournament design where combatants are pitted against multiple opponents in a series of contests. We still know relatively little about the extent to which morphology and performance traits are linked hormonally with aggression and its adaptive value in natural populations. Moreover, the performance traits that are usually examined are those that can be quantified using relatively rapid measurements of force and speed. Emphasis on such variables may bias consideration towards traits that are assumed to be under strong selection because they promote success during escalated fights. Although such fights are intensely interesting because they are spectacular and can often be readily provoked by introducing tethered intruders, selection often acts to reduce escalated fighting because it is costly, and instead promotes the use of less costly stereotypical displays to settle disputes. Evaluation of performance traits, therefore, should perhaps be expanded to include variation in the capacity of individuals to perform display patterns repetitively and/or conspicuously. Clearly, the value of future studies will be enhanced by observations, measurements and well-planned experiments that are firmly rooted in quantitative studies of how and when reptiles use aggression in their social interactions.

Acknowledgements

This work was partially supported by a grant from the office of Research and Grants at the University of Central Oklahoma. I greatly appreciate suggestions by Marie Babb, Gordon Shuett, Rick Shine and Brent Thomas during the preparation of this chapter. Dan Beck, Teresa Davis Baird, Maria Thacker and Martin Wikelski graciously provided permission to use their photographs. I give my sincere thanks to Teresa Davis Baird, Shirley Baird and Kathy McWilliams for their encouragement. This chapter is dedicated to the memory of John Lyle Baird, who, by his patient example, taught me the enduring value of perseverance and hard work.

References

Abell A (1997) Estimating paternity with spatial behaviour and DNA fingerprinting in the striped plateau lizard, *Sceloporus virgatus (Phrynosomatidae)*. *Behavioral Ecology and Sociobiology*, 41, 217–226.

Adkins-Regan E (2005) *Hormones and Animal Social Behavior: Monographs in Behavior and Ecology*. Princeton, NJ: Princeton University Press.

Alberts AC (1992) Constraints on the design of chemical communication systems in terrestrial vertebrates. *American Naturalist*, 139, 69–89.

Alberts AC (1994) Dominance hierarchies in male lizards: implications for zoo management programs. *Zoo Biology*, 13, 479–490.

Anderson RA & Vitt LJ (1990) Sexual selection versus alternative causes of sexual dimorphism in teiid lizards. *Oecologia*, 84, 145–157.

Anderson RA, McBrayer LD & Herrel A (2008) Bite force in vertebrates: opportunities and caveats for use of a nonpareil whole-animal performance measure. *Biological Journal of the Linnaean Society*, 93, 709–720.

Andrews RM (1971) Structural habitat and time budget of a tropical *Anolis* lizard. *Ecology*, 52, 262–270.

Angilletta MJ, Niewiarowski PH & Navas CA (2002) The evolution of thermal physiology in ectotherms. *Journal of Thermal Biology*, 27, 249–268.

Aragón P, López P & Martín J (2004) The ontogeny of spatio-temporal tactics and social relationships of adult Iberian rock lizards, *Lacerta monticola*. *Ethology*, 110, 1001–1019.

Babb ME (2004) Behavioral ecology and melanism in male red-eared sliders (*Trachemys scripta elegans*) in a central Oklahoma pond. Master's thesis, University of Central Oklahoma, Edmond.

Baird TA (2008) A growth cost of conspicuous coloration in first-year collared lizard males. *Behavioral Ecology*, 18, 1146–1154.

Baird TA, Baird TD & Shine R (2012) Aggressive transition between alternative male social tactics in a long-lived Australian dragon (*Physignathus lesueurii*) living at high density. *PLoS ONE*, e41819.

Baird TA & Curtis JL (2010) Context-dependent acquisition of territories by male collared lizards: the role of mortality. *Behavioral Ecology*, 21, 753–758.

Baird TA & Hews DK (2007) Hormone levels in territorial and non-territorial male collared lizards. *Physiology and Behavior*, 92, 755–763.

Baird TA & Sloan CL (2003) Interpopulation variation in the social organization of female collared lizards, *Crotaphytus collaris*. *Ethology*, 109, 879–894.

Baird TA & Timanus DK (1998) Social inhibition of territorial behaviour in yearling male collared lizards, *Crotaphytus collaris*. *Animal Behaviour*, 56, 989–994.

Baird TA, Acree MA & Sloan CL (1996) Age and gender-related differences in the social behavior and mating success of free-living collared lizards, *Crotaphytus collaris*. *Copeia*, 1996, 226–347.

Baird TA, Fox SF & McCoy JK (1997) Population differences in the roles of size and coloration in intra- and intersexual selection in the collared lizard, *Crotaphytus collaris*: influence of habitat and social organization. *Behavioral Ecology*, 8, 506–517.

Baird TA, Sloan CL & Timanus DK (2001) Intra- and interseasonal variation in the socio-spatial behavior of adult male collared lizards, *Crotaphytus collaris (Reptilia, Crotaphtidae)*. *Ethology*, 107, 15–32.

Baird TA, Timanus DK & Sloan CL (2003) Intra- and intersexual variation in social behavior: Effects of ontogeny, phenotype, resources, and season. In: SF Fox,

JK McCoy & TA Baird (eds.) *Lizard Social Behavior*, pp. 7–36. Baltimore, MD: Johns Hopkins.

Baird TA, Hranitz JM, Timanus DK, *et al.* (2007) Behavioral attributes influence annual mating success more than morphological traits in male collared lizards. *Behavioral Ecology*, 19, 589–593.

Beck DD (2005) *Biology of Gila Monsters and Beaded Lizards*. Berkeley, CA: University of California Press.

Belliure J & Clobert J (2004) Behavioral sensitivity to corticosterone in juveniles of the wall lizard *Podarcis muralis*. *Physiology and Behavior*, 81, 121–127.

Bels VL (1984) Ethological problems of anoline lizards in captivity. *Acta Zoologica et Pathologica Antverpiensa*, 78, 85–100.

Berry JF & Shine R (1980) Sexual selection in turtles (Order Testudines). *Oecologia*, 44, 185–191.

Berry KH (1974) The ecology and social behavior of the chuckwalla (*Sauromalus obesus obesus* Baird). *University of California Publications in Zoology*, 101, 1–60.

Braun CL, Baird TA & LeBeau JK (2010) Influence of substrate temperature and directness of approach on the escape responses of juvenile collared lizards. *Herpetologica*, 66, 418–424.

Bury RB & Wolfheim JH (1973) Aggression in free-living pond turtles (*Clemmys marmorata*). *Bioscience*, 23, 659–662.

Calsbeek R & Marnocha, E (2005) Context dependent territory defense: the importance of habitat structure in *Anolis sagrei*. *Ethology*, 112, 537–543.

Calsbeek R & Sinervo B (2002*a*) An experimental test of the ideal despotic distribution. *Journal of Animal Ecology*, 71, 513–523.

Calsbeek R & Sinervo B (2002*b*) The ontogeny of territoriality during maturation. *Oecologia*, 132, 468–477.

Carazo P, Font E & Desfilis E (2008) Beyond 'nasty neighours' and 'dear enemies'? Individual recognition by scent marks in a lizard (*Podarcis hispanica*). *Animal Behaviour*, 76, 1953–1963.

Carothers JH (1981) Dominance and competition in an herbivorous lizard. *Behavioural Ecology and Sociobiology*, 8, 261–266.

Carothers JH (1984) Sexual selection and sexual dimorphism in some herbivorous lizards. *American Naturalist*, 124, 244–254.

Carpenter CC (1977) Communication and displays of snakes. *American Zoologist*, 17, 217–223.

Carpenter CC & Ferguson GW (1977) Variation and evolution of stereotyped behavior in reptiles. In: C Gans & DW Tinkle (eds.) *Biology of the Reptilia. Vol. 7, Ecology and Behavior A*, pp. 335–403. London: Academic Press.

Carpenter GC (1995) Modeling dominance: the influence of size, coloration, and social experience on dominance relations in the tree lizards (*Urosaurus ornatus*). *Herpetological Monographs*, 9, 88–101.

Censky EJ (1995) Mating strategy and reproductive success in the teiid lizard, *Ameiva plei*. *Behaviour*, 132, 529–557.

Censky EJ (1997) Female mate choice in the non-territorial lizard Ameiva plei (Teidae). *Behavioral Ecology and Sociobiology*, 40, 221–225.

Cooper WE Jr (1999) Trade-offs between courtship, fighting, and anti-predatory behaviour by a lizard, *Eumeces laticeps*. *Behavioral Ecology and Sociobiology*, 47, 54–59.

Cooper WE Jr (2003) Social behavior and anti-predator defense in lizards. In: SF Fox, JK McCoy & TA Baird (eds.) *Lizard Social Behavior*, pp. 107–141. Baltimore, MD: Johns Hopkins.

Cooper WE Jr (2004) Adaptive chemosensory behaviour by lacertid lizards. In: V Pérez-Mellado, N Riera & A Perera (eds.) *The Biology of Lacertid Lizards: Evolutionary and Ecological Perspectives*, pp. 83–118. Institut Menorquí d'Estudís. Recerca, 8. Maó, Menorca.

Cooper WE Jr & Burns N (1987) Social significance of ventrolateral coloration in the fence lizard, *Sceloporus undulatus*. *Animal Behaviour*, 35, 526–532.

Cooper WE Jr & Greenberg N (1992) Reptilian coloration and behaviour. In: Gans C & Crews D (eds.) *Biology of the Reptilia, Physiology E – Hormones, Brain and Behavior, Vol. 18*, pp. 298–422. Chicago, IL: University of Chicago Press.

Cooper WE Jr & Vitt LJ (1987) Deferred agonistic behavior in a long-lived scincid lizard *Eumeces laticeps*. *Oecologia*, 72, 321–326.

Cooper WE Jr & Vitt LJ (1988) Orange head coloration of the male broad-headed skink (*Eumeces laticeps*), a sexually-selected social cue. *Copeia*, 1988, 1–6.

Cooper WE Jr, Mendonca MT & Vitt LJ (1987) Induction of orange head coloration and activation of courtship and aggression by testosterone in the male broad-headed skink (*Eumeces laticeps*). *Journal of Herpetology*, 21, 96–101.

Cote J, Boudsocq S & Clobert J (2008) Density, social information, and space use in the common lizard (*Lacerta vivipara*). *Behavioral Ecology*, 19, 163–168.

Crowley SR & Pietruszka RD (1983) Aggressiveness and vocalization in the leopard lizard (*Gambelia wislizennii*): the influence of temperature. *Animal Behaviour*, 31, 1055–1066.

Curtis JL (2010) Social modulation of androgens and glucocorticoids in territorial male collared lizards. Master's thesis, University of Central Oklahoma, Edmond.

Decourcy KR & Jenssen TA (1994) Structure and use of male territorial headbob signals by the lizard *Anolis carolinensis*. *Animal Behaviour*, 47, 251–262.

DeNardo DF & Sinervo B (1994) Effects of steroid hormone interaction on activity and home-range size in male lizards. *Hormones and Behavior*, 28, 273–287.

Díaz-Uriarte R (1999) Anti-predator behaviour changes following an aggressive encounter in the tree lizard *Tropidurus hispidus*. *Proceedings of the Royal Society London B*, 266, 2457–2464.

Douglass J (1986) Patterns of mate-seeking and aggression in a Southern Florida population of gopher tortoise, *Gopherus polyphemus*. *Proceedings of the Symposium of the Desert Tortoise Council*, 1986, 155–199.

Duvall D, Graves BM & Carpenter GC (1987) Visual and chemical composite signaling effects of *Sceloporus* lizard fecal boli. *Copeia*, 1987, 1028–1031.

Earley RL, Attum O & Eason P (2002) Varanid combat: perspectives from game theory. *Amphibia-Reptilia*, 23, 469–485.

Echelle AA, Echelle AF & Fitch HS (1971) A comparative analysis of aggressive display in nine species of Costa Rican *Anolis*. *Herpetologica*, 27, 271–288.

Font E & Desfilis E (2002) Chemosensory recognition of familiar and unfamiliar conspecifics by juveniles of the Iberian wall lizard (*Podarcis hispanica*). *Ethology*, 108, 319–330.

Fox SF (1983) Home range quality and aggression in *Uta stansburiana*. In: RB Huey, EE Pianka & TW Shoener (eds.) *Lizard Ecology: Studies of a Model Organism*, pp. 149–168. Cambridge, MA: Harvard University Press.

Fox SF & Baird TA (1992) The dear-enemy phenomenon in collared lizards, *Crotaphytus collaris*, with a cautionary note on experimental methodology. *Animal Behaviour*, 44, 780–782.

Fox SF & Shipman PA (2003) Social behavior at high and low elevations. In: SF Fox, JK McCoy & TA Baird (eds.) *Lizard Social Behaviour*, pp. 310–355. Baltimore, MD: Johns Hopkins.

Fox SF, Heger NA & Delay LS (1990) Social cost of tail loss in *Uta stansburiana*: lizard tails as status-signalling badges. *Animal Behaviour*, 39, 549–554.

Fox SF, McCoy JK, & Baird TA (2003). *Lizard Social Behavior*. Baltimore: Johns Hopkins.

Fraser S (1985) Variability of heating and cooling rates during radiant heating in a scincid lizard, *Egernia cunninghami*. *Comparative Biochemistry and Physiology*, 80, 281–286.

Garland T Jr, Hankins E & Huey RB (1990) Locomotor capacity and social dominance in lizards. *Functional Ecology*, 4, 243–250.

Garrick LD & Lang JW (1977) Social signals and behaviors of adult alligators and crocodiles. *American Zoologist*, 17, 225–239.

Gier PJ (2003) The interplay among environment, social behavior, and morphology: iguanid mating systems. In: SF Fox, JK McCoy & TA Baird (eds.) *Lizard Social Behavior*, pp. 278–309. Baltimore, MD: Johns Hopkins.

Gillingham JC (1987) Social behavior. In: RA Siegel, JT Collins & SS Novak (eds.) *Snakes: Ecology, and Evolutionary Biology*, pp. 184–209. New York, NY: McGraw-Hill.

Gillingham JC, Carmichael C & Miller T (1995) Social behavior of the tuatara, *Sphenodon punctatus*. *Herpetological Monographs*, 9, 5–16.

Glaes A (1982) Role of the vomeronasal organ in the reproductive behavior of the adder, *Vipera berus*. *Copeia*, 1982, 148–157.

Greenberg N & Crews D (1990) Endocrine and behavioral responses of aggression and social dominance in the green anole lizard, *Anolis carolinensis*. *General Comparative Endrocrinology*, 77, 246–255.

Grigg GC, Seebacher F, Beard LA, *et al.* (1998) Thermal relations of large crocodiles, *Crocodylus porosus*, free-ranging in a naturalistic situation. *Proceedings of the Royal Society of London B*, 265, 1793–1799.

Gvoždík L & Van Damme R (2003) Evolutionary maintenance of sexual dimorphism in head size in the lizard *Zootoca vivipara*: a test of two hypotheses. *Journal of Zoology London*, 259, 7–13.

Hare KM & Miller KA (2009) What dives beneath: diving as a measure of performance in lizards. *Herpetologica*, 65, 227–236.

Herrel A, Spithoven L, Van Damme R, *et al.* (1999) Sexual dimorphism of head size in *Gallotia galloti*: testing the niche divergence hypothesis by functional analysis. *Functional Ecology*, 13, 289–297.

Hertz PE, Huey RB & Nevo E (1982) Body temperature influences defensive responses of lizards. *Animal Behaviour*, 30, 676–679.

Hews DK (1990) Examining hypotheses generated by field measures of sexual selection on male lizards, *Uta palmeri*. *Evolution*, 44, 1956–1966.

Hews DK (1993) Food resources affect female distribution and male mating opportunities in the iguanian lizard *Uta palmeri*. *Animal Behaviour*, 46, 279–291.

Stapley J & Keogh JS (2004) Exploratory and antipredator behaviours differ between territorial and non-territorial male lizards. *Animal Behaviour*, 68, 641–646.

Stone PA, Snell HL & Snell HM (2003) Island biogeography of morphology and social behavior in the lava lizards of the Galápagos Islands. In: SF Fox, JK McCoy & TA Baird (eds.) *Lizard Social Behaviour*, pp. 190–239. Baltimore, MD: Johns Hopkins.

Stuart-Fox D (2006) Testing game theory models: fighting ability and decision rules in chameleon contests. *Proceedings of the Royal Society of London B*, 273, 1555–1561.

Stuart-Fox D & Johnston GR (2005) Experience overrides colour in lizard contests. *Behaviour*, 142, 329–350.

Stuart-Fox D, Firth D, Moussalli A, *et al.* (2006) Multiple signals in chameleon contests: desiging and analyzing animal contests as a tournament. *Animal Behaviour*, 71, 1263–1271.

Telemeco RS & Baird TA (2011) Capital energy drives production of multiple clutches whereas income energy fuels growth in female collared lizards *Crotaphytus collaris*. *Oikos*, 120, 915–921.

Thompson CW & Moore MC (1991) Throat colour reliably signals status in male tree lizards, *Urosaurus ornatus*. *Animal Behaviour*, 42, 745–753.

Tokarz RR (1985) Body size as a factor determining dominance in staged agonistic encounters between male brown anoles (*Anolis sagrei*). *Animal Behaviour*, 33, 746–753.

Tokarz RR (1987) Effects of corticosterone treatments on male aggressive behavior in a lizard (*Anolis sagrei*). *Hormones and Behavior*, 21, 358–370.

Trillmich F & Trillmich KGK (1984) The mating system of pinnepeds and marine iguanas: convergent evolution of polygyny. *Biological Journal of the Linnaean Society*, 21, 209–216.

Tucker JK, Maher J & Theiling CH (1995) Melanism in the red-eared slider (*Trachemys scripta elegans*). *Journal of Herpetology*, 29, 291–296.

Van Dyk DA & Evans CS (2007) Familiar–unfamiliar discrimination based on visual cues in the jacky dragon, *Amphibolurus muricatus Animal Behaviour*, 74, 33–44.

Vanhooydonck B, Herrel AY, Van Damme R, *et al.* (2005) Does dewlap size predict male bite performance in Jamaican *Anolis* lizards? *Functional Ecology*, 19, 38–42.

Vercken E & Clobert J (2008) Ventral color polymorphism correlates with alternative behavioral patterns in female common lizards (*Lacerta vivipara*). *Ecoscience*, 15, 320–326.

Vercken E, Clobert J & Sinervo B (2010) Frequency-dependent reproductive success in female common lizards: a real-life Hawk–Dove-bully game? *Oecologia*, 162, 49–58.

Vidal MA, Ortiz JC & Labra A (2008) Intraspecific variation in a physiological thermoregulatory mechanism: the case of the lizard *Liolaemus tenuis* (Liolaeminae). *Revista Chilena de Historia Natural*, 81, 171–178.

Vitt LJ & Caldwell JP (2009) *Herpetology: An Introductory Biology of Amphibians and Reptiles*, 3rd edn. San Diego, CA: Academic Press.

Vitt LJ & Cooper WE Jr (1985) The evolution of sexual dimorphism in the skink, *Eumeces laticeps*: an example of sexual selection. *Canadian Journal of Zoology*, 63, 995–1002.

Vitt LJ, Congdon JD, Hulse AC, *et al.* (1974) Territorial aggressive encounters and tail breaks in the lizard *Sceloporus magister*. *Copeia*, 1971, 990–993.

Vliet KA (1989) Social displays of the American alligator (*Alligator mississippiensis*). *American Zoologist*, 29, 1019–1031.

Watkins GG (1996) Proximate causes of sexual dimorphism in the Iguanian lizard *Microlophus occipitalis*. *Ethology*, 77, 1473–1482.

Watt MJ, Forster GL & Joss JMP (2003) Steroid correlates of territorial behavior in male jacky dragons, *Amphibolurus muricatus*. *Brain, Behavior and Evolution*, 61, 184–194.

Whiting MJ (1999) When to be neighborly: differential agonistic responses in the lizard *Platysaurus broadleyi*. *Behavioral Ecology and Sociobiology*, 46, 210–214.

Whiting MJ, Nagy KA & Bateman PW (2003) Evolution and maintenance of social status-signalling badges: experimental manipulations in lizards. In: SF Fox, JK McCoy & TA Baird (eds.) *Lizard Social Behavior*, pp. 47–82. Baltimore, MD: Johns Hopkins.

Whiting MJ, Stuart-Fox DM, O'Connor DO, *et al.* (2006) Ultraviolet signals ultra-aggression in a lizard. *Animal Behaviour*, 72, 353–363.

Wikelski M, Steiger SS, Gall B, *et al.* (2005) Sex, drugs and mating role: testosterone-induced phenotype-switching in Galapagos marine iguanas. *Behavioral Ecology*, 16, 260–268.

Wingfield JC, Hegner RE, Dufty AM Jr, *et al.* (1990) The 'challenge hypothesis': theoretical implications for patterns of testosterone secretion, mating systems and breeding strategies. *American Naturalist*, 136, 829–846.

Woodley SK & Moore MC (1999) Female territorial aggression and steroid hormones in mountain spiny lizards. *Animal Behaviour*, 57, 1083–1089.

Wyk JH van & Mouton P le FN (1998) Reproduction and sexual dimorphism in the montane viviparous lizard, *Pseudocordylus capensis* (Sauria: Cordylidae). *South African Journal of Zoology*, 33, 156–165.

Yang EJ, Phelps SM, Crews D, *et al.* (2001) The effects of social experience on aggressive behavior in the green anole lizard (*Anolis carolinensis*). *Ethology*, 107, 777–793.

Ydenberg RC, Giraldeau LA & Falls JB (1988) Neighbours, strangers and the asymmetric war of attrition. *Animal Behaviour*, 36, 343–347.

Zamudio KR (1998) The evolution of female-biased sexual size dimorphism: a population-level comparative study in horned lizards (*Phrynosoma*). *Evolution*, 52, 1821–1833.

Zamudio KR & Sinervo B (2003) Ecological and social contexts for the evolution of alternative mating strategies. In: SF Fox, JK McCoy & TA Baird (eds.) *Lizard Social Behavior*, pp. 83–106. Baltimore, MD: Johns Hopkins.

Zucker N (1989) Darkening and territoriality in a wild population of the tree lizard *Urosaurus ornatus*. *Journal of Herpetology*, 23, 389–398.

Zucker N (1994) A dual status-signalling system: a matter of redundancy or differing roles? *Animal Behaviour*, 47, 15–22.

Zucker N & Murray L (1996) Determinants of dominance in the tree lizard *Urosaurus ornatus*: the relative importance of mass, previous experience and coloration. *Ethology*, 102, 812–825.

Zug GR, Vitt LJ & Caldwell JP (2001) *Herpetology, An Introductory Biology of Amphibians and Reptiles*. San Diego, CA: Academic Press.

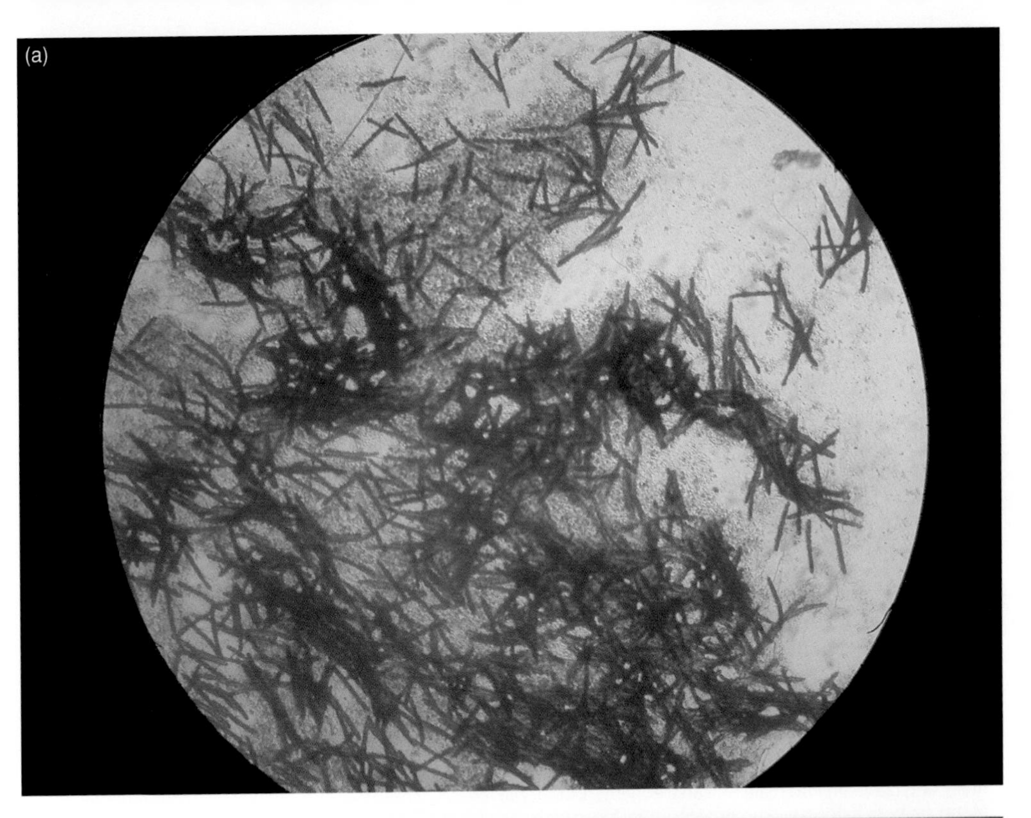

Figure 4.1 (**a**) Nematocysts, the stinging harpoon-like organelles of cnidarians, stained and viewed under light magnification. These nematocysts were from a sample of approximately 1 mm^2 of acrorhagial epidermis. (**b**) *Actinia equina* in combat: the acrorhagi are the blue bud-like tentacles visible on the individual on the left. The individual on the right has received a series of acrorhagial contacts and blue acrorhagial peels are visible on its epidermis. (Photograph credit (a) Mark Briffa, (b) Fabian Rudin.)

Figure 4.2 Agonistic interaction between two male African elephants, *Loxodonta africana*. These individuals, known to researchers as Dionysus (left) and Iain (right) are both in musth. (Photograph reproduced with permission © Joyce Poole, ElephantVoices.)

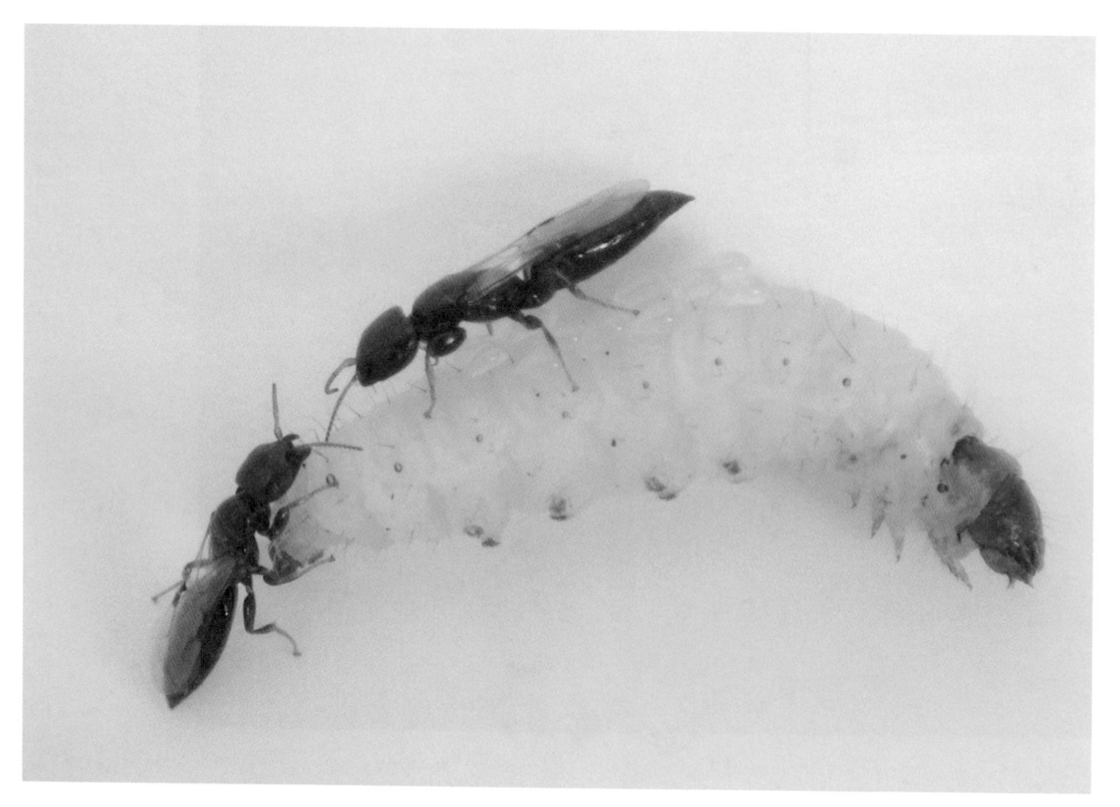

Figure 8.1 Contest interaction between two female bethylid wasps. Initial stage of an owner–intruder contest between two *Goniozus legneri* females. The owner (wasp on right) has already laid a clutch and the eggs have hatched into first instar larvae and started to feed on the paralysed host caterpillar. Note the intruder's flared mandibles. (Photograph: Sonia Dourlot.)

Figure 5.1 Examples of decapod crustaceans from groups that have been the focus of studies on aggression. (**a**) A common Eurpoean hermit crab, *Pagurus bernhardus* (Anomura) occupying a *Littorina littorea* shell. Note the asymmetrically sized chelipeds; (**b**) a male fiddler crab, *Uca mjoebergi* (Brachyura: Ocypodidae), waving its enlarged and conspicuously coloured major cheliped, an adaptation that is used to attract females and during fights against rival males; (**c**) a velvet swimming crab *Necora puber* (Brachyura: Portunidae), an example of a 'true crab' that engages in agonistic encounters; (**d**) a pair of Australian freshwater crayfish or 'yabbies', *Cherax dispar* (Astacidae), grappling with their chelipeds. (Photo credits: (**a**) Sophie Mowles, (**b**) Tanya Detto, (**c**) Thomas Guest, (**d**) Anthony O'Toole.)

Figure 5.5 (**a**) Use of fluorescein to visualise chemical communication via urine release in blindfolded crayfish, *A. leptodactylus*. Reproduced from Berry & Breithaupt (2010).

Figure 12.1 Territorial male collared lizard beginning a full show display characterised by lateral compression and elevation of the torso and dewlap extension (photograph by Teresa D. Baird).

Figure 12.3 (**b**) Male Australian water dragons (photograph by Teresa D. Baird).

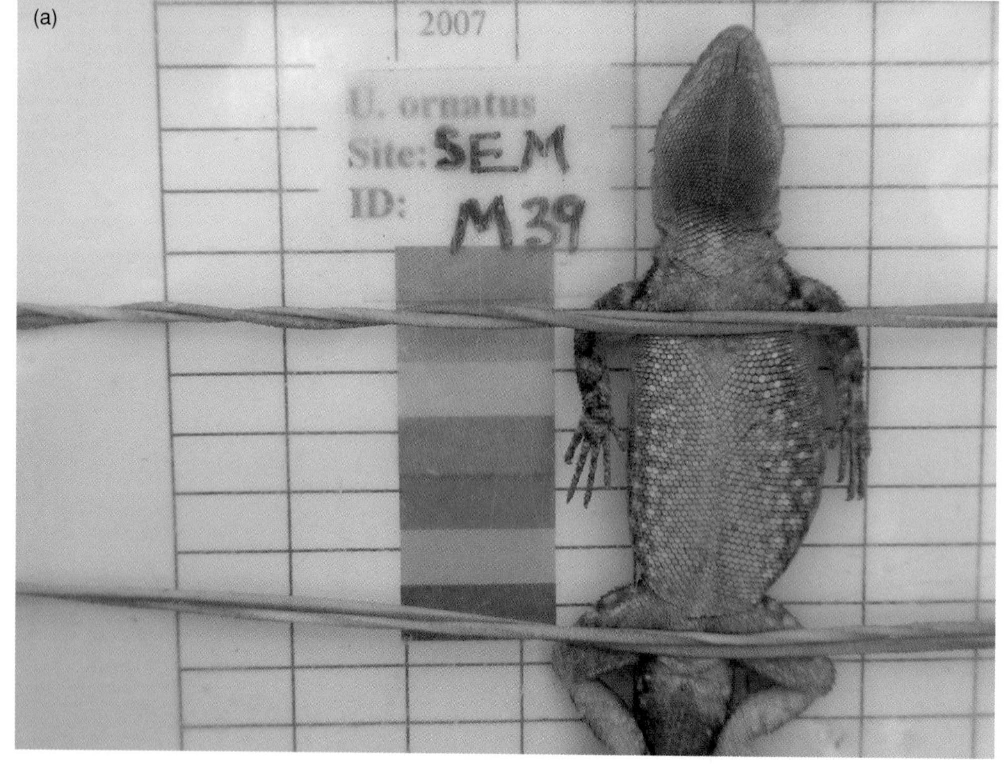

Figure 12.5 (**a**) Orange–male colour morph in male *Urosaurus ornatus* (photograph by Maria Thacker).

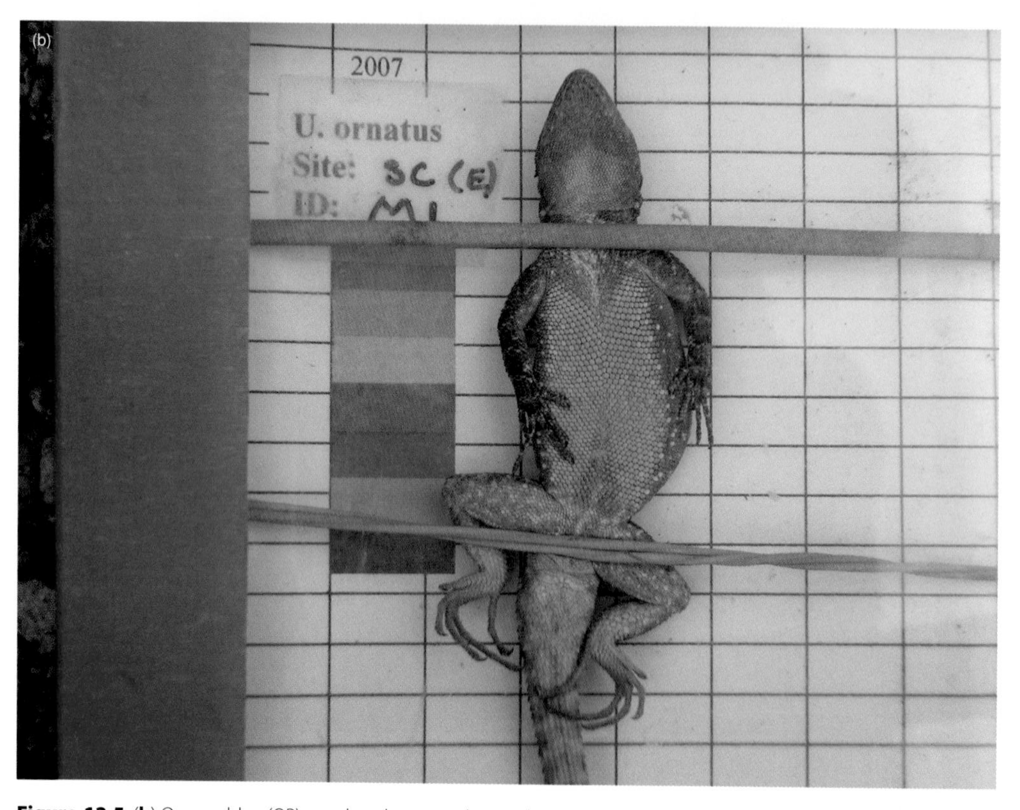

Figure 12.5 (**b**) Orange blue (OB) – male colour morph in male *Urosaurus ornatus* (photograph by Maria Thacker). The blue colouration is visible as a lighter patch in the centre of the dewlap, surrounded by the orange colouration. Note that both morphs, shown here and in figure 12.5(a), have a striking blue colouration on each side of the abdomen.

Figure 14.1 Fallow deer buck (*Dama dama*). ©iStockphoto.com/Damian Kuzdak.

Figure 14.3 Red deer stags *Cervus elaphus* preparing to fight. Early morning, but red deer stags in the rutting season start work at break of day. This sort of parallel walk allows the stags to size each other up, and frequently occurs at the boundary between two territories. A serious fight may follow, especially if there are hinds (females) close by. ©iStockphoto.com/Roger Whiteway.

Figure 14.4 Male Ibex fighting in the Gran Paradiso National Park, Italy. (Photograph: Patrick Bergeron.)

Chapter

13 Bird contests: from hatching to fertilisation

Sarah R. Pryke

13.1 Summary

Work on birds has prompted some key theoretical developments in the field of animal contests. Birds have an unusually conservative morphology, which constrains the ability of many species to fight physically, but many birds have evolved adaptations to aid in conflict resolution. These include the display of bright colours, elaborate vocalisations and physiological mechanisms. Birds also have a relatively conservative life history (parental care is ubiquitous), which influences the nature and extent of contests: for most birds, competition begins at birth for access to food provided by their carers. Intense resource competition continues into their adult lives, especially as many birds are territorial and/or reside in social flocks of related and/or unrelated conspecifics. At breeding, competition for access to mates is particularly severe, with birds competing both overtly and more cryptically to gain fertilisation precedence. Here I consider the diversity of avian contests.

13.2 Introduction

Bird contests can be highly conspicuous. An example is the ferocious fighting that occurs between two unfamiliar (or staged) jungle fowl (*Gallus gallus domesticus*) where males may fight to the death using sharp-pointed spurs as weapons (Mench 2009). Many avian contests are, however, less physical. Under more natural circumstances, for example within the jungle fowl's mixed-sex flock, disputes between flock members are unlikely to escalate to physical fighting (Ligon *et al.* 1990, Collias & Collias 1996). Agonistic encounters generally involve stereotyped behaviours, a range of complex vocalisations, and variation in conspicuously coloured morphological traits (e.g. Ligon *et al.* 1990, Cornwallis & Birkhead 2008). Nevertheless, the lack of physical contests does not necessarily indicate an absence of competition. Indeed, birds compete for the same things that cause and define competition in other taxa. However, there are a number of general avian morphological and life-history adaptations that are likely to have influenced the nature of competition in birds.

With only a few notable exceptions, birds have (1) forelimbs that are elegantly adapted solely for powered flight, (2) lightweight skeletons that are constrained by size and weight limits, (3) a beak that lacks teeth and (4) a generally much higher visual acuity than other vertebrates. Together, these highly specialised adaptations place considerable constraints on the evolution of contest competition in birds. For example, there has been no runaway selection on increased male body size or armaments to the extremes seen in highly polygynous mammals (e.g. elephant seals, red deer, Chapter 14) or insects (e.g. stalk-eyed flies, goliath beetles, Chapter 9). It is notable that the most extreme level of avian sexual size dimorphism recorded occurred in a flightless Moa (Bunce *et al.* 2003). Likewise, many birds lack specialised weapons, such as gripping, clawed forelimbs and toothed jaws, which has often constrained the physical nature of contests (exceptions include the well-developed talons and beaks of eagles and vultures which are used in prey capture and food handling but can also function in contest interactions).

In addition to their conserved morphological structure, birds also have a relatively conservative life history, which has important consequences for the nature of avian contests. Unlike many other vertebrates (e.g. fish, reptiles, mammals), birds are exclusively oviparous (egg-bearing). With the exception of the Megapodes, eggs and hatchlings are provided with

Animal Contests, ed. I.C.W. Hardy and M. Briffa. Published by Cambridge University Press.

direct parental care, either from only one parent (e.g. most polygynous and polyandrous birds), from both (e.g. most monogamous systems), or from a more extensive social group (e.g. cooperatively breeding systems). As a consequence of this close contact with carers during their development, young birds learn sex-specific features (e.g. complex song development: Soma *et al.* 2005, Catchpole & Slater 2008) that help them in future competition for mates. The overlap of generations, and extended contact with relatives and conspecifics, also provides new arenas in which competition can take place. Most species produce multiple eggs within the same clutch. For such species, competition starts from the moment the young hatch. In highly altricial species, the young hatch at an early stage of development (e.g. naked, blind and defenceless) and are confined to the nest for many weeks, completely dependent upon food brought to the nest by their carers. In such species, competition among the young can be very intense. In a few predatory birds, for example, if parents are unable to supply adequate food, the smallest chicks will be unable to compete against their older siblings and may eventually be killed and eaten (Mock *et al.* 1990, Bortolotti *et al.* 1991, Negro *et al.* 1992). Even though such extreme outcomes appear relatively rare, intense intra-brood competition could affect not only the relative development of different chicks (e.g. body size) but also potentially determine their relative fighting ability later in life (Chapter 8). The conflict between developing siblings is also a situation in which differences between the sexes become apparent (i.e. broods are often mixed sex), which has implications for the outcome of later competition and sexual conflict (Badyaev *et al.* 2001).

Although sexual size dimorphism is generally less extreme in birds than in other vertebrates (Székely *et al.* 2007), similar differences in the roles of the two sexes underpin intra-sexual competition. For birds, as well as many other groups of animals, the fundamental disparity in the number and size of gametes produced by females and males means that females (or more specifically, their eggs) are a limiting resource that is directly related to male reproductive success. Most birds are socially monogamous (> 90%: Ligon 1999); however, recently developed molecular techniques have revolutionised the traditional view of fidelity in avian mating systems, revealing that less than 14% are genetically monogamous (Hasselquist & Sherman 2001, Griffith *et al.* 2002). In birds, this variation in reproductive success, together with differences in parental contributions, and the often male-biased adult operational sex ratio (even in socially monogamous species: Breitwisch 1989), further exaggerates the disparity between the sexes. As a result, competition among males for access to females or resources important for breeding is intense, as males unable to compete successfully will have limited opportunity for reproduction. The exceptions to this rule are classically polyandrous species, in which females are more competitive and provide no parental care: for example, the jacanas and phalaropes (Ligon 1999). The general discrepancy in reproductive investment between the sexes sets the stage for the development of dimorphism in many morphological and behavioural secondary sexual characteristics related to competition, ranging from the evolution of conspicuous traits (e.g. ornamental plumage and long tails), male aggression and even the return dates of migratory species to their breeding grounds in northern and southern temperate zones. Birds may also compete directly by disrupting copulations (Trail 1985) or by destroying the attractive signal of another individual. Competition is evident, for example, when a male bowerbird has his bower dismantled and decorations stolen by another male (Pruett-Jones & Pruett-Jones 1994). Indeed, it may be that supposedly female-selected ornaments, such as bowers and display arena position, which are the focus of intense male competition, have evolved to allow females readily to identify the most competitive males (Hovi *et al.* 1995, Höglund *et al.* 1997). Not all males, however, compete so overtly for females. During copulation, males may use a number of cryptic behavioural and physiological adaptations to ensure that their sperm outcompetes their rivals and successfully fertilises the female.

Aspects of avian morphology, life history and sexual selection have all played a major role in the evolution and nature of contests in birds. Indeed, birds often use a variety of seemingly arbitrary traits, such as postural threat displays (reviewed in Hurd & Enquist 2001, Searcy & Nowicki 2005), elaborate and complex song (reviewed in Searcy & Nowicki 2000, Catchpole & Slater 2008) and colour signals (reviewed in Hurd & Enquist 2001, Searcy & Nowicki 2005, Senar 2006) to signal their intent or fighting ability. Because these mechanisms of avian competition have recently been reviewed from a variety of perspectives, this chapter will focus predominantly on the diversity of contests that occur throughout avian life (Box 13.1), referring to specific mechanisms where appropriate. This is by

Box 13.1 A classification of contexts for avian contests

Sibling rivalry occurs at hatching when multiple birds are reared together.

- **Precocial young.** Independent offspring compete with both siblings and carers for access to food (often limited).
- **Semi-precocial and altricial young.** Dependent or semi-dependent offspring compete with siblings for allocation of food delivered by carers (often limited).

Social rivalry occurs when multiple birds permanently or temporarily live together.

- **Cooperative breeders.** Related or unrelated birds live in small stable or unstable groups that compete primarily for breeding opportunities; dominant birds usually breed and subordinate birds help raise the dominants' offspring.
- **Social groups.** Typically unrelated birds live in largely stable groups where individuals compete within the group for access to all resources, including food, shelter and mates; dominant birds within the groups usually gain priority to resources.
- **Overwintering flocks.** Related and unrelated birds live in large unstable flocks (tens, hundreds or thousands of birds) that compete for access to non-reproductive resources, including food, shelter and roost position; dominant birds usually gain priority to resources.

Sexual rivalry occurs when birds compete for access to breeding resources and mates.

- **Pre-copulatory competition.** Birds compete directly for access to mates or indirectly to monopolise resources important for breeding or attracting mates; dominant birds usually gain access and/or high-quality mates or resources.
- **Post-copulatory competition.** Males compete directly by guarding females or indirectly through sperm competition (inside the female reproductive tract) for successful fertilisation.

no means an exhaustive review of the vast number of avian studies. Instead, the aim is to highlight some of the recent developments in a variety of contexts. I begin at the nest.

13.3 Competition between siblings

In birds, parental care is ubiquitous. Although precociality is the ancestral state (i.e. semi to fully independent at hatching), in most modern birds, hatchlings are completely dependent on their parents for their survival (i.e. altricial) (Starck & Ricklefs 1998). In this situation, sibling rivalry among nest-mates is intense because the effort put into competition directly affects food allocation from their parents (Mock & Parker 1997). Among some sea birds and predatory birds, this competition can take an extreme form. In the Australian kookaburra (*Dacelo novaeguineae*), for example, larger nestlings attack and may kill their siblings using a specialised beak hook (a weapon possessed only by young nestlings: Legge 2000, Figure 13.1). Although such elaborate morphological adaptations are very rare in birds, dominant individuals (typically older nestling) may stab a sibling to death (using sharp beaks) or push it out of the nest, thereby monopolising the food supplied by parents (Mock *et al.* 1990).

Even in species in which siblicide is absent (e.g. Passerines), the allocation of food to a nest of chicks is a classic example of scramble competition. The outcome of such competition is generally decided by the relative gape (mouth) size and colour, the extent to which they can reach towards the parents, the rate and intensity of begging noise, and by being in the best position in the nest (Kilner & Johnstone 1997, Wright & Leonard 2002, Soler & Avilés 2010), all of which will generally favour larger or faster-growing chicks. The factor, however, most likely to directly influence the intensity of sibling competition appears to be the level of relatedness within the brood. As relatedness decreases (e.g. due to extra-pair paternity or brood parasitism), selfishness within the brood and, hence sibling competition, is predicted to intensify (Ricklefs 1993, Briskie *et al.* 1994). Indeed, in a comparative study across 40 species, Royle *et al.* (1999) found that chicks grew faster in species where intra-brood relatedness was likely to be lower. Such intense intra-brood competition may also affect the relative rate of prenatal development (i.e. incubation of eggs: Lloyd & Martin 2003). These studies highlight the potential evolutionary consequences of contest competition that occurs at a very early stage of avian life.

Parents can also play a role in promoting or eliminating conflicts between their offspring. Sibling competition has obvious potential fitness costs to parents, such as brood reduction (e.g. siblicide) or wasteful investment (e.g. the energy cost of scrambling). However, sibling competition may maximise parental fitness in unpredictable conditions, as it

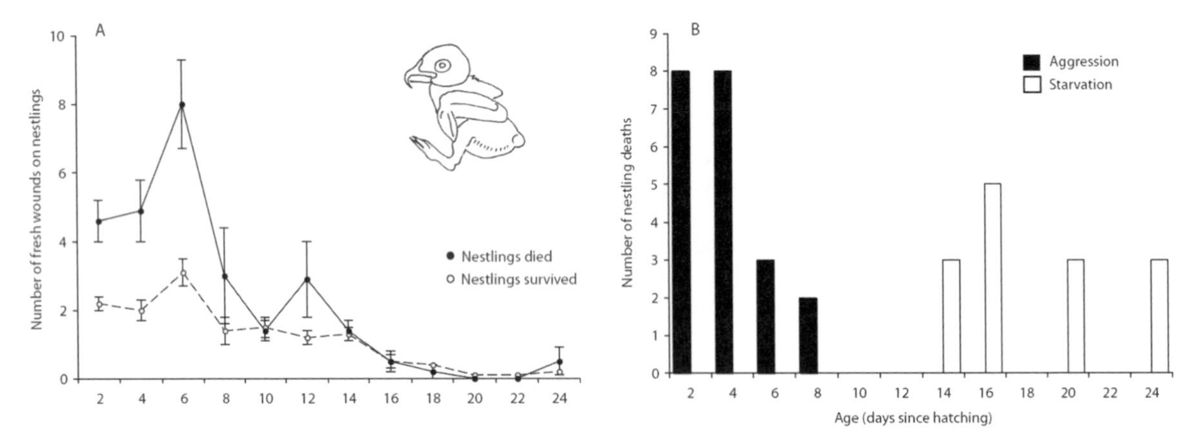

Figure 13.1 A rare example of morphological specialisation for sibling rivalry: Kookaburra (*Dacelo novaeguineae*) nestlings attack each other using a hook on the upper beak. (**A**) Nestlings that eventually died had more wounds than nestlings that survived to fledging. (**B**) Fresh wounds were more apparent early in the early nestling period (before 8 days old), and the direct effects of this aggression were primarily responsible for early nestling deaths. In contrast, later nestling deaths (> 14 days) were largely due to starvation (siblings monopolise food from parents without obvious signs of aggression). (From Legge 2000, with permission.)

provides a solution to the problem of which chick to feed if resources are limited (Bonabeau *et al.* 1998). Indeed, in many species of bird, parents preferentially feed the largest, most competitive chicks in a brood (Mock & Parker 1997); this characteristic is often exploited by brood parasites. Parents may further influence the outcome of sibling rivalry by manipulating competitive disparities between nestlings, for example by adjusting clutch size, egg size, hormone content (e.g. testosterone), nestling period and hatching asynchrony (Magrath 1992, Osorno & Drummond 1995, Schwabl *et al.* 1997, Gonzalez-Voyer *et al.* 2007). Alternatively, parents may use similar mechanisms to eliminate, rather than promote, competitive asymmetries within their broods. In canaries (*Serinus canaria*), for example, the last egg laid in the clutch typically receives twice as much testosterone as the first, which may accelerate chick development and allow later-hatched chicks to compete more effectively for food (Schwabl *et al.* 1997). Parents can therefore both directly and indirectly affect the dynamics of sibling competition within their broods.

It is likely that disparities in competitive abilities, learned behaviours and parental investment (whether direct or indirect) will not only affect offspring survival (e.g. body size), but may also set the precedence for further contests as adults. In the blue-footed boobie (*Sula nebouxii*), for example, chicks hatch asynchronously in the nest (Drummond & Osorno 1992). The older chick frequently pecks and jostles the younger for the first few weeks, even in the absence of food, and then remains dominant by less frequent but daily aggression. The younger chick responds to aggression with submissive gestures and rarely challenges the older one; even if it ultimately grows larger than the older chick. These initial aggressive effects may then continue through the life of these birds, with individuals remaining aggressive or submissive as adults (Drummond & Osorno 1992).

13.4 Competition in social groups

For many birds, communal living is adaptive because of the benefits to be gained through a number of social factors, including enhanced vigilance and protection from predators, increased foraging efficiency and limits on suitable breeding habitats (Jullien & Clobert 2000, Sridhar *et al.* 2009, Beauchamp 2010). However, intense competition between individuals within groups or flocks for critical resources may disrupt the social benefits gained from group living. To mitigate this, it is advantageous for birds to be able to assess dominance status rapidly while avoiding the risk of accidental injury, excessive energy consumption or elevated predation risk, which are associated with prolonged and intense fighting. Dominant birds receive obvious benefits from securing more or better resources (e.g. food and mates), but subordinates may also benefit from group living through increased foraging efficiency and improved vigilance. In addition, molecular-genetic studies have revealed that subordinate birds are able to access some alternative reproductive opportunities through extra-pair paternity and intraspecific brood parasitism (e.g. Richardson *et al.* 2002, Double & Cockburn 2003). Both dominants and subordinates also gain benefits from group augmentation, where a larger social group may hold better resources than a small one (Beauchamp 1998).

13.4.1 Competition in dominance hierarchies

Social stability in small groups of known (familiar) individuals may be established through a dominance hierarchy (Chapters 3 and 14), where each individual knows its relative status. Typically, dominance rank is initially determined through physical fighting or aggressive contests, and then a 'pecking order' is subsequently maintained or modified through more subtle daily interactions, such as displacements from feeding or resting areas, agonistic displays, or submissive behaviour (Crook & Butterfield 1970, Kalinoski 1975). Since Schjelderup-Ebbe (1922) first described intraspecific dominance relationships in flocks of domestic hens (*Gallus gallus domesticus*), this idea of a 'pecking order' has been the subject of much theoretical debate, both as a concept (reviewed in Drews 1993) and as a measurable individual attribute (e.g. Tufto *et al.* 1998, Jameson *et al.* 1999, de Vries & Appleby 2000).

In most group-living birds, the dominance hierarchy can be assessed by examining social behaviours. Alpha status tends to be readily discernible and stable because most alpha males exhibit (often extreme) aggression. In contrast, subordinate individuals typically avoid the dominant bird, although the relative rank of subordinates is not always clear or stable (Bayly *et al.* 2006). One individual (dominant) may control all of the others (i.e. despotically), or control may be exerted via a more linear hierarchy (Martin & Bateson 1993), such as those proposed for many fowl: each individual has a rank in the hierarchy (alpha, beta, gamma, etc.). The pecking order is also often dynamic and may be reordered, with a previously dominant bird being attacked if, for example, it gets old or sick.

The relative positions of individuals in a hierarchy are likely to incur differential costs. Primarily, dominance status may be costly because the competitive interactions with other members of the group can induce more stress (Goymann & Wingfield 2004). It is unclear, however, whether dominant or subordinate individuals suffer more from stress-related costs. Monopolising resources and enforcing dominance can be demanding for the dominant (alpha) bird, while being excluded from food sources or being harassed by dominants can also be stressful for subordinates (Hegner & Wingfield 1987, Poisbleau *et al.* 2005). Other costs linked to body size (Lindström *et al.* 2005), energy expenditure (Senar 1990, Bryant & Newton 1994) and immune responses or disease resistance (Apanius 1998, Zuk & Johnsen 2000, Lindström 2004) have also been suggested to affect dominance status, although there is surprisingly little direct experimental evidence to show that dominant (or alpha) birds do indeed incur higher costs than birds of lower rankings.

If the costly effort to achieve high status in a dominance hierarchy is adaptive, then high-ranking males should gain better resources or reproductive benefits. In times of extreme food shortage, for example, higher ranking birds may be more likely to survive than lower ranking ones (Koivula *et al.* 1996, Lahti 1998, Schubert *et al.* 2008). Another dominance-related benefit for many social groups is that the most dominant bird secures most of the matings (e.g. Collias *et al.* 1994, Pizzari & Birkhead 2000, Johnsen *et al.* 2001). Male boat-tailed grackles (*Quiscalus major*), for example, have a linear dominance hierarchy where males compete for relative position in a mating queue: alpha males at the top of the queue (notably older males) gain access to the colonies of females (Post 1992, Poston 1997).

The dynamics, and relative costs and benefits inherent in a dominance hierarchy, are perhaps best illustrated by the red junglefowl (*Gallus gallus*), an ancestor of the domestic chicken. In this system, both males and females form linear (or near-linear) dominance hierarchies. Dominance among hens is generally enforced by each female continually attacking the hen below it in rank (Collias & Collias 1990, Forkman & Haskell 2004). Among males, fights for rank and territories are often extremely aggressive, with the dominant male then asserting his dominance by stereotyped wingflapping and crowing (Leonard & Horn 1995). Although dominance hierarchies are generally stable, birds do occasionally change status both within and between years (Collias & Collias 1996, Cornwallis & Birkhead 2006). The dominant males provide no parental care but do provide females with food and vigilance against predators (Wilson *et al.* 2008) and females copulate almost exclusively with the

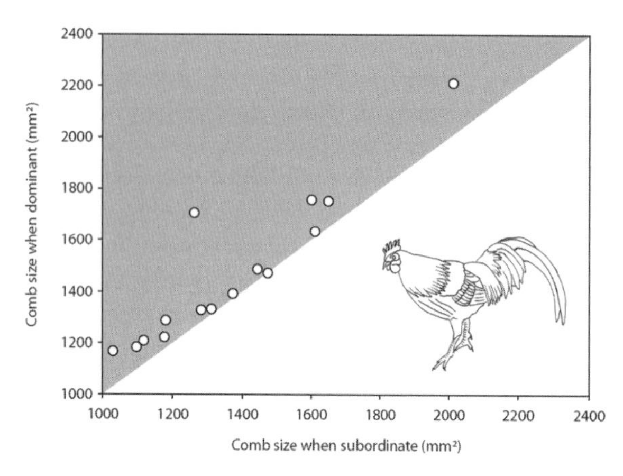

Figure 13.2 In male junglefowl (*Gallus gallus domesticus*) comb size is a plastic trait that changes when male status (dominant or subordinate) is experimentally manipulated. Points represent individual males. When individual males were dominant (*y*-axis) they had larger combs (mm^2) than when they were subordinate (*x*-axis). (From Cornwallis & Birkhead 2008, with permission.)

socially dominant male (Johnsen *et al.* 2001, Wilson *et al.* 2008). In a number of experimental studies, comb size has been consistently linked to dominance status (Ligon *et al.* 1990, Cornwallis & Birkhead 2008). The unique histological nature of the male's comb provides an honest indicator of the individual's physical condition, because comb size and colour are directly linked to current testosterone levels (Zuk *et al.* 1995), which in turn determines aggressiveness and fighting ability (Ligon *et al.* 1990). By experimentally reversing the dominance status of males (i.e. making dominant birds subordinate and vice versa), Cornwallis and Birkhead (2008) showed that comb size is a highly plastic trait, which is a consequence (rather than a predictor) of male status (Figure 13.2). Few systems have been studied in such depth, but overall it appears that social dominance hierarchies determine priority of access to resources, such as food, mates and roosting sites.

13.4.2 Competition with unfamiliar individuals and badges of status

Every winter, many passerines leave their breeding areas and form large unstable flocks of unfamiliar adults and juveniles (of both sexes). About one billion Palaeartic willow warblers (*Phylloscopus trochilus*), for example, migrate each year from Europe and Asia to warmer areas of Africa (Moreau 1972). In these large overwintering flocks, there are likely to be frequent encounters with unfamiliar rivals of unknown fighting ability. Therefore, if individuals that win escalated contests for food exhibited some type of badge reflecting their current fighting ability, this would allow potential fights to be settled without risking injury or wasting energy assessing the relative competitive ability of each and every opponent. The idea of a 'badge of status' was first proposed by Rohwer (1975, 1977), who demonstrated that the conspicuous variation in black head and throat patches of overwintering Harris's sparrows (*Zonotrichia leucophrys*), both within and between different age–sex classes, is linked to their relative fighting ability. Numerous correlative and experimental studies have subsequently confirmed that variation in plumage signals can be used to settle priority to resources (reviewed in Searcy & Nowicki 2005, Senar 2006).

Although the status-signalling hypothesis was initially invoked to resolve social competition over food or shelter in unstable winter flocks, this concept can be extrapolated to sexual conflicts among territorial birds (section 13.5.1), which may also benefit from displaying their status to the large number of unknown intruders (i.e. non-territorial birds) that frequently intrude onto occupied territories (e.g. Peek 1972, Røskaft & Rohwer 1987, Evans & Hatchwell 1992, Pärt & Qvarnström 1997, Pryke *et al.* 2001, 2002, Pryke & Andersson 2003*a*).

A number of, perhaps unintentional, perceptions about badges have arisen together with the increasing interest in status signalling. The first concerns whether badges should function within or between age and gender classes. For example, when status badges vary mainly by age and sex, some authors suggest that this challenges the status-signalling hypothesis (e.g. Whitfield 1987, Senar *et al.* 1993). However, if age and sex are the primary determinants of fighting ability (i.e. juveniles and/or females are inferior fighters), it seems that age and sex are what a status signal should provide information about (Searcy & Nowicki 2005). Dominance status is unlikely to be undermined unless individuals are able to assess age and sex independently of plumage variability (i.e. ignore plumage signals when assessing the competitive ability of opponents). Experiments using birds with delayed plumage maturation (i.e. age signalled through diagnostic colour patterns) typically show that birds primarily pay attention to differences in plumage (e.g. Muehter *et al.* 1997, VanderWerf & Freed 2003). Thus, badges advertising youth, and perhaps sex, may have evolved to reduce aggression in

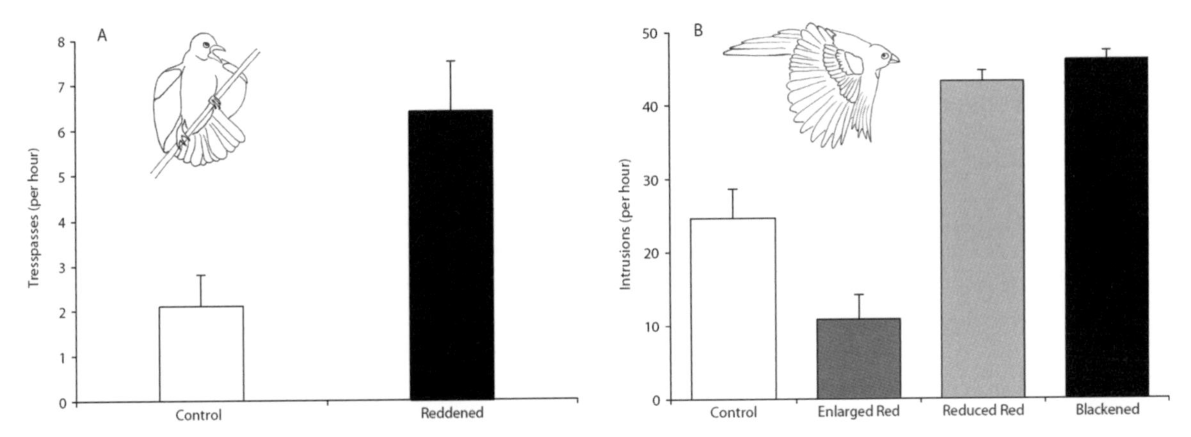

Figure 13.3 In the ecologically similar North American red-winged blackbirds (*Agelaius phoeniceus*) and the African red-shouldered widowbirds (*Euplectes axillaris*), both species display their yellow–red epaulets (on the wrist of the wing) during male threat displays and contests. (**A**) After epaulet manipulations, male red-winged blackbirds with experimentally reddened epaulets received more aggressive intrusions (mean ± SD) than males with control (sham-painted) epaulets. Therefore, males that exaggerate their social status provoke increased aggression from conspecifics, which is thought to prevent cheating in this system. (**B**) In contrast, male red-shouldered widowbirds with experimentally enlarged and reddened epaulets received fewer aggressive intrusions (mean ± SD) than males with experimentally reduced or orange epaulets. The contrasts in signal function between these two species probably relate to the coverable nature of the red-winged blackbird epaulet (i.e. black over-wing coverts can cover the red epaulet), which allows males to signal subordinance, whereas red-shouldered widowbirds are unable to cover their red epaulets; instead, the red epaulet functions as a signal of male dominance status (i.e. males with larger signals are rarely challenged). (**A**) From Yasukawa *et al.* (2009), with permission, (**B**) adapted from Pryke & Anderson (2003*a*), with permission.

otherwise monomorphic species in which there is limited information about age and sex.

Another common misconception is that badges should be melanin-based (i.e. brown and black). This generalisation is perhaps because initial studies specifically focused on the variation in the size or extent of melanin-based traits, presumably because black and brown plumage traits are particularly common in European and North American passerines (where most of the initial studies were conducted), and because melanin traits were presumed to be cheap to produce. This has probably been further exacerbated by the link between testosterone, melanin badge size and dominance (Bókony *et al.* 2008), as well as the divergent interests of avian mate choice studies focusing almost exclusively on variation in yellow–red carotenoid-based colouration (Hill & McGraw 2006). Nevertheless, together with recent technological advances in field spectrometry (i.e. objective quantification of colour variation), there is growing evidence that variation in carotenoid-based traits can play an important role in status signalling (e.g. Røskaft & Rohwer 1987, Evans & Hatchwell 1992, Pryke & Andersson 2003*a*, Crowley & Magrath 2004, Pryke & Griffith 2006, Pryke 2007, Figure 13.3). Recent studies have also suggested a status signalling function for variation in other colour types, such as ultraviolet and blue (Keyser & Hill 2000, Alonso-Alvarez *et al.* 2004, Siefferman & Hill 2005) and white structural plumage (Pärt & Qvarnström 1997, Dale *et al.* 1999), as well as haemoglobin-dependent colouration of bare (unfeathered) parts (Bamford *et al.* 2010). Overall, in the same way as females assess ornamental traits (Griffith & Pryke 2006), it seems likely that receivers will pay attention to the variation in a contest-related trait that reliably signals fighting ability, irrespective of the trait-type.

Indeed, as highlighted by theory (Maynard Smith & Harper 1988, Owens & Hartley 1991, Johnstone & Norris 1993), one of the biggest challenges to empirical studies is identifying what mechanism(s) prevent subordinates from developing large badges. By experimentally manipulating plumage traits, artificial 'cheaters' have been successfully produced in a number of species. However, for badges to aid in winning aggressive encounters in natural situations they need to signal an individual's fighting ability reliably. Studies investigating the reliability of badges currently fall into two categories. The first proposes that dominants are not always fitter than subordinates (i.e. that winning encounters is not always advantageous), perhaps because dominant birds have higher metabolic rates

(Røskaft *et al.* 1986, Hogstad 1987, but see Senar *et al.* 2000) or suffer high immunosuppression and stress responses (Pryke 2007). The second category suggests that exaggerated badges impose higher costs on individuals of low quality (i.e. reliability is maintained by signal costs) because they elicit higher retaliation costs (Rohwer 1977) or impose direct production or health-related costs on the signaller (Bókony *et al.* 2008). Given the large interest in status signalling in birds, it is perhaps surprising that few, if any, direct costs have yet to be unambiguously identified (reviewed in Searcy & Nowicki 2005, Senar 2006).

13.5 Competition for breeding resources

During the breeding season, many birds attempt to monopolise mates directly, or indirectly secure the resources that are important for attracting mates, such as food, nest sites and display arenas. As discussed above (section 13.2), intra-sexual competition is generally more pronounced among males. Males may directly compete for access to sexually receptive females, for example, by controlling female harems (Post 1992, Poston 1997), although such female-defence polygyny is very rare in birds (Ligon 1999). In fact, it appears that relatively few birds are able to directly access or to copulate with females without their consent. This is probably a consequence of choosy female birds (Griffith & Pryke 2006) and because most males are unable to overpower females physically. Although forced copulations have been observed in some species (e.g. Dunn *et al.* 1999), the general lack of an intromittant organ, the relatively weak sexual size dimorphism, and lack of gripping forelimbs, are likely to limit the effectiveness of forced copulations by males: most birds mate by touching their cloacae (vents) for a few seconds for males to transfer sperm to the female. Thus, slight female movement can disturb the close cloacal contact necessary for successful sperm transfer.

Most males, therefore, compete amongst themselves either to attract females or for access to limited breeding resources. Competition is usually thought to be most intense in polygynous and lek-mating species, due to the much greater variation among males in obtaining matings (Ligon 1999). However, the widespread prevalence of extra-pair copulations in most socially monogamous species may dramatically increase inter-male variation (Whittingham & Dunn 2005). This can be further inflated by differences in male competition for access to important breeding resources. For example, in the monogamous green woodhoopoe (*Phoeniculus purpureus*), variation in territory quality directly determines variance in lifetime reproductive success, with often less than 20% of males ever successfully breeding (Ligon & Ligon 1989).

13.5.1 Competition for territories

One of the most valued resources to a female is territorial space, whether it is a display arena to view males (e.g. lek), a nest site or extended space used for nesting and feeding. Indeed, for some species, it is the quality of the territory, rather than the quality of male territory-holder, that attracts the female (e.g. Alatalo *et al.* 1986). As a consequence of intense competition for high-quality territories, there often exists a considerable population of non-breeding 'floater' birds. Floaters frequently intrude onto occupied territories (Arcese 1987, Stutchbury 1991) and will rapidly fill vacant territories when residents are experimentally removed (Krebs 1982, Shutler & Weatherhead 1991, Pryke & Andersson 2003*b*). Territory owners are typically challenged by both neighbours (other territory owners) and floaters. Initial interactions with a new neighbour are likely to be intensely aggressive, but over time an owner develops a long-lasting and often stable association with its neighbours ('dear enemies', Chapters 2, 3 and 12). This is usually characterised by repeated low-level encounters, during which the owner signals its presence to neighbours (Stamps & Krishnan 1997). Floaters, on the other hand, are more of a constant threat because they are seeking to monopolise the exact resources that the owner is defending: the territory and females.

Such competition has led to extreme elaboration of territorial signalling behaviour between birds. Elaborate vocalisations (song) have been inextricably linked to territory establishment in birds, and many songbirds have made almost an art form of territorial song (Catchpole & Slater 2008). During competitive counter-calling, a process thought to be used to assess the relative stamina and aggressive motivation level of opponents, a bird's ability to learn fine features of their species-specific song gives them an opportunity to precisely match the details of their opponent's song (Searcy & Nowicki 2000). For example, song-type matching (i.e. match the specific notes of the song) is considered more aggressive than repertoire matching (i.e. match shared elements of the song), which in turn is more aggressive than non-matching, and birds may move between these to reveal their aggressive intentions (e.g. Beecher *et al.* 1996, Burt *et al.* 2001). Overall, long-term neighbours tend to engage in less aggressive counter-singing at a greater distance, while aggressive song-matching and escalated encounters are typically directed towards unknown floaters (Krebs *et al.* 1981, Beecher *et al.* 2000, Burt *et al.* 2001). Nevertheless, the often extensive song displays between neighbours may occasionally be used to challenge them, test their motivation and perhaps verify the exact position of the boundary (Catchpole & Slater 2008).

Typically, the territory owner wins conflicts with intruding floaters (Rohwer 1982), although it is often unclear why they do, especially as conflicts are often resolved rapidly (often within a matter of seconds) and without escalation to physical contact. One possibility is that the territory may be of greater value to the owner than intruder (the resource value asymmetry hypothesis, Chapter 2), perhaps because the resident has greater knowledge of the territory (Stamps 1987) and neighbours (Beletsky & Orians 1987), and therefore invests more in territorial contests (Krebs 1982). In European robins (*Erithacus rubecula*), for example, the relative time in tenure of (experimentally induced) territory replacements predicts the outcome of territorial contests when the original owner is released back onto the territory (Tobias 1997).

An alternative idea is that territory owners may have a competitive advantage over floaters, for example through body size and condition, that explains their consistent success in contests and fights (i.e. resource holding power or resource holding potential, RHP, Chapter 2). The superior RHP of owners may often be signalled by variation in conspicuous traits linked to their fighting ability. Variation in plumage, for example, can be used to signal relative status (section 13.4.2), and may also be used to advertise intrinsic qualities or fighting abilities of the territory holder (Pryke & Andersson 2003*a*,*b*, Figure 13.3). Variation in bird song has also been correlated with aspects of RHP (e.g. territory size, Krebs *et al.* 1978, Yasukawa 1981). However, the logistical problems of experimentally testing the function of specific, and often dynamic, features of aggressive song exchange (e.g. song-matching or switching) has left its direct role as an aggressive signal in territorial defence unresolved (reviewed in Catchpole & Slater 2008, Searcy & Beecher 2009). Furthermore, despite recognition that song can incur a number of condition-dependent costs (reviewed in Gil & Gahr 2002), studies are yet to directly test the potential costs or intrinsic qualities of aggressive songsters (e.g. song-matching and switching), and thus variation in bird song has mostly been associated with conveying information about ownership and aggression (e.g. Krebs *et al.* 1978, Searcy *et al.* 2006, Ballentine *et al.* 2008).

Overall, a number of studies have shown that competition for territories is intense and that territory owners may have superior competitive abilities (intrinsically and/or due to the benefits of ownership) or increased motivation for defending these important resources. However, the relative importance of these ideas in determining territorial status (i.e. owner–floater asymmetries) is often unclear. For example, results from the well-studied red-winged blackbirds (*Agelaius pheoniceus*) fail to show any consistent competitive differences between resident and floater males. The red epaulet (at the wrist of the wing) is thought to function as a badge of status in territorial defence, as territory owners display their red epaulets in response to male intruders, but trespassing and new establishing males keep their epaulets covered (Searcy & Yasukawa 1995). Although experimental studies have shown that variation in the size and colour of the epaulets displayed by territorial males affects trespassing rates and aggressive responses of intruders (Peek 1972, Røskaft & Rohwer 1987, Yasukawa *et al.* 2009, Figure 13.3), males with experimentally reduced epaulets can still defend their territories (Westneat 2006, Yasukawa *et al.* 2010), suggesting that territorial males have greater RHP. Other studies propose that occupancy time may be more important than variation in plumage traits to create perceived value asymmetries (Beletsky & Orians 1987, Shutler & Weatherhead 1991), while some suggest that arbitrary ownership conventions best explain territorial status (Shutler & Weatherhead 1992). Potentially, multiple effects may interact together, perhaps sequentially or dynamically, ultimately to determine territory ownership.

13.5.2 Competition among females

Intra-sexual competition tends to be regarded as a male domain (section 13.2) and female competition has consequently received relatively little attention. Female competition for mates may, however, be particularly widespread and intense. The vast majority of birds are socially monogamous with shared parental care, and a good-quality male with a high-quality territory (or breeding site) is a highly important resource to females (perhaps as important as male competition). Indeed, there is growing evidence that females may compete to monopolise male care, either by competing for high-quality males and territories (e.g. Dale *et al.* 1992), or by preventing secondary matings (e.g. Slagsvold *et al.* 1992, Sandell 1998). Recent studies have also revealed that female competition may be more common than initially assumed, with female birds competing for broadly the same functions as males do (reviewed in Langmore 2000, Amundsen & Pärn 2006). One difference between the sexes, however, is that males typically compete for access to females per se (whether social partners or extra-pair mates), while females generally compete with each other for male partners that are good parents and that possess good territories.

13.6 Post-copulatory competition

For most birds, competition for access to mates does not stop at copulation. Over the last decade, an increasing number of molecular studies has reported the incidence of extra-pair copulations and extra-pair paternity in a wide diversity of birds (Hasselquist & Sherman 2001, Griffith *et al.* 2002), emphasising the importance of post-copulatory competition (i.e. sperm competition). Indeed, there is increasing empirical evidence that sperm competition is a powerful evolutionary force, driving a large number of behavioural, morphological and physiological adaptations (Wigby & Chapman 2004, Pizzari & Parker 2009).

In many passerine birds, for example, males follow their partner very closely during the fertile period, guarding against the possibility that other males will be able to solicit and copulate with her during this crucial time (e.g. Birkhead 2010). Because males trade-off the benefits of mate guarding with other costly behaviours, such as foraging, territory defence and seeking extra-pair copulations themselves, males may adjust their investment in mate guarding relative to the risk of cuckoldry (e.g. Komdeur 2001). For species in which mate guarding is simply unfeasible, perhaps because one member of the pair has to remain to guard the nest (e.g. colonial birds and raptors: Mougeot 2003), males may copulate frequently with females in an attempt to swamp any extra-pair sperm that females have been inseminated with when unguarded.

Sperm competition has also driven considerable variation in sperm quantity (e.g. relatively larger testes: Møller & Briskie 1995, Pitcher *et al.* 2005) and quality (e.g. sperm size, morphology and motility: Calhim *et al.* 2007, Immler *et al.* 2007) both between and within bird species. In the case of the red junglefowl, for example, subdominant males with few opportunities to copulate with females produce 'aggressive' ejaculates with more and faster swimming sperm (Froman *et al.* 2002, Cornwallis & Birkhead 2007, 2008), suggesting that they could have a fertilisation advantage over more dominant males that receive more frequent and better-timed insemination opportunities.

Because sperm competition occurs exclusively within the female reproductive tract, an interesting dynamic is the role that females play. Recent work suggests that the outcome of sperm competition may be largely determined by the level of genetic similarity between the female and competing males (Pizzari *et al.* 2004, Thuman & Griffith 2005, Brouwer *et al.* 2010, Pryke *et al.* 2010). An interesting possibility is that despite considerable behavioural and morphological adaptations by males to win contests to fertilise eggs, the female (or more specifically, physiological mechanisms within the female tract) may ultimately override male competition. The relative importance of this, however, remains to be directly tested.

13.7 Competition using alternative strategies

Dominant or territorial birds are generally expected to gain the greatest fitness benefits from their status (e.g. Ekman & Askenmo 1984, Lahti 1998). Not all subordinate individuals, however, may strive to be dominant (Rohwer 1982, Számadó 2000); individuals may instead adopt a subordinate strategy if it allows them to achieve equal success to dominants (Studd & Robertson 1985). One potential avenue is through delayed plumage maturation if the interactions provide mutually fitness-enhancing benefits for individuals in different age groups. In Lazuli's buntings (*Passerina amoena*), for example, dominant adult males show reduced aggression to dull yearling birds on neighbouring territories, which increases the reproductive success of both adult and yearling birds (Greene *et al.* 2000).

Other dominance-related strategies which are independent of age include alternative behaviours; here individuals in disfavoured mating roles (e.g. subordinates) compensate by utilising alternative behaviours or tactics to those employed by dominant birds. Such strategies can be conditional if birds are able to switch between different tactics. In superb fairy wrens (*Malurus cyaneus*), for example, subordinate males intercept females that are searching for dominant extra-group males, and can subsequently gain over one-fifth of all extra-group paternity (Double & Cockburn 2003). Such strategies may be further enforced by a physiological trade-off between relative dominance status and its associated costs, as has been argued for the lower sperm quality of dominant red junglefowl (Pizzari *et al.* 2007, Cornwallis & Birkhead 2008, section 13.6).

Strategies can also be genetically fixed and regulated by genetic polymorphisms. The Australian Gouldian finch (*Erythrura gouldiae*), for example, exhibits genetically determined behavioural and physiological strategies that are related to their distinctive plumage polymorphism (Pryke *et al.* 2007, 2012). Red-headed individuals (ca. 10–30% of wild populations) are intrinsically more aggressive (Pryke 2009), outcompeting the more common black-headed individuals (ca. 70–90%) for access to limited food (Pryke & Griffith 2006, Pryke 2007) and nest-sites (Pryke & Griffith 2009). However, despite defending higher-quality resources, when red-headed morphs became too common (i.e. reach high densities), they trade-off investment in parental effort with increased aggression, which severely compromises their fitness (i.e. hawk–dove game, Chapter 2; Pryke & Griffith 2009). Although genetic polymorphisms are rare in birds, a number of studies, including both white-crowned sparrows (*Zonotrichia leucophrys*: Tuttle 2003) and lek-breeding ruffs (*Philomachus pugnax*: Lank *et al.* 1995, Widemo 1998) have also shown dominance-related strategies between alternative colour morphs.

Overall, however, it remains to be shown whether these different strategies used by birds do indeed result in equal fitness benefits to dominants and subordinates (i.e. evolutionarily stable strategies), as predicted by theory (Maynard Smith 1982).

13.8 Future challenges

Over the past decades, growing interest in bird contests has prompted a proliferation of studies investigating their nature, outcome and mechanisms. A variety of seemingly arbitrary signals, behaviours and vocalisations can be used effectively to signal aggressive motivation, ownership and fighting ability, and to resolve escalated contests for valuable resources, such as food and mates. Recent advances in technology have also uncovered more cryptic ways in which birds may compete with each other to gain a fertilisation advantage. Clearly, contests and conflicts among birds are both widespread and intense.

Yet, despite the important role of avian contests in the ecology and evolutionary biology of birds, and the large number of studies on avian competition, there are still large gaps in our knowledge. One of the most pressing challenges is to determine directly, rather than assume, the fitness benefits that dominant males may (or may not) gain from their higher status. Few empirical studies have directly tested this, and even in well-studied systems, it is not always clear that dominant males are indeed more fit. For example, despite the importance of red epaulets in resolving territorial competition in red-winged blackbirds (section 13.5.1, Figure 13.3), males with larger and redder epaulets receive few, if any, fitness benefits (Westneat 2006, Yasukawa *et al.* 2010). In reality, taking into consideration the full range of costs and benefits to competing birds over their lifetimes, together with the recent molecular advances in paternity tests, it may be (at least in some systems) that subordinates achieve similar overall fitness benefits to those of dominants.

It is also important that studies begin to investigate the fitness effects of contest outcomes (e.g. dominance) throughout an individual's life. In providing a structured review of the different contexts in which contests occur, artificial divisions have been placed between different contest types. In reality, of course, these contexts may be highly correlated and interact with one another in complex and interesting ways. For example, contests between siblings in the nest are likely directly to affect not only their physiology and morphology (e.g. growth rates, body size), and in turn be affected by the relative dominance status of their carers, but learned behaviours during these contests are also likely to affect an individual's fighting ability (and fitness) later in life.

Lastly, together with an increasing recognition of the importance of behaviour in conservation (Caro 2007), understanding the nature and consequences of contests in birds may become increasingly important in a number of conservation strategies (Brazill-Boast *et al.* 2011). In the Seychelles magpie robin (*Copsychus sechellarum*), for example, intensive conservation effort (e.g. habitat restoration and supplemental feeding) has rapidly increased populations (~150 birds) from critically endangered (17–21 birds in 1988–1990) (Norris & McCulloch 2003), but recovery efforts have been seriously hampered by intense competition among birds for high-quality breeding habitat (Komdeur 1996, López-Sepulcre *et al.* 2009). Recent translocations of disruptive subordinates from occupied territories to empty habitat on nearby islands has reduced their adverse effect on breeders and allowed them to establish themselves as breeders (López-Sepulcre *et al.* 2009). Understanding the nature and outcome of contest competition is likely to become increasing important for conservation management interventions.

Acknowledgements

I thank Mark Biffa, Ian Hardy, Simon Griffith, Simone Immler, Bob Wong and two reviewers for helpful comments on this chapter. While this chapter was being written, I was supported by grants and a Fellowship from the Australian Research Council (DP0770889).

References

Alatalo RV, Lundberg A & Glynn C (1986) Female pied flycatchers choose territory quality and not male characteristics. *Nature*, 323, 152–153.

Alonso-Alvarez C, Doutrelant C & Sorci G (2004) Ultraviolet reflectance affects male–male interactions in the blue tit (*Parus caeruleus ultramarinus*). *Behavioral Ecology*, 15, 805–809.

Amundsen T & Pärn H (2006) Female coloration: review of functional and nonfunctional hypotheses. In: GE Hill & KJ McGraw (eds.) *Bird Coloration: Function and Evolution*, pp. 280–348. Cambridge, MA: Harvard University Press.

Apanius V (1998) Stress and immune defense. In: AP Møller, M Milinski & PJB Slater (eds.) *Stress and Behaviour. Advances in the Study of Behavior*, 27, 133–153.

Arcese P (1987) Age, intrusion pressure and defence against floaters by territorial male song sparrows. *Animal Behaviour*, 35, 773–784.

Badyaev AV, Whittingham LA & Hill GE (2001) The evolution of sexual size dimorphism in the house finch. III. Developmental basis. *Evolution*, 55, 176–189.

Ballentine B, Searcy WA & Nowicki S (2008) Reliable aggressive signalling in swamp sparrows. *Animal Behaviour*, 75, 693–703.

Bamford AJ, Monadjem A & Hardy ICW (2010) Associations of avian facial flushing and skin colouration with agonistic interaction outcomes. *Ethology*, 116, 1163–1170.

Bayly KL, Evans CS & Taylor A (2006) Measuring social structure: a comparison of eight dominance indices. *Behavioural Processes*, 73, 1–12.

Beauchamp G (1998) The effect of group size on mean food intake rate in birds. *Biological Reviews*, 73, 449–472.

Beauchamp G (2010) Relaxed predation risk reduces but does not eliminate sociality in birds. *Biology Letters*, 6, 472–274.

Beecher MD, Campbell SE, Stoddard PK, *et al.* (1996) Repertoire matching between neighbouring song sparrows. *Animal Behaviour*, 51, 917–923.

Beecher MD, Campbell SE, Burt JM, *et al.* (2000) Song type matching between neighbouring song sparrows. *Animal Behaviour*, 59, 21–27.

Beletsky LD & Orians GH (1987) Territoriality among male red-winged blackbirds. II. Removal experiments and site dominance. *Behavioural Ecology and Sociobiology*, 20, 339–349.

Birkhead TR (2010) Post-copulatory sexual selection and the zebra finch. *Emu*, 110, 189–198.

Bókony V, Garamszegi LZ, Hirschenhauser K, *et al.* (2008) Testosterone and melanin-based black plumage coloration: a comparative study. *Behavioral Ecology and Sociobiology*, 62, 1229–1238.

Bonabeau E, Deneubourg J-L & Theraulaz G (1998) Within-brood competition and the optimal partitioning of parental investment. *American Naturalist*, 152, 419–427.

Bortolotti GR, Wiebe KL & Iko WM (1991) Cannabilism of nestling American kestrels by parents and siblings. *Canadian Journal of Zoology*, 69, 1447–1451.

Brazill-Boast J, van Rooij E, Pryke SR & Griffith SC (2011) Interference from long-tailed finches constrains reproduction in the endangered Gouldian finch. *Journal of Animal Ecology*, 80, 39–48.

Breitwisch R (1989) Mortality patterns, sex ratios, and parental investment in monogamous birds. In: DM Power (ed.) *Current Ornithology, Vol. 6*, pp. 1–49. New York, NY: Plenum Press.

Briskie JV, Naugler CT & Leech SM (1994) Begging intensity of nestling birds varies with sibling relatedness. *Proceedings of the Royal Society of London B*, 258, 73–78.

Brouwer L, Barr I, Pol M van de, *et al.* (2010) MHC-dependent survival in a wild population: evidence for hidden genetic benefits gained through extra-pair fertilizations. *Molecular Ecology*, 19, 3444–3455.

Bryant DM & Newton AV (1994) Metabolic costs of dominance in dippers *Cinclus cinclus*. *Animal Behaviour*, 48, 445–447.

Bunce M, Worthy TH, Ford T, *et al.* (2003) Extreme reversed sexual size dimorphism in the extinct New Zealand moa *Dinornis*. *Nature*, 425, 172–175.

Burt JM, Campbell SE & Beecher MD (2001) Song type matching as threat: a test using interactive playback. *Animal Behaviour*, 62, 1163–1170.

Calhim S, Immler S & Birkhead TR (2007) Postcopulatory sexual selection decreases variation in sperm morphology. *PLoS ONE*, 2, e413.

Caro T (2007) Behavior and conservation: a bridge too far? *Trends in Ecology and Evolution*, 22, 394–400.

Catchpole CK & Slater PJB (2008) *Bird Song: Biological Themes and Variations*. Cambridge: Cambridge University Press.

Collias NE & Collias EC (1990) Social organization of a high density population of red junglefowl (*Gallus gallus*). In: *Acta XX Congressus Internationalis Ornithologici*, Christchurch, New Zealand, pp. 466–467.

Collias NE & Collias EC (1996) Social organization of a red junglefowl, *Gallus gallus*, population related to evolution theory. *Animal Behaviour*, 51, 1337–1354.

Collias NE, Collias EC & Jennrich RI (1994) Dominant red junglefowl (*Gallus gallus*) hens in an unconfined flock rear the most young over their lifetime. *Auk*, 111, 863–872.

Cornwallis CK & Birkhead TR (2006) Social status and availability of females determine patterns of sperm allocation in the fowl. *Evolution*, 60, 1486–1493.

Cornwallis CK & Birkhead TR (2007) Changes in sperm quality and numbers in response to experimental manipulation of male social status and female attractiveness. *American Naturalist*, 170, 758–770.

Cornwallis CK & Birkhead TR (2008) Plasticity in reproductive phenotypes reveals status-specific correlations between behavioral, morphological and physiological sexual traits. *Evolution*, 62, 1149–1161.

Crook JH & Butterfield PA (1970) Gender role in the social system of Quelea. In: JH Crook (ed.) *Social Behaviour in Birds and Mammals: Essays on the Social Ethology of Animals and Man*, pp. 211–248. London: Academic Press.

Crowley CE & Magrath RD (2004) Shields of offence: signalling competitive ability in the dusky moorhen, *Gallinula tenebrosa*. *Australian Journal of Zoology*, 52, 463–474.

Dale S, Rinden H & Slagsvold T (1992) Competition for a male restricts mate search of female pied flycatchers. *Behavioral Ecology and Sociobiology*, 30, 165–176.

Dale S, Slagsvold T, Lampe HM, *et al.* (1999) Population divergence in sexual ornaments: the patch of Norwegian pied flycatchers is small and unsexy. *Evolution*, 53, 1235–1246.

Double MC & Cockburn A (2003) Subordinate superb fairy-wrens (*Malurus cyaneus*) parasitize the reproductive success of attractive dominant males. *Proceedings of the Royal Society of London B*, 270, 379–384.

Drews C (1993) The concept and definition of dominance in animal behaviour. *Behaviour*, 125, 283–313.

Drummond H & Osorno JL (1992) Training siblings to be submissive losers: dominance between booby nestlings. *Animal Behaviour*, 44, 881–893.

Dunn P, Afton A, Gloutney M, *et al.* (1999) Forced copulation results in few extrapair fertilizations in Ross's and lesser snow geese. *Animal Behaviour*, 57, 1071–1081.

Ekman JB & Askenmo CEH (1984) Social rank and habitat use in willow tit groups. *Animal Behaviour*, 32, 508–514.

Evans MR & Hatchwell BJ (1992) An experimental study of male adornment in the scarlet-tufted malachite sunbird: I. The role of pectoral tufts in territorial defence. *Behavioral Ecology and Sociobiology*, 29, 413–419.

Forkman B & Haskell MJ (2004) The maintenance of stable dominance hierarchies and the pattern of aggression: support for the suppression hypothesis. *Ethology*, 110, 737–744.

Froman DP, Pizzari T, Feltmann AJ, *et al.* (2002) Sperm mobility: mechanisms of fertilising efficiency, genetic variation and phenotypic relationship with male status in the fowl, *Gallus g. domesticus*. *Proceedings of the Royal Society of London B*, 269, 607–612.

Gil D & Gahr M (2002) The honesty of bird song: multiple constraints for multiple traits. *Trends in Ecology and Evolution*, 17, 133–141.

Gonzalez-Voyer A, Székely T & Drummond H (2007) Why do some siblings attack each other? Comparative analysis of aggression in avian broods. *Evolution*, 61, 1946–1955.

Goymann W & Wingfield JC (2004) Allostatic load, social status and stress hormones: the costs of social status matter. *Animal Behaviour*, 67, 591–602.

Greene E, Lyon BE, Muehter VR, *et al.* (2000) Disruptive sexual selection for plumage coloration in a passerine bird. *Nature*, 407, 1000–1003.

Griffith SC & Pryke SR (2006) Benefits to females of assessing color displays. In: GE Hill & KJ McGraw (eds.) *Bird Coloration: Function and Evolution*. Cambridge, MA: Harvard University Press.

Griffith SC, Owens IPF & Thuman KA (2002) Extra-pair paternity in birds: a review of interspecific variation and adaptive function. *Molecular Ecology*, 11, 2195–2212.

Hasselquist D & Sherman PW (2001) Social mating systems and extrapair fertilisations in passerine birds. *Behavioral Ecology*, 12, 457–466.

Hegner RE & Wingfield JC (1987) Social status and circulating levels of hormones in flocks of house sparrows, *Passer domesticus*. *Ethology*, 76, 1–14.

Hill GE & McGraw KJ (2006) *Bird Coloration: Function and Evolution*. Cambridge, MA: Harvard University Press.

Höglund J, Johansson T & Pelabon C (1997) Behaviourally mediated sexual selection: characteristics of successful male black grouse. *Animal Behaviour*, 54, 255–264.

Hogstad O (1987) It is expensive to be dominant. *Auk*, 104, 333–336.

Hovi M, Alatalo RV & Siikamaki P (1995) Black grouse leks on ice: female mate sampling by incitation of male competition? *Behavioral Ecology and Sociobiology*, 37, 283–288.

Hurd PL & Enquist M (2001) Threat display in birds. *Canadian Journal of Zoology*, 79, 931–942.

Immler S, Saint-Jalme M, Lesobre L, *et al.* (2007) The evolution of sperm morphometry in pheasants. *Journal of Evolutionary Biology*, 20, 1008–1014.

Jameson KA, Appleby MC & Freeman LC (1999) Finding an appropriate order for a hierarchy based on probabilistic dominance. *Animal Behaviour*, 57, 991–998.

Johnsen TS, Zuk M & Fessler EA (2001) Social dominance, male behaviour and mating in mixed-sex flocks of red jungle fowl. *Behaviour*, 138, 1–18.

Johnstone RA & Norris K (1993) Badges of status and the cost of aggression. *Behavioral Ecology and Sociobiology*, 32, 127–134.

Jullien M & Clobert J (2000) The survival value of flocking in neotropical birds: reality or fiction? *Ecology*, 81, 3416–3430.

Kalinoski R (1975) Intra- and interspecific aggression in house finches and house sparrows. *Condor*, 77, 375–384.

Keyser A & Hill GE (2000) Structurally based plumage coloration is an honest signal of quality in male blue grosbeaks. *Behavioral Ecology*, 11, 202–209.

Kilner R & Johnstone RA (1997) Begging the question: are offspring solicitation behaviours signals of need? *Trends in Ecology and Evolution*, 12, 11–15.

Koivula K, Orell M & Rytkönen S (1996) Winter survival and breeding success of dominant and subordinate Willow Tits *Parus montanus*. *Ibis*, 138, 624–629.

Komdeur J (1996) Breeding of the Seychelles magpie robin *Copsychus sechellarum* and implications for its conservation. *Ibis*, 138, 485–498.

Komdeur J (2001) Mate guarding in the Seychelles warbler is energetically costly and adjusted to paternity risk. *Proceedings of the Royal Society of London B*, 268, 2103–2111.

Krebs JR (1982) Territorial defence in the great tit, *Parus major*: do residents always win? *Behavioral Ecology and Sociobiology*, 11, 185–194.

Krebs JR, Ashcroft R & Webber M (1978) Song repertoire and territory defence in the great tit. *Nature*, 271, 539–542.

Krebs JR, Ashcroft R & Orsdol K van (1981) Song matching in the great tit *Parus major* L. *Animal Behaviour*, 29, 918–923.

Lahti K (1998) Social dominance and survival in flocking passerine birds: a review with an emphasis on the Willow Tit *Parus montanus*. *Ornis Fennica*, 75, 1–17.

Langmore NE (2000) Why do female birds sing? In: Y Espmark, T Amundsen & G Rosenquist (eds.) *Animal Signals: Signalling and Signal Design in Animal Communication*, pp. 317–327. Trondheim: Tapir.

Lank DB, Smith CM, Hanotte O, *et al.* (1995) Genetic polymorphism for alternative mating behaviour in lekking male ruff *Philomachus pugnax*. *Nature*, 378, 411–415.

Legge S (2000) Siblicide in the cooperatively breeding laughing kookaburra (*Dacelo novaeguineae*). *Behavioral Ecology and Sociobiology*, 48, 293–302.

Leonard ML & Horn AG (1995) Crowing in relation to social status in roosters. *Animal Behaviour*, 49, 1293–1290.

Ligon JD (1999) *The Evolution of Avian Breeding Systems*. Oxford: Oxford University Press.

Ligon JD & Ligon SH (1989) Green woodhoopoes. In: I Newton (ed.) *Lifetime Reproduction in Birds*, pp. 219–232. London: Academic Press.

Ligon JD, Thornhill R, Zuk M, *et al.* (1990) Male–male competition, ornamentation and the role of testosterone in sexual selection in red jungle fowl. *Animal Behaviour*, 40, 367–373.

Lindström KM (2004) Social status in relation to Sindbis virus infection clearance in greenfinches. *Behavioral Ecology and Sociobiology*, 55, 236–241.

Lindström KM, Hasselquist D & Wikelski M (2005) House sparrows (*Passer domesticus*) adjust their social status position to their physiological costs. *Hormones and Behavior*, 48, 311–320.

Lloyd JE & Martin TE (2003) Sibling competition and the evolution of prenatal development rates. *Proceedings of the Royal Society of London B*, 270, 735–740.

López-Sepulcre A, Norris K & Kokko H (2009) Reproductive conflict delays the recovery of an endangered social species. *Journal of Animal Ecology*, 78, 219–225.

Magrath RD (1992) The effect of egg size on growth and survival of blackbirds: a field experiment. *Journal of Zoology*, 227, 639–653.

Martin P & Bateson P (1993) *Measuring Behaviour: An Introductory Guide*. Cambridge: Cambridge University Press.

Maynard Smith J (1982) *Evolution and the Theory of Games*. Cambridge: Cambridge University Press.

Maynard Smith J & Harper DGC (1988) The evolution of aggression: can selection generate variability? *Philosophical Transactions of the Royal Society B*, 319, 557–570.

Mench JA (2009) Behaviour of fowl and other domesticated birds. In: P Jensen (ed.) *The Ethology of Domesticated Animals*, pp. 121–136. Cambridge, MA: CABI.

Mock DW & Parker GA (1997) *The Evolution of Sibling Rivalry*. Oxford: Oxford University Press.

Mock DW, Drummond H & Stinson CH (1990) Avian siblicide. *American Scientist*, 78, 438–449.

Møller AP & Briskie JV (1995) Extra-pair paternity, sperm competition and the evolution of testis size in birds. *Behavioral Ecology and Sociobiology*, 36, 357–365.

Moreau RE (1972) *The Palaearctic–African Bird Migration System*. London: Academic Press.

Mougeot F (2003) Breeding density, cuckoldry risk and copulation behaviour during the fertile period in raptors: a comparative analysis. *Animal Behaviour*, 67, 1067–1076.

Muehter VR, Greene E & Ratcliffe L (1997) Delayed plumage maturation in Lazuli Buntings: tests of the female mimicry and status signalling hypotheses. *Behavioral Ecology and Sociobiology*, 41, 281–290.

Negro JJ, Donazar JA & Hiraldo F (1992) Kleptoparasitism and cannibalism in a colony of lesser kestrels (*Falco naumanni*). *Journal of Raptor Research*, 26, 225–228.

Norris K & McCulloch N (2003) Demographic models and the management of endangered species: case study of the critically endangered Seychelles magpie robin. *Journal of Applied Ecology*, 40, 890–899.

Osorno JL & Drummond H (1995) The function of hatching asynchrony in the blue footed booby. *Behavioral Ecology and Sociobiology*, 37, 265–273.

Owens IPF & Hartley IR (1991) 'Trojan sparrows': evolutionary consequences of dishonest invasion for the badges-of-status model. *American Naturalist*, 138, 1187–1205.

Pärt T & Qvarnström A (1997) Badge size in collared flycatchers predicts outcome of male competition over territories. *Animal Behaviour*, 54, 893–899.

Peek FW (1972) An experimental study of the territorial function of vocal and visual displays in the male red-winged blackbird (*Agelaius phoeniceus*). *Animal Behaviour*, 20, 112–178.

Pitcher TE, Dunn PO & Whittingham LA (2005) Sperm competition and the evolution of testes size in birds. *Journal of Evolutionary Biology*, 18, 557–567.

Pizzari T & Birkhead TR (2000) Female feral fowl eject sperm of subdominant males. *Nature*, 405, 787–789.

Pizzari T & Parker GA (2009) Sperm competition and sperm phenotype. In: TR Birkhead, DJ Hosken & S Pitnick (eds.) *Sperm Biology: An Evolutionary Perspective*, pp. 207–245. Burlington, MA: Academic Press.

Pizzari T, Løvlie H & Cornwallis CK (2004) Sex-specific, counteracting responses to inbreeding in a bird. *Proceedings of the Royal Society of London B*, 271, 2115–2121.

Pizzari T, Cornwallis CK & Froman DP (2007) Social competitiveness associated with rapid fluctuations in sperm quality in male fowl. *Proceedings of the Royal Society of London B*, 274, 853–860.

Poisbleau M, Fritz H, Guillon N, *et al.* (2005) Linear social dominance hierarchy and corticosterone responses in male mallards and pintails. *Hormones and Behavior*, 47, 485–492.

Post W (1992) Dominance and mating success in male boat-tailed grackles. *Animal Behaviour*, 44, 917–929.

Poston JP (1997) Dominance, access to colonies, and queues for mating opportunities by male boat-tailed grackles. *Behavioral Ecology and Sociobiology*, 41, 89–98.

Pruett-Jones S & Pruett-Jones M (1994) Sexual competition and courtship disruptions: why do male bowerbirds destroy each other's bowers? *Animal Behaviour*, 47, 607–620.

Pryke SR (2007) Fiery red heads: female dominance among head color morphs in the Gouldian finch. *Behavioral Ecology*, 18, 621–627.

Pryke SR (2009) Is red an innate or learned signal of aggression and intimidation?. *Animal Behaviour*, 78, 393–398.

Pryke SR & Andersson S (2003*a*) Carotenoid-based status signalling in red-shouldered widowbirds (*Euplectes axillaris*): epaulet size and redness affect captive and territorial competition. *Behavioral Ecology and Sociobiology*, 53, 393–401.

Pryke SR & Andersson S (2003*b*) Carotenoid-based epaulettes reveal male competitive ability: experiments with resident and floater red-shouldered widowbirds. *Animal Behaviour*, 66, 217–224.

Pryke SR, Astheimer LB, Buttemer WA & Griffith SC (2007) Frequency-dependent physiological tradeoffs between competing colour morphs. *Biology Letters*, 3, 494–497.

Pryke SR, Astheimer LB, Griffith SC & Buttemer WA (2012) Covariation in life-history traits: differential effects of diet on condition, hormones, behavior, and reproduction in genetic finch morphs. *American Naturalist*, 179, 375–390.

Pryke SR & Griffith SC (2006) Red dominates black: agonistic signalling among head morphs in the colour polymorphic Gouldian finch. *Proceedings of the Royal Society of London B*, 273, 949–957.

Pryke SR & Griffith SC (2009) Socially mediated trade-offs between aggression and parental effort in competing color morphs. *American Naturalist*, 174, 455–464.

Pryke SR, Lawes MJ & Andersson S (2001) Agonistic carotenoid signalling in male red-collared widowbirds: aggression related to the colour signal of both the territory owner and model intruder. *Animal Behaviour*, 62, 695–704.

Pryke SR, Andersson S, Lawes MJ, *et al.* (2002) Carotenoid status signaling in captive and wild red-collared widowbirds: independent effects of badge size and color. *Behavioral Ecology*, 13, 622–631.

Pryke SR, Rollins LA & Griffith SC (2010) Females use multiple mating and genetically loaded sperm competition to target compatible genes. *Science*, 329, 964–967.

Richardson DS, Komdeur J & Burke T (2002) Direct benefits and the evolution of female-biased cooperative breeding in Seychelles warblers. *Evolution*, 56, 2313–2321.

Ricklefs RE (1993) Sibling competition, hatching asynchrony, incubation period, and lifespan in altricial birds. *Current Ornithology*, 11, 199–276.

Rohwer SA (1975) The social significance of avian winter plumage variability. *Evolution*, 29, 593–610.

Rohwer SA (1977) Status signaling in Harris Sparrows: some experiments in deception. *Behaviour*, 61, 107–129.

Rohwer SA (1982) The evolution of reliable and unreliable badges of fighting ability. *American Zoologist*, 22, 531–546.

Royle NJ, Hartley IR, Owens IPF, *et al.* (1999). Sibling competition and the evolution of growth rates in birds. *Proceedings of the Royal Society of London B*, 266, 923–932.

Røskaft E & Rohwer S (1987) An experimental study of the red epaulettes and black body colour of male red-winged blackbirds. *Animal Behaviour*, 35, 1070–1077.

Røskaft E, Järvi T, Bakken M, *et al.* (1986) The relationship between social status and resting metabolic rate in great tits (*Parus major*) and pied flycatchers (*Ficedula hypoleuca*). *Animal Behaviour*, 34, 838–842.

Sandell MI (1998) Female aggression and the maintenance of monogamy: female behaviour predicts male mating status in European starlings. *Proceedings of the Royal Society of London B*, 265, 1307–1311.

Schjelderup-Ebbe T (1922) Observation on the social psychology of domestic fowls. *Zeitschrift für Vergleichende Physiologie*, 88, 225–252.

Schubert KA, Mennill DJ, Ramsay SM, *et al.* (2008) Between-year survival and rank transitions in male black-capped chickadees (*Poecile atricapillus*): a multistate modeling approach. *Auk* 125, 629–636.

Schwabl H, Mock DW & Gieg JA (1997) A hormonal mechanism for parental favouritism. *Nature*, 386, 231.

Searcy WA & Beecher MD (2009) Song as an aggressive signal in songbirds. *Animal Behaviour*, 78, 1281–1291.

Searcy WA & Nowicki S (2000) Intra- and intersexual selection on bird song. In: Y Espmark, T Amundsen & G Rosenquist (eds.) *Animal Signals: Signalling and Signal Design in Animal Communication*, pp. 301–315. Trondheim: Tapir Academic Press.

Searcy WA & Nowicki S (2005) *The Evolution of Animal Communication*. Princeton, NJ: Princeton University Press.

Searcy WA & Yasukawa K (1995) *Polygyny and Sexual Selection in Red-winged Blackbirds*. Princeton, NJ: Princeton University Press.

Searcy WA, Anderson RC & Nowicki S (2006) Bird song as a signal of aggressive intent. *Behavioral Ecology and Sociobiology*, 60, 234–241.

Senar JC (1990) Agonistic communication in social species: what is communicated? *Behaviour*, 112, 270–283.

Senar JC (2006) Bird coloration as an intrasexual signals of aggression and dominance. In: GE Hill & KJ McGraw (eds.) *Bird Coloration: Function and Evolution*, pp. 87–136. Cambridge, MA: Harvard University Press.

Senar JC, Camerino M, Copete JL, *et al.* (1993) Variation in black bib of the Eurasian siskin (*Carduelis spinus*) and its role as a reliable badge of dominance. *Auk*, 110, 924–927.

Senar JC, Polo V, Uribe F, *et al.* (2000) Status signalling, metabolic rate and body mass in the siskin: the cost of being a subordinate. *Animal Behaviour*, 59, 103–110.

Shutler D & Weatherhead PJ (1991) Owner and floater redwinged blackbirds: determinants of status. *Behavioral Ecology and Sociobiology*, 28, 235–241.

Shutler D & Weatherhead PJ (1992) Surplus territory contenders in male red-winged blackbirds: where are the desperados? *Behavioral Ecology and Sociobiology*, 31, 97–106.

Siefferman L & Hill GE (2005) UV-blue structural coloration and competition for nestboxes in male eastern bluebirds. *Animal Behaviour*, 69, 67–72.

Slagsvold T, Amundsen T, Dale S, *et al.* (1992) Female–female aggression explains polyterriorality in male pied flycatchers. *Animal Behaviour*, 43, 397–408.

Soler JJ & Avilés JM (2010) Sibling competition and conspicuousness of nestling gapes in altricial birds: a comparative study. *PLoS ONE*, 5, e10509.

Soma M, Hasegawa T & Okanoya K (2005) The evolution of song learning: a review from a biological perspective. *Cognitive Studies*, 12, 166–176.

Sridhar H, Beauchamp G & Shanker K (2009) Why do birds participate in mixed-species foraging flocks? A large-scale synthesis. *Animal Behaviour*, 78, 337–347.

Stamps JA (1987) The effect of familiarity with a neighbourhood on territory acquisition. *Behavioural Ecology and Sociobiology*, 21, 273–277.

Stamps JA & Krishnan VV (1997) Functions of fights in territory establishment. *American Naturalist*, 150, 393–405.

Starck JM & Ricklefs RE (1998) Patterns of development: the altricial-precocial spectrum. In: JM Starck & RE Ricklefs (eds.) *Avian Growth and Development*, pp. 59–88. New York, NY: Oxford University Press.

Studd MV & Robertson RJ (1985) Life span, competition, and delayed plumage maturation in male passerines: the breeding threshold hypothesis. *American Naturalist*, 126, 101–115.

Stutchbury BJ (1991) Floater behaviour and territory acquisition in male purple martins. *Animal Behaviour*, 42, 435–443.

Számadó SZ (2000) Cheating as mixed strategy in a simple model of aggressive communication. *Animal Behaviour*, 59, 221–230.

Székely T, Lislevand T & Figuerola J (2007) Sexual size dimorphism in birds. In: W Blanckenhorn, DJ Fairburn & T Székely (eds.) *Sex, Size and Gender Roles*, pp. 27–37. Oxford: Oxford University Press.

Thuman KA & Griffith SC (2005) Genetic similarity and the nonrandom distribution of paternity in a genetically highly polyandrous shorebird. *Animal Behaviour*, 69, 765–770.

Tobias J (1997) Asymmetric territorial contests in the European robin: the role of settlement costs. *Animal Behaviour*, 54, 9–21.

Trail PW (1985) Courtship disruption modifies mate choice in a lek-breeding bird. *Science*, 227, 778–780.

Tufto J, Solberg EJ & Ringsby TH (1998) Statistical models of transitive and intransitive dominance structures. *Animal Behaviour*, 55, 1489–1498.

Tuttle EM (2003) Alternative reproductive strategies in the white-throated sparrow. Behavioral and genetic evidence. *Behavioral Ecology*, 14, 425–432.

VanderWerf EA & Freed LA (2003) Elepaio subadult plumages reduce aggression through graded status-signaling, not mimicry. *Journal of Field Ornithology*, 74, 406–415.

Vries H de & Appleby MC (2000) Finding an appropriate order for a hierarchy: a comparison of the I&SI and the BBS methods. *Animal Behaviour*, 59, 239–245.

Westneat DF (2006) No evidence of current sexual selection on sexually dimorphic traits in a bird with high variance in mating success. *American Naturalist*, 167, E171–E189.

Whitfield DP (1987) Plumage variability, status signalling and individual recognition in avian flocks. *Trends in Ecology and Evolution*, 2, 13–18.

Whittingham LA & Dunn PO (2005) Effects of extra-pair and within-pair reproductive success on the opportunity for selection in birds. *Behavioral Ecology*, 16, 138–144.

Widemo F (1998) Alternative reproductive strategies in the ruff, *Philomachus pugnax*: a mixed ESS? *Animal Behaviour*, 56, 329–336.

Wigby S & Chapman T (2004) Sperm competition. *Current Biology*, 14, R100–R103.

Wilson DR, Bayly KL, Nelson XJ, *et al.* (2008) Alarm calling best predicts mating and reproductive success in ornamented male fowl, *Gallus gallus*. *Animal Behaviour*, 76, 543–554.

Wright J & Leonard ML (2002) *The Evolution of Nestling Begging: Competition, Cooperation and Communication*. Dordrecht: Kluwer Academic Press.

Yasukawa K (1981) Song and territory defense in the red-winged blackbird. *Auk*, 98, 185–187.

Yasukawa K, Enstrom DA, Parker GA, *et al.* (2009) Epaulet color and sexual selection in the red-winged blackbird: a field experiment. *Condor*, 111, 740–751.

Yasukawa K, Enstrom DA, Parker PG, *et al.* (2010) Male red-winged blackbirds with experimentally dulled epaulets experience no disadvantage in sexual selection. *Journal of Field Ornithology*, 81, 31–41.

Zuk M & Johnsen TS (2000) Social environment and immunity in the red jungle fowl. *Behavioral Ecology*, 11, 146–153.

Zuk M, Johnsen TS & Maclarty T (1995) Endocrine–immune interactions, ornaments and mate choice in red jungle fowl. *Proceedings of the Royal Society of London B*, 260, 205–210.

Chapter

14 Contest behaviour in ungulates

Dómhnall J. Jennings & Martin P. Gammell

14.1 Summary

This chapter reviews our current understanding of ungulate contest behaviour. Before proceeding to the meat and bones we offer a caveat: following Yeats' question as to whether we can separate the dancer from the dance,[1] we recognise that no single aggressive action employed by a contestant can be considered independent of all other actions. Nevertheless, for reasons of structure and economy we have presented an overview of the competitive process by portraying contests more as parts of their sum rather than vice versa. The reader should keep this in mind when considering the various sections on display behaviour, contest structure, assessment processes and opponent choice. Many ungulates vocalise and appear to show their body and weapon size to opponents during contests, and so we begin with a review of how ungulates communicate their competitive ability prior to physical confrontation. Following this, we consider factors that might affect how ungulates structure their fights: do body size, resource availability and the level of familiarity between opponents influence contest duration? What assessment processes might be driving the willingness of a competitor to engage in a fight and how might this influence contest structure and subsequent outcome? We conclude with a review of the literature concerning what factors might be involved in mediating the decision for one individual to escalate an interaction with a particular group member to fighting.

14.2 Introduction

Ungulates comprise approximately 257 species classified broadly within two different orders,[2] the even-toed (order Artiodactyla) and uneven-toed (order Perissodactyla) ungulates. They represent the majority of large herbivores and, with the exception of the Antarctic, are currently resident in all continental regions. The numerous and diverse members of these orders and their wide geographic distribution is reflected in a complex range of social systems that extend from monogamous pair bonds to a variety of large group polygamous breeding systems. Despite the range and complexity of ungulate societies, competition to secure or defend access to resources is a common feature. Much of what we know and understand about ungulate contest behaviour, however, derives from studies of captive group-living species, primarily deer. Solitary-living and monogamous pair-bonding species have provided far less detailed information.

While numerous studies have examined ungulate competitive behaviour from the rather broad perspective of whether a fight occurred in the context of resource access or defence, relatively little work has been carried out on other questions. Questions remain concerning how ungulates structure their contests, why individuals vary in their tendency to engage in competitive interactions, why levels of intensity vary, what form of assessment process is used and what decision rules are employed in determining contest duration and resolution. Our aim here is to shed some light on these issues.

Competitive interactions can be divided into those in which opponents aggressively engage with each other but do not make physical contact (non-contact interactions) and fighting that usually involves contestants locking horns or antlers and engaging in a vigorous pushing contest. Frequently these two types of interaction are observed within the same engagement; contests can potentially oscillate from non-contact interactions to fighting and back to non-contact interactions. Common types of non-contact interaction

Animal Contests, ed. I.C.W. Hardy and M. Briffa. Published by Cambridge University Press.

include vocalisation contests (e.g. Clutton-Brock & Albon 1979), displacement interactions, horn or antler displays (Alvarez 1993, Jennings *et al.* 2002) and parallel walks (e.g. Clutton-Brock & Albon 1979, Jennings *et al.* 2003). Similar to non-contact interactions, fights also contain a variety of distinct actions such as the jump clash, charge, slam, butt and push (e.g. Geist 1971, Clutton-Brock & Albon 1979, Mloszewski 1983, Estes 1991, Alvarez 1993).

Given the diverse range of competitive actions employed during interactions, behavioural sequences of contests tend to be complex affairs. Jennings *et al.* (2004) defined fighting as the part of a contest during which animals are physically in contact with each other, such as when deer lock antlers. On the other hand, contests were considered to be temporally extended versions of the interaction as a whole. Contests therefore include all the behavioural actions that individuals employ during an interaction; for example, in the case of certain deer species, a contest was defined as beginning when contestants engage in some form of display behaviour such as parallel walking or bouts of roaring which then lead to fighting between the contestants and ends when a winner is determined. We use these definitions here. Teasing out the significance of any one behavioural action can only be achieved within the context of the suite of actions employed, while taking into account potential confounding factors such as age, dominance rank and individual phenotypic characteristics. For example, vocal behaviour in ungulates is considered by many to be a display of quality with calling rates linked to body size, age and dominance. Furthermore, body size is itself considered a visual indicator of quality that can be advertised to potential rivals through lateral displays. Body size is, given the style of pushing and twisting contests in many species of ungulate, central to fighting. The same is true of weapon size; larger weapons, that are displayed to opponents and also employed against opponents during fights, are borne by physically larger and older individuals (weapons in other taxa are discussed in Chapters 5, 7 and 8). Older animals are likely to be bigger physically and to hold high dominance rank relative to younger individuals. These characteristics are, to a greater or lesser extent, important mediating factors in contests irrespective of whether the interaction involves fighting. With these factors in mind, we review the literature on vocal behaviour (section 14.3.1) and then on visual displays of the weapons (section 14.3.2) and the body (section 14.3.3). We then address fighting and focus particularly on fight structure (section 14.4). Finally, we consider factors that influence the frequency with which individuals competitively engage each other (section 14.5).

Figure 14.1 Fallow deer buck (*Dama dama*). ©iStockphoto.com/Damian Kuzdak. For colour version, see colour plate section.

14.3 Display behaviour

14.3.1 Vocalisations

In ungulates, male vocal behaviour is almost exclusively restricted to periods of sexual activity, such as the relatively short temporal window where the majority of matings occur: the annual rut (Figure 14.1). Males repeatedly emit loud calls that, because these periods are likely to result in higher than usual local densities of males and females, are presumably received by both sexes. This has led to the suggestion that vocalisations potentially have a dual function, acting as a signal of dominance to rivals and of fitness to potential mates. If we assume that males do in fact advertise their quality through vocal displays, then a basic requirement for such an investment is that they should be individually identifiable. This is particularly true within populations where rivals aggregate and the probability arises that several individuals may vocalise simultaneously. Acoustic analyses of ungulate vocalisations support this idea; at a basic level it has been shown in computational terms using sound analysis software that individuals can be reliably identified by their vocal signature (e.g. goat, *Capra hircus*: Ruiz-Miranda *et al.* 1994; fallow deer, *Dama dama*: Reby *et al.* 1998, Vannoni & McElligott 2009; sheep, *Ovis aries*: Searby & Jouventin 2003; elephant, *Loxdonta africana*: McComb et al. 2003; red deer, *Cervus elaphus*: Reby *et al.* 2006).

Consistent with this idea, female red deer are capable of discriminating between a familiar male and an alien male based on vocal signature alone (McComb 1991, Reby *et al.* 2001) and display a preference for the vocalisations of larger males (Charlton *et al.* 2007). Female elephants also display greater interest in rumble vocalisations by males in musth (a periodic condition that male elephants endure characterised by a considerable increase in testosterone and accompanied by high levels of aggression, Chapter 4) than in vocalisations by non-musth males (Poole 1999). Early descriptive studies highlighted that the tendency for males to vocalise was related to the dominance status of the producer (e.g. Geist 1971, Kitchen & Bromley 1974, Sinclair 1974). Dominant males are most likely to be vocal when females are present (Moore & Marchinton 1974, Carranza *et al.* 1990) and subordinate males are less likely to vocalise when approaching a dominant male (Bergerud 1974). Furthermore, roaring rate also increases when competitor males are present, suggesting that roaring also conveys information about individual quality (Clutton-Brock & Albon 1979, Carranza *et al.* 1990). Nevertheless, it remains to be established whether males in these studies were signalling quality, indicating ownership of a territory or merely announcing their presence in the vicinity. Male fallow deer may or may not vocalise in the presence of females, but they are generally vocal in the presence of rival males, leading to the suggestion that male vocal behaviour is directed predominantly at potential competitors as a possible threat signal (Bartoš *et al.* 2007, McElligott & Hayden 2001). Similarly, bison males may approach an opponent and initiate a bellowing contest to challenge a male that is tending females (Wolff 1998). When this occurs the tending male often bellows at the same time in an apparent attempt to drown out the challenger. Where the tending male fails to compete with a challenger in this manner, there is an increased probability of losing his female charges to the more vocal rival (Wolff 1998).

Perhaps the most complete account of vocal behaviour between competing male ungulates comes from the island of Rhum (Clutton-Brock & Albon 1979). Red deer stags engage in roaring contests with each other, a finding that suggests that males use these contests to assess the potential fighting ability of rivals. During contests, changes in the number of roars between males, as indexed by matched increases in the rate of vocalising, provided evidence that mature red stags actively monitor their opponents' behaviour (see also studies by Poole 1987, 1999, Komers *et al.* 1997). However, the data provided by Clutton-Brock and Albon (1979) indicate, at least on a theoretical level, that this idea of using vocal rates to advertise quality to rivals might be too simplistic. It has been pointed out that the escalating and de-escalating rates at which rivals vocalised would make it difficult for opponents accurately to assess opponent quality (Payne 1998). Rather, this distribution of vocalisations corresponds more to a process of self-assessment where contestants monitor their own rates of repetition and use this to determine whether to continue to fight or yield (Payne 1998, Briffa & Elwood 2009). Furthermore, although Komers *et al.* (1997) reported an increase in vocal rate of their target males, they did not report whether the observed increase resulted in the target surpassing the playback vocal rate (tentative evidence for a mutual-assessment process: Enquist *et al.* 1990), or whether the target male matched vocal rates with the taped recording (tentative evidence for self-assessment: Payne 1998). In addition to vocal rate, there is also variation in the vocal structure of individual calls within bouts of roaring; for example, it has been noted within a bout of roars that males appear to vary the loudness and pitch of calls (Reby & McComb 2003). Although this observation regarding call structure remains to be investigated properly with respect to assessment processes, it does support the idea that the roars of red deer stags are consistent with self- rather than mutual assessment. Determining which of these two opposing theoretical perspectives is correct is critical to our understanding of male competitive behaviour. It has been reported that even when contests escalate to fighting, males can remain vocal, suggesting that contestants persist in advertising their quality using loud vocalisations (Frädrich 1974, Bartoš *et al.* 2007). Fallow deer vocalise at a similar rate to red deer during contests and, furthermore, vocal rates appeared to be related only to opponent contest behaviour rather than based on the differences in contest behaviour (Jennings *et al.* 2012). The authors concluded that their data supported a form of 'opponent-only' assessment process rather than self- or mutual assessment (Arnott & Elwood 2009). Based on the studies presented above, it is clear that this important issue regarding how individuals gather and use information during contests needs considerable further work.

Where the opportunity to engage a rival physically is absent, it has been shown that male red deer are sensitive to changes in the acoustic structure of the vocalisations themselves. Reby *et al.* (2005) demonstrated that recipient males were more attentive and vocalised more when presented with calls that simulated a large (dominant) opponent than with calls that simulated a smaller (subordinate) opponent. Such findings indicate that size and possibly age are encoded within the structure of these calls (Reby & McComb 2003). Sensitivity to the vocalisations of potential rivals is by no means restricted to deer; for example, dominant horses are more likely to investigate the vocalisations of subordinate animals than those of other dominant animals (Rubenstein & Hack 1992). Furthermore, there were significant temporal differences in the duration and frequency of squeal vocalisations based on dominance rank in this species.

Vocal behaviour encodes information about sex, age, reproductive and dominance status and permits communication between members of the same species (reviewed by Andersson 1994) and perhaps even emotional state (e.g. Weary & Fraser 1995). From the studies mentioned above it is apparent that males are extremely vocal when sexually active; indeed, it has been suggested that the onset of vocal activity by mature males marks the beginning of the rutting season in many species. In the experimental studies we have considered (e.g. Komers *et al.* 1997, Poole 1999, Reby *et al.* 2005) the only information available to the males was the structure and rate of the calls to which they were exposed suggesting that vocalisations alone can be used to asses opponent quality. However, context can be critical and vocal behaviour studied in isolation of the context in which it occurs does not necessarily support the idea that competitors are attempting to communicate their quality with each other. Nevertheless, it is apparent that vocalisations can convey something about opponent quality and also that individuals are sensitive to this from an early age.

14.3.2 Visual displays: antlers and horns

Antlers and horns are bilateral structures usually borne by male ungulates, and in terms of general appearance, antlers tend to be more complex structures than horns (section 14.4). Moreover, the annual casting and regrowth of antlers imposes considerable costs on the individuals; the growth process has been described as osteoporotic in that the skeleton provides a substantial proportion of the minerals necessary for production (Muir *et al.* 1987, Bubenik 1998). In moose, for example, energy requirements increase by between 13 and 20% during antler growth (*Alces alces*: Moen & Pastor 1998) and investment in horn growth can account for up to 15% of body weight in male bighorn sheep (Blood *et al.* 1970). While rapid growth is associated with lower life expectancy (Loehr *et al.* 2007) it is also associated with greater reproductive success (Ciuti & Apollonio 2011). In young males, horn growth is apparently unaffected in larger individuals that may have surplus resources to invest in growth; however, as body mass decreases there are smaller incremental levels of horn growth and these smaller individuals appear to invest more in increasing body mass (Festa-Bianchet *et al.* 2004). A critical feature impacting on antler and horn growth, therefore, is resource availability (Moen *et al.* 1999, Mysterud *et al.* 2005, Landete-Castillejos *et al.* 2010).

Because only mature, high-quality individuals can afford the costs imposed by antler growth, and due to their complex structure, it has been argued that antlers, in particular, serve a secondary display function in addition to their use as weapons (Zahavi 1975). In line with this suggestion there is a positive effect of breeding system on antler length: as polygyny increases there is an increase in antler length (Plard *et al.* 2011). Antler and horn displays between competing individuals are conspicuous and highly stereotyped interactions. They have, therefore, received a considerable amount of attention by students of ungulate behaviour. An early review of their form and function described displays as a form of 'expressive behaviour' classifying them as being threat, dominance, courtship, submissive, space claim or excitement displays (Walther 1974). Taken

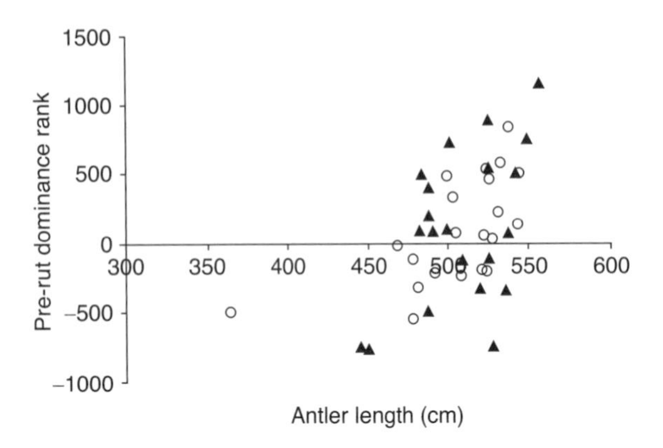

Figure 14.2 The relationship between dominance rank (calculated from pre-rut interactions, lower scores indicate higher dominance rank) and antler length in the fallow deer over two consecutive years. (Adapted from Jennings *et al.* 2006.)

together, Walther's classification of antler/horn displays is almost always invariably linked with the idea that males communicate their strength, dominance or aggressive intent through some aspect of the size of the structure (Figure 14.2). More recent work has shown that individuals that possessed larger antlers were also found to have larger testes and greater sperm velocity, thus, in addition to signalling fighting ability, antler size could also advertise physiological quality (Malo *et al.* 2005).

The relationship between body size and antler/horn size is complex, although as a general rule, large species tend to support large antlers (Andersson 1994). Consistent with this is the observation that ungulates alter the form of display in relation to weapon size; males of species with small antlers or horns tend to emphasise their body size, whereas species that possess relatively large horns or antlers more usually emphasise their head (Geist 1966*b*, 1971). Consequently, it has been suggested that lateral displays, similar to vocal displays, function as a means by which rival males can assess each other before they commit to physical combat (Clutton-Brock 1982). Outside of the rut, contests rarely escalate to fighting and most aggressive interactions are decided by low-level displacements. The idea that antlers serve as badges of quality during this period has received some support; for example, a study by Lincoln *et al.* (1970) indicated that dominance relations change from less well-structured relationships during antler growth (velvet stage) to a more structured linear hierarchy when the antler velvet has been shed. Furthermore, during the antler growth period dominance status at different periods of antler growth correlates with antler size at that particular time (Bartoš & Losos 1997).

Although males that suffer antler breakage or removal still actively compete with each other, they often decline in dominance rank because of a reduced ability to compete (Lincoln *et al.* 1970). The decline in rank observed by Lincoln and colleagues was apparently temporary; the following year, males re-established their original rank within the hierarchy. Antler loss does not, however, necessarily mean that individuals will decline in dominance rank. For example, when the antlers were removed from all members of a group of red stags, thus eliminating any potential advantage to those individuals in possession of large antlers, subordinate males soon began to challenge dominant males; however, these challenges were not successful and the hierarchy was maintained (Suttie 1980). A similar outcome was observed in a population of tule elk (*C. elaphus nannodes*: Johnson *et al.* 2007); although some males suffered fracturing and breakage of their antlers they retained their dominance rank, leading to the suggestion that body rather than antler size is the key variable in determining contest success.

Evidence that antler or horn displays serve to convey individual quality during contests, independent of factors such as age or body size, has remained somewhat elusive (e.g. Vanpé *et al.* 2010). Moreover, although the length of the structure is often related to dominance (Jennings *et al.* 2006, Figure 14.2), the contention that these structures serve to reduce fighting by facilitating assessment (e.g. Geist 1966*a*, 1991, Lincoln 1972, Barrette & Vandal 1986) has received only limited support (e.g. Wahlström 1994, Jennings *et al.* 2006). While it has been argued that species possessing large antlers frequently display them to opponents (reviewed by Geist 1991), studies that have directly investigated the display function of antlers or horns are rare. One exception is a study on mountain sheep (*Ovis dalli* and *Ovis canadensis*), where males were observed to respond to the size of their opponents' horns (Geist 1966*a*); in behavioural terms, males displayed their horns to each other prior to fighting and only males that were evenly matched proceeded to fight. This form of pre-contest display behaviour is also used by caribou (*Rangifer tarandus*), where antler length acts as a visual indicator of quality (Barrette & Vandal 1986). Similar to mountain sheep, the difference in antler size between opponents appeared to determine whether individuals engaged in a contest; a finding that is supported by studies on other ungulate species (e.g. Jennings *et al.* 2006, Hoem *et al.* 2007, Bergeron *et al.* 2010). The likelihood that two opponents will escalate to fighting is determined by how small the difference in antler length is (Hoem *et al.* 2007, Bergeron *et al.* 2010); this also influences the frequency with which two individuals will repeatedly engage each other in fights (Jennings *et al.* 2006).

In deer, a form of generalised ritualistic behaviour accompanies presentation of the antlers: males hold their heads high and erect before their opponents, frequently turning their heads as they present their antlers (Geist 1991). For example, male fallow deer present the antler palms to opponents during agonistic encounters, encouraging the idea that the antler (or palm) display represented a form of threat signal emphasising the size of the antler (Alvarez 1993, 1995). However, a detailed investigation of this behaviour indicated that the antler display did not serve to convey individual quality (Jennings *et al.* 2002). On the contrary, it was the eventual loser of the fight that was more likely to initiate an antler display irrespective of antler length or symmetry. Studies on the pronghorn (*Antilocapra americana*: Kitchen & Bromley 1974) and white-tailed deer (*Odocoileus virginianus*: Hirth 1977) have shown that staring at an opponent is an aggressive act whereas avoidance of eye contact signifies submission. In the context of the results of Jennings *et al.* (2002), the lateral display of the antlers of the fallow deer would appear to be a visually conspicuous act of submission rather than a display of antler size or structure. More than four decades since Geist (1966*a*) argued that horn displays are used to convey individual quality in sheep, further evidence that supports this view has not been forthcoming. A potential explanation for this is the fact that antlers and horns are fixed structures and as such, it is unlikely that they would indicate short-term changes in individual quality (Clutton-Brock 1982).

14.3.3 Visual displays: lateral body presentation

In ungulates, displaying the antlers or horns often coincides with a lateral presentation of the body such that the flanks are also oriented towards the opponent (Figure 14.3). Although presentation of the body appears to be used preferentially by species that possess small antlers (Geist 1966*b*, 1971), it also seems to

Figure 14.3 Red deer stags *Cervus elaphus* preparing to fight. Early morning, but red deer stags in the rutting season start work at break of day. This sort of parallel walk allows the stags to size each other up, and frequently occurs at the boundary between two territories. A serious fight may follow, especially if there are hinds (females) close by. ©iStockphoto.com/Roger Whiteway. For colour version, see colour plate section.

be part of the behavioural repertoire of most species of ungulate irrespective of the presence or absence of conspicuous weaponry. A common explanation for lateral displays is that they serve as a threat to potential opponents indicating a willingness to escalate to fighting (Kitchen & Bromley 1974) and are, therefore, a prelude to fighting itself (Geist 1971). Initiation of a lateral body display frequently evokes a reciprocal display, resulting in both animals standing within one or two metres parallel to each other (Alvarez 1993). A temporally extended form of the parallel display, the parallel walk, has been noted in several species of ungulate (e.g. fallow deer: Alvarez 1993; red deer: Clutton-Brock *et al.* 1979; wild boar, *Sus scrofa* and *Phacochoerus aethiopicus*: Frädrich 1974, Barrette 1986; *Cervus canadensis nannodes* tule elk: Wahome 1995). In general, as with the parallel display, the parallel walk appears to be associated with fighting. For example, fights between red deer stags and wild boar are typically preceded by parallel walking (Clutton-Brock & Albon 1979, Barrette 1986) and in fallow deer the parallel walk occurs either before, during or after approximately half of all recorded fights (Jennings *et al.* 2003, Bartoš *et al.* 2007). In the case of wild boar (Frädrich 1974) and red deer (Clutton-Brock *et al.* 1982), the parallel walk is often accompanied by bristling of the body hair (piloerection) which enhances the body profile. Such behaviour is consistent with the idea that lateral displays serve to transmit information about body size and possibly the motivational state of the contestants.

Because parallel walks usually occur between opponents that are in close physical proximity to each other, they may permit contestants to assess their opponent's body or antler/horn size (e.g. Geist 1966*a*, Clutton-Brock & Albon 1979) and initiating a lateral display increases the chances of victory (Kitchen & Bromley 1974). A direct test of this hypothesis in fallow deer indicated that the parallel walk did not serve to communicate information concerning antler length or body size, suggesting that the parallel walk does not appear to facilitate opponent assessment (Jennings *et al.* 2003). It has been shown that parallel walks were most likely to be observed between evenly matched individuals, that longer parallel walks were less likely to be followed by fights, and that if a fight did occur, it was likely to be longer if it had not been preceded by a parallel walk (Clutton-Brock & Albon 1979; see also Mattiangeli *et al.* 1998). However, because it is a temporally extended activity, parallel walking will naturally serve to increase overall contest duration. When the duration of the parallel walk was controlled for, the time spent fighting was not related to the time spent in parallel walk nor did opponents spend longer fighting with each other based on the presence of a parallel walk (Jennings *et al.* 2003). Nevertheless, one factor that appears to influence the duration of parallel walking concerns the location of the individual during the display: left or right side (Figure 14.3). Individuals that present their right flank are more likely to spend longer before terminating the parallel walk than if they present their left (Jennings 2012). One explanation for this apparent side preference centres on the idea that the brain hemispheres evolved to process different types of information more efficiently (Rogers & Andrew 2003). Therefore, when on the preferred right-hand side during parallel walks, males spend longer gathering and processing information about their opponent than they do when on the less-preferred left-hand side because they cannot process information about opponent quality as efficiently (Jennings 2012): this is an explanation that rests on the idea that individuals do (or at least attempt to) assess opponent quality.

Lateral display of the body during contests, therefore, appears to serve different functions in different species. It is a theoretical prediction that individuals should be sensitive to a potential opponent's body mass because weight is generally an important predictor of fighting ability (Chapter 2). However, not all empirical evidence accords with this expectation; for example, differences in body weight are not related to the frequency with which fallow deer engage in fights (Jennings *et al.* 2006). Nevertheless, there is a consistent relationship between agonistic behaviour and lateral display. When parallel walking occurs, there is a high probability that the contest will involve fighting. This is perhaps unsurprising given that, unlike vocal contests, parallel walking brings the protagonists into close proximity and thus, it is often viewed as an increase in contest intensity.

A consistent theme in published accounts of ungulate contest behaviour is that displays of weaponry and body size facilitate opponent assessment. Nevertheless, direct tests of this hypothesis are extremely rare: the effect of weapon size and body mass on parallel walking in fallow deer indicated that these factors are not involved in the assessment process (Jennings *et al.* 2003), although the side that the individual is located on is important for processing information about opponent quality (Jennings 2012). Ungulate behaviour can be highly flexible within and between populations (i.e. mating systems: Thirgood 1990) and there may also be a similar level of flexibility in the use of lateral displays. Consequently, there is a need for further investigation of the function of this behaviour in ungulates.

14.4 Fighting: structure, outcome and assessment processes

Fighting represents the most intense competitive phase of ungulate contests. In comparison to the more frequently observed low-level displacement interactions, fights are often quite spectacular sequences involving repeated clashing of the antlers and horns (Figure 14.4). The body mass of mature males, in combination with their weaponry means that intraspecific fighting is dangerous and on occasion lethal (e.g. Geist 1971, Wilkenson & Shank 1976, Clutton-Brock *et al.* 1979, Poole 1989). Therefore, it is unsurprising that both weapon size and body weight play important roles in determining how fights are structured. However, these are not the sole factors involved; the value of the contested resource, the familiarity of the opponents with each other and the form of assessment process are also predicted to shape contests. We now address each of these issues, focussing on fight duration and the frequency of behavioural actions within contests.

Figure 14.4 Male Ibex fighting in the Gran Paradiso National Park, Italy. (Photograph: Patrick Bergeron.) For colour version, see colour plate section.

14.4.1 Weapon use

The manner in which weapons are used during fighting is dependent on the physical structure of the weapon borne by different species. Species that possess horns with ridges use this structural feature to provide grip during fights, effectively holding the opponents together (Geist 1966*a*, Walther 1974, Lundrigan 1996). Moreover, it has been argued that the shape of the structure has developed in line with the style of combat; for example, bovids with inward-facing tips to their horns fight by wrestling, smooth horns are used for stabbing at opponents and antlers with more than five points are adapted for a fencing style of combat (Caro *et al.* 2003). Longer horns or antlers are closely associated with wrestling-type contests such as the twisting, pushing contests of deer (Lundrigan 1996). Recurved horns such as those of the buffalo (*Syncerus caffer*) are more compatible with a ramming style of fighting (Mloszewski 1983). Species such as the wild boar that do not possess horns or antlers use their sharp upper and lower canine teeth during fights (Barrette 1986). During fights, wild boar lean and push against each other in an attempt to unbalance their opponent which, if successful, leads in turn to goring attempts by the victor. Although rare, some species possess tusks and antlers (e.g. Muntjac, *Muntiacus reevesi* and *M. muntjak*) and both are employed during fights; males repeatedly clash with their antlers, twisting as they do so with the aim of engineering an opening by which they can use their tusks to gore their opponent (Barrette 1977). However, by far the most common form of fighting involves the use of horns and antlers in both offensive and defensive behaviour.

A number of studies have reported that there is a relationship between particular antler characteristics and contest success; for example, the number of antler points is related to fight success in red deer and mule deer (Clutton-Brock *et al.* 1979, Bowyer 1986). In the fallow deer, territory holders have longer antlers than non-territorial males (Ciuti & Apollonio 2011), and in roe deer it has been shown that if territorial males lose a contest they are also more likely to have shorter antlers than their opponent (Hoem *et al.* 2007). In nyala and eland (*Tragelaphus angasii* and *T. oryx*) it is horn shape rather than length that is deemed an important factor in fighting (Estes 1991). It has been reported that the right antler of fallow deer is more developed than the left and that deer make preferential use of this antler during fights (Alvarez 1995). Furthermore, larger male fallow deer initiate fighting by preferentially using risky tactics such as a jumping clash (one contestant jumps towards his opponent with his antlers lowered) against smaller opponents, suggesting that possession of longer weapons promotes the use of risky behaviour during contests (Alvarez 1993, Jennings *et al.* 2004). However, no relation between contest duration and differences in antler length was found (Jennings *et al.* 2004).

14.4.2 Body size

As noted in section 14.3.2, the question as to whether body size is related to fighting success independent of weapon size is difficult to answer given the allometric relationship between these two characters. However, when individuals suffer a break in the antler or horn and still retain the ability to defeat opponents it suggests that body size is a critical factor driving contest success (Johnson *et al.* 2007). A general finding is that asymmetries in body mass are a contributing factor to both outcome and duration of fights (reviewed by Riechert 1998). Although weapons have an important part to play in ungulate contests, body weight should also be important given the prevalence of ramming and pushing-style contests. In terms of contest outcome, findings concerning body weight have been variable: it has been shown in several species that the heavier of the two animals generally has an advantage over the lighter one (e.g. Bergerud 1974, Clutton-Brock *et al.* 1982, Mloszewski 1983, Wahlström 1994, Jensen & Yngvesson 1998); however, weight advantage has not been found to be important in determining fight outcome in all species studied (e.g. Lott 1979, Poole 1989, Wolff 1998, Jennings *et al.* 2004). Despite these findings, body size plays a central role in fight structure; for example, a decline in body weight and neck girth over the rut is related to tactical behaviour during fights, and fight duration is negatively related to pre-rut neck girth in fallow deer (Jennings *et al.* 2010).

14.4.3 Familiarity

Body weight influences, but is not the sole predictor of, fight duration in ungulates. For example, while fight duration in pigs is affected by asymmetries in body weight, fights are also longer when contestants are unfamiliar with each other (Rushen 1988, Jensen & Yngvesson 1998). Verification of the effects of body weight on contest duration in field populations has proven to be more difficult: Jennings *et al.* (2004) failed to show a relationship between body weight and either duration or intensity in fallow deer fights, but did find that contestant familiarity influenced duration. Specifically, as the number of fights that pairs of males engaged in increased, there was a corresponding decline in fight duration. These data suggest that opponents retained a memory of prior interactions with each other. It should be noted that fight duration was negatively related only to the number of previous fights between pairs of males and not to the overall interaction rate (Jennings *et al.* 2004). It would, therefore, appear that it is the type of interaction that individual dyads have previously engaged in (fight as opposed to low-intensity displacement) that influences contest structure. Moreover, although body weight might initially influence contest outcome it does not necessarily predict the outcome of subsequent encounters. For example, it has been shown that previous experience of being dominant is a stronger predictor of subsequent outcomes even when the body sizes of opponents were reversed (Taillon & Côté 2006).

14.4.4 Competition for resources

It is not always the size of the weapon or body of the protagonist, or familiarity with the other contestant, that influences contest structure or duration. Because they are expensive to the contestant on more than one level (immediate and more long-term energy requirements, possibility of injury, time spent distracted from tending the resource), we assume that fights will take place for a reason. In ungulates, fighting tends to increase as the number of matings increase (e.g. Floody & Arnold 1975, Clutton-Brock & Albon 1979, Komers *et al.* 1994, but see Festa-Bianchet *et al.* 1990). In some species there is a marked increase in contest duration and intensity and in others this does not happen. Roe deer appear to be more willing to escalate to fighting as the number of oestrous females increases (Hoem *et al.* 2007) and these fights are longer and more intense when females are likely to conceive (Clutton-Brock *et al.* 1979). However, this appears not to be the case in all ungulate species. For example, male ibex decrease the amount of time spent in aggressive interactions during the rut compared with the pre-rut (Willisch & Neuhaus 2010). In fallow deer, fights are of similar duration prior to and during the rut and, furthermore, fight duration is similar irrespective of the number of oestrous females in the population (Jennings *et al.* 2004, Fričová *et al.* 2007). It would appear, therefore, that while female presence and reproductive state promotes aggression between males, the effects of resource availability on contest structure are more equivocal. This may be related to the mating strategy adopted in the population: the red deer studied by Clutton-Brock *et al.* (1979) were harem-holding males while the males studied by Hoem *et al.* (2007) were territory-holders and the majority of males in the population studied by Jennings *et al.* (2004) held neither territories nor harems but rather moved with the female herd. Differences in contest structure could, therefore, be related to whether males hold the resource at stake.

14.4.5 Assessment processes during contests

One of the earliest studies into the decision-making processes used by ungulates during contests was conducted on red deer (Clutton-Brock & Albon 1979, Clutton-Brock *et al.* 1979). During the rut, contests often begin following the approach of a male towards a rival that possesses a harem. In this case, escalation to dangerous fighting generally does not occur immediately. Rather, males initially engage in a roaring contest (section 14.3.1); if the males are mismatched in terms of ability they will roar at each other at different rates and the weaker individual will eventually yield. Should the roaring contest fail to identify a winner, rivals may then move closer to allow inspection of opponent weapon and body size by parallel walking (section 14.3.3) and if this display should again fail to result in a winner then fighting is highly likely to occur. The common explanation for the structure of red deer contests is that males are mutually assessing each other, starting out from a relatively low energetic cost base and only increasing the intensity and cost of the interaction when they are unable to settle at the lower level. Nevertheless, as we have seen, suggestions that roaring contests and lateral displays are strategies that serve to communicate quality have not always been supported and have been challenged by alternative explanations (sections 14.4.1 to 14.4.4).

Since these studies on red stag contests were published, several assessment models that emphasise the mechanism by which individuals assess their probability of defeating an opponent have been developed. These models variously predict different forms of behaviour during fighting focussing on features such as fight duration, winner–loser disparity, repetition of behaviour and outcome (Arnott & Elwood 2009, Briffa & Elwood 2009). Nevertheless, in theoretical terms, addressing these aspects of fighting should enable us to determine whether assessment rules governing contest behaviour are self-assessment (e.g. energetic war of attrition, EWOA: Mesterton-Gibbons *et al.* 1996, Chapters 1, 2 and 4) or mutual assessment (e.g. sequential assessment, SAM: Enquist & Leimar 1983, Chapters 1, 2 and 4).

There are two general approaches to investigating assessment processes during contests. The first is based on examining the repetition rate of behavioural acts within contest phase (Payne 1998) and the second on a comparison of the resource holding potential, RHP (Chapter 1), of the contestants with contest duration (Taylor & Elwood 2003, Gammell & Hardy 2003, Briffa & Elwood 2009, Chapter 4). Jennings *et al.* (2004) used Taylor and Elwood's framework to investigate the effect of the body weight and antler length of both contestants on fight duration and intensity. In the case of body weight it was found that the correlation coefficient between the weight of the lighter contestant and both fight duration and intensity was greater than for the heavier contestant. These data support the idea that individuals fight and escalate contest intensity based on a form of self-assessment. The data concerning antler length did not follow predictions for either self-assessment or mutual assessment, consistent with the argument that antlers represent a relatively poor measure of current RHP (e.g. Clutton-Brock 1982, Jennings *et al.* 2006, Bartoš *et al.* 2007).

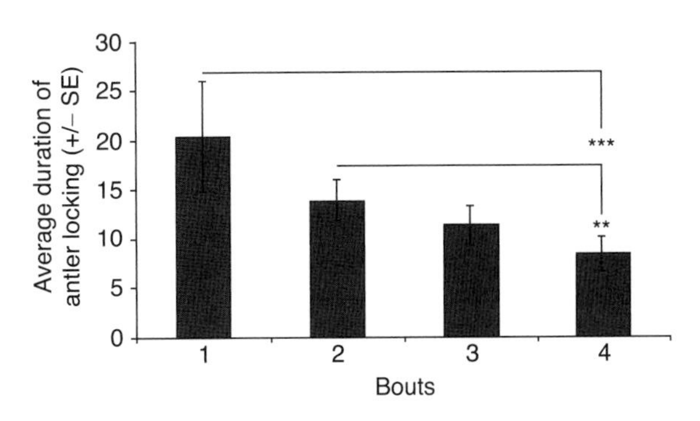

Figure 14.5 The average time (seconds) spent fighting by fallow bucks during the first four bouts of antler locking (± standard error). The first and second bouts of fighting are longer than the fourth bout. (Adapted from Jennings *et al.* 2005.)

Despite the apparent mutually exclusive categories of self-assessment and mutual assessment, recent empirical studies conducted with non-ungulate species have suggested that one form of assessment might be operating between contest phases and another form of assessment within contest phases (Killifish, *Kryptolebias marmoratus*: Hsu *et al.* 2008) or that assessment processes are a continuum with elements of more than one process active (e.g. freshwater amphipod, *Gammarus pulex*: Prenter *et al.* 2006; fiddler crab, *Uca mjoebergi*: Morrell *et al.* 2005). There is limited evidence that different assessment strategies are used between and within phases of the same contests in ungulates. An exception is a study conducted on pigs (Jensen & Yngvesson 1998): prior to fighting, a greater duration of pre-exposure between contestants resulted in a rapid escalation to the more costly biting phase as predicted by mutual assessment models. However, within this biting phase there was no difference in biting between the eventual winner and loser, a pattern of behaviour only accounted for by self-assessment models. In order to investigate assessment processes during fallow deer fights, the rates of behavioural change within a single contest phase (the fighting phase) were investigated (Jennings *et al.* 2005). The results of this study were equivocal: bouts of fighting tended to become progressively shorter over the contest, as predicted by self-assessment models (Figure 14.5), and there were differences between winners and losers in attacking behaviour, as expected by mutual assessment models. Taken together these results provide within-phase support for both self-assessment and mutual assessment during fights.

14.5 Choice of opponent: assessment and dominance

A central issue in the study of contest behaviour is what drives the decision for two individuals to start fighting with each other. Although the evidence is not conclusive, there is some support for opponent assessment. Where opponent RHP is highly asymmetric, contests are not expected to escalate to fighting, in contrast to when opponents are closely matched (Enquist & Leimar 1983) as has been shown in rutting moose (Peek *et al.* 1986). However, other factors such as age and dominance rank could be important in mediating contest frequency, suggesting that opponent assessment could be based on some composite measure of quality rather than on any one single measure (Peek *et al.* 1986). In terms of age, mature individuals are generally both dominant to and bigger than younger group members (e.g. Appleby 1982, Peek *et al.* 1986, Hass & Jenni 1991, Greenberg-Cohen *et al.* 1994, Wolff 1998, Côté 2000). However, there is a degree of flexibility here; for example, early in the rut, older larger bison tend to fight with other older males, but by late in the rut they tend to interact more with younger males (Komers *et al.* 1994). Although it is prime-aged males that invest most in fighting, it is possible that young males alter their investment in fighting in order to improve mating opportunities later in the rut (Jennings *et al.* 2010, Tennenhouse *et al.* 2011). Choice of opponent could therefore be related to the current quality of the individual which, for actively rutting males, is expected to decline over the rut (Bobek *et al.* 1990). Individuals of similar rank and age in red deer tend to associate more with each other than with other group members although the effect is stronger for rank than age (Appleby 1983). In fallow deer, individual age, independent of rank, was not related to the number of fights recorded between pairs of males (Jennings *et al.* 2006). It therefore appears that when individuals have reached full body size it is increasingly difficult to determine the influence of size or age independently of dominance rank in regulating opponent choice.

For many group-living species, the dominance hierarchy provides a useful summary of the social structure within the group. If appropriately measured, and assuming that rank is stable over time, the position of each member in the hierarchy provides a useful measure by which to estimate the quality of each individual (e.g. Appleby 1983, Bartoš *et al.* 1988, Moore *et al.* 1995, Roden *et al.* 2005, Jennings 2007, but see Lott 1979). It has been suggested that differences in dominance rank minimise the frequency of fighting by allowing individuals to resolve disputes via low-level interactions (Chapter 13). However, as we have noted above, the eventual winner of a fight (and therefore the dominant member of the dyad) is more likely to have initiated the contest. Consistent with these studies, possession of high dominance rank in several species has been shown to be strongly related to both fight rate and number of different opponents indicating that dominant individuals are highly competitive (e.g. Greenberg-Cohen *et al.* 1994, Barroso *et al.* 2000, Jennings *et al.* 2006, Shi & Dunbar 2006). Nevertheless, whether individuals are able to assess their rank relative to other group members is still a matter for debate. While there is evidence of rank awareness in primates (e.g. Cheney & Seyfarth 2007) and other group-living species (e.g. Engh *et al.* 2005, Chapter 13), evidence for ungulates has proved to be more contentious. An initial investigation of the patterning of contests between red deer stags indicated that fights between individuals with disparate dominance rankings were uncommon (e.g. Clutton-Brock *et al.* 1982, Appleby 1983). A similar finding for fallow deer was reported by Jennings *et al.* (2006); however, in their study only prime-aged males were found to interact more frequently with individuals of similar rank while younger males displayed no such preference. Contest patterns between female ungulates show that individuals are more likely to interact if they are closely ranked, consistent with findings for red deer stags (Rutberg 1986, Côté 2000).

Using data originally published by Appleby (1983), the patterning of contests between dyads of young red deer stags was investigated (Freeman *et al.* 1992). After accounting for differences in contest rate, Freeman *et al.* showed that individuals displayed a tendency to fight with those located two rankings from themselves. These findings led to the suggestion that stags were aware of their relative positions and selectively fought with group members on this basis (Freeman *et al.* 1992). However, this argument has been challenged on the basis that hierarchies, although extremely useful tools for summarising social behaviour, are artificial constructs with little meaning to the animals themselves (Appleby 1993); rather, individuals are more likely to view their social group in much simpler terms where members are either dominants or subordinates. Nevertheless, tests of whether animals are aware of their relative places in a social group are more easily carried out using triadic- rather than dyadic-level contests (Freeman *et al.* 1992). In the case of several primate species, an intimate knowledge of dominance relations within the group can be inferred by support offered during fights by third-party interveners allied to one or other competitor (e.g. Silk *et al.* 2004, Cheney & Seyfarth 2007). It has been suggested that ungulates do not engage in third-party interventions, adding to the argument that they are not aware of rank relations (e.g. Freeman *et al.* 1992, Ceacero *et al.* 2007). However, third-party intervention behaviour has been noted in some species of ungulate although rarely subjected to formal investigation (e.g. Schaller 1967, Keil & Sambraus 1998, Jennings *et al.* 2009, 2011). In partial agreement with observations on several primate species, interveners tend to be high-ranking although they do not selectively target one member of a competing dyad based on dominance rank (Jennings *et al.* 2009, 2011; see also Keil & Sambraus 1998). Rather, intervention appeared to be random with regard to the individual targeted, leading to the conclusion that intervention could serve to prevent either of the competing males from achieving a contest victory, rising up the hierarchy and becoming a threat to the intervening male (Dugatkin 1998*a*,*b* provides a theoretical overview). Although there was no evidence that interveners targeted individuals based on rank, they do in fact benefit from disrupting contests between other group members. Interveners are not just high-ranking individuals seeking to protect their position in the hierarchy; intervention rate is related to a gain in dominance status and, therefore, interveners actually advance in the hierarchy and enjoy enhanced reproductive success (Jennings *et al.* 2011). It would thus appear from the data available that there is no evidence for an awareness of relative rank positions in ungulate contest behaviour. However, we should also note that the surface of this fascinating aspect of contest behaviour in ungulates has barely been scratched.

14.6 Concluding remarks

We approached this chapter with the view that although ungulates represent a diverse and widespread group of animals, contest behaviour could be broken down into a series of questions concerning contest rates, contest structure, the probability of winning and the assessment strategies used. However, our understanding of ungulate contest behaviour has been hampered due to the fact that most species of ungulate have not been studied in detail, with analyses of their contests being particularly scarce (e.g. Geist 1971, Geist & Walther 1974, Estes 1991). Given their diversity and geographic distribution, this is hardly surprising. With some exceptions, most of the studies that we have considered have relied on populations resident in parks, observed over extended time periods and containing individually identifiable members. Based on current knowledge, it is apparent that ungulate contests are highly complex social interactions that include both vocal and visual components and range from low-level displacements to fights. Opponent assessment during contests is less likely to be based solely on self- or mutual-assessment processes than on a combination of both. The use of vocalisations during the parallel walks that often occur when there are intervals in fighting also suggest that the assessment processes employed during contests are flexible. The extent of this flexibility is but one of the outstanding questions that needs to be addressed. More recent investigation of triadic-level interactions has pointed to yet another level of complexity in ungulate competitive behaviour that requires further study. Future investigation into the evolutionary basis of competition should help to shed light on the mechanisms involved in structuring contests and maintaining aggression in ungulate societies.

Acknowledgements

We would like to express our thanks to our colleagues who have taken the time to read and comment on the contents of this chapter. We also thank the reviewers and editors for their constructive comments. We extend our thanks and appreciation to those of our colleagues who generously provided us with photographs of their study species (unfortunately space limitations meant that only a small number could be included).

Endnotes

1. W.B. Yeats – *Among School Children.*
2. Taxonomic description of the ungulata is currently in a state of flux. In this chapter we use the original description of ungulates as mammals with hooves. Apologies to those who expected cetaceans.

References

Alvarez F (1993) Risks of fighting in relation to age and territory holding in fallow deer. *Canadian Journal of Zoology*, 71, 376–383.

Alvarez F (1995) Functional directional asymmetry in fallow deer (*Dama dama*) antlers. *Journal of Zoology, London*, 236, 563–569.

Andersson M (1994) *Sexual Selection*. Princeton, NJ: Princeton University Press.

Appleby MC (1982) The consequences and causes of high social rank in red deer. *Behaviour*, 80, 259–270.

Appleby MC (1983) Competition in a red deer stag social group: rank, age and relatedness of opponents. *Animal Behaviour*, 31, 913–918.

Appleby MC (1993) How animals perceive a hierarchy: reactions to Freeman *et al. Animal Behaviour*, 46, 1232–1233.

Arnott G & Elwood, RW (2009) Assessment of fighting ability in animal contests. *Animal Behaviour*, 77, 991–1004.

Barrette C (1977) Fighting behavior of muntjac and the evolution of antlers. *Evolution*, 31, 169–176.

Barrette C (1986) Fighting behavior of wild *Sus scrofa*. *Journal of Mammalogy*, 67, 177–179.

Barrette C & Vandal D (1986) Social rank, dominance, antler size and access to food in snow-bound wild woodland caribou. *Behaviour*, 97, 118–146.

Barroso FG, Alados CL & Boza J (2000) Social hierarchy in the domestic goat: effect on food habits and production. *Applied Animal Behaviour Science*, 69, 35–53.

Bartoš L & Losos S (1997) Response of antler growth to changing rank of fallow deer buck during the velvet period. *Canadian Journal of Zoology*, 75, 1934–1939.

Bartoš L, Perner V & Losos S (1988) Red deer stags rank position, body weight and antler growth. *Acta Theriologica*, 33, 209–217.

Bartoš L, Fričová B, Bartošová-Víchová J, *et al.* (2007) Estimation of the probability of fighting in fallow deer (*Dama dama*) during the rut. *Aggressive Behavior*, 33, 7–13.

Bergeron P, Grignolio S, Apollonio M, *et al.* (2010) Secondary sexual characteristics signal fighting ability and determine social rank in Alpine ibex (*Capra ibex*). *Behavioral Ecology and Sociobiology*, 64, 1299–1307.

Bergerud AT (1974) Rutting behaviour of Newfoundland caribou. In: V Geist & F Walther (eds.) *The Behaviour of Ungulates and its Relation to Management*, pp. 395–435. Morges: IUCN.

Blood DA, Flook DR & Wishart WD (1970) Weights and growth of Rocky Mountain bighorn sheep in western Alberta. *Journal of Wildlife Management*, 34, 451–455.

Bobek B, Perzanowski K & Weiner J (1990) Energy expenditure for reproduction in red deer. *Journal of Mammalogy*, 71, 230–232.

Bowyer RT (1986) Antler characteristics as related to social status of male southern mule deer. *Southwest Naturalist*, 31, 289–298.

Briffa M & Elwood RW (2009) Difficulties remain in distinguishing between mutual and self-assessment in animal contests. *Animal Behaviour*, 77, 759–762.

Bubenik AB (1998) Evolution, taxonomy, and morphophysiology. In: AW Franzmann & CC Schwartz (eds.) *Ecology and Management of the North American Moose*, pp. 77–123. Washington, DC: Smithsonian Institution Press.

Caro TM, Graham CM, Stoner CJ, *et al.* (2003) Correlates of horn and antler shape in bovids and cervids. *Behavioral Ecology and Sociobiology*, 55, 32–41.

Carranza J, Alvarez F & Redondo T (1990) Territoriality as a mating strategy in red deer. *Animal Behaviour*, 40, 79–88.

Ceacero F, Landete-Castillejos T, García AJ, *et al.* (2007) Kinship discrimination and effects on social rank and aggressiveness levels in Iberian red deer hinds. *Ethology*, 113, 1133–1140.

Charlton BD, Reby D & McComb K (2007) Female red deer prefer the roars of larger males. *Biology Letters*, 3, 382–385.

Cheney DL & Seyfarth RM (2007) *Baboon Metaphysics: The Evolution of a Social Mind*. Chicago, IL: Chicago University Press.

Ciuti S & Apollonio M (2011) Do antlers honestly advertise the phenotypic quality of fallow buck (*Dama dama*) in a lekking population. *Ethology*, 117, 133–144.

Clutton-Brock TH (1982) The functions of antlers. *Behaviour*, 79, 108–125.

Clutton-Brock TH & Albon SD (1979) The roaring of red deer and the evolution of honest advertisement. *Behaviour*, 69, 145–170.

Clutton-Brock TH, Albon SD, Gibson RM, *et al.* (1979) The logical stag: adaptive aspects of fighting in red deer (*Cervus elaphus* L.). *Animal Behaviour*, 27, 211–255.

Clutton-Brock TH, Guinness FE & Albon SD (1982) *Red Deer: The Behaviour and Ecology of Two Sexes*. Chicago, IL: Chicago University Press.

Côté SD (2000) Dominance hierarchies in female mountain goats: stability, aggressiveness and determinants of rank. *Behaviour*, 137, 1541–1566.

Dugatkin LA (1998*a*) Breaking up fights between others: a model of intervention behaviour. *Proceedings of the Royal Society of London B*, 265, 433–437.

Dugatkin LA (1998*b*) A model of coalition formation in animals. *Proceedings of the Royal Society of London B*, 265, 2121–2125.

Engh AL, Siebert ER, Greenberg DA, *et al.* (2005) Patterns of alliance formation and postconflict aggression indicate spotted hyaenas recognize third-party relationships. *Animal Behaviour*, 69, 209–217.

Enquist M & Leimar O (1983) Evolution of fighting behaviour: decision rules and assessment of relative strength. *Journal of Theoretical Biology*, 102, 387–410.

Enquist M, Leimar O, Ljungberg T, *et al.* (1990) A test of the sequential assessment game: fighting in the cichlid fish *Nannacara anomala*. *Animal Behaviour*, 40, 1–14.

Estes RD (1991) *The Behavior Guide to African Mammals*. Berkeley, CA: University of California Press.

Festa-Bianchet M, Appllonio M, Mari F, *et al.* (1990) Aggression among lekking male fallow deer (*Dama dama*): territory effects and relationship with copulatory success. *Ethology*, 85, 236–246.

Festa-Bianchet M, Coltman DW, Turelli L, *et al.* (2004) Relative allocation to horn and body growth in bighorn rams varies with resource availability. *Behavioral Ecology*, 15, 305–312.

Floody DR & Arnold AP (1975) Uganda kob (*Adenota kob thomasi*) territory and spatial distribution of sexual and agonistic behaviours at a territorial ground. *Zeitschrift für Tierpsychologie*, 37, 192–212.

Frädrich H (1974) A comparison of behaviour in the Suidae. In: V Geist & F Walther (eds.) *The Behaviour of Ungulates and its Relation to Management*, pp. 133–143. Morges: IUCN.

Freeman LC, Freeman S & Romney AK (1992) The implications of social structure for dominance hierarchies in red deer (*Cervus elaphus* L.). *Animal Behaviour*, 44, 239–245.

Fričová B, Bartoš L, Bartošová J, *et al.* (2007) Female presence and male agonistic encounters in fallow deer, *Dama dama* during the rut. *Folia Zoologica*, 56, 253–262.

Gammell MP & Hardy ICW (2003) Contest duration: sizing up the opposition? *Trends in Ecology and Evolution*, 18, 491–493.

Geist V (1966*a*) The evolutionary significance of mountain sheep horns. *Evolution*, 20, 558–566.

Geist V (1966*b*) The evolution of horn-like organs. *Behaviour*, 27, 175–214.

Geist V (1971) *Mountain Sheep: A Study in Behaviour and Evolution*. Chicago, IL: University of Chicago Press.

Geist V (1991) Bones of contention revisited: did antlers enlarge with sexual selection as a consequence of neonatal security strategies? *Applied Animal Behaviour Science*, 29, 453–469.

Geist V & Walther F (1974). *The Behaviour of Ungulates and its Relation to Management*. Morges: IUCN Publications.

Greenberg-Cohen D, Alkon PU & Yom-Tov Y (1994) A linear dominance hierarchy in female Nubian ibex. *Ethology*, 98, 210–220.

Hass CC & Jenni DA (1991) Structure and ontogeny of dominance relationships among bighorn rams. *Canadian Journal of Zoology*, 69, 471–476.

Hirth DH (1977) Social behavior of white-tailed deer in relation to habitat. *Wildlife Monographs*, 53, 1–55.

Hoem SA, Melis C, Linnell JDC, *et al.* (2007) Fighting behaviour in territorial male roe deer *Capreolus capreolus*: the effects of antler size and residence. *European Journal of Wildlife Research*, 53, 1–8.

Hsu Y, Lee SP, Chen MH, *et al.* (2008) Switching assessment strategy during a contest: fighting in killifish *Kryptolebias marmoratus*. *Animal Behaviour*, 75, 1641–1649.

Jennings DJ (2007) The effect of chemical immobilization on dominance rank in the fallow deer. *Animal Behaviour*, 74, 1107–1110.

Jennings DJ (2012) Right-sided bias in the decision to terminate fallow deer parallel walks. Evidence for lateralization during a lateral display. *Animal Behaviour*, 83, 1427–1432.

Jennings DJ, Gammell MP, Carlin CM, *et al.* (2002) Does lateral presentation of the palmate antlers during fights by fallow deer (*Dama dama*) signify dominance or submission? *Ethology*, 108, 389–401.

Jennings DJ, Gammell MP, Carlin CM, *et al.* (2003) Is the parallel walk between competing male fallow deer (*Dama dama*) a lateral display of quality? *Animal Behaviour*, 65, 1005–1012.

Jennings DJ, Gammell MP, Carlin CM, *et al.* (2004) Effect of body weight, antler length, resource value and experience on fight duration and intensity in fallow deer. *Animal Behaviour*, 68, 213–221.

Jennings DJ, Gammell MP, Payne RJH, *et al.* (2005) An investigation of assessment games during fallow deer fights. *Ethology*, 111, 511–525.

Jennings DJ, Gammell MP, Carlin CM, *et al.* (2006) Is difference in body weight, antler length, age or dominance rank related to the number of fights between fallow deer (*Dama dama*)? *Ethology*, 112, 258–269.

Jennings DJ, Carlin CM & Gammell MP (2009) A winner effect supports third-party intervention behaviour during fallow deer, *Dama dama*, fights. *Animal Behaviour*, 77, 343–348.

Jennings DJ, Carlin CM, Hayden TJ, *et al.* (2010) Investment in fighting in relation to body condition, age and dominance rank in the male fallow deer, *Dama dama*. *Animal Behaviour*, 79, 1293–1300.

Jennings DJ, Carlin CM, Hayden TJ, *et al.* (2011) Third-party intervention behaviour during fallow deer fights: the role of dominance, age, fighting and body size. *Animal Behaviour*, 81, 1217–1222.

Jennings DJ, Elwood RW, Carlin CM, *et al.* (2012) Vocal rate as an assessment process during fallow deer contests. *Behavioural Processes*, 91, 152–158.

Jensen P & Yngvesson J (1998) Aggression between unacquainted pigs – sequential assessment and effects of familiarity and weight. *Applied Animal Behaviour Science*, 58, 49–61.

Johnson HE, Bleich VC, Krausman PR, *et al.* (2007) Effects of antler breakage on mating behavior in male tule elk (*Cervus elaphus nannodes*). *European Journal of Wildlife Research*, 53, 9–15.

Keil NM & Sambraus HH (1998) 'Interveners' in agonistic interactions amongst domesticated goats. *International Journal of Mammalian Biology*, 63, 266–272.

Kitchen D & Bromley PT (1974) Agonistic behaviour of territorial pronghorn bucks. In: V Geist & F Walther (eds.) *The Behaviour of Ungulates and its Relation to Management*, pp. 365–381. Morges: IUCN.

Komers PE, Meissier F & Gates CC (1994) Plasticity of reproductive behaviour in wood bison bulls: on risks and opportunities. *Ethology, Ecology and Evolution*, 6, 481–495.

Komers PE, Pélabon C & Stenström D (1997) Age at first reproduction in male fallow deer: age-specific versus dominance-specific behaviors. *Behavioral Ecology*, 8, 456–462.

Ladete-Castillejos T, Currey JD, Estevez JA, *et al.* (2010) Do drastic weather effects on diet influence changes in chemical composition, mechanical properties and structure in deer antlers? *Bone*, 47, 815–825.

Lincoln GA (1972) The role of antlers in the behaviour of red deer. *Journal of Experimental Zoology*, 182, 233–250.

Lincoln GA, Youngson RW & Short RV (1970) The social and sexual behaviour of the red deer stag. *Journal of Reproduction and Fertility*, 11(Suppl.), 71–103.

Loehr J, Carey J, Hoefs M, *et al.* (2007) Horn growth rate and longevity: implications for natural and artificial selection in thinhorn sheep (*Ovis dalli*). *Journal of Evolutionary Biology*, 20, 818–828.

Lott DF (1979) Dominance relations and breeding rate in mature male American bison. *Zeitschrift für Tierpsychologie*, 49, 418–432.

Lundrigan B (1996) Morphology of horns and fighting behavior in the family Bovidae. *Journal of Mammalogy*, 77, 462–475.

Malo AF, Roldan ERS, Garde J, *et al.* (2005) Antlers honestly advertise sperm production and quality. *Proceedings of the Royal Society of London B*, 272, 149–157.

Mattiangeli V, Mattiello S & Verga M (1998) Factors affecting the duration of fights in fallow deer (*Dama dama*) during the rut. *Ethology, Ecology and Evolution*, 10, 87–93.

McComb KE (1991) Female choice for high roaring rate in red deer, *Cervus elaphus. Animal Behaviour*, 41, 79–88.

McComb K, Reby D, Baker L, *et al.* (2003) Long-distance communication of acoustic cues to social identity in African elephants. *Animal Behaviour*, 65, 317–329.

McElligott AG & Hayden TJ (2001) Postcopulatory vocalizations of fallow bucks: who is listening? *Behavioral Ecology*, 12, 41–46.

Mesterson-Gibbons M, Marden JH & Dugatkin LA (1996) On wars of attrition without assessment. *Journal of Theoretical Biology*, 181, 65–83.

Mloszewski MJ (1983) *The Behaviour and Ecology of the African Buffalo*. Cambridge: Cambridge University Press.

Moen RA & Pastor J (1998) A model to predict nutritional requirements for antler growth in moose. *Alces*, 34, 59–74.

Moen RA, Pastor J & Cohen Y (1999) Antler growth and extinction of Irish elk. *Evolutionary Ecology Research*, 1, 235–249.

Moore NP, Kelly PF, Cahill JP, *et al.* & Hayden TJ (1995) Mating strategies and mating success of fallow (*Dama dama*) bucks in a non-lekking population. *Behavioral Ecology and Sociobiology*, 36, 91–100.

Moore WG & Marchington L (1974) Marking behaviour and its social function in white-tailed deer. In: V Geist & F Walther (eds.) *The Behaviour of Ungulates and its Relation to Management*, pp. 447–456. Morges: IUCN.

Morrell LJ, Backwell PRY & Metcalfe NB (2005) Fighting in fiddler crabs *Uca mjoebergi*: what determines duration? *Animal Behaviour*, 70, 653–662.

Muir PD, Sykes AR & Barrell GK (187) Calcium metabolism in red deer (*Cervus elaphus*) offered herbages during antlerogenesis: kinetic and stable balance studies. *Journal of Agricultural Science*, 109, 357–364.

Mysterud A, Meisingset E, Langvtan R, *et al.* (2005) Climate-dependent allocation of resources to secondary sexual traits in red deer. *Oikos*, 111, 245–252.

Payne RJH (1998) Gradually escalating fights and displays: the cumulative assessment model. *Animal Behaviour*, 56, 651–662.

Peek JM, Vanballanberghe V & Miquelle DG (1986) Intensity of interactions between rutting bull moose in central Alaska. *Journal of Mammalogy*, 67, 423–426.

Plard F, Bonefant C & Galliard JM (2011) Revisiting the allometry of antlers among deer species: male–male competition as a driver. *Oikos*, 120, 601–606.

Poole JH (1987) Rutting behaviour in African elephants: the phenomenon of musth. *Behaviour*, 102, 283–316.

Poole JH (1989) Announcing intent: the aggressive state of musth in African elephants. *Animal Behaviour*, 37, 140–152.

Poole JH (1999) Signals and assessment in African elephants: evidence from playback experiments. *Animal Behaviour*, 58, 185–193.

Prenter J, Elwood RW & Taylor PW (2006) Self assessment by males during energetically expensive contests over precopula females in amphipods. *Animal Behaviour*, 72, 861–868.

Reby D & McComb K (2003) Anatomical constraints generate honesty: acoustic cues to age and weight in the roars of red deer stags. *Animal Behaviour*, 65, 519–530.

Reby D, Joachim J, Lauga J, *et al.* (1998) Individuality in the groans of fallow deer (*Dama dama*) bucks. *Journal of Zoology, London*, 245, 79–84.

Reby D, Izquierdo M, Hewison AJM, *et al.* (2001) Red deer (*Cervus elaphus*) hinds discriminate between the roars of their current harem holder stag and those of neighbouring stags. *Ethology*, 107, 951–959.

Reby D, McComb K, Cargnelutti B, *et al.* (2005) Red deer stags use formants as assessment cues during intrasexual agonistic interactions. *Proceedings of the Royal Society of London B*, 272, 941–947.

Reby D, André-Obrecht R, Galinier A, *et al.* (2006) Cepstral coefficients and hidden Markov models reveal idiosyncratic voice characteristics in red deer (*Cervus elaphus*) stags. *Journal of the Acoustical Society of America*, 120, 4080–4089.

Riechert SE (1998) Game theory and animal conflict. In: LA Dugatkin & HK Reeve (eds.) *Game Theory and Animal Behavior*, pp. 64–93. New York, NY: Oxford University Press.

Roden C, Vervaecke H & Elsacker L van (2005) Dominance, age and weight in American bison males (*Bison bison*) during non-rut in semi-natural conditions. *Applied Animal Behaviour Science*, 92, 169–177.

Rogers LJ & Andrew RJ (2003) *Comparative Vertebrate Lateralization*. Cambridge: Cambridge University Press.

Rubenstein DI & Hack M (1992) Horse signals: the sounds and scents of fury. *Evolutionary Ecology*, 6, 254–260.

Ruiz-Miranda CR, Szymanski MD & Ingals JW (1994) Physical characteristics of the vocalization of domestic goat does *Capra hircus* in response to their offspring's cries. *Bioacoustics*, 5, 99–116.

Rushen J (1988) Assessment of fighting ability or simple habituation – what causes young pigs (*Sus scrofa*) to stop fighting? *Aggressive Behaviour*, 14, 155–167.

Rutberg AT (1986) Dominance and its fitness consequences in American bison cows. *Behaviour*, 96, 62–91.

Schaller GB (1967) *The Deer and the Tiger*. Chicago, IL: University of Chicago Press.

Searby A & Jouventin P (2003) Mother–lamb acoustic recognition in sheep: a frequency coding. *Proceedings of the Royal Society of London B*, 270, 1765–1771.

Shi J & Dunbar RIM (2006) Feeding competition within a feral goat population on the Isle of Rhum, NW Scotland. *Journal of Zoology*, 24, 117–124.

Silk JB, Alberts SC & Altmann J (2004) Patterns of coalition formation by adult female baboons in Amboseli, Kenya. *Animal Behaviour*, 67, 573–582.

Sinclair ARE (1974) The social organization of the East African buffalo. In: V Geist & F Walther (eds.) *The Behaviour of Ungulates and its Relation to Management*, pp. 676–689. Morges: IUCN.

Suttie JM (1980) The effect of antler removal on dominance and fighting behaviour in farmed red deer stags. *Journal of Zoology, London*, 190, 217–224.

Taillon J & Côté SD (2006) The role of previous social encounters and body mass in determining social rank: an experiment with white-tailed deer. *Animal Behaviour*, 72, 1103–1110.

Taylor PW & Elwood RW (2003) The mismeasure of animal contests. *Animal Behaviour*, 65, 1195–1202.

Tennenhouse E, Weladji RB, Holland Ø, *et al.* (2011) Mating group composition influences somatic costs and activity in rutting dominant male reindeer (*Rangifer tarandus*). *Behavioral Ecology and Sociobiology*, 65, 287–295.

Thirgood SJ (1990) Alternative mating strategies and reproductive success in fallow deer. *Behaviour*, 116, 1–10.

Vannoni E & McElligott AG (2009) Fallow bucks get hoarse: vocal fatigue as a possible signal to conspecifics. *Animal Behaviour*, 78, 3–10.

Vanpé C, Gaillard JM, Kjellander P, *et al.* (2010) Assessing the intensity of sexual selection on male body mass and antler length in roe deer *Capreolus capreolus*: is bigger better in a weakly dimorphic species? *Oikos*, 119, 1484–1492.

Wahlström LK (1994) The significance of male–male aggression for yearling dispersal in roe deer (*Capreolus capreolus*). *Behavioral Ecology and Sociobiology*, 35, 409–412.

Wahome JM (1995) Changes in interactions among tule elk bulls, *Cervus elaphus nannodes*. PhD thesis, University of California, Berkeley.

Walther FR (1974) Some reflections on expressive behaviour in combats and courtship of certain horned ungulates. In: V Geist & F Walther (eds.) *The Behaviour of Ungulates and its Relation to Management*, pp. 56–98. Morges: IUCN.

Weary DM & Fraser D (1995) Signalling need: costly signals and animal welfare assessment. *Applied Animal Behaviour Science*, 44, 159–169.

Wilkenson PF & Shank CC (1976) Rutting fight mortality among musk oxen on Banks Island, Northwest Territories, Canada. *Animal Behaviour*, 24, 756–758.

Willisch CS & Neuhaus P (2010) Social dominance and conflict reduction in rutting male ibex, *Capra ibex*. *Behavioral Ecology*, 21, 372–380.

Wolff JO (1998) Breeding strategies, mate choice, and reproductive success in American bison. *Oikos*, 83, 529–544.

Zahavi A (1975) Mate selection – a selection for a handicap. *Journal of Theoretical Biology*, 53, 205–214.

Chapter 15

Human contests: evolutionary theory and the analysis of interstate war

Scott A. Field & Mark Briffa

15.1 Summary

The past decade has seen a marked convergence between evolutionary models of animal contests and the analysis of interstate war, or 'militarised interstate disputes' (MID). Since James Fearon's landmark paper in 1995 on war as a bargaining problem, the literature on 'rationalist' approaches to modelling war has burgeoned and become increasingly sophisticated. It has moved from a 'Costly Lottery' approach (in which the decision to cease bargaining and fight is a game-ending move with a costly, probabilistic outcome) to a 'Costly Process' approach, in which states continue to accumulate information on relative strength and motivation while fighting, and use this to inform their strategic decisions about whether to continue fighting or revert to bargaining. The Costly Process approach has much in common with the evolutionary analysis of animal conflict, and may stand to gain from incorporating some of its theoretical insights and approaches. The actors in evolutionary models are in a very similar strategic situation to those of rationalist models: they are unitary actors with imperfect information who have a range of behavioural options to facilitate mutual assessment and may have incentives to resolve conflicts short of lethal combat. The concept of rational utility maximisation is analogous to the assumption that, over evolutionary time, natural selection has honed behaviour such that it represents a game-theoretic equilibrium. Most importantly, the expectation that signallers will misrepresent their capabilities and intentions means that costly, inefficient actions will usually be required to stabilise the reliability of the signalling system. We discuss two key evolutionary models of conflict, comparing them with recent Costly Process models of war and suggesting how they could stimulate new theoretical and empirical research. We argue that the ultimate utility of models of war will turn on their ability to make qualitatively distinct predictions about state behaviour, particularly with respect to the temporal organisation of costly actions, and how they process and respond to information on the costs incurred and imposed in the course of war.

15.2 Introduction: warfare as a Costly Process

The quest to develop a formal modelling framework for understanding the causes of war has spawned a rich and productive research programme over the past two decades (Powell 2002). Widely known as the 'rationalist' approach to war, it draws heavily on bargaining theory to address the central puzzle of war: why, when war is so self-evidently risky, costly and destructive, has it occurred with such regularity throughout history? As formulated by Fearon (1995) in his seminal paper on the rationalist approach, this reduces to answering the question why, when it can be shown that there always exists a bargain that both sides prefer to make than to fight each other, might they choose to fight nonetheless? His answer revolved around two pervasive constraints capable of preventing states from locating their bargaining range: informational asymmetries that produce uncertainty about the costs of fighting and the likelihood of winning; and commitment problems that make credible deals hard to strike. Most modelling work on warfare has focused on informational problems (e.g. Fearon 1995, Powell 1996*a*,*b*, 1999*a*, 2004, Wagner 2000, Filson & Werner 2002). Less attention has been paid to commitment problems, the main contributions being from Fearon (1995) and Powell (2006). In the process of exploring these stimuli

Animal Contests, ed. I.C.W. Hardy and M. Briffa. Published by Cambridge University Press.

to war, modellers have made a number of salutary contributions to security studies, clarifying the strategic logic of conflict and illuminating basic flaws in some of the most fundamental and longstanding debates in international relations theory. For example, Fearon (1995, pp. 384–385) discusses the roles of anarchy, the 'security dilemma' and 'preventive war', Powell (1999*b*) discusses absolute versus relative gains and the offence–defence balance and Powell (2002) discusses the balance versus preponderance of power debate and arms races. Thus, the rationalist analyses of warfare focus on the processes that lead to the outbreak of war, following an interaction that involves 'bargaining' between the protagonists.

Despite their contribution to the field of security studies, most early rationalist models shared a questionable assumption: that the decision to go to war effectively ended the bargaining game and the adversaries then entered into a 'costly lottery', in which the outcome was decided by a mixture of relative power and chance. This behaviour has also been referred to as a 'methodological bet' (Lake & Powell 1999). In the bargaining literature this notion is referred to as an 'outside option' or a 'game-ending' move, in which warfare is triggered when one of the parties resorts to an option outside the rules of the bargaining game (but see Wittman 1979, for an early exception). A more recent 'second wave' (also called the 'Costly Process' approach to modelling war: Powell 2004, p. 345) of rationalist models have sought to relax this assumption, treating war as an 'inside option' (i.e. a part of the bargaining game) and allowing the parties to continue interacting strategically once fighting begins. Such models take account of the well-documented fact that leaders do in fact modify their beliefs, expectations and bargaining or fighting strategy as a war unfolds, and this approach has proven capable of generating novel insights and predictions about the onset, duration and outcome of fighting. More generally, these costly process models of war address what numerous scholars have identified as an outstanding problem in security studies: the challenge of conceptually integrating different temporal phases of conflict. Levy (1995), for example, suggested that 'scholars must begin to focus more explicitly on the dynamic processes through which international conflicts progress from one stage to another', a sentiment later echoed in Brecher's (1999) call for a 'unified model of crisis' and more recently by Diehl (2006), who asked: 'If we know that conflicts go through different phases or stages, and these are likely to be interdependent, why do scholars insist on looking at only one moment...?'

This endeavour shows a striking parallel with developments in another scientific discipline that largely shares the same basic philosophy, methods, assumptions and subject matter as the rationalist approach: the evolutionary analysis of animal conflict. Evolutionary models of animal conflict (Chapters 1, 2 and 3) share the same basic assumptions of rationalist models of human interstate warfare: unitary actors with imperfect information who have a range of behavioural options to facilitate mutual assessment and incentives to settle conflicts short of lethal combat. Note that the term 'evolutionary' has also been used in a less-formal sense in international security, to refer to the developmental trajectory of long-term interstate rivalries (e.g. Hensel 1999). Scholars studying such rivalries have also borrowed terms such as 'punctuated equilibrium' from evolutionary biology (e.g. Cioffi-Revilla 1998), but these models from international security studies are distinct from the biological models discussed here.

The 'methodological bet' of rational utility-maximisation is also somewhat analogous to the 'phenotypic gambit' in evolutionary biology (Grafen 1990, 1991); the assumption that, over evolutionary time, natural selection has honed behavioural strategies such that they approximate a fitness-maximising optimum. Although this assumption is often criticised as unrealistic in both disciplines (e.g. Walt 1999 in international security; Pierce & Ollason 1987 in evolutionary ecology) and is indeed likely to be an approximation of reality, this is far from a fatal flaw. In both cases, the motivation of constructing models based on such strong assumptions is not that they are likely to be true; it is to gain understanding of how a simplified model of the system behaves, to serve as a reference point from which to explore the more realistic behaviour that results when these assumptions are relaxed. As Wagner (2000, p. 483) put it, when discussing decisions to go to war: 'It is common to attribute these decisions...to the irrationality of individual decision-makers...Such explanations may often be correct, but we cannot be sure without knowing what rationality would lead one to expect'. Formal models also serve the very important function, often neglected in informal verbal arguments, of making underlying assumptions quantitatively explicit, which provides structure and discipline to subsequent debates. Fearon (1995) and Powell

(1999*b*) make this point forcefully when showing how seemingly endless debates on fundamental concepts in international security have been clarified by making their logic explicit in formal models. Nonetheless, it remains an open question as to whether animal or human behaviour can attain or even approximate the equilibria predicted by models. For an argument that both biological and cultural phenomena will often exist out of equilibrium, see Enquist *et al.* (2002).

Rationalist and evolutionary approaches to conflict are also united as part of the broader transdisciplinary study of the strategic aspects of communication and signalling (Morrow 1999, Maynard Smith & Harper 2004, Hurd & Enquist 2005). A common theme in both literatures, and one which we will emphasise below, is the central role of costly signalling in shaping conflict dynamics (e.g. Fearon 1995, Maynard Smith & Harper 2004). Fearon (1995) explicitly recognised that because states have an incentive to misrepresent private information, they are compelled to force one another to make costly (and therefore more informative) signals, which could substantially increase the impetus towards war. However, in his model (an example of the literature on war as an outside option) costs could not be used to explain the dynamics of fighting, as this was simply a 'costly lottery' that occurred after bargaining had failed. Costs were incurred but did not determine the course of events, which were left to a mixture of the balance of power and chance. In contrast, the second wave of models of war as an inside option have begun to grapple with how the costs of fighting influence the trajectory, duration and outcome of conflict.

In this chapter, we argue that this is where certain evolutionary models can be particularly useful. The fact that animals are unable to bargain explicitly reverses the essential puzzle facing evolutionary biologists and scholars of international relations: the task for evolutionary biologists has been to explain why combatants do not immediately launch into escalated fighting, whereas for international relations scholars it is to explain why they fight at all. As a result, while rationalist models initially focused on bargaining and relegated the fighting to a lottery, evolutionary theorists have long been interested in how to model fighting as a process. Since evolutionary game theory started to be applied to animal contests in the 1970s (see the *Foreword*), four main models have been developed, each of which deals with the issues of costs differently, and each of which leads to distinct predictions about contest dynamics. Our purpose in this chapter is to compare and contrast two of these approaches with recent process models of war, to highlight unanswered questions that persist in both disciplines and to propose new avenues of empirical research designed to address them.

This chapter proceeds in four main sections. In section 15.3 we describe concepts derived from the evolutionary analysis of animal conflict that seem especially pertinent to understanding the process of interstate warfare in humans. We consider the variables thought to determine contest outcome, discuss the fundamental role of costly signalling in stabilising both animal and human communication systems, and discuss two evolutionary models of fighting behaviour, the sequential assessment model (SAM) (Enquist & Leimar 1983, 1990) and the cumulative assessment model (CAM) (Payne 1998), that we consider to be potentially relevant to interstate warfare. In section 15.4 we compare and contrast the most recent wave of 'Costly Process' models of war with the SAM and the CAM. This analysis highlights some key unresolved issues in the study of war as a process and suggests some novel directions for future empirical research, which we briefly explore in section 15.5. Specifically, we suggest that documenting the temporal sequencing of actions of various cost during crisis and war, and studying the decision rules states used to track and respond to costs incurred and costs imposed during fighting will be important to advancing the field. In section 15.6, we conclude with an overview of the common ground between evolutionary and rationalist models of conflict, emphasising the need to focus on extracting diagnostic predictions from the broad array of models now available, and subjecting them to empirical tests.

15.3 Evolutionary analysis of animal conflict: basic parallels with interstate warfare

15.3.1 The numbers game: a note on attrition laws versus strategic decisions

Although we discuss interstate warfare, i.e. aggressive encounters between multiple human opponents, we focus on parallels with models that have been devised to explain the evolution of strategic decisions in dyadic animal contests (Chapter 2) rather than those that focus on multi-party contests (Chapter 3). Multi-party contest theory can, of course, also be applied to the case of human conflict. Indeed, it was first devised during and following the First World War (Lanchester 1956, reprinted from a paper originally published in 1916) in order to model the outcomes of mechanised warfare between opposing armies. Rather than being concerned with the evolution of strategic decisions during warfare, this body of theory ('Lanchester's attrition models') was devised in order to predict attrition rates in the field and is therefore concerned with questions about the optimal level of investment in troop numbers and equipment such as aircraft and artillery, rather than the decision of whether to 'go to war'. These attrition laws have, in turn, been adapted to the case of inter-group aggression in animals, in a series of models that deal with the evolution of trade-offs in investment between the number of individuals and individual fighting ability (Chapter 3). These developments have been tested empirically using non-human animals, most notably ants: for an example of an attempt simultaneously to analyse multi-party contests in ants from the perspectives of strategic and tactical decisions and attrition rates, see Batchelor and Briffa (2010, 2011, Chapter 8).

15.3.2 Variables influencing contest outcome

Early theoretical work on animal conflict centred on the idea that there must exist asymmetries between the parties that could be used as a basis for settling the conflict short of lethal combat (Maynard Smith & Price 1973, Maynard Smith 1974, Parker 1974, Maynard Smith & Parker 1976). As discussed in Chapters 1 and 2, this led to the underlying proposition that has informed much subsequent work on animal contests: that animals use the interactions during a contest for the purpose of either 'mutual' (e.g. as assumed by the SAM) or 'self' assessment (e.g. as assumed by the CAM or energetic war of attrition, EWOA), and use this to decide whether to continue fighting or give up. As discussed in Chapter 2, and in many of the chapters based on empirical studies, these decisions might be based on asymmetries in various correlates of fighting ability (resource holding potential, RHP) or resource value, *V*. Four additional variables, assumed to act as modulators of fighting ability and resource value, are also of potential importance. The first is ownership status, whether one of the animals is in possession of the resource and better able or motivated to defend it from an intruder. Second is previous fighting experience, which may give an individual a prior expectation of success that influences its starting estimate of relative fighting ability. The third is the 'shadow of the future', or expectations of future fitness rewards subsequent to the current contest. This is of particular importance if the individual is nearing the end of its lifespan, in which case it may become highly motivated to engage in escalated or fatal fighting (Enquist *et al.* 1990). A fourth factor is inherent 'aggressiveness', a willingness to escalate a contest independent of fighting ability and *V*. Although the evolutionary significance of this trait has been questioned (Hurd 2006), its presence has been demonstrated empirically in studies on fish (Earley & Hsu 2008) and recent work on aggressiveness has been examined from the perspective of 'animal personalities' (Wilson *et al.* 2011, Rudin & Briffa 2012).

15.3.3 Deception and the importance of costly signalling

A fundamental problem common to both animal communication and international relations is the possibility of deception, or bluffing. In a range of biological interactions, and particularly in mating and fighting behaviour, the interests of senders and receivers may differ radically and therefore attempts by senders to manipulate receivers by making deceitfully exaggerated signals of traits such as genetic quality or fighting ability were expected to be widespread (Maynard Smith 1974, Dawkins & Krebs 1978). Subsequent empirical work shows that while bluffing by exaggerating RHP can occur (e.g. in crustaceans, Chapter 5) it appears to be rare compared to the use of 'honest signals'. The theoretical possibility of bluffing, however, made the widespread existence of honest signals something of a puzzle. A major part of the answer was provided by the theory of costly signalling: the notion that the biological meaning of signals could be evolutionarily stabilised by the imposition of costs. The development of this theory is reviewed in depth in Chapter 2. Briefly, natural selection, it was hypothesised, would act upon receivers to respond preferentially to signals that were costly for the sender to make, thus reducing the scope for bluffing. At evolutionary equilibrium, signals would be costly to the sender in a manner directly related to the quality they were attempting to signal, and would thus be 'honest' or reliable. As noted above, this is not to say that cheating would be eliminated entirely; it may still exist as a kind of 'tax' on the reliability of the signal. However, it should occur at sufficiently low frequency that the signal would remain, on average, honest (Grafen 1990). Consequently, honest signals may also be quite extravagant and wasteful, as suggested by Zahavi in his famous 'handicap principle' (Zahavi 1975). For example, a male's ability to produce an elaborate costly ornament and survive despite its cumbersome effect could reliably signal genetic quality to a female. Similarly, an individual conspicuously wasting food might be signalling that it is a highly capable forager, or an individual conspicuously wasting energy signalling that it is in excellent condition. As Grafen (1990) put it in his mathematical vindication of Zahavi's verbal arguments, 'signalling systems require waste to ensure honesty'.

By considering the possibility of bluffing and the evolution of honest signals, what Fearon called the 'central puzzle of war', namely that it recurs despite being costly and inefficient (Fearon 1995), looks much less enigmatic. State leaders find themselves in much the same strategic situation as the actors in evolutionary signalling systems; knowing their adversaries have an incentive to misrepresent their capabilities and intentions through signals, they seek to make these signals costly to produce. Fearon (1995), of course, independently arrived at much the same conclusion as to how a state could enforce honesty; by choosing actions that 'run a real risk of (inefficient) war in order to signal that it will fight if not given a good deal in bargaining'. The delicate task of a statesman is to control the cost of these actions such that they achieve this task without becoming embroiled in a confrontation whose costs exceed the bargaining benefits it was intended to extract. In a sense, they are in the unenviable position of having to try and solve a problem that natural selection has been working on for millions of years. With that in mind, it may be instructive to review the ways in which natural selection has organised behaviour in response to these same pressures, and how evolutionary biologists have sought to model them.

15.3.4 Animal contests and evolutionary game-theory models

It is evident from many chapters in this book that although the full spectrum of animal contests includes a staggering diversity of forms and behaviour patterns, one of the most robust generalisations is that agonistic encounters do not necessarily begin with dangerous fighting, but with a series of actions that involve little or no physical contact. Such early phases are usually 'ritualised'. This term refers to the stereotyping and amplification (e.g. through exaggerated repetition) of behavioural acts (Huntingford & Turner 1987, p. 40, Maynard Smith & Harper 2004, Chapter 5). The exact switching point from ritualised display to un-ritualised fighting may often be difficult to define, as the key evolutionary models focused on the puzzle of how such restrained behaviour could have evolved (e.g. the Hawk–Dove game) and then subsequent models (e.g. SAM, CAM, EWOA) focussed on how non-injurious (but often costly) behaviour could induce one opponent to give up.

Thus, evolutionary modellers have come up with a number of different explanations for how the giving-up decision that terminates a contest is made. As discussed in Chapter 2, each of these models, subsequent to the initial Hawk–Dove game and its derivatives, assumes a different underlying decision rule for how the opponent is assessed and how decisions about whether to continue fighting are made. Here we consider two that appear most relevant to modelling human war: the sequential assessment model, SAM, and the cumulative assessment model, CAM. A full discussion of these models is given in Chapter 2, and they are summarised in Chapter 1, but their key relevance to the field of interstate warfare is that they make qualitatively different predictions that are amenable to testing with available data on warfare; sources of such data are discussed below.

Briefly, the SAM treats contests as a series of repeated bouts of fighting through which the combatants update their estimate of their relative strength, in a process akin to statistical sampling. Players engage each other in bouts of fighting that allow estimates to be made but, because there is a random error in that estimate, they must increase their 'sample size' by repeating the bouts numerous times. Several behavioural options are available, and when one option begins to deliver diminishing informational returns, the players switch to a new one. They continue to engage the opponent until they have a sufficiently precise estimate that they are either inferior, or are so evenly matched that the contest is likely to continue beyond what is justified by the subjective value of the resource, V. Thus they will only give up if and when their estimate of relative fighting ability falls below an 'evolutionarily stable switching line' (Chapter 2 provides general discussion of evolutionarily stable strategies or 'ESSs') whose shape reflects the increasing precision of the estimate as information is acquired throughout the contest (Figure 15.1; see also Figure 6.2). Contests are divided into 'phases', which may be defined by a clean switch from one behaviour to another, but the model also allows for phases to contain more than one behaviour, in which case they are characterised by the relative proportions of each behaviour used. Figure 15.1 shows that early in the contest the animal is predicted to give up only if it has a negative estimate of fighting ability that is very large, because that estimate is also very imprecise and its negative value may be due to random sampling error. Later in the contest, however, even a very slight negative estimate will cause it to capitulate, because the animal is by now highly certain of the estimate's accuracy and that it will be a good predictor of the eventual fight outcome. The position of the switching line varies according to three factors: (1) the value of the resource; (2) the costs of fighting; and (3) the accuracy of the sampling process. Combatants will have a higher switching line (and thus accept defeat earlier) if the resource is of low value, the costs of fighting are high, or if there is a high degree of noise associated with the type of signalling being used. In terms of contest dynamics, the SAM makes several key predictions (see also Chapter 2).

1. The contest should be organised into phases characterised by a single behaviour or a mix of behaviours in constant proportions, and the final phase should be dangerous, unrestrained fighting.
2. Each player should fight with constant intensity, to ensure maximum accuracy of the signal given and received.
3. When contestants are more evenly matched, the contests should include more phases, last longer on average and be more variable in duration.
4. When resource value is higher, the contests should last longer on average and be more costly to both parties.
5. High resource value will favour the owner of the resource over a challenger.

While the SAM conceptualises informational updating during fighting, it does not deal with the impacts of cost accumulation. It assumed that the sole purpose of cost expenditures incurred by combatants was for the exchange of information, whereas in the case of

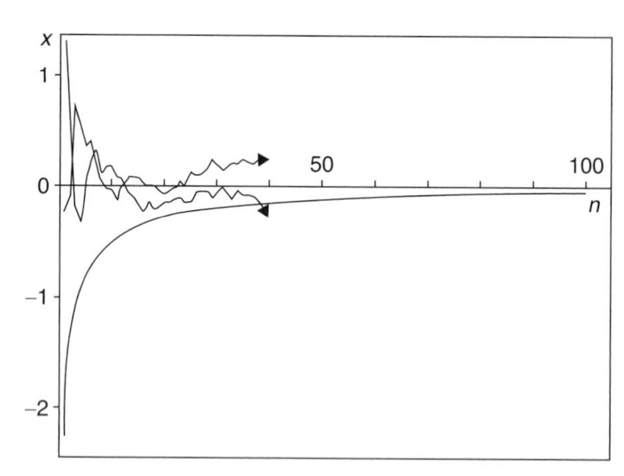

Figure 15.1 An example of an animal contest as conceptualised in the sequential assessment model. The *x*-axis represents the number of repetitions of a behavioural action (and thus the time-course of the contest, '*n*') and the *y*-axis the sampling average ('*X*'), that is each player's estimate of relative fighting ability. The ESS switching line (below which a player gives up) is the smooth solid curve and the two traces represent the separate estimates of the two contestants, one of whom gives up after crossing the switching line after refining its estimate 39 times. (Reproduced, with permission, from Enquist & Leimar 1983.)

interstate conflict costly action may also be designed to inflict damage on the opponent. Indeed, a recent model of interstate warfare by Kadera and Morey (2008) shows how the specific targeting of industrial capacity should lead to conflicts of shorter duration. Certainly, these two aims – to intimidate the opponent into making a giving-up decision (i.e. to capitulate) and to inflict damage that will reduce the opponent's ability to persist – are not easy to disentangle in the case of interstate warfare. For example, during the Second World War the Royal Air Force pursued an 'area bombing' campaign against the major industrial centres of Germany. Rather than restricting the targets to industrial and transport infrastructure the campaign also targeted the associated civilian populations. According to Air Chief Marshall Arthur Harris, the Commander in Chief of RAF Bomber Command (1942–1945), these attacks were primarily designed to degrade the morale of the enemy civil population (Harris 1995). However, this (in the reasoning of Bomber Command) entailed inflicting significant damage. Thus, the strategy ('why') was closely bound with the tactics ('how') employed: 'The ultimate aim of an attack on a town area is to break the morale of the population which occupies it. To ensure this, we must achieve two things: first, we must make the town physically uninhabitable and, secondly, we must make the people conscious of constant personal danger. The immediate aim is, therefore, twofold, namely, to produce (i) destruction and (ii) fear of death' (Harris 1995).

In the context of interstate warfare 'morale' seems similar to the concept of 'motivation' as applied to animal contests: in both cases the term pertains to the 'willingness to fight' and experimental studies on animal contests show that motivation varies with access to information about the costs and benefits of persistence (e.g. Chapter 5). It is clear that a key assumption made by Bomber Command is that the informational state of the enemy would influence their strategic decision-making. However, this information could not be supplied without inflicting significant costs. One form of the SAM, the 'SAM with fatal fighting' (Enquist *et al.* 1990), does include damage costs, but only in the form of a random chance of being injured or killed as the fight progresses. In terms of interstate warfare, this would equate to a random chance of an entire belligerent state being somehow rendered incapable of persisting. Therefore, although the role of information may be central in determining the protagonist's decision-making during interstate warfare, the accumulation of actual damage is also important.

In models of contest behaviour, the challenge of incorporating cost accumulation was taken up in the energetic war of attrition, EWOA (Payne & Pagel 1997), and the cumulative assessment model, CAM (Payne 1998). In both cases the decision about whether to keep fighting is determined not by an averaged estimate of relative strength, but by the total physical costs that an individual is able to sustain. These costs come from two sources: an individual's own energetic costs from performing actions of a chosen type and intensity and damage costs inflicted by the other party. The key difference between the CAM and the EWOA is that in the CAM damage costs are not under an individual's own control (Chapters 1 and 2). This means each player is expected to adjust its rate of escalation in response to that of its opponent. Unlike the SAM, which predicts combatants' fighting intensities to be stable within phases, the CAM predicts that each player could change its intensity, with escalation predicted in costly or damaging phases. This opens up greater possibilities for variation in contest dynamics. The most interesting result is that a disparity in escalation rates between weaker and stronger fighters is expected. In short contests the encounter will be resolved without escalation while in long contests the stronger opponent is expected to escalate more rapidly and, in medium-length contests, weaker individuals start at lower intensity but then escalate more rapidly than their opponents (Figure 15.2). This result appears to hold whether the quality of an individual is determined more by its ability to tolerate cost accumulation, its defensive skill, or its ability to inflict damage on its opponent. The issue of 'quality' is discussed in more detail in section 15.4.

15.4 Costly Process models of war in relation to evolutionary models

As mentioned in section 15.2, unpacking the 'Costly Lottery' assumption of war and exploring the dynamics generated by bargaining while fighting has recently opened up as a major research frontier. A spate of new Costly Process models have approached the problem in different ways and revealed new insights. While animal conflict models introduced here have some basic differences and would be in need of revision and/or extension before their predictions could be seriously tested, they do suggest different ways of looking at the problem and may yield useful insights. In what follows, we concentrate on connecting the SAM and the CAM to the rationalist literature and suggesting a research agenda to explore their consequences.

15.4.1 War as information-gathering

One striking feature of several of the Costly Process models of war is that they share a fundamental assumption with the SAM: namely, war is viewed as an information acquisition process that reduces uncertainty about relative strength. Wagner, using a formulation that precisely mirrors that of the SAM's authors and evolutionary game-theory models in general, suggested that war is 'an experiment that allows states to test competing hypotheses about the outcome of a war fought to the finish' (Wagner 2000, p. 478). In Filson and Werner's model, 'the war ends when the attacker's and the defender's beliefs about the defender's power converge sufficiently' (Filson & Werner 2002, p. 832). Slantchev (2003*a*), in whose model war ends 'after opponents learn enough about their prospects in war to decide that continuation is unprofitable', has called this the 'principle of convergence'. These models agree that fighting shifts players' estimates of relative strength and narrows the uncertainties around those estimates, in principle much as

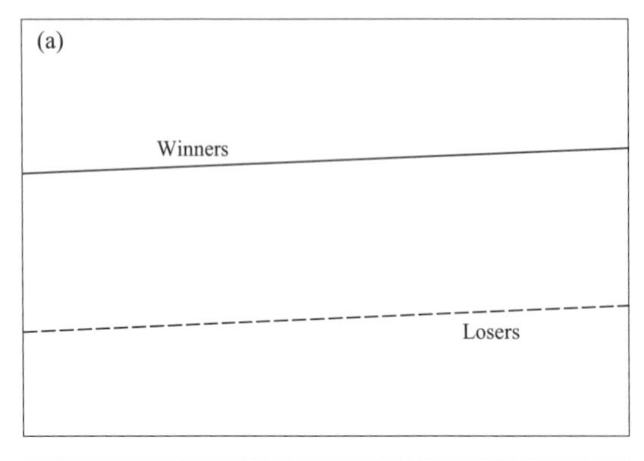

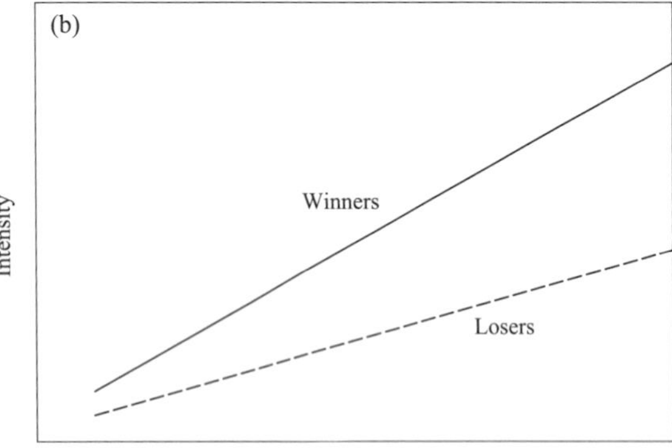

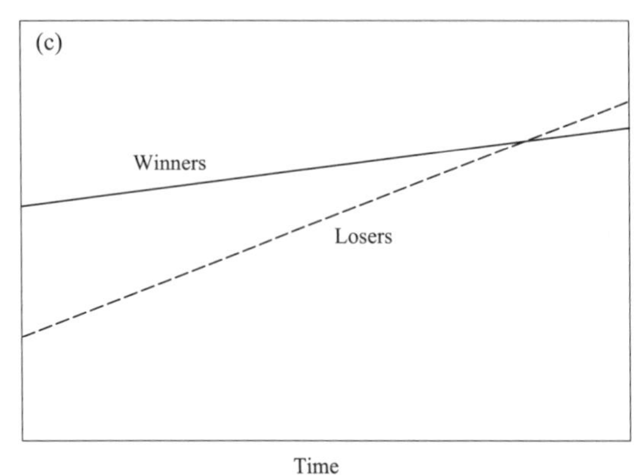

Figure 15.2 Rates of escalation of eventual winners and losers during contests of different duration, as predicted by the cumulative assessment model for contests of short duration (**a**), long duration (**b**) and intermediate duration (**c**). In short contests, individuals with high RHP perform at a greater intensity than those with low RHP and win the encounter quickly without needing to escalate. If the weaker opponent persists for a significant period, the initial rates of aggression become irrelevant. The opponents are forced to escalate and the individual with greater RHP will win the contest by escalating more rapidly than the individual of lower RHP. In intermediate-length contests, the escalation rates are determined by the initial levels of performance. The weaker individual will try to compensate for the initial performance of the opponent by escalating more rapidly. (Reproduced, with permission, from Payne 1998.)

depicted in the ESS switching curve derived from the SAM in Figure 15.1.

However, a fundamental difference between war and animal contests is the presence in war of explicit bargaining, in which language is used to propose specific, detailed suggestions about the terms of settlement. This is a fundamental part of all of the Costly Process models of war discussed here, which allow players to alternate between bargaining and fighting. Bargaining provides an alternative, parallel channel for information transmission and coordination of expectations, which, via the nuance and sophistication of language, should contribute disproportionately to clarifying the informational situation. As Fearon puts it, 'One might expect that... state leaders who disagree on some issue could simply tell each other what they would be willing to accept rather than fight, and then choose a mutually acceptable bargain' (Fearon 1994, p. 578). Slantchev (2003*a*) also emphasises the greater informational content of bargaining over fighting, saying: 'Fighting does reduce uncertainty, but the battlefield is a noisy source of information and not the only one'.

Even the powerful communicative advantage of language does not, however, provide an escape from the universal risk of deception in signalling, whose potency evolutionary theorists have long grappled with. Fearon (1994) goes on to say that, given the incentives to misrepresent, 'quiet diplomatic exchanges may be rendered uninformative' and 'normal forms of diplomatic communication may be worthless'. Taking the example of the Cuban Missile Crisis, even with the potentially catastrophic consequence of escalation to a nuclear exchange between the USA and the USSR, this dilemma rendered it pointless for President Kennedy to ask General Secretary Khruschev what he would do in the event of a US blockade of Cuba (Fearon 1994).

The point is that the most informative parts of bargaining may not be the words spoken but the symbolic effect of the opening and closing actions of the session: sitting down (representing the willingness to offer a settlement rather than fight) and standing up before a deal is struck (representing the willingness to return to combat). Furthermore, negotiators do not have infinite flexibility in the range of deals they bring to the table, the upper (most generous) limit of which represents the point at which they will walk away and return to fighting. These limits are set by the most recent round of fighting, which, as mentioned above, has modified their point estimates of relative strength and motivation and the associated degree of uncertainty. The bargain they present will simply reflect their expectations about the outcome of continued combat. In Slantchev's (2003*a*) formulation, bargaining is a means of attempting to coordinate these expectations and thus put an end to fighting.

15.4.2 War as a balance between incurring and inflicting costs

While the SAM fits very well with the perspective of war as an information-acquisition process, it does not deal very effectively with the issue of cost accumulation. Costs influence fight dynamics only in the sense that the players choose progressively more costly actions as the fight progresses, in order to reveal information about themselves and their opponent. It does not account for the fact that the players may have a limited ability to bear the costs incurred by fighting, and the costs imposed on them opponent, and that this might influence the trajectory of the contest and its outcome. Similarly, Wagner's and Powell's Costly Process models of war, while they include costs of fighting, do so as a fixed per-period cost that is not directly under the players' control and thus not able to influence fight dynamics directly (Wagner 2000, Powell 2004). Filson and Werner make some headway on this problem by making states' ability to fight dependent on their resource levels, which can rise and fall during war as battles are won and lost (Filson & Werner 2002, 2004). In a second model, they advance a little further by allowing states to vary in their sensitivity to the risk of losing resources, western-style liberal democracies being the most cost-sensitive and thus readily prepared to make concessions rather than fight. However, in all these cases it is only the ability to bear the costs of fighting that influence the course and outcome of the contest, and not the ability to impose costs. As we have already seen in the CAM, the relationship between the two may be critical.

The only Costly Process model to address this point is that of Slantchev (2003*b*), who points out that the abilities to bear costs and inflict costs jointly determine the bargaining range that emerges through fighting. Thus he suggests 'the process of war can be usefully viewed as a contest, in which both sides attempt to reduce the opponent's ability to impose costs on them while simultaneously trying to impose costs on the opponent, thereby improving their bargaining position'. By assuming complete information and that war brings only costs and no benefits, he convincingly makes the case that variation in the abilities to inflict and sustain costs can of themselves lead to the inefficient outcome of fighting. Citing a number of historical examples, he claims that the 'power to hurt', which depends on the relative magnitude of these two costs, is a key variable that can explain otherwise anomalous behaviour, especially prolonged cases of asymmetric conflict where the weaker side is vastly inferior in conventional military terms, but its 'power to hurt' is significant, nonetheless.

While this approach succeeds in highlighting how the intriguing interaction between costs incurred and imposed can cause fighting, the CAM goes one step further and asks how it might affect the dynamics of fighting. As Slantchev (2003*b*) correctly notes: 'the costs associated with these actions [bearing and imposing costs] are analytically distinct and not necessarily related in a straightforward manner'. It is precisely this issue that Payne addresses when he says: 'Having a higher rate of performance reduces one's own give-up time, but also increases the opponent's damage costs, which reduce the opponent's give-up time. It is therefore not necessarily clear how individuals of various qualities are likely to differ in their optimal values of r and ρ' (Payne 1998). Payne's (1998) analysis revolves around these two quantities, what he calls the 'performance parameters': the initial rate of performance; and the rate of escalation. As discussed above, lower quality individuals may escalate more rapidly because they are accumulating costs more quickly and therefore have greater urgency to bring the conflict to an end (Figure 15.2). Given their lower relative efficiency in inflicting cost on their opponent, their only hope of doing so is to escalate rapidly, hoping that this might overwhelm the opponent before they are themselves exhausted.

This model also raises a critical concept that encompasses the two kinds of costs discussed above, and will be pivotal in future work in this area: quality. By focusing on the complex relationship between the ability to incur and inflict costs, it illuminates the fact that the traditional surrogates for quality ('fighting ability' in animal contests and 'military power' in international relations) are in great need of being unpacked and critically examined, particularly with respect to their interactions with the two different types of costs discussed above. As Payne (1998) points out, 'whichever phenotypic trait has the major role in causing the costs to differ [between the opponents] is the "quality" upon which success hangs'. He suggests various possibilities: 'the energetic efficiency of performing the actions, or defensive skill, or ability to persist despite heavy costs, and so on'. To the extent the information-acquisition view of war is correct, it is dimensions such as these that states are likely to be assessing their opponent on during combat. All of Payne's suggestions have obvious analogues in warring states, and exploring their interactions with traditional notions of power promises to be an illuminating way to advance the study of war.

15.5 New directions for empirical research on interstate warfare

As mentioned in section 15.2, the degree to which formal modelling can enlighten us about the dynamics of war hinges on the extent to which a given model accurately captures the underlying decision rules being used by participants. In turn, our ability to distinguish which model performs better depends on showing how the different decision rules translate into behaviour that is both qualitatively different and measurable. Payne and Pagel (1997) provide an example with evolutionary models, showing that indicators such as the number of repetitions of particular behaviours and whether the intensity of performance increases, decreases or stays the same over time can help distinguish which model better captures the decision rules driving the contest. Other indicators are whether contest duration correlates better with relative strength or the strength of the weaker contestant only (Chapter 4), whether information transfer is highly error-prone and whether the time costs to continuing in the contest accrue linearly, sublinearly or superlinearly. Analysis of time costs can prove particularly informative, because under models that assume a switching line based on cumulative cost, rather than on information about the opponent (i.e. self-assessment rather than mutual assessment, as described in Chapter 4), the rate of accrual is expected to drive the pattern of escalation in intensity (Payne & Pagel 1997, Figure 15.3).

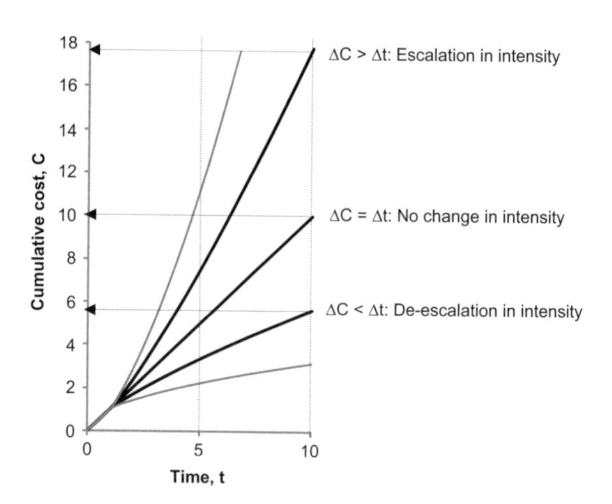

Figure 15.3 Rates of cost-accrual in animal contests. Δt = change in time and ΔC = change in the cost of performing agonistic behaviour. If costs accumulate linearly with time there will be no change in the intensity of agonistic behaviour. If the rate of cost accrual is superlinear ($\Delta C > \Delta t$) such that a doubling of duration leads to more than a doubling of costs, then the intensity will increase. Therefore, in the scenarios with the greatest rates of cost accrual ($\Delta C > \Delta t$) escalation in intensity is predicted. If the rate of cost accrual is sublinear ($\Delta C < \Delta t$), and a doubling of time units produces less than a doubling of costs, then the intensity will decrease (de-escalation). Therefore, the pattern of change in intensity is driven by the second derivative of the costs that accumulate over time (d^2C/dt^2). The rates of escalation and de-escalation for interactions represented by grey lines would be greater than for those represented by the black lines. Drawn from the EWOA model described in Payne & Pagel (1997), reproduced from Briffa & Sneddon (2010).

A precise statement of how such reasoning could motivate an empirical research programme to test rationalist models of war must await the theoretical work of adapting the mechanics or logic of models such as the CAM and SAM, but in what follows we offer some general suggestions for how it might proceed. A first step might be to attempt a comprehensive survey of the temporal sequencing of actions taken by states between the onset of crisis and the conclusion of a war. A clear and basic prediction of the SAM is that they should be ordered into a sequence of increasingly costly phases. In theory, the different 'phases' of the SAM map onto the 'rounds' of fighting that alternate with bargaining in Costly Process models of war. Aside from increasing in cost, the SAM also predicts that if multiple options are employed during a phase, those options should be applied with constant intensity and in constant proportions to one another. The Codebook for the Dyadic Militarised Interstate Incident Data from the Correlates of War data set (Ghosn *et al.* 2004) provides a starting point from which to categorise and set about collecting and analysing the relevant data. For example, variable number 14 (HiAct) in the codebook is the highest-level hostile action taken in a particular dispute and consists of 21 different actions broken into five 'Hostility Levels' (Table 15.1). One caution, however, is that these categories encompass actions generally included under the heading 'crisis', which is much broader than war. Although Costly Process models are generally aimed at modelling only war, there is no reason in principle they could not be generalised. This is the key to meeting the challenge of developing a 'unified model of crisis' discussed in section 15.2. Nevertheless, historical narratives of interstate conflict could be coded using these categories and used to search for temporal patterns.

An obvious problem is that the costs of different actions could be very difficult to quantify, and the assumption that states were cognizant of their relative costs similarly hard to justify. However, simply documenting the sequencing patterns could be useful in that it might form a basis for inferring how states have historically ranked the costs of different coercive policy options in practice. Documenting whether these patterns have changed over time, and if so in what direction and why, could prove an illuminating exercise in itself and a useful first step.

Following from the CAM's focus on the balance between costs incurred and imposed, empirical work on how states measure and respond to these variables will also be important. Some such work has already been attempted, and has shown that states indeed attempt to measure precisely the kind of variables that underpin the model (Gartner 1997). For example, evidence from the two World Wars and the Vietnam War shows that decision-makers explicitly based decisions on the costs incurred (such as merchant shipping lost or battle deaths) and costs imposed (U-boats sunk, or communist forces killed and weapons captured) and, more interestingly, also on the rates of change in these variables. Decision-makers were thus explicitly calculating some of the key parameters in the CAM, and, according to this analysis, reacting to them in a 'rational' manner. Where future work in this area should lie is in defining exactly what is meant by 'rational', i.e. quantifying the strategic response to changes in performance indicators, and attempting to deduce the underlying decision rule as a means of distinguishing between alternative models.

Table 15.1 Military actions at five different levels of hostility. Hostility levels are as listed in the *Codebook for the Dyadic Militarised Interstate Incident Data* in the *Correlates of War* data set (Ghosn *et al.* 2004).

Level 1: No militarised action	
	0 No militarised action
Level 2: Threat to use force	
	1 Threat to use force
	2 Threat to blockade
	3 Threat to occupy territory
	4 Threat to declare war
	5 Threat to use chemical, biological or radiological (CBR) weapons
	6 Threat to join war
Level 3: Display of force	
	7 Show of force
	8 Alert
	9 Nuclear alert
	10 Mobilisation
	11 Fortify border
	12 Border violation
Level 4: Use of force	
	13 Blockade
	14 Occupation of territory
	15 Seizure
	16 Attack
	17 Clash
	18 Declaration of war
	19 Use of CBR weapons
Level 5: War	
	20 Begin interstate war
	21 Join interstate war

15.6 Conclusion

In this chapter we have argued that when the fields of evolutionary ecology and international relations analyse conflict, they stand on remarkably similar theoretical foundations. Both animal conflict and war are examples of signalling systems in which the possibility of deception compels the parties to engage in costly actions in order to stabilise the meaning of the signals transmitted. Following this insight, both theoretical traditions have emphasised the strategic significance of information deficits, and have sought to model fighting as a process of information acquisition that clarifies combatants' relative strength and motivation and may thus end the conflict. This view of fighting has a strong cooperative dimension: the parties are in a very real sense agreeing to a coordinated, restrained process of information exchange that allows them to distribute the benefits in dispute equitably without paying the costs of undue escalation. This view predicts conflict to consist of a sequence of actions that are increasingly costly, but also increasingly likely to yield a convergence of expectations that ends the fighting.

However, both fields have also recognised that this cooperative information exchange is conjoined with an evil twin: a highly competitive process in which the parties aim primarily to inflict costs on one another and which may generate very different dynamics. In this view, the actors pay the costs of conflict not as a signal of strength or resolve but in a straightforward attempt to drive the costs incurred by their opponent above some maximum sustainable threshold. The outcome of the contest then turns largely on the fighting efficiency of each party: their ability to maximise the cost inflicted on the opponent while minimising the performance and damage costs they incur themselves. The critical difference here is that cost-incurring behaviour may be used strategically: rather than inflicting costs in a consistent manner designed to reflect strength accurately, an actor may increase and decrease its performance depending on how long the contest has been going on and as a best response to the performance of the opponent thus far.

The issue of such dynamics generated by strategic manipulation of the ratio of 'cost incurred to cost imposed' during fighting has not been adequately dealt with thus far in rationalist models. To the extent that this parameter is under the control of participants, they should obviously strive to minimise it, and their attempts to do so should have a profound impact on both the choice of actions and their temporal patterning during fighting. As theorists from both disciplines have noted, how this will play out is by no means straightforward and capturing its essence in a model that makes clear predictions is a major unsolved problem that only formal game-theoretic modelling is capable of addressing. This should be a prime target of future work.

The issue of deriving testable predictions from models is a difficult issue that bedevils both disciplines. The models developed thus far have unarguably done much to clarify the strategic logic of conflict, which in itself is a major contribution, but it is as yet unclear whether they can go beyond generalities and yield predictions sufficiently detailed to propel a productive programme of empirical research. In evolutionary ecology the extreme difficulty of performing quantitative tests of models has been largely abandoned in favour of using diagnostic qualitative predictions from competing models to try to pinpoint the decision rules underlying behaviour. The rapidly expanding variety of formal models of bargaining and war suggest that a similar endeavour is needed in international security, as Powell advocated in his review of the field (Powell 2002). As Wagner put it, 'clearly understanding the relation between war and bargaining is crucial to explaining the great variety of military conflicts that have occurred' (Wagner 2000, p. 483), and this is a task that has barely begun. Clarifying the essential differences in decision rules among competing models and identifying mutually exclusive predictions will be an important way to progress towards this goal, as it will anchor the models to an empirical research programme and help ensure that rigour, to use the metaphor of one prominent critic (Walt 1999), does not turn into rigor mortis.

Acknowledgements

We are grateful to Sarah Collins and Tim Batchelor for their referees' comments on this chapter and to Ian Hardy for his editorial comments.

References

Brecher M (1999) International studies in the twentieth century and beyond: flawed dichotomies, synthesis, cumulation: ISA presidential address. *International Studies Quarterly*, 43, 213–264.

Batchelor TP & Briffa M (2010) Influences on resource-holding potential during dangerous group contests between wood ants. *Animal Behaviour*, 80, 443–449.

Batchelor TP & Briffa M (2011) Fight tactics in wood ants: Individuals in smaller groups fight harder but die faster. *Proceedings of the Royal Society of London B*, 278, 3243–3250.

Briffa M & Sneddon LU (2010) *Contest behavior*. In: DF Westneat & CW Fox (eds.) *Evolutionary Behavioral Ecology*, pp. 246–265. New York, NY: Oxford University Press.

Cioffi-Revilla C (1998) The political uncertainty of interstate rivalries: a punctuated equilibrium model. In: P Diehl (ed.) *The Dynamics of Enduring Rivalries*, pp. 64–97. Urbana and Chicago, IL: University of Illinois Press.

Dawkins R & Krebs JR (1978) Animal signals: information or manipulation? In: JR Krebs & NB Davies (eds.)

Behavioural Ecology: An Evolutionary Approach, pp. 282–309. Blackwell: Oxford.

Diehl P (2006) Just a phase? Integrating conflict dynamics over time. *Conflict Management and Peace Science*, 23, 199–210.

Earley RL & Hsu Y (2008) Reciprocity between endocrine state and contest behavior in the killifish, *Kryptolebias marmoratus*. *Hormones & Behavior*, 53, 442–451.

Enquist M & Leimar O (1983) Evolution of fighting behaviour: decision rules and assessment of relative strength. *Journal of Theoretical Biology*, 102, 387–410.

Enquist M & Leimar O (1990) The evolution of fatal fighting. *Animal Behaviour*, 39, 1–9.

Enquist M, Leimar O, Ljungberg T, *et al.* (1990) A test of the sequential assessment game: fighting in the cichlid fish *Nannacara anomala*. *Animal Behaviour*, 40, 1–14.

Enquist M, Arak A, Ghirlanda S, *et al.* (2002) Spectacular phenomena and the limits to rationality in genetic and cultural evolution. *Philosophical Transactions of the Royal Society B*, 357, 1585–1594.

Fearon JD (1994) Domestic political audiences and the escalation of international disputes. *American Political Science Review*, 88, 577–592.

Fearon JD (1995) Rationalist explanations for war. *International Organisation*, 49, 379–414.

Filson D & Werner S (2002) A bargaining model of war and peace: anticipating the onset, duration, and outcome of war. *American Journal of Political Science*, 46, 819–838.

Filson D & Werner S (2004) Bargaining and fighting: the impact of regime type on war onset, duration, and outcomes. *American Journal of Political Science*, 48, 296–313.

Gartner S (1997) *Strategic Assessment in War*. New Haven, CT: Yale University Press.

Ghosn F, Palmer G & Bremer S (2004) The MID3 Data Set, 1993–2001: procedures, coding rules, and description. *Conflict Management and Peace Science*, 21, 133–154.

Grafen A (1990) Biological signals as handicaps. *Journal of Theoretical Biology*, 144, 517–546.

Grafen A (1991) Modelling in behavioural ecology. In: JR Krebs & NB Davies (eds.) *Behavioural Ecology*, pp. 5–31. Oxford: Blackwell Scientific Publications.

Harris AT (1995) Despatch on war operations 23rd February 1942 to 8th May 1945. In: S Cox (ed.) *Studies in Air Power 3*. London: Frank Cass & Co. Ltd.

Hensel P (1999) An evolutionary approach to the study of interstate rivalry. *Conflict Management and Peace Science*, 17, 175–206.

Huntingford F & Turner A (1987) *Animal Conflict*. London: Chapman and Hall.

Hurd PL (2006) Resource holding potential, subjective resource value, and game theoretical models of aggressiveness signalling. *Journal of Theoretical Biology*, 241, 639–648.

Hurd PL & Enquist M (2005) A strategic taxonomy of biological communication. *Animal Behaviour*, 70, 1155–1170.

Kadera KM & Morey DS (2008) The trade-offs of fighting and investing: a model of the evolution of war and peace. *Conflict Management and Peace Science*, 25, 152–170.

Lake D & Powell R (1999) International relations: a strategic-choice approach. In: D Lake & R Powell (eds.) *Strategic Choice in International Relations*, pp. 3–38. Princeton, NJ: Princeton University Press.

Lanchester FW (1956) Mathematics in warfare. In: JR Newman (ed.) *The World of Mathematics, Vol. 4*, pp. 2138–2157. New York, NY: Simon & Schuster.

Levy J (1995) On the evolution of militarized interstate conflicts. In: S Bremer & T Cusak (eds.) *The Process of War: Advancing the Scientific Study of War*, pp. 219–226. Amsterdam: Overseas Publishers Association.

Maynard Smith J (1974) The theory of games and the evolution of animal conflicts. *Journal of Theoretical Biology*, 47, 209–221.

Maynard Smith J & Harper D (2004) *Animal Signals*. Oxford: Oxford University Press.

Maynard Smith J & Parker GA (1976) The logic of asymmetric contests. *Animal Behaviour*, 24, 159–175.

Maynard Smith J & Price GR (1973) The logic of animal conflict. *Nature*, 246, 15–18.

Morrow J (1999) The strategic setting of choices: signalling, commitment, and negotiation in international politics. In: D Lake & R Powell (eds.) *Strategic Choice and International Relations*, pp. 77–114. Princeton, NJ: Princeton University Press.

Parker GA (1974) Assessment strategy and the evolution of fighting behaviour. *Journal of Theoretical Biology*, 47, 223–243.

Payne RJH (1998) Gradually escalating fights and displays: the cumulative assessment model. *Animal Behaviour*, 56, 651–662.

Payne RJH & Pagel M (1997) Why do animals repeat displays? *Animal Behaviour*, 54, 109–119.

Pierce GJ & Ollason JG (1987) Eight reasons why optimal foraging theory is a complete waste of time. *Oikos*, 49, 111–118.

Powell R (1996*a*) Bargaining in the shadow of power. *Games and Economic Behavior*, 15, 255–289.

Powell R (1996*b*) Stability and the distribution of power. *World Politics*, 48, 239–267.

Powell R (1999*a*) *In the Shadow of Power: States and Strategies in International Politics*. Princeton, NJ: Princeton University Press.

Powell R (1999*b*) The modelling enterprise and security studies. *International Security*, 24, 97–106.

Powell R (2002) Bargaining theory and international conflict. *Annual Review of Political Science*, 5, 1–30.

Powell R (2004) Bargaining and learning while fighting. *American Journal of Political Science*, 48, 344–361.

Powell R (2006) War as a commitment problem. *International Organization*, 60, 169–203.

Rudin FS & Briffa M (2012). Is boldness a Resource Holding Potential trait? Fighting prowess and changes in startle response in the sea anemone *Actinia equina*. *Proceedings of the Royal Society of London B*, 279, 1904–1910.

Slantchev B (2003*a*) The principle of convergence in wartime negotiations. *American Political Science Review*, 97, 621–632.

Slantchev B (2003*b*) The power to hurt: costly conflict with completely informed states. *American Political Science Review*, 97, 123–133.

Wagner R (2000) Bargaining and war. *American Journal of Political Science*, 44, 469–484.

Walt S (1999) Rigor or rigor mortis? Rational choice and security studies. *International Security*, 23, 5–48.

Wilson AJ, Boer M de, Arnott G, *et al.* (2011) Integrating personality research and animal contest theory: aggressiveness in the green swordtail *Xiphophorus helleri*. *PLoS ONE*, 6, e28024.

Wittman D (1979) How a war ends: a rational model approach. *Journal of Conflict Resolution*, 23, 743–763.

Zahavi A (1975) Mate selection: a selection for a handicap. *Journal of Theoretical Biology*, 53, 205–214.

Chapter 16

Prospects for animal contests

Mark Briffa, Ian C.W. Hardy & Sophie L. Mowles

16.1 Repeated patterns in animal contest behaviour research

Among studies of contest behaviour that have been conducted within the framework of evolutionary theory, one can discern distinct phases of activity that have been associated with developments in an underpinning body of theory. As recounted in Geoff Parker's *Foreword* to this volume, the initial period of intense activity that occurred in the early to mid 1970s involved the laying down of a fundamental body of theory. During this time, contest behaviour provided the original context for the biological application of evolutionary game theory (as opposed to economic game theory, from which it derives). Game theory still acts as a cornerstone for behavioural ecology research and it is testament to its explanatory power that the Hawk–Dove game, wars of attrition and other examples of 'Evolutionarily Stable Strategy, or ESS, thinking' (Davies *et al.* 2012) still dominate undergraduate curricula in the subject. These early models stimulated empirical studies that provided evidence for ESSs in contests in diverse study systems including scorpionflies (Thornhill 1984), butterflies (Davies 1978) and red deer (Clutton-Brock *et al.* 1979). Studies such as these provided the early foundation for the cross-taxon approach to contests that we have continued in this book.

As in behavioural ecology as a whole, studies of contest behaviour show how sound evolutionary reasoning can explain (and enable us to recognise) similar phenomena displayed by very different animals. For example, the concept of 'self-assessment' of resource holding potential (RHP; Chapter 4), which is a new term for an assumption made in some war of attrition (WOA) models, now seems equally useful for explaining giving up decisions in examples of animals with vastly different nervous systems and information-gathering adaptations (contrast, for instance, sea anemones: Rudin & Briffa 2011, with amphipods: Prenter *et al.* 2006). The next phase in the development of contest theory was Enquist and Leimar's (1983) sequential assessment model (SAM), which stimulated another round of empirical research. In the late 1990s the addition of two new models for repeated signals, the energetic war of attrition (EWOA: Payne & Pagel 1997) and the cumulative assessment model (CAM: Payne 1998), produced a renewed interest in contests and collectively these 'assessment rules' have set the agenda for contest research through to the current time of writing. A characteristic of this phase has been an interest in investigating the physiological mechanisms that underpin aggression, particularly those that might relate to the costs of performing demanding agonistic behaviour (e.g. studies on crustaceans, Chapter 5, and fish, Chapter 10). Alongside 'assessment-orientated' research there has been a continuation of interest in evaluating predictions of some of the original models concerning factors determining contest outcomes (Chapter 2). For instance, in hymenopterans, a series of studies have shown that asymmetries in both RHP and resource value, V, influence the probability of winning and also that these influences can have many subcomponents; for instance, there are at least five aspects of V that can influence parasitoid contests, many apparently having a physiological or biochemical basis (Chapter 8). As with studies on assessment, we encourage research (in a range of taxa) that further couples physiological and biochemical variation with aspects of RHP and V. A very recent example is the finding of winner effects without concomitant

Animal Contests, ed. I.C.W. Hardy and M. Briffa. Published by Cambridge University Press.

loser effects in female parasitoids, which are thought to be generated by winning influencing physiological state and thus the value of the resource, *V*, to the winner but not to the loser (Goubault & Decuignière 2012, Chapter 8). In this instance, there has as yet been no formal theory developed for *V*-based winner–loser effects; current models are all based on influences of contest outcomes on competitors' subsequent RHPs. More broadly, as new theory is formulated (e.g. eco-genetic models, Chapter 2), we can look forward to a continued stimulus for new empirical research, perhaps taking population structure into account as much as measures of individual RHP, as proposed in Chapter 10 for future studies of contests in fish.

The chapters in this book, and the references sections for many of the papers cited therein, support the assertion that interest in animal contests has closely tracked key theoretical developments. Just like those models this view is, of course, a simplification of reality. It is not the case that those using empirical approaches have felt the need to wait for a new formally derived hypothesis in order to find inspiration for their work. A Web of Science search[1] covering 1981 to June 2012 results in 554 papers on the topic of animal contest behaviour. In 1981, one study was published; in 1997, halfway through this sample period, there were 24 papers, and in 2011 there were 65. By this fairly basic measure (which is undoubtedly incomplete as it does not encompass all keywords, such as 'aggression', for reasons discussed in Chapter 1), we can see that interest in animal contests, while certainly linked to theory, shows an overall pattern of escalation that does not appear to be entirely coupled to the periodic publication of key theoretical papers. As noted in Chapter 1, many authors seem to be motivated as much by a desire to learn more about their particular study systems as by wanting to solve a given problem that could apply generally across diverse species. Among the chapters in this book we can find examples that show how progress in the field has benefited from authors who have incorporated elements of both approaches into their work (for an interesting historical perspective on the positive tension between these two approaches in behavioural ecology, see Birkhead & Monaghan 2010).

One consequence of this duality of approaches has been that those working on contest behaviour are interested in questions that range more widely than what many will see as the core body of contest theory. In some cases this has meant that questions that span across different areas of behavioural ecology have been addressed. In crustaceans (Chapter 5), for example, the large amount of intraspecific variation in the appearance of weapons has led several authors to investigate the possibility of bluffing during fights, providing tests of Zahavi's (1975) handicap principle (Chapters 4 and 5). In other cases, questions that pertain more specifically to contests have been investigated. In conditions of high population density an individual might engage in a sequence of encounters, and in studies on fish in particular the influence of 'winner and loser' effects has been studied extensively (Chapters 3, 4 and 10). Moreover, winner and loser effects might be modulated by individual recognition; hence, the 'dear enemy effect' has been investigated in fiddler crabs (Chapter 5), salamanders (Chapter 11) and lizards (Chapter 12). Many contests will involve an intruder and an owner and the effects of residency and prior ownership have also featured heavily in studies of contest behaviour (e.g. Chapters 5, 7, 8 and 9). Therefore, studies of contest behaviour are based on a mix of theory that is often assumed to be specific to contest behaviour (assessments and decision rules, but see section 16.3), theory that is definitely considered to be relevant to many aspects of animal behaviour (handicaps and the fundamentals of ESSs) and a plethora of other issues that are specific to contests and which arise from the natural history of the study systems concerned. Below we turn to possible future directions for contest research, which are potentially amenable to all of these different study approaches.

16.2 Opportunities for integrating contest studies

16.2.1 Integration with other areas of behavioural ecology

In several examples contained in this book, the focus has been on analysing continuous variation in specific RHP traits among individuals that are otherwise similar. For example, in Carolina anole lizards, *Anolis carolinensis*, of similar size the individual with the stronger bite force is more likely to win an encounter (Chapter 12), and in big-clawed shrimp, *Alpheus heterochaelis*, variance around the relationship between overall body size and cheliped size leads to the possibility of bluffing in the use of agonistic displays (Chapter 5), and there is a similar example of dishonest signalling in a species of fig wasp in which males with atypically large mandibles do not fare well in fights but their large mandibles encourage opponents to retreat prior to escalation (Chapter 8). In terms of behaviour, as well as morphology, individuals that are otherwise similar might vary in ways that contribute to the chance of winning a contest. In fish, for example, individuals of the same size can differ in aggressiveness (Chapters 4 and 10) and in the beadlet anemone, *Actinia equina*, individuals showing short startle response durations demonstrate enhanced RHP (Rudin & Briffa 2012). While the presence of such continuous variation in RHP traits has been analysed at length (Chapter 4), the reasons for its presence have received less attention. Consistent between-individual variation in behaviour has been called 'animal personality' and recent studies have focussed on how to quantify this and test for different aspects of its presence (e.g. Martin *et al.* 2011, Stamps *et al.* 2012) as well as how to explain it. One explanation for the presence of animal personalities is that, given an initial presence of alternative behavioural strategies (e.g. thorough versus superficial exploration), disruptive selection then promotes the evolution of stable polymorphisms in other traits such as boldness and aggressiveness, which are also correlated to form a 'behavioural syndrome' (Wolf *et al.* 2007). There is clear potential for studies of contest behaviour to enhance our understanding of animal personality by providing empirical tests for such predictions. For example, recent studies have demonstrated the presence of consistent variation in aggressiveness in swordtail fish *Xiphophorus helleri* (Wilson *et al.* 2011) and the presence of a stable behavioural syndrome involving aggression, boldness and inquisitiveness in the hermit crab *Pagurus bernhardus* (Mowles *et al.* 2012). There is also the opportunity for personality research to inform our understanding of the evolution of contest behaviour. While the Hawk–Dove game shows how a mixed ESS, containing two types of contest behaviour and hence the presence of non-injurious contests, could arise (Chapter 2), theories about the evolution of personality could explain why we see continuous between-individual variation in aggressiveness. For example, in the Hawk–Dove game the different roles can either be maintained by competitors switching between strategies, or by individuals maintaining consistent strategies. Until the relatively recent research into behavioural syndromes, it was thought that the likely solution was that individuals switched strategies in a frequency-dependent manner. However, as consistent individual differences are now known to exist in nature, an interpretation of the Hawk–Dove game based on consistent behavioural types is possible (Dall *et al.* 2004).

Another explanation for the presence of behavioural syndromes is the presence of constraints that link the causal mechanisms underpinning different behaviours. Developmental constraints (pleiotropies) might be particularly important and study organisms such as beetles (Chapter 9), fish (Chapter 10) and birds (Chapter 13), where the links between contest behaviour and development have been analysed, offer the potential for investigating the role of constraints in promoting behavioural syndromes. 'Omic' approaches have already started to shed light on the links between development and behaviour (e.g. Aubin-Horth *et al.* 2012) and they could also complement the study of contests by identifying mechanisms associated with different levels of RHP and by matching these with different sets of developmental and experiential circumstances in which they are expressed. For example, juvenile swordtails exposed to poor diet undergo catch-up growth if their diet improves (Chapter 4). Although they attain the same adult size as individuals that have experienced normal growth, this appears to be at the cost of reduced RHP. Little is known, however, about why this is the case. Comparing transcription profiles between the two groups of individuals could shed light on the mechanisms involved and explain the adaptive value of this trade-off that leads to differential fighting abilities in similarly sized individuals.

Another area of behavioural ecology that has experienced a recent surge of interest is the analysis of social relationships within groups. Analysis of group structure is interesting for many reasons and when contests occur within social groups (i.e. when the resource in question is a position in a dominance hierarchy) analysing the links between group structure and contests can be insightful. Tools for analysing intra-group relationships have undergone a significant amount of recent development (Croft *et al.* 2011) and measures such as connectivity could be very complementary to traditional approaches from the contest literature, such as the dominance-ranking techniques discussed in Chapter 4.

16.2.2 Integration with other contest studies

A striking feature of the different study systems contained in this book is, in addition to the taxonomic diversity, the diversity of approaches that have been used. In a few cases the approaches used will be suitable only to the specific study system, but we suspect that there is also huge potential for the cross-fertilisation of experimental approaches between study systems. One example is the analysis of metabolites in order to understand the proximate mechanisms involved in giving-up decisions. These analyses were extensively applied in crustaceans (Chapter 5) and are now being applied in parasitoid wasps, initially to understand variation in host quality, *V*, in terms of biochemical composition (Chapter 8), but also with the clear potential to be applied to the contestants themselves to understand components of RHP. Another approach that could be shared between study systems is that of 'whole-body' performance measures, which have been extensively used in reptiles (Chapter 12) but have only recently been applied to crustacean contests (Mowles *et al.* 2010, Chapter 5). A related issue to that of performance capacities is variation in fighting 'skill' between opponents. As discussed in Chapter 10, this concept has been applied to courtship behaviour and could soon be applied to contest behaviour in fish. A possible development that could come from a more integrated approach across study systems is therefore a more precise definition of RHP where the concept of 'fighting prowess' is partitioned into components defined by measures of size, strength, stamina and skill. There are many other approaches, from developmental studies (e.g. beetles, Chapter 9) to analysis of colouration (e.g. birds, Chapter 13), that could usefully cross species boundaries and we hope that the taxonomic chapters of this book will help to promote this process.

16.3 Securing the future of contest research

Despite the upwards trajectory of publications on contest behaviour, other topics, such as sexual selection, have probably garnered greater interest among behavioural ecologists in recent years. Perhaps one fundamental biological reason for this is that directly sexually selected traits can be more immediately tied to proxies of fitness than non-sexually selected RHP traits. For instance, if a male is chosen by a female and gets to copulate, the first stage to achieving fitness happens relatively soon after the behavioural interaction of interest, while the male that wins against another male in a contest for access to a female is not necessarily assured to become her mate. Obviously, we know that even a successful copulation is but one step on the path to achieving fitness (Birkhead & Pizzari 2002) and there is no reason that other more immediate measures could not be taken, especially in laboratory animals that may be followed for large proportions of their lifetimes under controlled conditions. Female crustaceans, for example, fight over territory and shelters and carry egg masses externally, while in parasitoids, females often directly contest vital reproductive resources (Chapter 8). The number and volume of eggs and the quality of eggs can be assessed readily. It should therefore be relatively straightforward to look for links between fighting success and fecundity in female crustaceans (similarly, male crustaceans invest in spermatophores and spermatophore investment could serve as a valid proxy of fitness). In some model organisms that are short-lived, one need not rely on measuring such proxies of fitness. *Drosophila* have already been proposed as model organisms for investigating the mechanisms of aggression (Baier *et al.* 2002) and populations of freely interacting individuals have been subjected to automatic high-throughput behavioural measures of social behaviours ('ethomics': Branson *et al.* 2009) and microsatellite markers for paternity analysis are available.

Therefore, a challenge to those working on animal contests is to communicate to others the importance of fighting in the lives of most animals. While the theoretical underpinnings of our discipline are a definite strength (adding rigour), they also have the potential to be a weakness in terms of engaging with the widest possible audience. The finer points of the key assessment models (Chapters 1 and 2) form the central hypotheses of many papers in contest research, and for those unfamiliar with this body of theory the questions addressed may seem rather intangible. A related issue is that many empirical studies are concerned with measuring behaviour; for example, the changes in rates of behaviour predicted by the assessment models. It is less common for assessment-orientated studies of contests to focus on traits, such as plumage in birds, which often feature strongly in studies of sexual selection, that are more familiar to readers. This is curious, because contest behaviour involves some of the most dramatic behavioural interactions that we can observe (although admittedly in some cases, such as fights in sea anemones, one might have to view filmed encounters at high speed in order to fully appreciate the inherent drama of the situation), as well as some of the most striking morphological traits such as weapons (e.g. horns in beetles, antlers in deer, claws in crustaceans, jaws in ants and wasps). An example of how greater emphasis on visually appealing aspects of contests can bring this research to a wider audience is given by recent work on paper wasps, where analysis of varied facial patterns has demonstrated their use as signals of quality as well as the social costs of bluffing (Tibbetts & Dale 2004, Tibbetts & Lindsay 2008, Tibbetts *et al.* 2010).

Thus one way that the appeal of contest behaviour could be broadened is to augment the focus on a fairly specialised area of theory with approaches that are common in other areas of behavioural ecology. Conversely, it has long been recognised that those supposedly specialised areas of theory, are not really all that specialised at all. Sequential assessment, replacement of previous signals and energetic wars of attrition are fundamentally explanations for why animals might repeat displays. In their seminal 1997 paper (entitled 'Why do animals repeat displays?'), which contrasted the various features of these models, Payne and Pagel pointed out that although developed with fighting in mind, these explanations for repeated displays could be equally applicable to other behavioural contexts such as courtship. Thus, although interest in contests might have been 'taken over' by interest in sexual selection, the continued development of contest theory would bear attention by those interested in mate choice. Mowles and Ord (2012) explore these possibilities in detail, describing how dynamic repeated courtship displays may be better understood following the application of models analogous to those used to understand repetitive signalling during animal contests (e.g. in hermit crabs, Chapter 5 and see Chapters 1 and 2). The essential link is that choosy potential reproductive partners have to make a choice based upon the repeated display produced by the signalling individual in order to try to impress them. Moreover, by analysing signal costs and the structure of the repeated signal, as is necessary for the identification of the model of repeated signalling in an agonistic interaction (Chapter 4), investigators interested in sexual selection and mate choice can identify the mechanisms governing the intensity of courtship displays as well as the attributes that the choosy sex is using to assess the quality of the signaller. This application of contest theory to other signalling systems need not end with courtship, however, and may be equally applicable to analysing the honesty of anti-predatory signals used by species in order to demonstrate their quality (and therefore unsuitability as a target of attempted predation) in predator–prey interactions.

The ultimate means of securing the survival of a research strand is to convince others of its utility (or, in the parlance of research funding organisations, its 'impact'). An obvious application of contest research is to the field of animal welfare. For example, domesticated pigs can be aggressive when groups are newly established and finding ways of reducing aggression would improve the husbandry of farmed pigs. Similarly, studies of contests in fish might have applications to the aquaculture industry and to the welfare of fish kept as pets. For instance, Oldfield (2011) has recently shown that in the Midas cichlid fish, *Amphilophus citrinellus*, aquaria of sizes typically used by hobbyists do not provide optimal welfare conditions and that less of a cichlid's time is spent behaving aggressively when kept in larger aquaria provided with more complex habitat. Further, we have seen how understanding the fighting behaviour of insects might aid the strategic biocontrol of crop pests and the invasiveness of ecologically damaging ants (Chapter 8).

Because behaviour is the link between individuals and population ecology (e.g. Sutherland 1995, Candolin & Wong 2012), studies of contest behaviour could help us to predict wider ecological impacts of pollution. For example, Sopinka *et al.* (2010) showed how exposure to heavy metal contaminants disrupted the contest behaviour of the round goby, *Neogobius melanostomus*, leading to changes in their social structures. Exposure to pollution and climate change can disrupt a wide range of behaviours through routes including reduced metabolic scope and info-disruption (e.g. Briffa *et al.* 2012). Animal contests have the potential to provide an ideal context in which to study such anthropogenic impacts on behaviour; as can be seen throughout this book, many experiments on contest behaviour (and their associated statistical analyses) have been devised precisely in order to investigate the links between metabolic costs, information-gathering and behaviour.

Finally, it seems traditional to end books on animal behaviour by attempting to make the case that what we know about the behaviour of non-human animals can be applied to *Homo sapiens*, to better understand our own behaviour (e.g. Danchin *et al.* 2008, Westneat & Fox 2010, Bradbury & Vehrencamp 2012) and perhaps to rise above the 'tyranny of the selfish replicators' that shape its evolution (Dawkins 1976). As aggression certainly occurs within the human behavioural repertoire, studies of animal contests might indeed shed light on aspects of human behaviour. In *Animal Conflict*, Huntingford and Turner (1987) discussed human aggression in detail and Chapter 15 of this book deals with this from the perspective of organised interstate warfare. Similarities between inter-group conflict in ants and humans have long been recognised, although in some cases authors may have been influenced by the anxieties of the day as much as by biology; for instance, Haskins (1939) considers at length whether ant societies are primarily fascist or communist in nature. Nevertheless, just as models of human conflict have been adapted to explain multi-party contests in other animals, models (Chapter 3) and perhaps data (Chapter 8) on multi-party animal contests could well shed light on multi-party contests in humans.

In summary, the body of theory that has been developed to explain the evolution of contest behaviour, and the range of approaches that have been devised to investigate contests empirically ought to be, and indeed is, relevant to questions that are far broader than determining why hermit crabs rap their shells and why larger wasps tend to win.

Acknowledgement

We thank Geoff Parker for comments on an earlier version of this chapter.

Endnote

1. WoS search parameters: Topic = (animal*) AND Topic = (contest*). Refined by: Web of Science Categories = (Zoology or Behavioral Sciences or Ecology or Biology or Evolutionary Biology or Ornithology or Marine Freshwater Biology or Entomology) [6/6/12].

References

Aubin-Horth N, Deschênes M, Cloutier S (2012) Natural variation in the molecular stress response is correlated to a behavioural syndrome. *Hormones and Behavior*, 61, 140–146.

Baier A, Britta W & Brembs B (2002) *Drosophila* as a new model organism for the neurobiology of aggression? *Journal of Experimental Biology*, 205, 1233–1240.

Birkhead TR & Monaghan P (2010) Ingenious ideas – the history of behavioral ecology. In: DF Westneat & CW Fox (eds.) *Evolutionary Behavioral Ecology*, pp. 3–15. Oxford: Oxford University Press.

Birkhead TR & Pizzari T (2002) Postcopulatory sexual selection. *Nature Reviews*, 3, 262–273.

Bradbury JW & Vehrencamp SL (2012) *Principles of Animal Communication*, 2nd edn. Sunderland, MA: Sinauer Associates.

Branson K, Robie AA, Bender J, *et al.* (2009) High-throughput ethomics in large groups of *Drosophila*. *Nature Methods*, 6, 451–457.

Briffa M, Haye K de la & Munday PL (2012) High CO_2 and marine animal behaviour: potential mechanisms and ecological consequences. *Marine Pollution Bulletin*, 64, 1519–1528.

Candolin U & Wong BBM (eds.)(2012) *Behavioural Responses to a Changing World*. Oxford: Oxford University Press.

Clutton-Brock TH, Albon RM & Guinness FE (1979) The logical stag: adaptive aspects of fighting in red deer (*Cervus elaphus* L.) *Animal Behaviour*, 27, 211–225.

Croft DP, Madden J, Franks D, *et al.* (2011) Hypothesis testing in animal social networks. *Trends in Ecology and Evolution*, 26, 502–507.

Dall SRX, Houston AI & McNamara JM (2004) The behavioural ecology of personality: consistent individual differences from an adaptive perspective. *Ecology Letters*, 7, 734–739.

Danchin E, Giraldeau L-A & Cézilly F (eds.) (2008) *Behavioural Ecology*. Oxford: Oxford University Press.

Davies NB (1978) Territorial defence in the speckled wood butterfly (*Pararge aegeria*): the resident always wins. *Animal Behaviour*, 26, 138.

Davies NB, Krebs JR & West SA (2012) *An Introduction to Behavioural Ecology*, 4th edn. Chichester: John Wiley & Sons Ltd.

Dawkins R (1976) *The Selfish Gene*. Oxford: Oxford University Press.

Enquist M & Leimar O (1983) Evolution of fighting behaviour: decision rules and assessment of relative strength. *Journal of Theoretical Biology*, 102, 387–410.

Goubault M & Decuignière M (2012) Prior experience and contest outcome: winner effects persist in absence of evident loser effects in a parasitoid wasp. *American Naturalist*, 180, 364–371.

Haskins CP (1939) *Of Ants and Men*. New York, NY: Prentice-Hall.

Huntingford FA & Turner A (1987) *Animal Conflict*. London: Chapman & Hall.

Martin JGA, Nussey DH, Wilson AJ, *et al.* (2011) Measuring individual differences in field and experimental studies: a power analysis of random regression models. *Methods in Ecology and Evolution*, 2, 362–374.

Mowles SL & Ord TJ (2012) Repetitive signals and mate choice: insights from contest theory. *Animal Behaviour*, 84, 295–304.

Mowles SL, Cotton PA & Briffa M (2010) Whole-organism performance capacity predicts resource holding potential in the hermit crab *Pagurus bernhardus*. *Animal Behaviour*, 80, 277–282.

Mowles SL, Cotton PA & Briffa M (2012) Consistent crustaceans: the identification of stable behavioural syndromes in hermit crabs. *Behavioral Ecology and Sociobiology*, 66, 1087–1094.

Oldfield RG (2011) Aggression and welfare in a common aquarium fish, the Midas cichlid. *Journal of Applied Animal Welfare Science*, 14, 340–360.

Payne RJH (1998) Gradually escalating fights and displays: the cumulative assessment model. *Animal Behaviour*, 56, 651–662.

Payne RJH & Pagel M (1997) Why do animals repeat displays? *Animal Behaviour*, 54, 109–119.

Prenter J, Elwood RW, Taylor PW (2006) Self-assessment by males during energetically costly contests over precopula females in amphipods. *Animal Behaviour*, 72, 861–868.

Rudin SF & Briffa M (2011) The logical polyp: assessments and decisions during contests in the beadlet anemone *Actinia equina*. *Behavioral Ecology*, 22, 1278–1285.

Rudin FS & Briffa M (2012) Is boldness a resource holding potential trait? Fighting prowess and changes in startle response in the sea anemone *Actinia equina*. *Proceedings of the Royal Society of London B*, 279, 1904–1910.

Sopinka NM, Marentette JR & Balshine S (2010) Impact of contaminant exposure on resource contests in an invasive fish. *Behavioral Ecology and Sociobiology*, 64, 1947–1958.

Stamps JA, Briffa M & Biro PA (2012) Unpredictable animals: individual differences in intra-individual variability (IIV). *Animal Behaviour*, 83, 1325–1334.

Sutherland WJ (1995) *From Individual Behaviour to Population Ecology*. Oxford: Oxford University Press.

Thornhill R (1984) Fighting and assessment in *Harpobittacus* scorpionflies (Insecta: Mecoptera). *Evolution*, 38, 204–214.

Tibbetts EA & Dale J (2004) A socially enforced signal of quality in a paper wasp. *Nature*, 432, 218–222.

Tibbetts EA & Lindsay R (2008) Visual signals of status and rival assessment in *Polistes dominulus* paper wasps. *Biology Letters*, 4, 237–239.

Tibbetts EA, Mettler A & Levy S (2010) Mutual assessment via visual status signals in *Polistes dominulus* wasps. *Biology Letters*, 6, 10–13.

Westneat DF & Fox CW (2010) *Evolutionary Behavioral Ecology*. Oxford: Oxford University Press.

Wilson AJ, Boer M de, Arnott G, *et al.* (2011) Integrating personality research and animal contest theory: aggressiveness in the green swordtail *Xiphophorus helleri*. *PLoS ONE*, 6, e28024.

Wolf M, Doorn GS van, Leimar O, *et al.* (2007) Life-history trade-offs favour the evolution of animal personalities. *Nature*, 447, 581–584.

Zahavi A (1975) Mate selection – a selection for a handicap. *Journal of Theoretical Biology*, 53, 205–214.

Index